T0398196

Handbook of RAMS in Railway Systems

Theory and Practice

Handbook of RAMS in Railway Systems

Theory and Practice

Edited by
Qamar Mahboob
Enrico Zio

CRC Press
Taylor & Francis Group
Boca Raton London New York

CRC Press is an imprint of the
Taylor & Francis Group, an **informa** business

CRC Press
Taylor & Francis Group
6000 Broken Sound Parkway NW, Suite 300
Boca Raton, FL 33487-2742

CRC Press is an imprint of Taylor & Francis Group, an Informa business

Printed by CPI Group (UK) Ltd, Croydon, CR0 4YY on acid-free paper

International Standard Book Number-13: 978-1-138-03512-6 (Hardback)

Library of Congress Cataloging-in-Publication Data

Names: Mahboob, Qamar, editor. | Zio, Enrico, editor.
Title: Handbook of RAMS in railway systems / edited by Qamar Mahboob, Enrico Zio.
Other titles: Handbook of reliability, availability, maintainability, and safety in railway systems
Description: Boca Raton : Taylor & Francis, CRC Press, 2018. | Includes bibliographical references and index.
Identifiers: LCCN 2017044301| ISBN 9781138035126 (hardback) | ISBN 9781315269351 (ebook)
Subjects: LCSH: Railroads--Safety measures--Handbooks, manuals, etc. |
Railroads--Security measures--Handbooks, manuals, etc. |
Railroads--Equipment and supplies--Reliability--Handbooks, manuals, etc.
Classification: LCC TF610 .H36 2018 | DDC 625.10028/9--dc23
LC record available at https://lccn.loc.gov/2017044301

Visit the Taylor & Francis Web site at
http://www.taylorandfrancis.com

and the CRC Press Web site at
http://www.crcpress.com

Thinking ahead to the smart railway systems of the future, for reliable and safe transportation.

Contents

Section 1

Section 2

Preface

Modern railway systems are complex, integrate various technologies, and operate in an environment where the identification of exact system response and behaviors has limitations. Complexity in railway systems has significantly increased due to the use of modern technologies in relation to computers, microprocessors, communication, and information technologies, in combination with the historically developed electromechanical components. The resulting inter- and intradependencies and redundancies among technologies extend the boundaries of the railway network system. Unwanted and unknown system states may emerge due to unidentified behaviors, which cannot be predicted and, thus, eliminated. It then becomes a target for the designers, operators, maintainers, and approvers to consider the acceptable limits of system states. To do so, railway-related standards have been introduced in terms of reliability, availability, maintainability, and safety (RAMS).

Modeling, assessment, and demonstration of RAMS require careful and combined handling of the following:

- Railway engineering
- System analysis
- Development and application of mathematical and statistical methods, techniques, and tools
- Compliance with international and local standards and laws and specific project requirements
- Financial and economic analysis and, management

Today, a main objective for carrying out RAMS-related tasks is to obtain a safe, highly reliable and available, innovative and sustainable railway system. Within this, RAMS activities are also fundamental for increasing the lifetime of railway systems. Railway RAMS-related standards provide specifications and require the railway manufacturers and operators to implement a RAMS management system and demonstrate particular safety standards and RAM requirements. The standards mainly provide general guidelines on different RAMS issues but do not provide details on how to proceed in real-world projects. Consequently, endless discussions, disagreements, and rework of RAMS activities are experienced in many real-world projects. This is partly due to the fact that there is lack of purposeful understanding of the standards themselves and of the mathematical and statistical methods applicable for RAMS and their use. The topics in this handbook aim to improve the understanding and application of RAMS-related standards and the theory, methods, tools and techniques and related background. To this aim, dedicated efforts have been coordinated worldwide in writing this handbook.

This is the first-ever comprehensive reference handbook that deals with the many unique RAMS issues involved in the difficult environment of an operating railway system. This is of concern given that the European Union and some other parts of the world have already mandated the use of RAMS in legislation. The implementation of the RAMS requirements then extends to system and equipment designers and manufacturers, who supply their products. The handbook provides detailed guidance for those involved in the integration of the highly complex and multitechnology systems that must seamlessly perform to guarantee a safe and reliable railway transport system. It can be used as guidance to get the RAMS tasks successfully accomplished, especially in complex railway projects. It focuses on the several topics of risk, safety, reliability, and maintenance in railway systems and provides state-of-the-art knowledge on the issues therein. The handbook includes 38 chapters authored by key senior experts from industry and renowned professors and researchers from academia. The handbook is divided into two major sections: Section 1 on basic concepts, prediction, and estimation techniques and the second section on RAMS in practice and special topics. Highlights of the topics covered under each section are presented in the following.

Highlights of the topics in the handbook of RAMS in Railway Systems: Theory and Practice	
Section 1	**Section 2**
01. Introduction to RAMS requirements	21. Methodology and application of RAMS management along the railway rolling stock lifecycle
02. Basic methods for RAM and decision making	22. RAMS and security
03. Advance methods for RAM and decision making	23. Modeling reliability analysis European Train Control System (ETCS)
04. Safety integrity concept	24. Decision support systems for railway RAMS: An application from the railway signaling subsystem
05. SIL apportionment and SIL allocation	
06. Prognostics and health management	25. Fuzzy reasoning approach and fuzzy hierarchy process for expert judgement capture and process in risk analysis
07. Human factors and their applications in railways safety	
08. Individual and collective risk and F–N curves for risk acceptance	26. Independent safety assessment process and methodology—a practical walk through a typical public transport system/railway system or subsystem/product/project
09. Practical demonstrations of reliability growth in railway projects	
10. Methods for RAM demonstration in railway projects	27. IFF-MECA: Interfaces and functional failure mode effects and criticality analysis for railway RAM and safety assessment
11. A guide for preparing comprehensive and complete case for safety for complex railway products and projects	28. RAMS as an integrated part of the engineering process and the application for railway rolling stock
12. Reliability demonstration tests: Decision rules and associated risks	29. Model based HAZOP to identify hazards for modern train control system
13. The hazard log: Structure and management in complex project	
14. Life cycle cost (LCC) in railways RAMS management with example applications	30. Application of risk analysis methods for railway level crossing problems
15. System assurance for the railways electromechanical products	31. Human reliability and RAMS management
16. Software reliability in RAMS management	32. A standardized approval process for a public transport system/railway system/or subsystem/product/project
17. Safety software development for the railway applications	33. Importance of safety culture for RAMS management
18. Practical statistics and demonstration of RAMS in projects	34. Railway security policy and administration in the USA: Reacting to the terrorists attack after Sep. 11, 2011
19. Proven in use for software—assigning a SIL based on statistics	
20. Target reliability for new and existing railway engineering structures	35. Introduction to IT transformation of safety and risk management systems
	36. Formal reliability analysis of railway system using theorem proving technique
	37. Roles and responsibilities for new built, extension or modernization of public transport system—a walk through the life cycle
	38. A holistic view of charm and challenges of CENELEC standards for railway signaling

The authors of the individual chapters have devoted quite some efforts to include current and real-world applications, recent research findings, and future works in the area of RAMS for railway systems and to cite the most relevant and latest references in their chapters. Practical and insightful descriptions are presented in order to nourish RAMS practitioners of any level, from practicing and knowledgeable senior RAMS engineers and managers to beginning professionals and to researchers. Given the multidisciplinary breadth and the technical depth of the handbook, we believe that readers from different fields of engineering can find noteworthy reading in much greater detail than in other engineering risk and safety-related publications. Each chapter has been written in a pedagogical style, providing the background and fundamentals of the topic and, at the same time, describing practical issues and their solutions, giving real-world examples of application and concluding with a comprehensive discussion and outlook. The content of each chapter is based on established and accepted practices, publications in top-ranked journals, and conferences. The comprehensive contents and the involvement of a team of multidisciplinary experts writing on their areas of expertise provide the editors confidence that this handbook is a high-quality reference handbook for students, researchers, railway network operators and maintainers, and railway safety regulators, plus the associated equipment designers and manufacturers worldwide. We believe that this handbook is suited to the need of RAMS in railways.

This book contains information obtained from authentic and highly regarded sources. Reasonable efforts have been given to publish reliable data and information, but the authors, editors, and publisher cannot assume responsibility for the validity of all materials or consequences of their use. The authors, editors, and publisher have attempted to trace the copyright holders of all material reproduced in this publication and apologize to copyright holders if permission to publish in this form has not been obtained. If any copyright material has not been acknowledged please write and let us know so we may rectify in any future reprint.

Dr. Engr. Qamar Mahboob
Erlangen, Germany, and Lahore, Pakistan

Prof. Dr. Enrico Zio
Milano, Italy, and Paris, France

Editors

Dr. Engr. Qamar Mahboob has more than 15 years of project experience and several scientific publications in the field of reliability, availability, maintainability, and safety (RAMS).

Qualification: Degrees: PhD in railway risk, safety and decision support from Technische Universität Dresden (TU Dresden), Dresden, Germany; MSc in transportation systems from the Technical University of Munich (TU Munich), Munich, Germany; MS in total quality management from Punjab University, Lahore, Pakistan; BSc in mechanical engineering from the University of Engineering and Technology Lahore, Lahore, Pakistan; and B-Tech from the (GCT) Government College of Technology, Railway Road, Lahore, Pakistan. Experience: maintenance engineer for rolling stock for Pakistan railways; scientific researcher for engineering risk analysis group of TU Munich; scientific researcher of Railway Signalling and Transport Safety Technology of TU Dresden; RAMS consultant for CERSS.com; key senior expert for RAMS for Siemens AG, Germany; technical lead for HI-TEK Manufacturing Pvt. Ltd., Lahore; services manager for KKPower International Pvt. Ltd., Lahore; and director and CTO (Core Technology Officer), SEATS (Science, Engineering and Technology for Systems), Lahore.

Prof. Dr. Enrico Zio (M'06–SM'09) received his MSc degree in nuclear engineering from Politecnico di Milano in 1991 and in mechanical engineering from the University of California, Los Angeles, in 1995, and his PhD degree in nuclear engineering from Politecnico di Milano and the Massachusetts Institute of Technology (MIT) in 1996 and 1998, respectively. He is currently the director of the chair on systems science and the energetic challenge of the Foundation Electricite' de France at CentraleSupélec, Paris, France; full professor and president of the Alumni Association at Politecnico di Milano; adjunct professor at University of Stavanger, Norway, City University of Hong Kong, Beihang University, and Wuhan University, China; codirector of the Center for REliability and Safety of Critical Infrastructures, China; visiting professor at MIT, Cambridge, Massachusetts; and distinguished guest professor of Tsinghua University, Beijing, China. His research focuses on the modeling of the failure–repair–maintenance behavior of components and complex systems, for the analysis of their reliability, maintainability, prognostics, safety, vulnerability, resilience, and security characteristics and on the development and use of Monte Carlo simulation methods, soft computing techniques and optimization heuristics. He is the author or coauthor of seven international books and more than 300 papers on international journals.

Contributors

Waqar Ahmad
School of Electrical Engineering
  and Computer Sciences
National University of Sciences
  and Technology
Islamabad, Pakistan

Allegra Alessi
Alstom
Saint-Ouen, France

Min An
University of Salford
Greater Manchester, United Kingdom

Kyoumars Bahrami
Siemens Mobility (Rail Automation)
Melbourne, Australia

Jens Braband
Siemens AG
Braunschweig, Germany

Mehdi Brahimi
Alstom
Saint-Ouen, France

Yao Chen
Siemens AG
Munich, Germany

Attilio Ciancabilla
Rete Ferroviaria Italiana
Bologna, Italy

Pierre Dersin
Alstom
Saint-Ouen, France

Dimitris Diamantidis
Ostbayerische Technische Hochschule
  Regensburg
Regensburg, Germany

Georg Edlbacher
Bombardier Transportation
Zürich, Switzerland

Alessandro Fantechi
University of Florence
Florence, Italy

Alessio Ferrari
Istituto di Scienza e Tecnologie
  dell'Informazione
Pisa, Italy

Miguel Figueres-Esteban
University of Huddersfield
Huddersfield, United Kingdom

Olga Fink
Zürcher Hochschule für angewandte
  Wissenschaften,
Winterthur, Switzerland

Simone Finkeldei
Schweizerische Bundesbahnen SBB
Bern, Switzerland

Lance Fiondella
University of Massachusetts Dartmouth
North Dartmouth, Massachusetts

Heinz Gall
TÜV Rheinland
Cologne, Germany

Stefania Gnesi
Istituto di Scienza e Tecnologie
  dell'Informazione
Pisa, Italy

Gary A. Gordon
Massachusetts Maritime Academy
Buzzards Bay, Massachusetts

Stephan Griebel
Siemens AG
Braunschweig, Germany

Thomas Grossenbacher
Schweizerische Bundesbahnen SBB
Bern, Switzerland

Coen van Gulijk
University of Huddersfield
Huddersfield, United Kingdom

Malcolm Terry Guy Harris
Topfield Consultancy Limited
London, United Kingdom

Osman Hasan
School of Electrical Engineering
    and Computer Sciences
National University of Sciences
    and Technology
Islamabad, Pakistan

Ali Hessami
Vega Systems
London, United Kingdom

Milan Holicky
Czech Technical University in Prague
Prague, Czech Republic

Peter Hughes
University of Huddersfield
Huddersfield, United Kingdom

Lei Jiang
Southwest Jiaotong University
Chengdu, China

Andreas Joanni
Siemens AG
Munich, Germany

Karel Jung
Czech Technical University in Prague
Prague, Czech Republic

Benjamin Lamoureux
Alstom
Saint-Ouen, France

Yiliu Liu
Norwegian University of Science
    and Technology
Trondheim, Norway

Andrei Loukianov
University of Huddersfield
Huddersfield, United Kingdom

Cristian Maiorano
Ansaldo STS
Genoa, Italy

Birgit Milius
Technische Universität Braunschweig
Braunschweig, Germany

Vidhyashree Nagaraju
University of Massachusetts Dartmouth
North Dartmouth, Massachusetts

Alban Péronne
Alstom
Saint-Ouen, France

Jeremy F. Plant
Pennsylvania State University
Harrisburg, Pennsylvania

Hendrik Schäbe
TÜV Rheinland
Cologne, Germany

Eric J. Schöne
Traffic Sciences Department
Dresden Technical University
Dresden, Germany

Joerg Schuette
Technische Universität Dresden
Dresden, Germany

Holger Schult
Siemens AG
Erlangen, Germany

Rohan Sharma
TÜV Rheinland
Cologne, Germany

Miroslav Sykora
Czech Technical University in Prague
Prague, Czech Republic

Sofiène Tahar
Department of Electrical and Computer
  Engineering
Concordia University
Montreal, Canada

René Valenzuela
Alstom
Saint-Ouen, France

Peter Wigger
TÜV Rheinland
Cologne, Germany

Xiaofei Yao
Casco Signal
Shanghai, China

Richard R. Young
Pennsylvania State University
Harrisburg, Pennsylvania

Datian Zhou
Centrale Supélec
Laboratoire Génie Industriel
Paris, France

and

Beijing Jiaotong University
Beijing, China

Ruben Zocco
TÜV Rheinland
Dubai, United Arab Emirates

Section 1

1

Introduction to the Requirements of Railway RAM, Safety, and Related General Management

Qamar Mahboob, Enrico Zio, and Pierre Dersin

CONTENTS

1.1 Introduction and Background

Railway-related standards introduce terms used in reliability, availability, maintainability, and safety (also called RAMS) and require railway suppliers, operators, maintainers, and duty holders to implement a comprehensive RAMS management system (EN 50126 [CEN 2003], EN 50129 [CEN 2003], IEC 61508 [IEC 2000], IEC-DTR-62248-4 [IEC 2004], IEC 62267-2 [IEC 2011], and IEC 62278-3 [IEC 2010]). This chapter explains the general requirements toward railway RAMS management throughout the life cycle of a technology or system for railway application. The topics of RAMS management and its process are covered in this chapter a way that a reliable, safe, cost-optimal, and improved quality of railway systems may be achieved. To achieve all these, a life cycle approach needs to be adopted. The life cycle applicable to the RAMS management is shown in Figure 1.1 and is adopted from EN 50126. This life cycle approach provides basic concepts and structure for planning, managing, implementing, controlling, and monitoring of all aspects of a railway project, incorporating RAMS as well, as the project proceeds through the life cycle phases. The general RAMS management process consists of three major areas shown in Figure 1.1.

This life cycle and three major areas are applicable to any railway product or subsystem under consideration regardless of its level or position within the complete railway system. In other words, each considered subsystem level can be combined and integrated into the superior system until the top level of the complete railway system has been obtained. The life cycle process can be simplified depending on the applicable project phases. For example, in railway projects, suppliers usually commit the demonstration of RAMS performance-related targets until the "trial run and system acceptance" phase. It is important to mention that warranty periods are also defined within the life cycle phases of a project, and the warranty demonstration period may overlap with more than one phase of the life cycle.

According to railway-related RAMS standards, safety is "freedom from unacceptable risks, danger and injury from a technical failure in railways." The focus of the RAMS

3

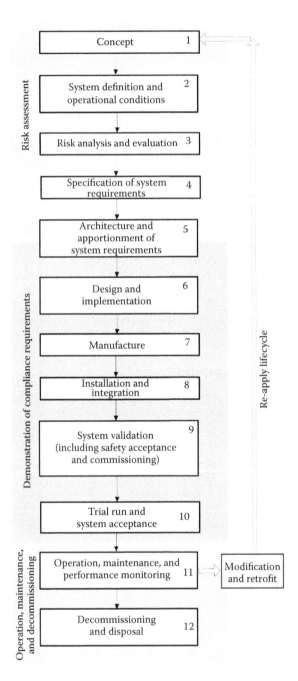

FIGURE 1.1
System life cycle applicable for RAMS management process. (Based on CEN, EN 50126. *Railway Applications—The Specification and Demonstration of Reliability, Availability, Maintainability and Safety (RAMS)*, CEN, Brussels, 2003.)

management process is either to identify and then reduce the safety relevant failures or to eliminate the consequences of the failures throughout the life cycle. The objective is always to minimize the residual risk from the safety-related failures (Birolini 2014; Rausand and Hoyland 2004). The risk assessment process, defined in applicable standards such as EN 50129 (CEN 2003), should be performed in order to identify the degree of safety required for each particular situation. The tolerability of safety risk of a railway system is dependent upon the safety criteria set by the legal authorities or by the railway duty holder in accordance with the rules given by legal authorities (HSE 2001).

After a concept for a project has been set up, the life cycle process consists of three major steps:

- Risk assessment that includes risk analysis and risk evaluation on the basis of the system definition including the specification of system requirements
- Demonstration (includes theoretical and practical ways) that the system fulfils the specified requirements
- Operation, maintenance, and decommissioning

In addition to the process flow—starting from the "Concept" and ending at the "Decommissioning and Disposal"—within the life cycle phases, the process flow involves a so-called feedback loop. The risk needs to be reevaluated in case additional information on safety risk is obtained during the related phases of the project. Consequently, some phases of the life cycle have to be revaluated from the risk point of view. The logical flow of information and associated decisions in project phases are more important than the time-based flow of the phases. It requires the risk assessment to be confirmed at the end of the life cycle. Here the aim is to have complete visualization of the risk picture, at the end of the lifetime of a technology/system, in order to confirm whether the "risks expectations" were met during the whole lifetime. This reassessment, at the end of the lifetime, will help in updating/improving risk-based decisions in the future (in reference to a particular system/subsystem), based on the lifetime considerations. RAMS tasks contribute to the general project tasks for each phase, and requirements for RAMS tasks are detailed in the succeeding sections of this chapter. The process flow in Figure 1.1 shows life cycle-related RAMS tasks as components of general project tasks. The next sections will explain the phases and RAMS requirements in each phase of the project, considering EN 50126. For the applications of the RAMS, we refer the reader to Mahboob (2014) and references therein.

1.2 RAMS Management Requirements

A V representation, also provided in EN 50126 and other Comité Européen de Normalisation Électrotechnique (CENELEC) standards, of the life cycle is widely used in the RAMS management. Please refer to Figure 1.2. The top–down branch on the left side of the V-shaped diagram is generally called the development branch and begins with the concept and ends with the manufacturing of the subsystem components of the system. The bottom–up branch on the right side is related to the installation or assembly, the system handover, and then the operation of the whole railway system.

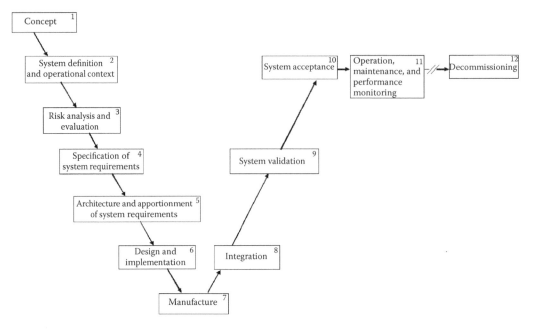

FIGURE 1.2
V representation of system life cycle. (From DIN [Deutsches Institut für Normung] and VDE [Verband der Elektrotechnik, Elektronik und Informationstechnik], DIN EN 50126 (VDE 0115-103), edition 2000-03. *Bahnanwendungen—Spezifikation und Nachweis von Zuverlässigkeit, Verfügbarkeit, Instandhaltbarkeit und Sicherheit (RAMS)—Teil 1: Generischer RAMS Prozess; Deutsche Fassung prEN 50126-1:2015)*, Beuth Verlag and VDE Verlag, Berlin, 2003. With permission No. 12.017 of DIN and VDE, conveyed by VDE.*)

For the roles and responsibilities as well as the general activities throughout, the applicable life cycle, e.g., of the system-level considerations, will be described in another chapter of this handbook. The life cycle phases in the V diagram are briefly explained in the following and are in line with the prevailing EN 50126.

1. Concept: Contracts are signed, agreements are made, and transmittals of the project are drawn up.
2. System definition and operational context: System characteristics and functions are described; interfaces and requirements are clarified; system inputs and outputs are recorded; intended operational conditions, maintenance, and environment are stated; and the RAMS performance-related parameters of the attached subsystems and components are derived. The RAMS management (e.g., using RAMS plan) and organization are established.
3. Risk assessment: Risk assessment includes risk analysis and risk evaluation. The risk assessment involves steps such as hazard identification, identification

---

* It should be noted that the latest edition be always used. This can be obtained from Beuth Verlag, Berlin, Germany (http://www.beuth.de) and from VDE Verlag, Berlin, Germany, (http://www.vde-verlag.de). The English edition, named BS EN 50126, can be obtained from British Standards Institution, London (http://www.bsigroup.com).

of events leading to hazards, and determination of risks associated with hazards (requires consequence analysis). The process for ongoing risk management should be established and then followed to decide if a risk is tolerable. This requires risk acceptance criteria to be in place. Risk analysis is a continuous and iterative step and goes in parallel with subsequent phases. There can be a condition that can lead to defining further safety system requirements induced by the risk acceptance criteria in order to reduce the risk to an acceptable level. Based on the risk assessment, system requirements can be derived.

4. Specification of system requirements: Detailing the initial system requirements (expected functions including their RAMS requirements) and the ones derived from risk assessment in phase 3 as well as defining criteria for acceptance and specifying the overall demonstration of compliance.

5. Architecture and apportionment of system requirements: Definition and allocation of RAMS requirements for subsystems. This phase might be a part of a demonstration of compliance, which can be possibly achieved in theoretical ways such as design simulations and software-based demos. Subsystems and their component requirements can be directly allocated if they are already available up to this point or are apportioned by deriving them from system-level requirements.

6. Design and implementation: During this phase, subsystems and components should be developed according to the RAMS requirements. Furthermore, plans for future life cycle tasks have to be established.

7. Manufacture: The components (and subsystems) of the railway system are manufactured, and RAMS-specific assurance plans have to be established and applied in the later project phases.

8. Integration: All subsystems should be assembled and installed to form the complete railway system in order to achieve the system-level mission.

9. System validation: It must be validated that the system and associated processes fully comply with the RAMS requirements, and the external risk reduction measures are considered.

10. System acceptance: This refers to the demonstration of compliance of complete railway system with overall contractual RAMS requirements and provides evidence that the system is now acceptable for entry into service.

11. Operation, maintenance, and performance monitoring: It is required to operate, maintain, and support the product through performance monitoring such that compliance with system RAMS requirements is consistent.

12. Decommissioning: In case of decommissioning, the system risk is controlled during the transition phase.

1.3 Life Cycle-Based RAM, Safety, and General Management Tasks

Table 1.1, which is based on EN 50126:2003, provides a summary of the main RAM, safety, and general project-related tasks in different phases of the project life cycle.

TABLE 1.1

Summary of the Project Phase-Related Requirements, Management, and Processes

Life Cycle Phases	RAM Tasks	Safety Tasks	General Tasks
1. Concept	• Consider previously achieved RAM performance of similar projects • Consider and define RAM implications of new project • Review RAM targets	• Consider previously achieved safety performance of similar project and application conditions • Consider and define safety implications of new project • Safety policy and safety targets are reviewed	• Establish and define the scope and purpose of project • Define project concept • Carry out financial analysis and feasibility studies • Set up management
2. System definition and application conditions	• Perform preliminary RAM analysis, based on historical data of RAM • Define RAM policy • Identify life cycle-based operation and maintenance conditions • Identification of the influences on RAM of existing interfaces of infrastructure and other constraints	• Perform preliminary hazard analysis, based on the historical data of safety • Create safety plan • Define risk acceptance criteria • Identification of the influences on safety of existing interfaces of infrastructure and further constraints	• Define system mission profile • Prepare system-level technical description • Identify operation and maintenance strategies • Identify operating and maintenance conditions • Identify influence of existing interfaces of infrastructure and local constraints
3. Risk analysis	• Not relevant	• Perform systematic hazard analysis and safety risk analysis on system level • Set up central hazard log • Make complete risk assessment (= risk analysis + risk evaluation)	• Project-level risk analysis (may have to be repeated at several stages)
4. System requirements	• Specify system RAM requirements • Define RAM acceptance criteria • Define system functional concept and structure • Establish RAM program on system level • Establish RAM management on system level	• Specify system safety requirements • Define safety acceptance criteria • Define safety-related functional concept and requirements • Establish safety management on system level	• Requirements analysis • System specific • Specify local environment • Define system assurance, demonstration, and acceptance criteria • Establish verification and validation plan • Establish management, quality, integration, and organization requirements • Introduce and implement change control procedure

(Continued)

TABLE 1.1 (CONTINUED)

Summary of the Project Phase-Related Requirements, Management, and Processes

Life Cycle Phases	RAM Tasks	Safety Tasks	General Tasks
5. Apportionment of system requirements	• Apportionment of system RAM requirements to the specific subsystem and component RAM requirements • Define subsystem and components RAM acceptance criteria	• Apportionment of system safety targets and requirements for specific subsystems and components • Define subsystem and components safety acceptance criteria • Update system safety plan, if necessary	• Apportionment of system requirements • Define subsystem and components requirements and acceptance criteria
6. Design and implementation	• Implement RAM program by review, analysis, testing and data assessment, reliability, availability, maintainability and maintenance, and analysis logistic support • Control programs: RAM program management, control of suppliers and contractors	• Implement safety plan by review, analysis, testing, and data assessment. It includes: • Hazard log • Hazard analysis and risk assessment • Undertake program control for safety management and supplier control • Preparation of generic safety case • Preparation of generic application safety case, if required	• Planning • Design and development • Design analysis and testing • Design certification • Implementation and validation • Design of logistic support resources
7. Manufacturing	• Requires environmental stress screening • Requires RAM improvement testing • Initiate failure reporting and corrective action system (FRACAS)	• Implement safety plan (by following review, analysis, testing and data assessment) • Use and update hazard log	• Production planning • Manufacture • Manufacture and test subassembly of components • Documentation management • Design associated trainings
8. Installation	• Start trainings for the maintenance people from a maintainer • Establish spare parts and tool provision-related (inventory) lists	• Establish installation program • Implement installation program	• Subsystem assembly and system-level integration • Multiple subsystem installations

(Continued)

TABLE 1.1 (CONTINUED)

Summary of the Project Phase-Related Requirements, Management, and Processes

Life Cycle Phases	RAM Tasks	Safety Tasks	General Tasks
9. System validation	• RAM demonstration (and evaluation in reference to the penalty criteria, e.g., trip losses during operation times)	• Establish and then commission program • Preparation of application-specific safety case begins	• System-level commissioning • Perform transition or probationary period of operation • Carry out related trainings
10. System acceptance	• Assess RAM demonstration in reference to the acceptance criteria	• Assess application-specific safety case in reference to the given acceptance criteria	• Observe acceptance procedures, based on acceptance criteria • Documented evidence for acceptance • System bringing into service • Continue transition or probationary period of operation, if necessary
11. Operation and maintenance	• Procurement of spare parts and tools • Apply reliability-centered maintenance, logistic support	• Safety-centered maintenance • Safety performance monitoring • Continuous management and maintenance of hazard log	• System operation for long-term basis • Maintenance activities based on system-level considerations
12. Performance monitoring	• Through FRACAS (and SCADA, if relevant) collect, analyze, evaluate, and use performance and RAM statistics	• Through FRACAS (and SCADA, if relevant) collect, analyze, evaluate, and use performance and safety statistics	• Through FRACAS (and SCADA, if relevant) collect operational performance statistics and analyze and evaluate collected data
13. Modification and retrofit	• Reconsider RAM implications for modification and retrofit and revise and update the necessary requirements	• Reconsider safety implications for modification and retrofit and revise and update the necessary requirements	• Take care of change request procedures • Take care of modification and retrofit procedures
14. Decommissioning and disposal	• Not relevant	• Safety plan for the decommissioning and disposal and its implementation • Decommissioning and disposal-based hazard analysis and risk assessment	• Planning and procedure of decommissioning and disposal

Source: CEN, EN 50126. *Railway Applications—The Specification and Demonstration of Reliability, Availability, Maintainability and Safety (RAMS)*, CEN, Brussels, 2003.

1.4 Summary

In general, the responsibilities for the tasks in the various life cycle phases depend on the contractual and, sometime, legal relationship between the stakeholders involved. It is important that the related responsibilities are defined and agreed upon. The RAMS management process shall be implemented under the control of an organization, using competent personnel assigned to specific RAMS-related roles. Selection, assessment, and documentation of personnel competence, including technical knowledge, qualifications, relevant experience, skill, and appropriate training, shall be carried out in accordance with given requirements to be defined by the project-specific safety management organization (involves subsystem [e.g., suppliers of signaling, rolling stock, platform screen doors, rail electrification, supervisory control and data acquisition [SCADA], communication, auxiliaries, and civil works], transit system, consortium, operator, maintainer, and customer or public authorities levels). Several roles within an organization, such as ISA (Independent Safety Assessor) verifier, and validator, are viewed in reference to the RAMS performance of what has been produced by other specialists in the project such as design engineers for different subsystems. It is worth noting that the series of RAMS CENELEC standards (EN 50126, EN 50128, and EN 50129) are undergoing a drastic revision. In parallel, a CENELEC ad hoc group, AHG9, has been drafting a technical report which aims at strengthening the RAM contents of EN 50126 by introducing the concept of "RAM risk," whereas EN 50126 mainly concentrated so far on safety risks.

References

Birolini, A. *Reliability Engineering: Theory and Practice*. Heidelberg: Springer, 2014.

CEN (European Committee for Standardization). *EN 50126: Railway Applications—The Specification and Demonstration of Reliability, Availability, Maintainability and Safety (RAMS)*. Brussels: CEN, 2003.

CEN. *EN 50129: Railway Applications—Communications, Signalling and Processing Systems—Safety-Related Electronic Systems for Signalling*. Brussels: CEN, 2003.

HSE (UK Health and Safety Executive). *Reducing Risks, Protecting People (R2P2)*. Liverpool: HSE, 2001.

IEC (International Electrotechnical Commission). IEC 61508: *Functional Safety of Electrical/Electronic/Programmable Electronic Safety-Related Systems (IEC 61508-1 to 7)*. Geneva: IEC, 2000.

IEC. *IEC-DTR-62248-4: Railway Applications—Specification and Demonstration of Reliability, Availability, Maintainability and Safety (RAMS), Part 4: RAM Risk and RAM Life Cycle Aspects*. Geneva: IEC, 2004.

IEC. *IEC TR 62278-3-2010: Railway Applications—Railway Applications—Specification and Demonstration of Reliability, Availability, Maintainability and Safety (RAMS)—Part 3: Guide to the Application of IEC 62278 for Rolling Stock RAM*. Geneva: IEC, 2010.

IEC. *IEC TR 62267-2-2011: Railway Applications—Automated Urban Guided Transport (AUGT)—Safety Requirements—Part 2: Hazard Analysis at Top System Level*. Geneva: IEC, 2011.

Mahboob, Q. *A Bayesian Network Methodology for Railway Risk, Safety and Decision Support*. Technische Universität Dresden: Dresden, PhD Thesis, 2014.

Rausand, M., and A. Hoyland. *System Reliability Theory: Models, Statistical Methods, and Applications*. Hoboken, NJ: John Wiley & Sons, 2004.

2

Basic Methods for RAM Analysis and Decision Making

Andreas Joanni, Qamar Mahboob, and Enrico Zio

CONTENTS

2.1   Introduction

This chapter is the first of two chapters describing basic and advanced methods for reliability, availability, and maintainability (RAM) analysis and decision making, respectively. In practice, we refer to these methods as "basic" because they are the most frequently used in practice for specification, assessment, and possible improvement of the RAM characteristics of products and systems throughout all phases of the life cycle of a railway project. Moreover, these methods are used in the context of functional safety for identifying, assessing, and (if necessary) mitigating potentially hazardous events that may emerge in relation to the operation of these technical systems. For railway applications, the definition of reliability, availability, maintainability, and safety (RAMS) and of the systematic processes for specifying the requirements for RAMS and demonstrating that these requirements are achieved are given in the standard EN 50126-1. This European standard provides railway system operators, suppliers, maintainers, and duty holders with a process which will make them able to implement a systematic and consistent approach to the management of RAMS. The system-level approach given in EN 50126-1 facilitates assessment of the RAMS interactions between elements of railway applications, regardless of their complexity.

The methods described in the present chapter have certain limitations when dealing with complex systems characterized by dependencies, time effects, etc. Some of these limitations can be overcome by applying more advanced methods described in the next chapter, such as Markov models and Monte Carlo simulation. However, this comes at the cost of increased complexity of the analysis methods, modeling efforts, and computational costs.

In the following sections of this chapter, most of the basic mathematical concepts, definitions, and terminology related to RAM analysis can be found in IEC 61703:2016 and IEC 60050-192:2015 (for terminology, see also http://www.electropedia.org). The examples presented in this chapter are for academic and illustration purposes only, for instance, by considering limited failure scenarios only.

2.2   Failure Modes and Effects Analysis

A failure modes and effects analysis (FMEA) is a frequently applied, systematic method for analyzing an item in order to identify its potential failure modes, their likelihood of occurrence, and their effects on the performance of the respective item and of the system that embeds it. This may done in order to document the current status of the item (e.g., for regulatory purposes), as well as in order to derive measures that may lead to improved characteristics of the item such as higher reliability, reduced frequency and/or severity of hazardous events, or optimized maintenance strategies. Moreover, the results of an FMEA may serve as input for other analyses, such as fault tree analysis (FTA). For a more detailed reference of the FMEA method, the reader may consult IEC 60812:2006, on which the following contents are partly based, as well as textbooks such as that by Zio (2007).

An FMEA can be conducted in the design and planning phases of an item, as well as in later stages of development such as during manufacturing and operation phases. It can be performed at any level of detail and may be iteratively applied for the same item in order to refine previous analyses with more detailed information or to assess the effectiveness of improvement measures. The specific approach taken for an FMEA, including the scope and level of detail, usually needs to be adapted to the specific item or process, to the life cycle phase, as well as to the overall objectives of the analysis. It is customary to distinguish between

- System FMEA, which usually considers an item concept and analyzes the failure modes and effects with respect to the key risks in the system design. The level of detail is typically restricted to subsystems, their interfaces, and corresponding functions. The goal is to assess and develop a promising product concept, which is feasible within given boundary conditions.
- Design FMEA, which considers either a previously existing design (which may be due to design changes, changes in environmental conditions, etc.) or a new item design and helps determine potential failure modes of the item. The level of detail of the analysis is typically as low as the component level, including the functions. The goal is to develop a flawless and robust design of the item.
- Process FMEA, which considers an existing or planned process and helps determine potential failure modes of the process. The analysis considers the process steps and their functions (purposes) to the desired level of detail. The goal is typically to develop a faultless and robust manufacturing, maintenance, or installation process for a given item.

If the failure modes are to be prioritized according to some criticality measure, then the process is called failure modes, effects, and criticality analysis (FMECA). In a safety-related context, an FMEA may be focused on safety-relevant failure modes with corresponding failure rates, where the possibility of diagnosis of the failure modes is included and quantified. This is then referred to as a failure modes, effects, and diagnostic analysis (FMEDA), as demonstrated, e.g., in IEC 61508:2010.

2.2.1 Procedure of an FMEA

The FMEA procedure is usually composed of the steps of planning, performing, and documentation. It is typically a team effort composed of an analyst/facilitator who is an expert in the FMEA methodology and involved in all steps (including planning and communicating with management and other stakeholders) as well as domain experts who contribute applicable domain knowledge and experience for the item or process under consideration. Moreover, it also usually requires support by management for defining of the objectives of the FMEA and for approving the implementation of the resulting improvement measures and recommendations.

In the following, the typical tasks for carrying out an FMEA are described in more detail:

- Planning phase: This usually involves the definition of the objectives of the analysis and the identification of the analysis boundaries and operating conditions. For large and complex systems, this may also mean that the overall system needs to be divided into subsystems, and an FMEA is performed on each of them, so that the individual analyses remain reasonably sized. Also, the appropriate level of detail at which the analysis is to be carried out must be determined. Required input for the planning phase may include drawings and flowcharts and details of the environment in which the system operates. Before performing the analysis, it must be decided whether or not a criticality analysis will be part of the procedure, and criteria for deciding which failure modes require improvement measures should be defined (see also criticality analysis in the following). Finally, the way in which the FMEA is documented should be decided in the planning phase.

- Performing phase: Normally, a team of experts guided by the analyst/facilitator divides the item or process into individual elements. As a minimum for each element, (a) the functions, (b) the potential failure modes (i.e., the deviations from the designed function), and (c) the consequences (effects) of failure modes at the local and global levels are identified and recorded. If it helps to achieve the objective of the analysis, and in order to support decisions regarding if and how to implement improvement measures, then the mechanisms that may produce these failure modes (failure causes), as well as detection methods and possible mitigation measures, are identified. For a criticality analysis, it is required to evaluate the relative importance of the failure modes using the chosen criticality analysis method. Finally, the improvement measures that were decided on are recorded. These may change the likelihood of a given failure mode, mitigate the effects, or both. For an FMEDA, it should be recorded whether the failure mode is potentially safe or dangerous.

- Throughout the FMEA process, the documentation of all relevant information should be done (including the results of the planning phase) and updated if necessary. The subsequent analysis should then be documented and reported as initially decided (which may be achieved by means of electronic spreadsheet documents or information in a database, if a specialized tool is used). An example of the documented analysis as a spreadsheet document is given in Figure 2.1, which is based on the study by Dinmohammadi et al. (2016). As a general rule, all information should be documented in a clear and comprehensive manner so that the analysis and all resulting conclusions can be understood by a person not involved in the FMEA.

2.2.2 Criticality Analysis

When performing an FMECA, failure modes are prioritized according to some criticality measure that results from a combination of multiple parameters (normally, the severity of the global failure effect of a failure mode, its likelihood of occurrence, and possibly some measure of detectability). If only one of these parameters is used for ranking the failure modes, then this is still commonly considered an FMEA rather than an FMECA.

The possible range of values of the derived criticality measure, as well as the scales that are used for each of the parameters and the method for deriving the criticality measure, should have been determined during the planning phase. Prioritization can then be carried out by the experts and analysts, either as part of the analysis for each failure mode or following the identification of all failure modes and their derived criticality measure. The resulting prioritized failure modes are then either assigned improvements measures, or alternatively, the failure mode is accepted as is depending on the decision criteria agreed upon during the planning phase. These may be influenced by the efficacy of available improvements measures, the required effort to implement them, and their potential impact on the overall system.

There are at least two common methods of combining the individual parameters in order to yield a criticality measure:

- A criticality matrix, in which qualitative, quantitative, or semiquantitative scales for each of the parameters (usually, the severity of a failure mode and its likelihood of occurrence) describe the rows and columns of a matrix; a criticality rank is then

No.	Characteristic (Part/Step)	Function	Failure mode	Potential		Current state				
				Effect of failure (Local/Global)	Cause of failure	Prevention/ detection measures	O	S	D	RPN
1	Door drive—motor and gearbox	Drive spindle and locking shaft	No drive to spindle and locking shaft	Door will not open or close automatically but can be moved manually	Electrical failure or wire rupture	Detectable at railway station	4	7	2	56
				Door blocked in open position	Mechanical failure and fracture	Detectable at railway station	2	8	2	32
				Door blocked in closed position, also in case of emergency	Mechanical failure and fracture	Not detectable between railway stations in case of emergency	2	9	6	108
2	Door open push button	Transmit opening signal after passenger action	No transmission	Door cannot be opened by passenger, door must be locked manually	Sensor defect or wire rupture	Detectable at railway station	3	7	2	42
			Permanent transmission	Door always open at railway stations	Sensor defect or permanent short circuit	Detectable at railway station	3	7	2	42
			Unintended transmission	Spurious opening at railway stations when door release is present	Spurious short circuit	Detectable at railway station	3	5	2	30

FIGURE 2.1
Spreadsheet documentation of the results of an FMEA. (Based on F. Dinmohammadi et al., *Urban Rail Transit*, 2, 128–145, 2016.)

		Severity			
		Insignificant	Minor	Major	Catastrophic
Likelihood of occurrence	High	1	2	3	3
	Medium	1	2	3	3
	Low	A	1	2	3
	Remote	A	A	1	2

FIGURE 2.2
Example of a criticality matrix.

assigned to each of the cells within the matrix. The criticality rank is associated with specific improvement measures for the respective failure mode, or no action is required. An example of a criticality matrix is shown in Figure 2.2, where the criticality may assume the values A (acceptable), 1 (minor), 2 (major), and 3 (unacceptable).

- A risk priority number (RPN), which is a method that multiplies values derived from semiquantitative assessments made on ordinal scales for the severity of a failure mode, its likelihood of occurrence, and possibly some measure of detectability (a failure mode is given a higher priority if it is more difficult to detect), respectively. The range of the resulting RPN values depends on the scales for the parameters. Usually, ordinal rating scales from 1 to 10 are used, therefore resulting in RPN values between 1 and 1000 if three parameters are combined (as illustrated in the spreadsheet in Figure 2.1). The failure modes are then ordered according to their RPN values, and higher priority is commonly assigned to higher RPN values. This approach has some limitations if not used cautiously, and therefore, alternative RPN methods have been developed in order to provide a more consistent assessment of criticality by using addition of the parameter values instead of multiplication, provided that the parameters can be quantified on a logarithmic scale; see, e.g., Braband (2003).

FMECA usually provides a qualitative ranking of the importance of the individual failure modes, but can yield a quantitative output if suitable failure rate data and quantitative consequences are available.

2.2.3 Strengths and Limitations

The strengths of the FMEA/FMECA methodology are that (a) it is widely applicable to human operator, equipment, and system failure modes and to hardware, software, and processes and it is (b) a systematic approach to identify component failure modes, their causes, and their effects on the system and to present them in an easily readable format. For instance, it helps avoid the need for costly equipment modifications in service by identifying problems early in the design process.

Among others, the limitations of the FMEA/FMECA methodology are that (a) it can generally only be used to identify single failure modes, not combinations of failure modes (in which case other methods like FTA are more appropriate) and (b) an analysis can be time consuming and, therefore, costly, if it is not properly planned and focused. Even if this is the case, it can involve large efforts for complex systems.

To overcome some of the limitations of the FMEA, researchers have introduced modified forms of FMEA. For example, the approach called interface and function failure modes effects and criticality analysis (IFF-MECA) provides a framework to handle the system properties, functions, interfaces, components, and combination of all these together with external events for assessing RAM and safety in railways; see Mahboob, Altmann, and Zenglein (2016).

2.3 Reliability Block Diagrams and Boolean Methods

A reliability block diagram (RBD) is an effective means to visually represent and quantitatively assess the behavior of a system with regard to functioning or failure and, thus, may help decide upon possible improvement measures in the system logic configuration. An RBD shows the logical connection of (functioning) components needed for successful operation of the system, which essentially makes it a graphical representation of Boolean expressions linking the success state (up state) of a system (i.e., the overall RBD) to the success states (up states) of its components.

The RBD methodology is analogous to FTA in the sense that the underlying mathematical relations are the same; however, an RBD is focused on system success while the FTA is focused on system failure. A good reference for RBDs is IEC 61078:2016 as well as textbooks such as that by Birolini (2007).

2.3.1 Basic Models for RBDs

The basic assumptions of the RBD methodology are the following:

- The system, as well as the individual blocks that it is made of, have only two states (success/up state or failure/down state).
- The RBD represents the success state of a system in terms of connections of the success states of its components (success paths).
- The individual components (blocks) are assumed statistically independent.

Essentially, an RBD is constructed by drawing the various success paths between the input and output of the diagram that pass through the combinations of blocks that need to function in order for the system to function. The elementary structures that form an RBD are shown in Figure 2.3 and explained in the following. Hereby, the input and output of the diagram are represented by the ends of the connection, which are usually interchangeable.

- Series structures imply that all the blocks need to function in order for the system to function. In terms of Boolean expressions, if S represents the event that the system is in an up state, then we have $S = A \cap B \cap C \ldots \cap Z$, where $A, B, C, \ldots$ represent the events that the corresponding blocks are in an up state.
- For a parallel structure, only one out of a given number of blocks needs to function in order for the system to function. We then have the Boolean expression $S = A \cup B \cup C \ldots \cup Z$ (where the same symbols as before are used).
- For M/N structures, it is assumed that at least M out of N blocks need to function in order for the system to function, which is sometimes referred to as a "majority

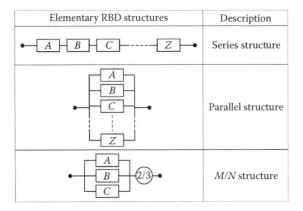

FIGURE 2.3
Elementary structures of RBDs.

vote" structure. It should be noted that this is different from the K/N gate used for FTA. The latter is expressed in terms of failed components, where K/N would be equal to $(N - M + 1)/N$. The Boolean expression corresponding to the 2/3 structure shown in Figure 2.3 is $S = (A \cap B) \cup (A \cap C) \cup (B \cap C)$.

It is quite evident that the elementary structures presented earlier can be combined to form more complicated RBDs (see upper part of Figure 2.4). Other structures, for instance with common blocks as shown in the lower part of Figure 2.4, cannot be expressed in terms of the preceding elementary structures. For instance, the Boolean expression for the combination of series and parallel structures is $S = D \cap [(A1 \cap B1 \cap C1) \cup (A2 \cap B2 \cap C2)]$. For the structure with the common block, the Boolean expression amounts to $S = (B1 \cap C1) \cup (B2 \cap C2) \cup [A \cap (C1 \cup C2)]$.

For large RBDs, a system can be divided into smaller RBDs and then connected via so-called *transfer gates*.

2.3.2 Qualitative and Quantitative Aspects

In principle, the following two approaches can be used for a qualitative analysis of an RBD. One can identify either of the following:

- All minimal combinations of blocks in up states that cause the system being in the up state (minimal tie sets)
- All minimal combinations of the blocks in down states that cause the system being in the down state (minimal cut sets)

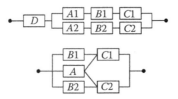

FIGURE 2.4
More complicated structures of RBDs.

In both cases, the Boolean expression for the system success and system failure event is equal to the union over the minimal tie and cut sets, respectively, where the event of a block being in a down state is denoted by, say, $\overline{A}$. For instance, for the structure with the common block in Figure 2.4, the minimal tie sets are given by $(B1 \cap C1)$, $(B2 \cap C2)$, $(A \cap C1)$, and $(A \cap C2)$, while the minimal cut sets are given by $(\overline{C1} \cap \overline{C2})$, $(\overline{A} \cap \overline{B1} \cap \overline{B2})$, $(\overline{A} \cap \overline{B2} \cap \overline{C1})$, and $(\overline{A} \cap \overline{B1} \cap \overline{C2})$. Generally, more insight is gained by considering minimal cut sets, as those with the lowest number of events may represent potential weak points of the system (or even single points of failure).

If an RBD is to be evaluated quantitatively, the following situations can be generally distinguished for simple series and parallel structures:

- For the case of constant probability P_i of a block i being in an up state, or a time-dependent survival probability $R_i(t)$ for nonrepairable blocks, the formulas as given in Table 2.1 apply.
- For the case of repairable blocks with simultaneous repair, the respective equations are given in Table 2.2. The steady-state availability values in the Boolean formulas could be replaced by instantaneous block availabilities $A_i(t)$, which may result from periodically tested blocks. The latter is useful for calculating measures such as the average probability of failure on demand in a functional safety context; see, e.g., IEC 61508:2010.

The equations given earlier describe only the simplest structures. For more complicated structures such as RBDs with dependent blocks, or large RBDs, more sophisticated

TABLE 2.1

Formulas for the System Probability Measures for Nonrepairable Blocks

System Reliability Measure	Series Structure	Parallel Structure (Hot Standby)
Survival probability $R_S(t)$	$\prod_i R_i(t)$	$1 - \prod_i \left(1 - R_i(t)\right)$
Survival probability $R_S(t)$ (constant failure rates λ_i)	$\prod_i \exp(-\lambda_i t)$	$1 - \prod_i \left(1 - \exp\left(-\lambda_i\, t\right)\right)$
Mean operating time to failure MTTF$_S$ (constant failure rates λ_i)	$\left(\sum_i \lambda_i\right)^{-1}$	$\int_0^\infty R_S(t)\, dt$

TABLE 2.2

Formulas for the System Probability Measures for Repairable Blocks

System Reliability Measure	Series Structure	Parallel Structure (Hot Standby)
Steady state availability A_S	$\prod_i A_i$	$1 - \prod_i (1 - A_i)$
Mean operating time between failure MTBF$_S$ (constant failure rates λ_i and repair rates μ_i)	$\left(\sum_i \lambda_i\right)^{-1}$	$\left(\prod_i \frac{\lambda_i}{\mu_i} \sum_i \lambda_i\right)^{-1}$ for $\lambda_i \ll \mu_i$
Mean down time MDT$_S$ (constant failure rates λ_i and repair rates μ_i)	$\sum_i \frac{\lambda_i}{\mu_i} \left(\sum_i \lambda_i\right)^{-1}$ for $\lambda_i \ll \mu_i$	$\left(\sum_i \mu_i\right)^{-1}$

algorithms have to be used; see, e.g., IEC 61078:2016, which are implemented in commercial analysis packages. Dynamic RBDs allow the modeling of standby states where components fail at a lower rate, as well as shared spare components and functional dependencies.

2.3.3 Strengths and Limitations

The strengths of RBDs lie in the fact that most system configurations can be described in the form of a compact, easily understandable diagram, which can often be derived from functional diagrams of the system. Moreover, qualitative and quantitative assessments can be obtained from RBDs, as well as certain importance measures (see Section 2.7). The limitations are that the application of RBDs cannot describe systems where the order of failures is to be taken into account or complex repair strategies are to be modeled (e.g., repaired blocks do not behave independently from each other). In these cases, other techniques such as Monte Carlo simulations, Markov models, or stochastic Petri nets may be more appropriate (see the discussions in the accompanying Chapter 3).

2.4   Fault Tree Analysis

Fault tree analysis (FTA) is an established methodology for system reliability and availability analysis, as well as safety analysis. It was originally developed in 1962 at Bell Laboratories; see Ericson (1999). Based on a logical representation of a technical system, it provides a rational framework for modeling the possible scenarios that contribute to a specified undesired event, referred to as the *top event* (TE) of the tree. These scenarios originate from so-called *basic events* (BE) at the bottom of the tree and are described by a series of logical operators and intermediate events leading to the TE. The system is analyzed in the context of its operational and safety requirements, in addition to its environment, to find all combinations of BEs that will lead to the occurrence of the TE. It is, hence, a deductive method that investigates a system by understanding in which ways a system can fail, rather than looking at a system in terms of how it can successfully operate. A graphical representation is called a *fault tree* (FT) and describes the relationship between TE, intermediate events, and BEs using symbols for events, for gates that describe the relationship between events using Boolean logic, as well as transfer symbols. A comprehensive reference for FTA is, e.g., IEC 61025:2006 or textbooks like that by Zio (2007); for an overview of classical as well as modern techniques related to FTA, the reader is referred to, e.g., Xing and Amari (2008).

2.4.1   Standard FTA Methodology and Construction Guidelines

The basic assumptions of the standard FTA methodology are the following:

- The FTA is based on events. Thus, it can only represent random variables with binary states.
- BEs are assumed to be statistically independent.
- The relationship between events is represented by the simple logical gates given in the following.

The most basic constituents of the standard FTA are shown in Figure 2.5 and explained in the following:

- The TE represents the undesired event, which is usually a defined system failure mode or accident.
- BEs represent the basic causes that contribute to the undesired event. These could be failures of hardware components or software failures, human errors, or any other events. No further investigation of failure causes leading to a BE failure is done.
- Intermediate events are events other than TEs or BEs, which are defined through logical gates that represent combinations of BEs and/or other intermediate events. In particular,
  - OR gate: The intermediate event occurs if one of the subordinate events occur.
  - AND gate: The intermediate event occurs if all of the subordinate events occur.
  - *K/N* gate: The intermediate event occurs if at least *K* of the *N* subordinate events occur.

In constructing an FT, the TE is analyzed by first identifying its immediate and necessary causes. Each of those is then investigated and further refined until the basic causes of failure, i.e., the BEs, are reached and not further investigated. In doing so, it is essential to define boundary conditions for the analysis, including the physical boundaries and environmental conditions, as well as the level of resolution of the model. Moreover, the TE should be defined in a clear and unambiguous manner. Finally, the FT should be developed in levels, and each level should be completed before investigating the next lower level. More complex FTs can be dealt with in an easier way, if they are split into subtrees and combined via transfer gates.

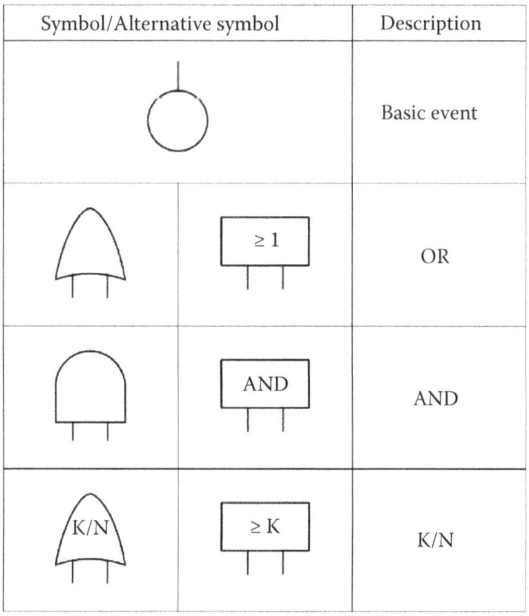

Symbol/Alternative symbol		Description
		Basic event
	≥ 1	OR
	AND	AND
K/N	≥ K	K/N

FIGURE 2.5
Basic symbols used for standard FTA methodology.

2.4.2   Qualitative and Quantitative Analyses; Advanced FTA Methodology

An FTA can be used in order to perform a qualitative or quantitative analysis. A qualitative analysis usually has the objective to identify potential causes and pathways to the TE. This can be done by calculating the minimal cut sets, that is, the sets of BEs whose simultaneous occurrence leads to occurrence of the TE and which are minimal in the sense that if one of the BEs is removed from the minimal cut set, then the TE does not occur any longer (see also Section 2.3).

The objective of a quantitative FTA is, in the case of nonrepairable systems, to calculate the probability of the TE. For repairable systems, the FTA is used to calculate measures such as mean operating time between failure, mean down time, and availability. As static, coherent FTs can be converted into RBDs, the results from Section 2.3 can be analogously applied, noting that the blocks in an RBD represent functioning components, while events in an FT represent failure events.

FTs can be broadly classified into coherent and noncoherent structures. Roughly speaking, coherent structures do not improve after failure of a component, i.e., they do not use inverse gates, which would require other analysis techniques (see, e.g., Xing and Amari [2008]). Noncoherent FTs can include inverse gates such as NOT gates or Exclusive-OR gates.

Coherent FTs may be further classified into static and dynamic FTs. Static FTs use only static coherent gates such as the ones described earlier, and the occurrence of the TE is independent of the order of occurrence of the BE. Dynamic FTs allow for taking into account the sequence of occurrence of BEs, which is useful for modeling more sophisticated systems and maintenance strategies using gates such as functional dependency gates, cold spare gates, and Priority-AND gates. An overview of dynamic FTs including appropriate numerical techniques is given e.g., by Xing and Amari (2008). For a recent survey of the current modeling and analysis techniques, see Ruijters and Stoelinga (2015). These references also discuss other topics such as dependent BEs.

2.4.3   Maintenance Strategies

As a result of an FTA that involves nonrepairable components and redundancies, the time-dependent failure time (occurrence of the TE) distribution function $F(t) = 1 - R(t)$ corresponds to a failure rate $\lambda(t) = \mathrm{d}F(t)/(R(t) \cdot \mathrm{d}t)$ that is usually increasing, which means that it describes some kind of aging (by successive failures of redundant components). In this case, one might ask what the effect of performing maintenance actions is on the mean time to failure, because failure of a system during operation may be costly or dangerous.

A possible maintenance strategy is the so-called *age replacement* strategy; i.e., the system is always and immediately replaced or restored to a state as good as new after failure or after a scheduled time T, whichever comes first. Then, the mean operating time to failure depending on T is obtained from

$$\mathrm{MTTF}(T) = \frac{\int_0^T R(t)\,\mathrm{d}t}{F(T)}.$$

If $F(t)$ follows an exponential distribution with constant failure rate λ (i.e., no ageing), then, consequently, MTTF $(T) = 1/\lambda$, and there is no effect of T. Another possible maintenance strategy is called the *block replacement* strategy, where the system is replaced or restored to a state as good as new at regular intervals irrespective of the failure history. For a comprehensive treatment of inspection and maintenance strategies, see, e.g., Barlow and Proschan (1965), Aven and Jensen (1999), or Pham and Wang (1996).

2.4.4 Applications in Railway Industry

An example of an FT developed in the manner described earlier is shown in Figure 2.6, which is adapted from Mahboob (2014). It is mentioned here that the example is created for academic purposes and considers limited failure scenarios only. For example, there can be other reasons for train derailment such as broken rail, broken wheel, broken axle, deteriorated track superstructure, and uneven or heavy shifted loads (in case of goods trains). A complete analysis of functional safety in railways or accident analysis such as any train derailment accident is beyond the scope of this chapter.

The possible scenarios leading to the TE are as follows: the intermediate event of a signal passed at danger (SPAD) will occur when a train is approaching a red signal, and there are (a) combined failures of train protection and warnings with errors toward brake

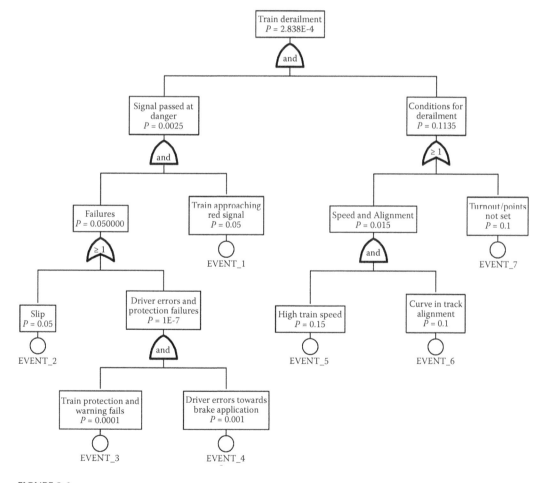

FIGURE 2.6
Example of a static FT for a train derailment accident. (Adapted from Q. Mahboob, *A Bayesian Network Methodology for Railway Risk, Safety and Decision Support*, PhD thesis, Technische Universität Dresden, Dresden, 2014.)

application or (b) slip due to poor adhesion between the rail and wheels. The TE representing train derailment can occur when the SPAD is followed by a (a) a turnout/point with prevented route or (b) a curve in track alignment as well as high train speed. It is assumed that the driver has no information on a slippery track conditions. Therefore, the driver is unable to take care of slip-related aspects during brake application.

2.4.5 Strengths and Limitations

The strengths of FTA are that it is a systematic, deductive approach that can account for a variety of failure causes, including human interactions, and it is especially useful for analyzing systems with many interfaces and complex combinations that lead to system failure. Moreover, the graphical representation makes it easy to understand the system behavior and the factors included. Its limitations include that sequences of occurrence of events are not addressed, and the FTA deals only with binary states (in the latter case, one may resort to methods described, e.g., by Lisnianski, Frenkel, and Ding (2010). Further, FTAs consider only one TE or failure mode at a time, respectively. However, there are recent extensions of classical FTA such as component fault trees (CFTs) to overcome this limitation; see, e.g., Adler et al. (2010). Moreover, CFTs allow the creation of FTAs in a compositional way and allow their elements to be systematically reused.

2.5 Event Tree Analysis

Event tree analysis (ETA) is an inductive technique that, starting with an initiating event, considers the mutually exclusive sequences of events that could arise. These ultimately lead to different possible outcomes, some of which undesired or unsafe. Moreover, the probabilities or frequencies of occurrence of the outcomes, given the initiating event, can be quantified by means of basic mathematical operations.

ETAs can be used for risk analyses, where the initiating event represents a potentially hazardous event and the sequence of events following the initiating event represents mitigating factors (protection layers such as human actions or safety systems interventions). The outcomes are then distinguished into the ones where the unsafe consequences could be prevented and the ones ultimately leading to unsafe consequences. In the process industry, the so-called *layer of protection analysis* constitutes an adaption of an ETA for the assessment of potential hazards.

The basic principles of ETA are described, e.g., in IEC 62502:2010 and in textbooks such as that by Zio (2007).

2.5.1 Construction of an Event Tree

The following steps are usually taken for construction of an event tree:

- Definition of the system under consideration including system boundaries and systematic identification of the list of initiating events that require detailed investigation by means of an ETA. Initiating events that could be excluded may be those which are considered sufficiently unlikely or of negligible consequences.

- Identification of the mitigating factors that could potentially reduce the probability or frequency of occurrence of the initiating events. These could include technical devices such as monitoring devices, protection devices, human interaction, and physical phenomena. The effect of each of these factors is characterized by possible events, e.g., successful and unsuccessful prevention of harmful consequences due to a protection device. Usually, two or more events describe the effect of each factor, which leads to branching of the subsequent event tree into two or more branches, respectively. The probabilities of these events are normally conditional probabilities, conditional on the initiating event, and conditional on the assumed effects of the preceding mitigating factors. It is an important thing to note that the effectiveness of the factors depends on the assumed previous events along the branch of the event tree.

- Identification of a suitable (temporal or logical) ordering of the potentially mitigating factors and drawing of the branches of the event tree considering the assumed ordering.

- Qualitative or quantitative analysis of the resulting event tree (as demonstrated in the following), which results in probabilities or frequencies of occurrence for each of the outcomes. The individual outcomes may be categorized into degree of severity or damage (as a simple example the two categories safe or dangerous) and possibly combined in order to obtain the probabilities or frequencies of occurrence for each category.

The results of the qualitative or quantitative analysis may support decisions regarding the acceptability of the resulting risk, the need for additional mitigating factors, the need for changes in the design, and the like.

2.5.2 Qualitative and Quantitative Analyses

The goal of a qualitative analysis by means of an event tree includes gaining a deeper understanding of the possible sequences of events that lead to various outcomes and identifying dependencies among the initiating events and potential mitigating factors (including functional, structural, and physical dependencies, such as common causes).

For the principles of a quantitative analysis, consider the event tree shown in Figure 2.7, which was adapted from IEC 62502:2010.

The tree is typically drawn from left to right, starting with the initiating event *I* and nodes that indicate the effects of certain potential factors with associated events that lead to branching of the tree. In the figure, there are two mitigating factors *A* and *B* (possibly protection devices) and four outcomes α, β, γ, and δ that are each the result of a given sequence of events.

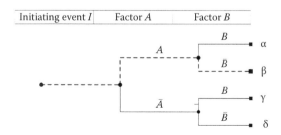

FIGURE 2.7
Event tree with possible outcomes and a given sequence of events highlighted. (Adapted from IEC 62502:2010. *Event tree analysis (ETA),* IEC, Geneva, 2010.)

Consider the sequence of events that leads to outcome β. Mathematically, the probability of this outcome is expressed as

$$P(\beta) = P(I \cap A \cap \bar{B}) = P(I) \bullet P(A|I) \bullet P(\bar{B}|I \cap A),$$

where the effects of each of the factors is described here by the successful or unsuccessful functioning of the protection device, say, A and $\bar{A}$ (always conditional on the previous events). As the sequences of events that lead to the outcomes are mutually exclusive, we have $P(\alpha) + P(\beta) + P(\gamma) + P(\delta) = P(I)$. If the initiating event is characterized by a frequency of occurrence $f(I)$ instead of a probability, then

$$f(\beta) = f(I) \bullet P(A \cap \bar{B}|I) = f(I) \bullet P(A|I) \bullet P(\bar{B}|I \cap A).$$

It was mentioned before that the ordering of potentially mitigating factors should be determined based on temporal or logical arguments. Consider the case where the protection device A shown in Figure 2.7 requires a functioning protection device B. In other words, $P(A|\bar{B}) = 0$ and $P(\bar{A}|\bar{B}) = 1$. This implies, by virtue of Bayes's theorem, that $P(\beta) = P(I \cap A \cap \bar{B}) = 0$ and $P(\delta) = P(I \cap \bar{A} \cap \bar{B}) = P(I \cap \bar{B})$. While the evaluation of the event tree from Figure 2.7 would still yield a correct result, in this case, it would be better to swap the ordering of factors A and B, as shown in Figure 2.8, again adapted from IEC 62502:2010.

On the contrary, if the probability of successful or unsuccessful functioning of the protection devices are independent of the other factors, as well as independent of the initiating event, then, for instance,

$$P(\beta) = P(I \cap A \cap \bar{B}) = P(I) \bullet P(A) \bullet P(\bar{B}).$$

In cases where the effects of potentially mitigating factors can be described by more events than just binary events, we would get more than two branches at a node and, for instance, deal with events such as $P(B_1|A)$, $P(B_2|A)$, and $P(B_3|A)$.

An ETA can be combined with other modeling techniques, for instance, FTA. In this case, FTA may be used to calculate the probability or frequency of the initiating event and the conditional probabilities.

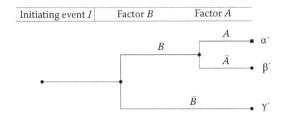

FIGURE 2.8
Event tree from previous figure with different ordering. (Adapted from IEC 62502:2010. *Event tree analysis (ETA)*, IEC, Geneva, 2010.)

2.5.3 Applications in Railway Industry

As a continuation of the train derailment example from the previous section on FTAs, consider the event tree shown in Figure 2.9, which is a simplified example from Mahboob (2014) that starts with the potentially hazardous event of train derailment and models the sequence of potentially mitigating factors following the initiating event. It is mentioned again that the example is created for academic purposes and considers limited failure scenarios only. Complete analysis of functional safety in railways or accident analysis such as any train derailment accident is beyond the scope of this chapter.

The occurrence or non-occurrence of four mitigating factors included in the ETA can lead to the seven outcomes with different degrees of severity. For example, accidents α, β, and ε do not involve hitting a line side structure and can be regarded as minor. In contrast, accidents δ and θ, where the train hits a line side structure and individual vehicles collapse, are catastrophic events with many fatalities.

2.5.4 Strengths and Limitations

The strengths of ETAs are that they provide a structured and systematic investigation and quantification of potential scenarios following an initiating event as well as the influence of subsequent, potentially mitigating factors, which eventually lead to various outcomes. The applicability of ETAs is limited if the probabilities of the considered events are dependent on time, in case of which, Markov models could be more appropriate. Also, event trees can be difficult to construct for large, complex systems.

2.6 Bowtie Method

The bowtie method (also called a *cause–consequence diagram*) is a risk evaluation method that is used to assist in the identification and management of risks (or hazards in a safety context). It visualizes the causes and consequences of a hazardous event by means of a bowtie diagram, as well as control measures and escalation factors. The right side of the diagram (see Figure 2.10) is similar to an event tree describing the possible consequences that may result from a hazardous event, while the left side of the diagram resembles a simple FT explaining the possible causes of the event.

The bowtie method is frequently used for the analysis of major hazards and to ensure that these are properly managed and mitigated by control measures. Moreover, it identifies how these control measures may fail (so-called *escalation factors*). It can be a reactive or proactive approach, by identifying scenarios that have already occurred or which have not occurred yet. Due to its visual and intuitive nature, the bowtie method facilitates a common understanding of hazards by all parties involved.

For an application of the bowtie method to railway applications, see, e.g., Trbojevic (2004).

2.6.1 Construction of a Bowtie Diagram

Bowtie diagrams are most often developed in a team effort. Its construction usually involves the following initial steps (see also Figure 2.10):

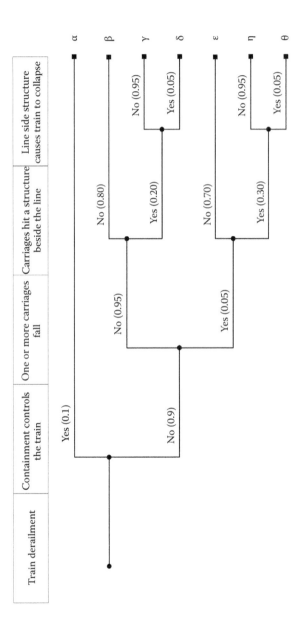

FIGURE 2.9
Exemplary event tree for a train derailment accident. (Modified from Q. Mahboob, *A Bayesian Network Methodology for Railway Risk, Safety and Decision Support*, PhD thesis, Technische Universität Dresden, Dresden, 2014.)

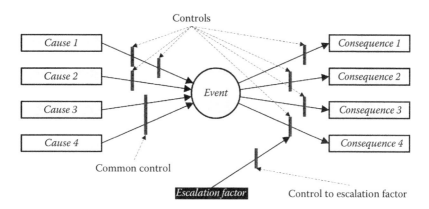

FIGURE 2.10
Exemplary bowtie diagram.

- Identification of the risk or hazardous event that may potentially cause damage: It is represented by the central node of a bowtie diagram.
- Identification of the causes that contribute to the hazardous event: These are listed at the left-hand side of the central node and connected to the node by lines, which indicate that these are the different mechanisms by which causes lead to the event.
- Identification of the potential consequences resulting from the hazardous event, i.e., the results of the hazardous event directly ending in loss or damage: These are listed on the right-hand side and connected to the central node by lines.

Afterwards, the bowtie diagram is extended by identifying controls (or barriers) that influence the various scenarios from the causes of the hazardous event to the potential consequences. These could be technological controls (e.g., protection devices) or human interaction. One can distinguish between the following:

- Controls that modify the likelihood of each cause leading to the hazardous event: These are depicted as vertical bars across the lines from the respective cause to the central node.
- Controls that interrupt the evolution of the hazardous event into the consequence: These are shown as vertical bars across the lines emanating from the central node to the respective consequence.

Finally, the bowtie diagram is completed by identifying the following:

- Factors that might cause the controls to fail (escalation factors): These could, in turn, be extended by additional controls that influence the escalation factors.
- Management functions which support the effectiveness of controls: These can be shown below the bowtie diagram and connected to the respective control.

Although some degree of quantification by means of a bowtie diagram could be possible in simple cases, one would rather resort to methods such as FTA or ETA for determination of the probability of frequency of occurrence of the hazardous event or its consequences.

2.6.2   Strengths and Limitations

Among the strengths of bowtie analysis is that it is an easy-to-use and visual tool that clearly illustrates a hazardous event as well as its causes and consequences. Hereby, it draws attention to controls which are supposed to be in place and their effectiveness. Its limitations include that it cannot adequately reflect the situation where combinations of causes lead to consequences. Further, it is not well suited for quantification purposes.

2.7   Component Importance Measures

Component importance measures (CIMs) are useful in ranking components in a system according to their impact on the probability of success (or failure) of the overall system. For instance, by making use of CIMs, one can prioritize the components according to their safety importance for careful protection and maintenance; see Birnbaum (1969), Fussell (1975), and Borgonovo et al. (2003). To this end, CIMs (a) enable the designers to identify the components that should be improved, (b) helps maintenance planners to improve maintenance plans for important components, and (c) facilitates decision making such as utilization of engineering budgets by prioritizing the components. For application of CIMs to transport industry, readers are referred to Chen, Ho, and Mao 2007; Zio, Marella, and Podofillini (2007); and Mahboob et al. (2012a).

A number of CIMs have been developed, and there are at least two ways of computing them within system modeling frameworks; namely, FTA and Bayesian networks; see e.g., Mahboob et al. (2012b). In the following, the classical method of computing several CIMs for system components by means of FTA is expounded.

2.7.1   Definitions of CIM

The CIMs are classified as follows: The structural importance measure (SIM) considers the importance of a component due to its position in the logic of the system and does not consider the reliability of the component. Hence, a SIM cannot distinguish components that occupy the same structural position but have different reliabilities. Alternatively, the reliability importance measure (RIM) considers both the structural position and the reliability of the component in a system. Various importance measures corresponding to the RIM are shown in Table 2.3.

2.7.2   Computation of CIM by Using FTA

Importance measures are difficult to calculate by hand on realistic system models. The classical method of computing CIM for system components is within the FTA.

For the conditional probability (CP) and risk achievement worth (RAW), the conditional probability of system failure given failure of component i can be obtained by evaluating the FT under the assumption $P(F_i) = 1$. Similarly, for the risk reduction worth (RRW) one can evaluate the FT, assuming $P(F_i) = 0$.

The definition of the diagnostic importance factor (DIF) from Table 2.3 can be extended as follows:

$$\text{DIF}_{(i)} = P\left(F_i|F_s\right) = \frac{P(F_s \cap F_i)}{P(F_s)} = \frac{P\left(F_s|F_i\right) \cdot P(F_i)}{P(F_s)} = \text{CP}_{(i)}\frac{P(F_i)}{P(F_s)},$$

and the evaluation is straightforward.

TABLE 2.3

Definitions of CIM

CIM	Description	Mathematical Definition
CP	CP gives the probability of system failure given the failure of an individual component i	$CP_{(i)} = P\left(F_s\middle|F_i\right) = \dfrac{P(F_s \cap F_i)}{P(F_i)}$
RAW	RAW measures the value of component i in achieving the present level of system reliability	$RAW_{(i)} = \dfrac{P\left(F_s\middle|F_i\right)}{P(F_s)} = \dfrac{CP_{(i)}}{P(F_s)} = \dfrac{P(F_s \cap F_i)}{P(F_i) \cdot P(F_s)}$
RRW	RRW measures the decrease of the system unreliability by increasing the reliability of component i	$RRW_{(i)} = \dfrac{P(F_s)}{P\left(F_s\middle|\overline{F_i}\right)}$
DIF	DIF gives the failure probability of component i conditional on system failure	$DIF_{(i)} = P\left(F_i\middle|F_s\right) = \dfrac{P(F_s \cap F_i)}{P(F_s)}$
BM	BM measures the sensitivity of the system unreliability with respect to the changes in the failure probability of component i	$BM_{(i)} = \dfrac{\partial P(F_s)}{\partial P(F_i)}$
FV importance	The standard FV failure importance is the contribution of the failure probability of component i to the probability of system failure	$FV_{(i)} = \dfrac{P(F_s) - P\left(F_s\middle|\overline{F_i}\right)}{P(F_s)}$
CIF	CIF gives the probability that component i has caused system failure given that the system has already failed	$CIF_{(i)} = \dfrac{P(F_i)}{P(F_s)} \cdot BM_{(i)} = \dfrac{P(F_i)}{P(F_s)} \cdot \dfrac{\partial P(F_s)}{\partial P(F_i)}$
IP	IP gives the improvement potential in system reliability if component i is replaced by a component that is perfectly working	$IP_{(i)} = P(F_i) \cdot BM_{(i)} = P(F_i) \cdot \dfrac{\partial P(F_s)}{\partial P(F_i)}$

Birnbaum's measure (BM) is usually computed numerically by taking partial derivatives or assigning zero to the reliability value of the component, which is readily done by means of an FTA tool. The Fussell–Vesely (FV) measure can be obtained with the conditional probability also used for the RRW. Finally, the criticality importance factor (CIF) and the improvement potential (IP) can also be calculated using the value for BM.

2.8 Decision Tree Analysis

Decision tree (DT) analysis attempts to structure decisions on the expected financial loss or gain or, more generally, on expected utility. It helps to provide a logical basis to justify reasons for a decision that needs to be taken under uncertainty. Using a DT, one can graphically organize a sequential decision process, for instance, whether to initiate project A or project B. As the two hypothetical projects proceed, a range of events might occur and further decisions will need to be made. These are represented in a tree format, similar to an event tree. The expected utility of any decision can then be computed on the basis of the weighted summation of all branches of the tree, from the initial decision to all leaves from that branch. The decisions are then taken to maximize the expected utility.

DT analysis is described, e.g., by Jensen and Nielsen (2007) and Barber (2012). For a classical treatment of statistical decision theory, see Raiffa and Schlaifer (1961).

2.8.1 Construction and Evaluation of a DT

An example of a simple DT is shown in Figure 2.11.

It contains nodes which are either decision nodes (rectangular boxes) or random nodes (circles). The diamond-shaped leaves at the end of the branches represent utility nodes. The branches usually start with a root decision node and the labeled links emanating from the decision node indicate which decision has been taken. Similarly, the links emanating from random nodes indicate the possible realizations of the corresponding random variable. As for event trees, the possible realizations of the random variable, and the corresponding probabilities may be conditional on all previous decisions taken and all previous realizations of random variables from the root node to the current position in the DT. This is sometimes referred to as *nonforgetting* assumption.

The quantitative information of a DT consists of the (conditional) probabilities, as well as a utility value attached to the leaves that represents the utility of the decision scenario identified by the path from the root node to the leaf under consideration. Some hints on the interpretation of uncertainties and probabilities in engineering decision making can be found in the study by Faber and Vrouwenvelder (2008).

The evaluation of a DT is best explained by means of the simplified railway example shown in Figure 2.11. Consider the case where a project developer has to decide whether to build a large or small railway station, under the uncertainty of future demand by passengers. The cost for the large railway station is 100, and if future demand by passengers is high, then the revenue amounts to 200. Therefore, the utility associated with this decision scenario is −100 + 200 = +100. If the future demand by passengers turns out to be low, then the revenue is 120, and consequently, the utility for this scenario equals to −100 + 120 = +20.

If the decision was made to build a smaller railway station at a cost of 50, then it could not completely satisfy a high passenger demand, and the revenue would amount to 150. Hence, the utility for this scenario is −50 + 150 = +100. For low passenger demand, the size of the station would suffice and the revenue is, as earlier, 120 with a utility for the latter scenario of −50 + 120 = +70.

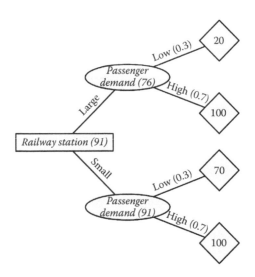

FIGURE 2.11
Example of a DT.

Assume that the probability of high and low demand is 0.7 and 0.3, respectively, which is, in this case, independent from the size of the railway station. Then, the expected utility for a large railway station amounts to $0.7 \cdot 100 + 0.3 \cdot 20 = 76$, whereas for a small railway station it would be $0.7 \cdot 100 + 0.3 \cdot 70 = 91$. Consequently, the expected utility is maximized by building a small railway station.

It would also be possible to consider further decisions, for instance, whether the railway station should be extended given that a small one has been built, and it turned out that the passenger demand is high. In this case, one would start the evaluation backward with the last decision to be made and perform the maximization step for this decision, conditional on the considered decision scenario up to that point in the tree.

Mathematically, the procedure described earlier can be formulated as

$$\mathbf{d}_{opt} = \arg \max_{\mathbf{d}} E\left[u(\mathbf{d}, \boldsymbol{\theta}) \middle| \mathbf{d} \right],$$

where the optimal set of decisions is sought out of a set of decisions $\mathbf{d}$, which maximizes the expected utility given the set of decisions $\mathbf{d}$ and uncertain factors $\boldsymbol{\theta}$. It is quite clear that a DT can become quite complex even for a moderate number of possible decisions and uncertain factors. It is sometimes possible to reduce the complexity by coalescing the tree; see Jensen and Nielsen (2007). A more compact modeling such kinds of decision problems is possible using so-called *influence diagrams*, which are an extension of Bayesian networks described in the accompanying Chapter 3.

It should be clear that the results depend on possibly subjective estimates of the conditional probabilities. Nevertheless, DTs help decision makers to overcome inherent perception biases. Depending on the type of decision to be made, it can be used for operational as well as strategic decisions.

2.8.2 Strengths and Limitations

The strengths of a DT analysis are that it provides a clear graphical representation of the details of a decision problem, and developing the tree yields insights into the problem and an optimal decision strategy under the assumptions that have been made. Among the limitations are that its application is limited to relatively simple decision problems.

2.9 Summary

Several frequently used methods have been presented in this chapter, which are relevant for specification, assessment, and possible improvement of the RAM and safety characteristics of products and systems throughout all phases of the life cycle of a railway project. More specifically, an FMEA is preferably applied for identification of risks in early phases of the life cycle, but should be continually updated. Similarly, the bowtie method and ETA help analyze risks and controls, which should also be proactively performed. RBDs and FTAs, along with CIMs, belong to the class of assessment techniques that are often quantitative and more useful the more information about the product or system is available. Finally, DT analysis supports decision making at any time a decision needs to be taken.

References

R. Adler, D. Domig, K. Höfig et al. Integration of component fault trees into the UML, *Models in Software Engineering. MODELS 2010. Lecture Notes in Computer Science*, Dingel J., Solberg A., Eds., vol 6627, Springer, Berlin, Heidelberg, pp. 312–327.

T. Aven and U. Jensen, *Stochastic Models in Reliability*, Springer, Berlin, 1999.

D. Barber, *Bayesian Reasoning and Machine Learning*, Cambridge University Press, Cambridge, UK, 2012.

R. E. Barlow, and F. Proschan, *Mathematical Theory of Reliability*, Wiley, Hoboken, NJ, 1965.

Z. W. Birnbaum and P. R. Krishanaiah, On the importance of different components in a multi-component system, *Multivariable Analysis II*, P. R. Krishnaiah, Ed., Academic Press, New York, 1969, pp. 581–592.

A. Birolini, *Reliability Engineering: Theory and Practice, Fifth Edition*, Springer, Berlin, 2007.

E. Borgonovo, G. E. Apostolakis, S. Tarantola, and A. Saltelli, Comparison of global sensitivity analysis techniques and importance measures in PSA, *Reliability Engineering & System Safety*, 79(2), pp. 175–185, 2003.

CEN (European Committee for Standardization), EN 50126-1, *Railway Applications—The Specification and Demonstration of Reliability, Availability, Maintainability and Safety (RAMS) Part 1: Basic Requirements and Generic Process*, CEN, Brussels, 2003.

J. Braband, Improving the risk priority number concept, *Journal of System Safety*, 3, 2003, pp. 21–23.

S. Chen, T. Ho, and B. Mao, Reliability evaluations of railway power supplies by fault-tree analysis, *IET Electric Power Applications*, 1(2), pp. 161–172.

F. Dinmohammadi, B. Alkali, M. Shafiee, C. Berenguer, and A. Labib, Risk evaluation of railway rolling stock failures using FMECA technique: A case study of passenger door system, *Urban Rail Transit*, 2(3–4), 2016, pp. 128–145.

C. A. Ericson. Fault tree analysis—A history, *Proceedings of the 17th International Systems Safety Conference*, Orlando, FL. pp. 1–9, 1999.

M. H. Faber and A. C. W. M. Vrouwenvelder, *Interpretation of Uncertainties and Probabilities in Civil Engineering Decision Analysis*, Background Documents on Risk Assessment in Engineering, No. 2, Joint Committee of Structural Safety, http://www.jcss.byg.dtu.dk/, 2008.

J. Fussell, How to hand-calculate system reliability and safety characteristics, *IEEE Transactions on Reliability*, R-24(3), 1975, pp. 169–174.

IEC (International Electrotechnical Commission) 60812:2006: *Analysis Techniques for System Reliability—Procedures for Failure Mode and Effect Analysis (FMEA)*, IEC, Geneva.

IEC 60050-192:2015: *International Electrotechnical Vocabulary—Part 192: Dependability*, IEC, Geneva.

IEC 61025:2006: *Fault Tree Analysis (FTA)*, IEC, Geneva.

IEC 62502:2010: *Event Tree Analysis (ETA)*, IEC, Geneva.

IEC 61508:2010: *Functional Safety of Electrical/Electronic/Programmable Electronic Safety-Related Systems*. IEC, Geneva.

IEC 61703:2016: *Mathematical Expressions for Reliability, Availability, Maintainability and Maintenance Support Terms*, IEC, Geneva.

IEC 61078:2016: *Reliability Block Diagrams*, IEC, Geneva.

F. V. Jensen and T. D. Nielsen, *Bayesian Networks and Decision Graphs*, Springer, Berlin, 2007.

A. Lisnianski, I. Frenkel, and Y. Ding, *Multi-state System Reliability Analysis and Optimization for Engineers and Industrial Managers*, Springer, Berlin, 2010.

Q. Mahboob, B. Altmann, and S. Zenglein, Interfaces, functions and components based failure analysis for rail electrification RAMS assessment, *ESREL 2016*. CRC, Glasgow, 2016. 2330–2334.

Q. Mahboob, *A Bayesian Network Methodology for Railway Risk, Safety and Decision Support*, PhD thesis, Technische Universität Dresden, Dresden, 2014.

Q. Mahboob, E. Schöne, M. Kunze, J. Trinckauf, and U. Maschek, Application of importance measures to transport industry: Computation using Bayesian networks and fault tree analysis, *International Conference on Quality, Reliability, Risk, Maintenance, and Safety Engineering (ICQR2MSE)*, 2012a.

Q. Mahboob, F. M. Zahid, E. Schöne, M. Kunze, and J. Trinckauf, Computation of component importance measures for reliability of complex systems, *10th International FLINS Conference on Uncertainty Modeling in Knowledge Engineering and Decision Making*, 2012b. 1051–1057.

H. Pham and H. Wang. Imperfect maintenance, *European Journal of Operational Research*, 94, 1996, pp. 425–438.

H. Raiffa and R. Schlaifer, *Applied Statistical Decision Theory*, Harvard Business School, Boston, MA, 1961.

E. Ruijters and M. Stoelinga, Fault tree analysis: A survey of the state-of-the-art in modeling, analysis and tools, *Computer Science Review* 15–16, 2015, pp. 29–62.

V. Trbojevic, Linking risk analysis to safety management, *PSAM7/ESREL04*, 2004.

L. Xing and S. V. Amari, Fault tree analysis, *Handbook of Performability Engineering*, K. B. Misra (ed.), Springer, Berlin, 2008, pp. 595–620.

E. Zio, *An Introduction to the Basics of Reliability and Risk Analysis*, World Scientific, Singapore, 2007.

E. Zio, M. Marella, and L. Podofillini, Importance measures-based prioritization for improving the performance of multi-state systems: Application to the railway industry, *Reliability Engineering & System Safety*, 92(10), 2007, pp. 1303–1314.

3

Advanced Methods for RAM Analysis and Decision Making

Andreas Joanni, Qamar Mahboob, and Enrico Zio

CONTENTS

3.1 Introduction

For the complex systems used in the railway industry, the basic methods of reliability, availability, and maintainability (RAM) analysis, and the subsequent decision making as illustrated in the previous chapter may not be sufficient. Thus, in this chapter, we introduce advanced techniques for modeling such complex systems. For the application of these methods to railway systems, please refer to other relevant chapters in this handbook and references therein. Some of the basic mathematical models, expressions, definitions, and terminology related to RAM analysis can be found in IEC 61703:2016 and IEC 60050-192:2015 (for terminology, see also http://www.electropedia.org). The examples presented in this chapter are for academic and illustration purposes only, for instance, by considering limited failure scenarios only.

3.2 Markov Models

When applied in the context of reliability and safety analyses, Markov models provide a quantitative technique to describe the time evolution of a system in terms of a set of discrete states and transitions between them, given that the current and future states of the system do not depend on its state at any time in the past but only on the present state; see, e.g., Zio (2009). Markov models are particularly useful for analyzing systems with redundancy, as well as systems where the occurrence of system failure depends on the sequence of occurrence of individual component failures. They are also well suited for analyzing systems with complex operation and maintenance strategies such as cold standby components, prioritized repair actions, and limited resources for corrective maintenance activities. Considering the system as a whole, the system states can be classified into states with various degrees of degraded performance, therefore allowing the modeling of multistate systems as described, e.g., by Lisnianski, Frenkel, and Ding (2010) and Podofillini, Zio, and Vatn (2004).

Markov models can be visualized by means of state-transition diagrams as shown in the following. Depending on the system and the type of analysis, various transient and stationary reliability and availability measures can be obtained by solving the corresponding model, such as instantaneous and steady-state availability, average availability, mean operation time to (first) failure, and mean down time. In the context of functional safety, Markov models allow the computation of measures such as dangerous failure rate and average probability of failure on demand.

A detailed discussion of Markov models can be found, e.g., in the books by Zio (2009), Birolini (2007), and Rausand and Høyland (2004). Guidelines for the application of Markov models as well as some theoretical background are also described in IEC 61165:2006.

In the following section, we will confine attention to Markov processes with finite state space (where the possible states of the system are numbered 1, 2, ..., n) and continuous time parameter t. Such a process can be characterized by means of the so-called *transition probabilities* $p_{i,j}(t, s)$, where

$$p_{i,j}(t,s) := P\left\{X(t) = j \,\middle|\, X(s) = i\right\} \tag{3.1}$$

with $0 \leq s \leq t$ and $i, j = 1, ..., n$. $X(t)$ denotes the state of the system at time t. The preceding relation is a sufficient description of a Markov process due to the assumption that the current and future states of the system do not depend on its state at any time in the past (i.e., at any time before s).

In general, a system can be described by means of a homogeneous Markov process provided that all random failure and repair times of its constituent components are independent and exponentially distributed. Other kinds of distributions can be approximated with a homogeneous Markov process; however, this requires an appropriate state space expansion that significantly increases the number of states.

3.2.1 Theoretical Background

For a homogeneous Markov process with a finite state space and continuous time parameter, we have, instead of Equation 3.1,

$$p_{i,j}(t) := P\{X(t+s) = j \mid X(s) = i\} = P\{X(t) = j \mid X(0) = i\}$$

with $s \geq 0$, $t \geq 0$, and $i, j = 1, \ldots, n$. Noting that

$$p_{i,j}(0) = \left\{ \begin{array}{ll} 1 & \text{for } i = j \\ 0 & \text{for } i \neq j \end{array} \right\}$$

for $i, j = 1, \ldots, n$ and under certain conditions, it can be shown that the transition probabilities are the solution of the following system of differential equations:

$$\frac{\mathrm{d}}{\mathrm{d}t} p_{i,j}(t) = \sum_{k=1, k \neq j}^{n} p_{i,k}(t) a_{k,j} + p_{i,j}(t) a_{j,j},$$

with the so-called *transition rates* $a_{k,j} := \lim_{t \downarrow 0} \dfrac{p_{k,j}(t)}{t}$, where $i, j = 1, \ldots, n$, $i \neq j$, and $a_{j,j} := -\sum_{l=1, l \neq j}^{n} a_{j,l}$.

The so-called *state probabilities* $p_j(t) := P\{X(t) = j\}$, with $t \geq 0$ and $j = 1, \ldots, n$, are obtained from the transition probabilities as

$$p_j(t) = \sum_{i=1}^{n} p_i(0) p_{i,j}(t).$$

Hence, it follows that the state probabilities are the solution of the system of differential equations

$$\frac{\mathrm{d}}{\mathrm{d}t} p_j(t) = \sum_{k=1, k \neq j}^{n} p_k(t) a_{k,j} + p_j(t) a_{j,j}, \tag{3.2}$$

which is equivalent to

$$\frac{\mathrm{d}}{\mathrm{d}t} \mathbf{p}(t) = \mathbf{A}\, \mathbf{p}(t) \tag{3.3}$$

with $\mathbf{p}(t) = (p_1(t), p_2(t),..., p_n(t))^T$ and

$$
\mathbf{A} =
\begin{bmatrix}
-\sum_{l=2}^{n} a_{1,l} & a_{2,1} & \cdots & a_{n,1} \\
a_{1,2} & -\sum_{l=1,l\neq2}^{n} a_{2,l} & \cdots & a_{n,2} \\
\vdots & \vdots & \ddots & \vdots \\
a_{1,n} & a_{2,n} & \cdots & -\sum_{l=1}^{n-1} a_{n,l}
\end{bmatrix}.
$$

The sums over the individual columns are equal to zero, which reflects the fact that the change in probability over all states must always cancel out (independent of the current state probabilities). The preceding system of differential equations for the state probabilities has the formal solution

$$
\mathbf{p}(t) = \exp(\mathbf{A}t)\mathbf{p}(0), \tag{3.4}
$$

which makes use of the matrix exponential function. In simple cases, the preceding equations can be solved in a closed form; for the numerical computation of the matrix exponential function, see Moler and Van Loan (2003).

One can visualize the preceding relations using a state-transition diagram, as demonstrated with the following examples. Consider first case (a) of a parallel system (hot standby) consisting of two components A and B with failure rates λ_A and λ_B, respectively. Suppose that there is no possibility of repair. The corresponding state-transition diagram is shown in Figure 3.1. The fact that the system is supposed to be in the initial state at $t = 0$ is modeled by assuming $\mathbf{p}(0) = (1,0,0,0)^T$ in Equation 3.4.

The set of possible states for the system is represented by four states: initial state 1, where both components A and B are functioning; state 2 where component A is failed ($\bar{A}$), given that no repair is possible after failure, and component B is functioning; state 3 where component A is functioning and component B is failed ($\bar{B}$); system failure state 4, where both component A and component B are failed ($\bar{A}\bar{B}$). State 4 is a so-called *absorbing state*, which

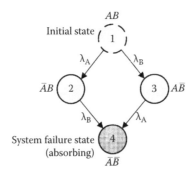

FIGURE 3.1
State-transition diagram for a parallel system of components A and B without repair. The overline negation sign means that a component is failed.

implies that the system remains in this state because there are no outgoing transitions. The corresponding transition matrix $\mathbf{A}$ of Equation 3.3 takes the following form:

$$\mathbf{A} = \begin{bmatrix} -(\lambda_A + \lambda_B) & 0 & 0 & 0 \\ \lambda_A & -\lambda_B & 0 & 0 \\ \lambda_B & 0 & -\lambda_A & 0 \\ 0 & \lambda_B & \lambda_A & 0 \end{bmatrix}.$$

The failure probability $F(t)$ for the system is equal to the probability $p_4(t)$ of the system being in state 4, which is—by definition—the failure state. The mean operating time to (first) failure is obtained as $\int_0^\infty [1 - F(t)]\,dt$.

As a variant of the previous case, consider the case (b) in which both components are repairable with repair rate μ in those configurations in which the system is functioning; on the contrary, once the system has failed the components remain failed. The corresponding state-transition diagram is shown in Figure 3.2.

The corresponding transition matrix $\mathbf{A}$ of Equation 3.3 now takes the following form:

$$\mathbf{A} = \begin{bmatrix} -(\lambda_A + \lambda_B) & \mu & \mu & 0 \\ \lambda_A & -(\lambda_B + \mu) & 0 & 0 \\ \lambda_B & 0 & -(\lambda_A + \mu) & 0 \\ 0 & \lambda_B & \lambda_A & 0 \end{bmatrix}.$$

State 4 is again an absorbing state, and the failure probability $F(t)$ for the system is equal to the probability $p_4(t)$.

Finally, as the last case (c), consider repair of components A and B with repair rate μ also when the system is in the failed state. The corresponding state-transition diagram is shown in Figure 3.3, and the transition matrix $\mathbf{A}$ equals

$$\mathbf{A} = \begin{bmatrix} -(\lambda_A + \lambda_B) & \mu & \mu & 0 \\ \lambda_A & -(\lambda_B + \mu) & 0 & \mu \\ \lambda_B & 0 & -(\lambda_A + \mu) & \mu \\ 0 & \lambda_B & \lambda_A & -2\mu \end{bmatrix}.$$

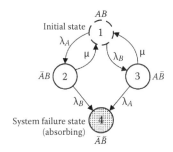

FIGURE 3.2
State-transition diagram for a parallel system of components A and B with partial repair.

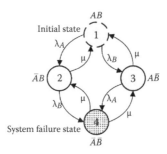

FIGURE 3.3
State-transition diagram for the parallel system of components *A* and *B* with full repair.

It should be noted that the probability $p_4(t)$ associated with the—by definition—failure state 4 here represents the *instantaneous unavailability* $U(t)$, i.e., the probability that an item is not in a state to perform as required at given instants. In order to calculate the mean operating time to first failure, one would have to remove the outgoing transitions from the failure state, which would effectively yield the transition matrix from case b.

The last case (c) of the example represents what is called an *irreducible* Markov process, i.e., every state can be reached from every other state (possibly through intermediate states). For an irreducible Markov process, it can be shown that there exist limiting distributions $p_i > 0$, $i = 1,\ldots, n$, such that

$$\lim_{t \to \infty} p_i(t) = p_i$$

independent of the initial distribution $\mathbf{p}(0)$. These limiting distributions are computed by noting that at stationarity $\dfrac{d}{dt} p_j(t) = 0$, $j = 1,\ldots, n$ and solving the linear system of equations derived from Equation 3.2. The state probabilities are uniquely determined by

$$-p_j \sum_{k=1, k \neq j}^{n} a_{j,k} + \sum_{k=1, k \neq j}^{n} p_k a_{k,j} = 0, \quad j = 1,\ldots, n,$$

where one (arbitrarily chosen) equation must be replaced by $\sum_{j=1}^{n} p_j = 1$. For the case (c) of the example, this could look like

$$
\begin{bmatrix}
-(\lambda_A + \lambda_B) & \mu & \mu & 0 \\
\lambda_A & -(\lambda_B + \mu) & 0 & \mu \\
\lambda_B & 0 & -(\lambda_A + \mu) & \mu \\
1 & 1 & 1 & 1
\end{bmatrix}
\begin{bmatrix}
p_1 \\
p_2 \\
p_3 \\
p_4
\end{bmatrix}
=
\begin{bmatrix}
0 \\
0 \\
0 \\
1
\end{bmatrix}.
$$

In general, the system states can comprise more than one failure state. That is, the system states are divided into two classes: one for the system being in an up state and one for the down state. For the calculation of corresponding values for the mean operating time between failures and mean down times for the stationary case, see, e.g., Birolini (2007).

3.2.2 Markov Model Development

The following steps are commonly followed when developing a Markov model:

- Definition of the objectives of the analysis, for instance, defining whether transient or stationary measures are of interest and identifying the relevant system properties and boundary conditions, e.g., whether the system is repairable or nonrepairable
- Verification that Markov modeling is suitable for the problem at hand; this includes verification whether the random failure and repair times of the system components can indeed be assumed independent and exponentially distributed
- Identification of the states required to describe the system for the objectives of the analysis, for instance, several failure states may be required depending on the possible modes a system can fail and ways it can recover from a failed state; moreover, system failure due to common causes may require additional failure states
- Verification whether the system states can be combined for model simplification without affecting the results; this may be the case if the system exhibits certain symmetries: for example, it may be sufficient to distinguish between the number of functioning identical components, instead of distinguishing each of the individual functioning components
- Setting up the transition matrix based on the identified system states; evaluating the model according to the objective of the analysis; and interpreting and documenting the results

3.2.3 Further Extensions

One extension of a homogeneous Markov process with finite state space is the so-called *semi-Markov* process, where the times the system resides in a given state can be arbitrarily distributed and not necessarily exponentially distributed. Another extension are Markov reward models that take into account reward functions for transitions between states as well as reward functions for the time the system is in a given state, hence allowing the computation of a wider class of performance measures, such as average maintenance cost.

3.2.4 Applications in Railway Industry

Frequently, Markov models are used to describe the deterioration and critical damage of railway track assets and to investigate the effects of an asset management strategy on a railway track section as described, e.g., by Prescott and Andrews (2015) and Podofillini, Zio, and Vatn (2004). Other uses include modeling of the reliability and availability of rolling stocks under complex maintenance strategies or the reliability and safety of railway transportation systems as in the study by Restel and Zajac (2015).

An example application of a homogeneous discrete-time Markov chain for modeling a railway facility is presented next. Discrete-time Markov chains differ from the continuous-time Markov chains discussed earlier in that state changes are assumed to occur at discrete time steps only. They are described by initial probabilities $\mathbf{p}_0$ and transition probabilities (instead of transition rates) represented through a transition matrix $\mathbf{\Pi}$. The transition probabilities indicate the probability that the system attains a certain state at time step i, given that the system was in a certain state a time step $i - 1$. Suppose that the initial probabilities of the four states of a nonrepairable railway facility are

$$\mathbf{p}_0 = \begin{bmatrix} 1 - \text{Safe and fully available} \\ 2 - \text{Safe and unavailable} \\ 3 - \text{Unsafe and fully available} \\ 4 - \text{Unsafe and unavailable} \end{bmatrix} = \begin{bmatrix} 0.96 \\ 0.02 \\ 0.01 \\ 0.01 \end{bmatrix}.$$

In the preceding equation, the system states 3 and 4 can cause risks on railways, and their probabilities are changing over time. The transition matrix $\mathbf{\Pi}$ that governs the changes of a state that occur within a system is given by

$$\mathbf{\Pi} = \begin{bmatrix} \pi_{11} & \pi_{12} & \pi_{13} & \pi_{14} \\ \pi_{21} & \pi_{22} & \pi_{23} & \pi_{24} \\ \pi_{31} & \pi_{32} & \pi_{33} & \pi_{34} \\ \pi_{41} & \pi_{42} & \pi_{43} & \pi_{44} \end{bmatrix} = \begin{bmatrix} 0.95 & 0 & 0 & 0 \\ 0.05 & 0.9 & 0.1 & 0 \\ 0 & 0.05 & 0.8 & 0 \\ 0 & 0.05 & 0.1 & 1 \end{bmatrix}.$$

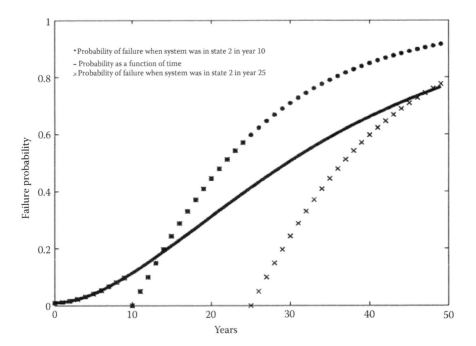

FIGURE 3.4
Failure probabilities of a nonrepairable railway facility.

In the preceding equation, π_{jk}, $j \neq k$ is the transition probability from state k to state j in successive time steps. For example, π_{21} is the transition probability in state 2 at time step i, given that it was in state 1 in time step $i - 1$. The probability of system failure as a function of time is shown in Figure 3.4.

3.2.5 Strengths and Limitations

The strengths of Markov modeling include the capability of describing complex redundancies and dynamic multistate systems (see, e.g., the book by Lisnianski, Frenkel, and Ding [2010]), providing various transient or stationary measures as results. The supporting state-transition diagrams provide a simple and intuitive means for visualizing and communicating the structure of the model. However, Markov modeling is complicated by the fact that the numerical evaluation requires more effort, and limited in that the number of states may explode for complex systems.

3.3 Monte Carlo Simulation

Monte Carlo simulation is a stochastic simulation technique in which multiple independent outcomes of a model are obtained by repeatedly solving the model for randomly sampled values of the input variables and events. The sample of outcomes thereby obtained is statistically treated to compute the system quantities of interest. The method inherently takes into account the effects of uncertainty and is free from the difficulty of solving the underlying model, which is why it is well suited for complex systems that are difficult to model using analytical techniques, see, e.g., the book by Zio (2013).

In the context of RAM, the Monte Carlo simulation has found many applications, for instance, in order to evaluate fault trees (FTs) or Markov models where the failure times of individual components or the times the system remains in a given state follow arbitrary distributions and measures such as the mean operating time between failures or the probability of failure within a given time interval are sought. Also, the Monte Carlo simulation is used to estimate cumulative density functions in risk assessment, e.g., when the average availability within the warranty period is critical and a value must be estimated that is only exceeded with a given probability. Finally, the Monte Carlo simulation is a quite flexible methodology that can be applied when changes in the state of a system occur at discrete time instants, which allows modeling complex operation and maintenance scenarios.

For a comprehensive treatment of the Monte Carlo simulation method, see, e.g., the books by Rubinstein and Kroese (2016), Kroese, Taimre, Botev (2011), and Zio (2013).

3.3.1 Generation of Random Variates

A key aspect of the Monte Carlo simulation is the generation of realizations of the value of the random variables, which are then input to the model to produce samples of outcomes. Two methods for generating realizations of random variables that follow an arbitrary distribution are described in the following.

Let X be a random variable that follows an arbitrary probability distribution with cumulative distribution function $F_X(x)$. The *inverse transform method* (see, e.g., the book by Zio [2013]) uses the inverse function of $F_X^{-1}(y)$, where

$$F_X^{-1}(y) = \inf\left\{x : F_X(x) \geq y\right\}, \quad 0 \leq y \leq 1.$$

If the random variable U is uniformly distributed between 0 and 1, then it can be shown that $X = F_X^{-1}(U)$ has the cumulative distribution function $F_X(x)$. This is also illustrated in Figure 3.5.

It may happen that it is not easy to compute the inverse of a complex cumulative distribution function $F_X(x)$. In this case, the so-called *acceptance–rejection method* may be used; see, e.g., the book by Zio (2013). Let $f_X(x)$ be the probability density function corresponding to $F_X(x)$, i.e.,

$$f_X(x) = \frac{\mathrm{d}}{\mathrm{d}x} F_X(x)$$

and let $f_G(x)$ be the probability density function of another random variable G, for which it is easy to generate realizations. Further, there must exist a constant $C \geq 1$ such that $f_X(x) \leq C \cdot f_G(x)$ for each $x \in \mathbb{R}$. Then, the following steps are carried out:

1. Generate two random variates: U, which is uniformly distributed between 0 and 1, and G, which is distributed according to $f_G(x)$.
2. If $U \cdot C \cdot f_G(G) < f_X(G)$, then accept G as a variate generated from $f_X(x)$. Otherwise, reject it and go to step 1.

This is illustrated in Figure 3.6, where the crosses indicate accepted values, and the circles, rejected values.

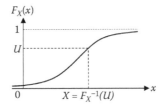

FIGURE 3.5
Illustration of the inverse transform method.

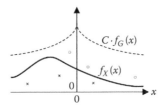

FIGURE 3.6
Illustration of the acceptance–rejection method.

Statistical software packages provide functions to generate random variables from a variety of parameterized univariate or multivariate probability distributions.

3.3.2 Basic Concepts of Monte Carlo Simulation

Typically, the Monte Carlo simulation is used in order to estimate the expected value $\mu = E[Y]$ of a random variable Y with probability density function $f_Y(y)$, where

$$\mu = E[Y] = \int_{-\infty}^{\infty} y \cdot f_Y(y) \, dy.$$

More generally, Y could be a function $Y = g(\mathbf{X})$ of a vector of random variables $\mathbf{X} \in \mathcal{D} \subseteq \mathbb{R}^d$ with joint probability density function $f_{\mathbf{X}}(\mathbf{x})$, so that

$$\mu = E[Y] = \int_{\mathcal{D}} g(\mathbf{x}) \cdot f_{\mathbf{X}}(\mathbf{x}) \, d\mathbf{x}.$$

(3.5)

The estimated value $\hat{\mu}_n$ of μ is obtained by generating n independent samples $Y_1, Y_2, \ldots, Y_n$ from Y and taking the average

$$\hat{\mu}_n = \frac{1}{n} \sum_{i=1}^{n} Y_i.$$

It should be noted that $\hat{\mu}_n$ is a random variable, too. Intuitively, the estimated value $\hat{\mu}_n$ approximates well the true value μ if the number of sample n is large enough. Indeed, if Y has a finite variance $\text{Var}(Y) = \sigma^2 < \infty$, then it can be shown that the expected value of $\hat{\mu}_n$ is

$$E[\hat{\mu}_n] = \mu,$$

and its standard deviation is

$$\sqrt{\text{Var}(\hat{\mu}_n)} = \frac{\sigma}{\sqrt{n}}.$$

(3.6)

Approximate confidence intervals for μ with confidence level $(1 - \alpha)$ can be computed based on the central limit theorem as

$$\hat{\mu}_n - z_{\alpha/2} \frac{s}{\sqrt{n}} \leq \mu \leq \hat{\mu}_n + z_{\alpha/2} \frac{s}{\sqrt{n}},$$

where $z_{\alpha/2}$ denotes the upper $\alpha/2$ quantile of the standard normal distribution and

$$s^2 = \frac{1}{n-1} \sum_{i=1}^{n} (Y_i - \hat{\mu}_n)^2, \quad n \geq 2.$$

An important special case of Equation 3.5 is when the function $Y = g(\mathbf{x})$ is the index function $g(\mathbf{x}) = \mathbb{I}_A(\mathbf{x})$, which is equal to 1 if $\mathbf{x} \in A$ and 0 otherwise. In this case, $E[Y] = E\left[\mathbb{I}_A(\mathbf{x})\right]$ is equal to the probability $P(\mathbf{X} \in A) = p$. The estimated value $\hat{p}_n$ of p is obtained as

$$\hat{p}_n = \frac{n_A}{n},$$

where n_A is the number of samples out of n for which $\mathbf{X} \in A$, i.e., the number of "successes" of falling in A. Approximate confidence intervals for p can also be computed for this case using

$$s^2 = \frac{\hat{p}_n(1 - \hat{p}_n)}{n - 1}, \quad n \geq 2,$$

but it is also possible to obtain exact confidence intervals by noting that n_A follows a binomial distribution, see, e.g., the book by Birolini (2007).

For instance, in the context of risk analysis where measures such as value at risk are sought, it may be required to compute the so-called *empirical cumulative distribution function* $\hat{F}_Y(y)$ from the samples $Y_1, Y_2, \ldots, Y_n$, where

$$\hat{F}_Y(y) = \frac{1}{n} \sum_{i=1}^{n} \mathbb{I}(Y_i \leq y).$$

One possibility to estimate the error is through pointwise confidence intervals, which may essentially be computed for each fixed point in the same way as for simple probabilities described in the previous paragraph. A better alternative is to construct *confidence bands* using the Dvoretzky–Kiefer–Wolfowitz inequality; see, e.g., the book by Wasserman (2006). Estimation procedures for the quantiles of a cumulative distribution function are also available as described by Guyader, Hengartner, and Matzner-Lober (2011).

3.3.3 Variance Reduction Techniques

Variance reduction techniques, loosely speaking, attempt to reduce the value of Equation 3.6 not by increasing n but by reducing σ. This can only be achieved by using some form of available information about the problem. Several variance reduction techniques have been devised, and it should be kept in mind that the effectiveness depends on how well the chosen technique fits to the problem at hand. Moreover, there is a trade-off between reducing σ and the increased computational cost per simulation run, in addition to the time required for programming and testing the simulation code.

Importance sampling biases the realizations by concentrating samples in the regions of interest, which are usually of rare occurrence, but introduces a correction factor in order to account for that. *Correlated sampling* is useful to estimate the effects of small changes in the system, e.g., for sensitivity analyses. Other methods such as *subset simulation* use information from previous samples to refine subsequent simulations. A detailed discussion of these and other methods can be found, e.g., in the books by Rubinstein and Kroese (2016); Kroese, Taimre, and Botev (2011); Guyader, Hengartner, and Matzner-Lober (2011); and Zio (2013).

3.3.4   Applications in Railway Industry

Monte Carlo simulation is typically used for the computation of correction factors applied to the nominal emergency brake deceleration curve within the European Train Control System/European Rail Traffic Management System requirement specifications framework; see UIC B 126/DT 414:2006. This is done in order to ensure a sufficient safety margin, given the statistical dispersion of the emergency braking performance. For instance, the correction factor K_{dry} takes into account the rolling stock characteristics and represents the statistical dispersion of braking efforts on dry rails. It depends on a given Emergency Brake Confidence Level (EBCL), which is transmitted by trackside and can range from $1 - 10^{-1}$ up to $1 - 10^{-9}$. Due to the high confidence levels, variance reduction techniques and estimation procedures suitable for extreme quantiles can prove useful.

Other applications of Monte Carlo simulation in the railway industry arise in the context of risk analysis, in order to quantify scenarios that involve uncertain events; see, e.g., the study by Vanorio and Mera (2013). Furthermore, the Monte Carlo simulation can prove useful in order to model uncertainties related to the reliability characteristics of the components. One might, for instance, assume uniform distributions for the failure rates assigned to the basic events of a FT model, which, then, propagate through the tree to yield a distribution for the failure rate of the top event of interest.

3.3.5   Strengths and Limitations

The strengths of Monte Carlo simulation are its flexibility to take into account any distributions of input variables and to provide solutions for complex systems that do not lend themselves to analytical solutions, while providing a measure of the accuracy of the results. The method has limitations as the accuracy of the solution depends upon the number of simulations that can be performed, where the rate of convergence is generally slow (but may be improved by variance reduction techniques). Also, the method must be used with consideration when simulating extreme events.

3.4   Petri Nets

Petri nets (PNs) are a technique for describing the global behavior of a system by modeling local states, local events (transitions), and their relations to each other. Their graphical notation, in the form of a directed graph with tokens, allows intuitive and compact modeling of deterministic or stochastic systems for various applications, such as concurrent processes in logistics and transportation, as well as distributed and parallel communication systems.

Very relevant for RAM purposes are the so-called *generalized stochastic PNs* (GSPNs), where random variables are introduced to represent random times until certain events occur. In this context, local events may represent failure or repair events, while local states may represent functioning states, failure states, or states with various degrees of degraded performance.

In case the random transitions in the model are exponentially distributed and the deterministic transitions have zero delay (immediate transitions), GSPNs can be represented as Markov models; however, the former usually have a much more compact representation.

A detailed description of PNs can be found, e.g., in IEC 62551:2012 and in the book by Marsan et al. (1996).

3.4.1 Basic Elements of GSPNs and Construction Guidelines

GSPNs consist of at least the following basic elements (Birolini 2007):

- Places, which are graphically indicated by circles and denoted by $P_1, P_2,\ldots,P_n$. They may contain zero, one or more tokens to describe the local state. The combination of the number of tokens in the places is referred to as the *marking* of a PN and represents information about the global system state at a given time
- Transitions denoted by $T_1, T_2,\ldots,T_m$, which are indicated by outlined rectangles in case of exponentially distributed random transitions and by solid bars in case of immediate transitions
- Directed arcs, which indicate the relationships between places and transitions. A so-called *input arc* is an arc from a (input) place to a transition, whereas an *output arc* is an arc from a transition to a (output) place. A transition is said to be *enabled*, i.e., any deterministic or random times are counted from that instant, when the input places connected to it each have the required number of tokens, which is determined by the multiplicity of the arc. Output arcs can have a multiplicity, too, which describes the number of tokens put into a place after a transition has occurred. If no multiplicity is indicated next to an arc, it defaults to one.

In addition, GSPNs may contain other elements such as inhibitor arcs, which indicate that a local state may disable a transition instead of enabling it. These can have multiplicities similar to ordinary arcs. Moreover, there exist testing arcs, which do not reduce the number of tokens in an input place IEC 62551:2012.

The dynamic behavior of the system described by means of GSPNs is determined by the following rules:

1. An initial marking is determined.
2. If a transition is enabled (when the input places connected to it each have the required number of tokens), it occurs ("fires") after the respective deterministic or random time. Only one transition can occur at a time. If several transitions at the same time are about to occur, then this is resolved by weighing the individual transitions.
3. As a consequence, after a transition has occurred, the respective numbers of tokens are removed from the input places according to the multiplicity of the input arcs, and a number of tokens are put into the output places according to the multiplicity of the output arcs. This is also illustrated in Figure 3.7.
4. The procedure is repeated from step 2.

As a simple example, consider the stochastic PN from Figure 3.8, which models a parallel system of two components with identical failure rates λ and repair rates μ. The (initial) marking shown in the figure indicates that both components are currently functioning.

Despite the similarity of the graph shown in the figure to an analogous Markov model, there is, generally, a fundamental notational difference between PNs and Markov models:

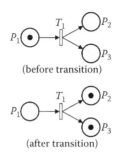

FIGURE 3.7
Transition in a stochastic PN.

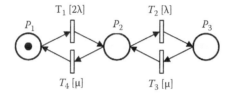

FIGURE 3.8
Example of a parallel system modeled with a stochastic PN.

in a Markov model, each state denotes one of the possible global states of the system; in a PN, the individual places with given numbers of tokens in that place denote local states, while the possible states of the system are represented by all possible combinations of tokens in the respective places (which may depend on the initial marking).

The following (iterative) steps should be taken when constructing and analyzing complex PN models IEC 62551:2012:

1. Identification of system boundaries, context, and environment, as well as main components and functions

2. Modeling the structure of the system in a modular fashion by creating submodels related to the functions of the system including dependability aspects, as well as submodels for the control aspects of the system and submodels for the dependability of the control of the system. These submodels should then preferably be linked by elements such as inhibitor arcs or testing arcs

3. Checking and refinement of the model until the required level of detail is reached

4. Analyzing the model according to the objectives of the analysis (qualitative or quantitative, stationary or transient)

5. Visualization, interpretation, and documentation of the results of the analysis

Extensions of PNs have been developed, for instance, where certain properties can be attached to the tokens of so-called *colored PNs*; see, e.g., the book by Jensen and Kristensen (2009). For guidelines on efficient construction of PNs, one may consult Signoret et al. (2013)

3.4.2 Applications in Railway Industry

As PNs are well known for describing distributed and concurrent complex systems, and analysis techniques such as simulation, testing, and model checking are available. There are numerous examples of applications to railway industry; see, e.g., the studies by Wu and Schnieder (2016), Giua and Seatzu (2008), and Zimmermann and Hommel (2003). For an assessment of proof-testing strategies for safety-instrumented systems, see the study by Liu and Rausand (2016). As an academic and illustrative example, a simplified PN model of the traffic at a railway level crossing based on IEC 62551:2012 is shown in Figure 3.9 and described next.

It is a model of two concurrent processes: one models a train which can either be outside the danger zone of the railway level crossing or be located within the danger zone. The transitions between the two places are governed by exponential distributions. The other process models cars that can be outside the danger zone, approaching, or be located within the danger zone. The transitions are again exponentially distributed except for transition T_1. The latter is an immediate transition between an approaching car and the car being in the danger zone, provided that there is no train in the danger zone. One should note the role of the inhibitor arc to transition T_1, which effectively enables the transition as long as there is no token in the corresponding place.

For the simplified model shown in Figure 3.9, it should also be noted that although cars do not enter the danger zone while a train is present, an accident might still occur because a train can enter the danger zone while a car is present. For more explanations and further refinements of the model, the reader is referred to IEC 62551:2012.

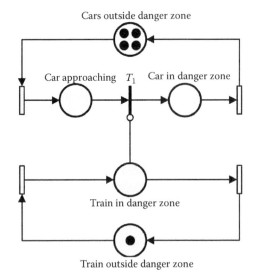

FIGURE 3.9
Simplified PN model of a railway level crossing. (Based on IEC 62551:2012. *Analysis Techniques for Dependability—Petri Net Techniques*, IEC, Geneva, 2012.)

3.5   Bayesian Networks and Influence Diagrams

3.5.1   Basic Concepts of Bayesian Networks

Bayesian networks (BNs) are directed, acyclic probabilistic graphical models. The nodes in BNs are random variables, and the directed links between them represent the dependence structure (or causal relation). The random variables represent a finite set of possible mutually exclusive states. The directed links are used to create dependencies among the random variables in the BNs.

A *conditional probability table* (CPT) is attached to child random variables, which is defined conditional on its parents. In Figure 3.10, X_3 and X_4 are the children of the parent X_2. The variable X_2 has no parents and therefore it has unconditional probabilities, called *prior probabilities*.

The BNs are based on the so-called *d-separation* properties of causal networks. In a serial connections (such as formed by random variables X_2, X_4, X_6 in Figure 3.10), the evidence is transmitted through the network if the state of the intermediate random variable is not known with certainty. In a diverging connection (such as formed by random variables X_2, X_3, X_4), the evidence is transmitted when the state of the common parent variable is not known with certainty. Random variables X_3, X_4, X_5 form a converging connection in which evidence is transmitted only if there is some information about the child variable or one of its descendants.

Suppose we have the BN from Figure 3.10 with random variables that are defined in a finite space $\mathbf{X} = [X_1, X_2,...,X_6]$. For the discrete case, the probability mass function (PMF) of each random variable is defined conditional on its parents, i.e., $p(x_1|x_2,...,x_6)$, and the joint PMF $p(\mathbf{x})$ of a set of random variables $\mathbf{X}$ is obtained by multiplying all the conditional probabilities:

$$p(\mathbf{x}) = p\left(x_6|x_4\right) \cdot p\left(x_5|x_4, x_3, x_2\right) \cdot p\left(x_4|x_2\right) \cdot p\left(x_3|x_2, x_1\right) \cdot p(x_2) \cdot p(x_1).$$

The joint PMF of the BN can be generalized as

$$p(\mathbf{x}) = p(x_1, x_2,...,x_6) = \prod_{i=1}^{6} p\left[x_i|\text{pa}(x_i)\right],$$

where pa(x_i) are realizations of the parents of variables X_i.

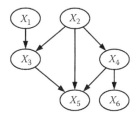

FIGURE 3.10
Exemplary BN.

BNs have the advantage that they can predict (using the total probability theorem, which is the denominator in Bayes's theorem) and estimate (using Bayes's theorem) states of random variables:

$$p(X|Y) = \frac{p(Y|X)p(X)}{\sum_X p(Y|X)p(X)}.$$

This equation essentially states that the posterior probability of X given Y is equal to the probability $p(Y|X)\,p(X) = p(X \cap Y)$ divided by the probability of Y.

The details on how random variables are dependent on each other and how information can flow in such networks are well explained, for instance, in the books by Jensen and Nielsen (2007) and Barber (2012). For the application of BNs in various industries, we refer to Lampis and Andrews (2009) and Oukhellou et al. (2008).

3.5.2 Inference in BNs

BNs are used for performing three tasks—*structural learning* (SL), *parameter learning* (PL), and *probabilistic inference* (PI). SL and PL are data-driven processes, whereas PI is used to compute the probability values for a set of variables, given some evidences in the network. In SL, algorithms determine the topology of the network model, such as the number of links and directions among the random variables in the network. In PL, unknown parameters of the joint or conditional probability distributions in BNs are determined from given data, using algorithms such as the expectation maximization algorithm. For more details on SL, PL, and PI, the reader is referred to the books by Jensen and Nielsen (2007), Koller and Friedman (2009), and Russell and Norvig (2009).

A number of questions related to the (conditional) probability of variables being in a given state can be answered using PI. Consider again the network in Figure 3.10, where all random variables have $m = 2$ states named "True" and "False." In order to answer questions such as what is the probability of random variable X_6 being in a specific state, say, "True", given that another (set of) random variables X_3 is observed to be equal to "True", one must compute the following:

$$p(X_6 = \text{"True"} | X_3 = \text{"True"}) = \frac{p(X_6 = \text{"True"} \cap X_3 = \text{"True"})}{\sum_{X_6} p(X_6 \cap X_3 = \text{"True"})}.$$

One has to solve the joint PMFs (numerator) and marginalized probability (denominator) in the preceding equation. A number of algorithms (including both exact and approximate) exist to answer probability questions in a BN. The exact inference methods such as inference by enumeration and variable elimination are used for relatively smaller and simpler BNs, which have discrete random variables. Large or complex BNs require simulation-based approximate algorithms such as direct sampling, rejection sampling, likelihood weighting, and Markov chain Monte Carlo. The major advantage of the approximate algorithms is that they allow handling continuous random variables. In other words, the discretization of continuous variables can be avoided. The main disadvantage of the approximate algorithms is that it is difficult to analyze and assess their accuracy. For details on approximate algorithms, one may consult Koller and Friedman (2009) and Russell and Norvig (2009).

3.5.3 Mapping of Fault Trees and Event Trees to BNs

A complete FT can be mapped to BNs by performing numerical and graphical related algorithmic tasks. The gates of the FT are managed by making use of CPTs attached to nodes in BNs. To this end, all the basic events and gates (i.e., intermediate events) are treated as random variables in BNs, and dependences among them are managed by introducing the required directed links. For example, the logical AND and OR gates, respectively, for the random variables X_1 and X_2 from Figure 3.10 are expressed as CPTs, as shown in Table 3.1.

It is easy to see that BNs represent a methodology with which some of the limitations of classical FT analysis (FTA) can be overcome. For instance, the modeling of more than only two binary states per variable, as well as dependent events and noncoherent systems (introduced by Exclusive-OR or NOT gates) is straightforward. More details regarding the mapping of FTA and BNs can be found, e.g., in the study by Bobbio et al. (2001).

Similar to FTs, event trees (ETs) can be mapped to BNs by performing numerical and graphical related algorithmic tasks. A graphical representation of the simplified mapping procedure for both FTA and ET analysis (ETA) based on the thesis by Mahboob (2014) is presented in Figure 3.11.

For understanding of the mapping procedure, the nodes in the BNs are further classified as follows:

- *Root nodes of BNs* are used to represent basic events of the FTA, and technical and operational barriers as well as neutralizing factors in the ETA. Barriers can actively mitigate the hazard evolution to an accident. They can be characterized by their functionality, frequency of occurrence, and effectiveness in reference to specific hazard occurrences. Neutralizing factors also mitigate the hazard propagation to an accident after the barriers have failed.

- *Intermediate nodes of BNs* are used to represent intermediate events, i.e., logical gates for FTA and intermediate scenarios for ETA.

- *Leaf node of BNs* are used to represent the output of network models, i.e., one or more top events for FTA and one or more consequences in ETA.

3.5.4 Influence Diagrams

Influence diagrams (IDs) are the extensions of BNs, where decision nodes (rectangular) and utility nodes (diamond shaped) are added (see Figure 3.12) to the regular nodes. The directed links (arrows) represent probabilistic dependencies among the system, decision, and utility variables in the IDs.

A decision analysis with given information (such as on system states and their probabilities, decision alternatives, and their utilities) is called *prior analysis*. In the IDs with

TABLE 3.1

CPTs Equivalent to Logical AND and OR Gates in FTs

AND Gate					OR Gate				
$X_1 =$	"True"		"False"		$X_1 =$	"True"		"False"	
$X_2 =$	"True"	"False"	"True"	"False"	$X_2 =$	"True"	"False"	"True"	"False"
$P(X_3 = $ "True")	1	0	0		$P(X_3 = $ "True")	1	1	1	0
$P(X_3 = $ "False")	0	1	1	1	$P(X_3 = $ "False")	0	0	0	1

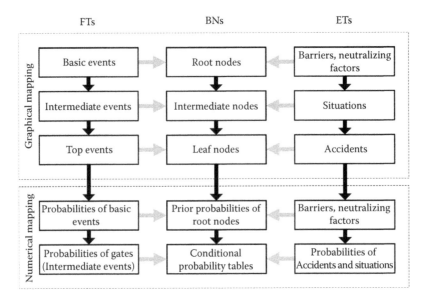

FIGURE 3.11
Simplified mapping procedure of FTs and ETs to BNs. (Based on Q. Mahboob, *A Bayesian Network Methodology for Railway Risk, Safety and Decision Support*, PhD thesis, Technische Universität Dresden, Dresden, 2014.)

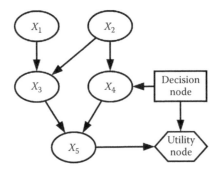

FIGURE 3.12
Exemplary ID.

given information, the decision nodes are parents to the system variables and the utility nodes. In other words, the decision nodes influence both the system variables and utility outcomes. Figure 3.12 is a simple representation of a prior decision analysis using IDs. In the prior analysis, the evaluation and decision optimization are made on the basis of probabilistic modeling and statistical values available prior to any decision. Guidelines on the interpretation of uncertainties and probabilities in engineering decision making can be found in the study by Faber and Vrouwenvelder (2008).

The optimal set of decisions d_{opt} is identified as the set of decisions that maximizes the expected utility. Mathematically, it is written as

$$\mathbf{d}_{opt} = \arg\max_{\mathbf{d}} E\big[u(\mathbf{d},\mathbf{x})\big] = \arg\max_{\mathbf{d}} \sum_{\mathbf{x}} u(\mathbf{d},\mathbf{x}) \cdot p(\mathbf{x}|\mathbf{d}).$$

In the equation above, $\mathbf{d} = (\mathbf{d}_1, \mathbf{d}_2, \ldots, \mathbf{d}_n)$ are sets of decision alternatives, $u(\mathbf{d}, \mathbf{x})$ is the utility function, $p(\mathbf{x}|\mathbf{d})$ is the probability, $E[u(\mathbf{d}, \mathbf{x})]$ is the expectation operator, and $\mathbf{x}$ is a vector of system variables (from BNs). By using prior analysis, the simple comparison of utilities associated with different system outcomes, which are influenced by the system variables and sets of decision alternatives, can be performed, and the sets of decision alternatives can be ranked and optimized.

IDs are described, e.g., by Jensen and Nielsen (2007) and Barber (2012). For a classical treatment of statistical decision theory, see the book by Raiffa and Schlaifer (1961).

3.5.5 Applications in Railway Industry

As a continuation of the risk models based on FTA and ETA for a train derailment accident from Chapter 2, a corresponding BN model is shown in Figure 3.13. For the sake of simplicity, shorter names of random variables in the BN are used, for instance, "Fall" instead of "One or more carriages fall," "SPAD" instead of "Signal passed at danger," and "TPWS" instead of "Train protection and warning system." It is mentioned here that the example is created for academic purposes and considers limited failure scenarios only. For example, there can be other reasons for train derailment such as broken rail, broken wheel, broken axle, deteriorated track superstructure, and uneven or heavy shifted loads (in case of goods trains). Complete analysis of functional safety in railways or accident analysis such as any train derailment accident is beyond the scope of this chapter.

As an example, the AND logic between the random variables "High train speed" (HTS) and "Curve in track alignment" (CTA) for "Speed and alignment" (SA) is modeled by a conditional probability table as shown in Table 3.1. The probability of the event SA = "Yes" is calculated as

$$p(\text{SA} = \text{``Yes''}) = \sum_{\text{CTA, HTS}} p\big(\text{SA} = \text{``Yes''}|\text{CTA} \cap \text{HTS}\big) \cdot p(\text{CTA} \cap \text{HTS}).$$

In the preceding equation, the basic events CTA and HTS are statistically independent, meaning that $p(\text{CTA} \cap \text{HTS}) = p(\text{CTA}) \cdot p(\text{HTS})$.

3.5.6 Strengths and Limitations

BNs and IDs are powerful techniques for modeling and handling uncertainties and complex dependencies in risk scenarios and decision problems. Moreover, the dependency structure among variables can be clearly visualized and communicated, and updating of probabilities is easily performed by entering new evidence. The limitations include that BNs and IDs are usually built with discrete variables, making it necessary to discretize continuous variables. Moreover, dynamic problems are harder to model and require discretization of the time axis into a limited number of discrete time slices (so-called *dynamic BNs*).

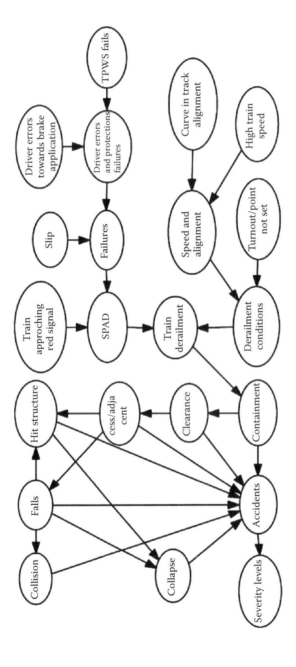

FIGURE 3.13
BNs model for train derailment accident equivalent to the models based on FTA and ETA from Chapter 2. (From Q. Mahboob, *A Bayesian Network Methodology for Railway Risk, Safety and Decision Support*, PhD thesis, Technische Universität Dresden, Dresden, 2014.)

3.6 Summary

A number of advanced modeling techniques for complex systems used in the railway industry have been described in this chapter. Particularly for modeling of systems with redundancy, concurrent behavior, and complex operation and maintenance strategies, Markov models are presumably more intuitive to develop and understand compared with PNs. However, PNs allow a more compact modeling and are therefore better suited for modeling complex systems encountered in practice. Monte Carlo simulation offers great modeling flexibility to take into account any uncertainties and to provide solutions for complex systems of any kind. This, however, comes at the cost of increased effort in developing the simulation models, oftentimes by means of writing programming code. Further, the convergence of the numerical results must always be assessed. Finally, BNs and IDs are powerful techniques for modeling and handling uncertainties and complex dependencies in risk scenarios and decision problems. However, BNs are less suitable for modeling problems in continuous time and for seeking stationary solutions.

References

D. Barber, *Bayesian Reasoning and Machine Learning*, Cambridge University Press, Cambridge, UK, 2012.

A. Birolini, *Reliability Engineering: Theory and Practice*, Fifth Edition, Springer, Berlin, 2007.

A. Bobbio, L. Portinale, M. Minichino, and E. Ciancamerla, Improving the analysis of dependable systems by mapping fault trees into Bayesian networks, *Reliability Engineering and System Safety* 71, pp. 249–260, 2001.

M. H. Faber and A. C. W. M. Vrouwenvelder, *Interpretation of Uncertainties and Probabilities in Civil Engineering Decision Analysis*, Background Documents on Risk Assessment in Engineering, No. 2, Joint Committee of Structural Safety, http://www.jcss.byg.dtu.dk/, 2008.

A. Giua and C. Seatzu, Modeling and supervisory control of railway networks using Petri nets, *IEEE Transactions on Automation Science and Engineering*, 5, 3, pp. 431–445, July 2008.

A. Guyader, N. Hengartner, and E. Matzner-Lober, Simulation and estimation of extreme quantiles and extreme probabilities, *Applied Mathematics & Optimization*, 64, 2, pp. 171–196, 2011.

IEC (International Electrotechnical Commission) 61165:2006: *Application of Markov techniques*, IEC, Geneva.

IEC 62551:2012: *Analysis Techniques for Dependability—Petri Net Techniques*, IEC, Geneva.

IEC 60050-192:2015: *International Electrotechnical Vocabulary—Part 192: Dependability*, IEC, Geneva.

IEC 61703:2016: *Mathematical Expressions for Reliability, Availability, Maintainability and Maintenance Support Terms*, IEC, Geneva.

International Union of Railways, UIC B 126/DT 414: *Methodology for the Safety Margin Calculation of the Emergency Brake Intervention Curve for Trains Operated by ETCS/ERTMS*, International Union of Railways, Paris, 2006.

K. Jensen and L. M. Kristensen, *Coloured Petri Nets; Modelling and Validation of Concurrent Systems*, Springer, Berlin, 2009.

F. V. Jensen and T. D. Nielsen, *Bayesian Networks and Decision Graphs*, Springer, Berlin, 2007.

D. Koller and N. Friedman, *Probabilistic Graphical Models—Principles and Techniques*, MIT Press, Cambridge, MA, 2009.

D. P. Kroese, T. Taimre, and Z. I. Botev, *Handbook of Monte Carlo Methods*, Wiley, Hoboken, NJ, 2011.

M. Lampis and J. D. Andrews, Bayesian belief networks for system fault diagnostics, *Quality and Reliability Engineering International*, 25, 4, pp. 409–426, 2009.

A. Lisnianski, I. Frenkel, and Y. Ding, *Multi-state System Reliability Analysis and Optimization for Engineers and Industrial Managers*, Springer, Berlin, 2010.

Y. Liu and M. Rausand, Proof-testing strategies induced by dangerous detected failures of safety-instrumented systems, *Reliability Engineering & System Safety*, 145, pp. 366–372, 2016.

Q. Mahboob, *A Bayesian Network Methodology for Railway Risk, Safety and Decision Support*, PhD thesis, Technische Universität Dresden, Dresden, 2014.

M. A. Marsan, G. Balbo, G. Conte et al., *Modelling with Generalized stochastic Petri Nets*, Wiley, Hoboken, NJ, 1996.

C. Moler and C. Van Loan, Nineteen dubious ways to compute the exponential of a matrix, twenty-five years later, *SIAM Review*, 45, 1, pp. 3–49, 2003.

L. Oukhellou, E. Côme, L. Bouillaut, and P. Aknin, Combined use of sensor data and structural knowledge processed by Bayesian network: Application to a railway diagnosis aid scheme, *Transportation Research Part C: Emerging Technologies*, 16, 6, pp. 755–767, 2008.

L. Podofillini, E. Zio, and J. Vatn, Modelling the degrading failure of a rail section under periodic inspection. In: Spitzer, C., Schmocker, U., and Dang V. N. (eds), *Probabilistic Safety Assessment and Management* (pp. 2570–2575). Springer, London, 2004.

D. Prescott and J. Andrews, Investigating railway track asset management using a Markov analysis, *Proceedings of the Institution of Mechanical Engineers, Part F: Journal of Rail and Rapid Transit*, 229, 4, pp. 402–416, 2015.

H. Raiffa and R. Schlaifer, *Applied Statistical Decision Theory*, Harvard Business School, Boston, MA, 1961.

M. Rausand and A. Høyland, *System Reliability Theory: Models, Statistical Methods, and Applications*, Second Edition, Wiley, Hoboken, NJ, 2004.

F. J. Restel, M. Zajac, Reliability model of the railway transportation system with respect to hazard states, In: *2015 IEEE International Conference on Industrial Engineering and Engineering Management (IEEM)*, pp. 1031–1036, Institute of Electrical and Electronics Engineers, Piscataway, NJ, 2015.

R. Y. Rubinstein, D. P. Kroese, *Simulation and the Monte Carlo Method*, Third Edition, Wiley, Hoboken, NJ, 2016.

S. Russell and P. Norvig, *Artificial Intelligence: A Modern Approach*, Pearson, London, 2009.

J.-P. Signoret, Y. Dutuit, P.-J. Cacheux, C. Folleau, S. Collas, and P. Thomas. Make your Petri nets understandable: Reliability block diagrams driven Petri nets, *Reliability Engineering and System Safety*, 113, pp. 61–75, 2013.

G. Vanorio and J. M. Mera, Methodology for risk analysis in railway tunnels using Monte Carlo simulation, pp. 673–683, *Computers in Railways XIII*, WIT Press, Ashurst, UK, 2013.

L. Wasserman, *All of Nonparametric Statistics*, Springer, Berlin, 2006.

D. Wu and E. Schnieder, Scenario-based system design with colored Petri nets: An application to train control systems, *Software & Systems Modeling*, pp. 1–23, 2016.

A. Zimmermann and G. Hommel. A train control system case study in model-based real time system design. In: *Parallel and Distributed Processing Symposium*, Institute of Electrical and Electronics Engineers, Piscataway, NJ, 2003.

E. Zio, *Computational Methods for Reliability and Risk Analysis*, World Scientific, Singapore, 2009.

E. Zio, *The Monte Carlo Simulation Method for System Reliability and Risk Analysis*, Springer, Berlin, 2013.

4

Safety Integrity Concept

Holger Schult

CONTENTS

4.1 Introduction

The safety integrity concept and the philosophy behind it are explained. Allocation and apportionment of safety integrity requirements due to CLC EN 50126 are outlined. With regard to safety-related electronic equipment, safety integrity levels (SILs), their application, allocation, and apportionment are described. This includes pitfalls as well as potential misuse of the integrity concept. The specific issues with regard to mechatronic systems in comparison to electronic systems are considered.

4.2 Integrity Concept

This chapter outlines the key elements of the *integrity concept*. In principle, this concept is applicable to functions implemented by use of computers as well as electrical, mechatronic, or mechanical functions. This basic concept is not restricted to the matter of safety but is valid for reliability and availability aspects as well. Nevertheless, this chapter focuses on the matter of safety.

Functional safety is ensured if there is sufficient confidence that the functions considered do not fail unsafe more often than specified by the respective requirements. This perspective applies in an analogous manner to the requirements of reliability or availability as well.

But how to get this confidence? Here the integrity concept provides an answer. The key message can be put in simple words: In order to ensure sufficient confidence, both random failures and systematic faults shall be covered and protected against when considering reliability, availability, or safety. With regard to safety, the measures to make a function as safe as specified by its requirements shall be balanced for both sources of malfunctioning.

Safety integrity is a generic concept, which should be applied to all kinds of technical systems, regardless of the technology used. In other words, the concept itself is not specific to electronic systems. The concept of safety integrity is not a synonym for *safety integrity levels.*

The more complex a system and its functions are and the higher the required level of safety to be achieved is, the more potential causes of systematic faults have to be considered as well. Ensuring the adequate balance of random failures and systematic faults means to ensure the required degree of safety integrity.

Random failures, as the name says, are failures which, if statistically analyzed, randomly occur due to physical, chemical, or other processes. Ideally, their causes are not of any systematic nature.

Systematic faults happen due to errors, which cause the product, system, or process to fail deterministically under a particular combination of inputs or under particular environmental or application conditions. Systematic faults are always caused by human beings or their interaction.

Examples of causes of systematic faults are human errors, e.g., in safety requirements specifications and design; in manufacturing, installation, and operation of hardware; or in software design and software implementation.

Modern-technology industrial equipment nowadays is very reliable. That means that random failures are well under control, and the focus for further improvement is more on the systematic fault aspect.

4.3   Safety Integrity Requirements

Random failures can be quantified, whereas no commonly accepted concept for the quantification of systematic faults exists. If random failures are observed statistically correct, they follow the respective failure distribution curve.

As mentioned earlier, integrity can be obtained by adequate balance of measures against both types of sources of malfunctioning. For random failures, the decision on adequate requirements is not that difficult since their occurrence can be described by a rate. For systematic faults, it is not possible to scientifically determine the degree of measures required in a comprehensible way. Therefore, a set of measures to be applied needs to be agreed upon between experts and assigned to categories or levels. If commonly accepted, this set of measures against systematic faults and their relation to the degree of random failures is considered as representing the state of the art. Consequently, if the measures and methods required for level x are correctly applied, it is not required to consider the systematic faults when demonstrating the dedicated tolerable hazard rate (THR) is achieved. Those faults are covered and deemed negligible in comparison to the occurrence of random failures.

With regard to safety-related protective electrical/electronic/programmable electronic functions, IEC 61508 [1] provides such measures assigned to levels. For railways, the command, control and signaling standards EN 50128 (safety-related software) [2] and EN 50129 (system perspective and safety-related electronic hardware) [3] as well as the rolling stock standard EN 50657 (onboard software) [4] provide those normative measures or requirements. The categories used to distinguish the strength of measures to be fulfilled are called SILs. If a certain SIL is to be achieved, the respective dedicated measures have to be fulfilled. The measures agreed are always specific to the respective technology.

4.4 Safety Integrity Requirements for Mechatronic Functions

Consequently, if SILs for, e.g., the mechatronic function "operational braking" or the "access and egress function" were intended to be introduced, those measures had to be proposed and agreed on. At the European Committee for Electrotechnical Standardization, a TC9X working group aimed to find an overall approach applicable to any technology, at least restricted to two classes ("levels of integrity"). To ensure this overall applicability, the measures compiled became so high level that the practical use has been questioned and the idea of creating a common classification and belonging measures which are standardized and applicable to any technology has been dismissed, which does not impair the concept itself.

Nevertheless, the concept of safety integrity has been applied to mechanical, pneumatic, or mechatronic functions in the industry for decades, even if commonly agreed SILs for those technologies are not explicitly available. Thus, companies often have internal rules on measures to be applied, avoiding systematic faults in the design and manufacturing process of the functions implemented in the equipment produced. The classes distinguished are for example "high level safety performance" or "normal safety performance." Analogously, this concept applies to the reliability and availability aspect as well. The "higher class" simply contains more measures against systematic faults than the other class(es). These can be, e.g., more stringent quality criteria for selecting subsuppliers, more experts involved in the deeper review of specifications, or more extensive application of fault tree analyses.

The concept has to be applied in a responsible manner and tailored with regard to the specific needs of the area of application.

4.5 Apportionment of Safety Targets and Allocation of Safety Integrity Requirements

If the risk-based approach is applied, a risk analysis has to be performed to achieve safe design. One of the results is a list of potential hazards which could occur.

Safety requirements have to be derived from the requirements and the safety targets at the higher system level. They have to be apportioned down (derived) to the functions at the next levels below. As far as one of these functions is implemented by electronic means, an SIL has to be allocated at the level of the last independent function to ensure that the part of the safety target apportioned to this function is covered.

Safety integrity requirements are expressed at the system level as tolerable hazard rate (THR) dedicated to the respective hazard (e.g., "train braking ability impaired or lost"). Depending on the system architecture chosen, all the functions potentially contributing to this hazard, in case they fail, have to fulfill the required tolerable functional unsafe failure rate (TFFR).

It should be noted that a component being part of a system cannot have a THR but a TFFR. This component might fail with a TFFR in an unsafe manner and would thus contribute to a hazard at the system level under consideration. The acceptable or tolerated level of this hazard caused by the TFFRs of the contributing components is given by the THR at system level. Each analysis shall clearly state what the system under consideration is.

If the relationship between hazard and the functions contributing in case of a safety-related failure is one to one, for each hazard with related quantitative target, the quantitative safety integrity requirement (expressed as THR) shall be completely allocated to the function that protects against that hazard.

In most cases the one-to-one relation is not given. Here apportionment has to be applied as described in EN 50126-2:2017 [5]: "Otherwise, when in case of combination of multiple functions, the THR shall be apportioned to sub-hazards and their tolerable functional failure rates down to the level of the last independent functions. This is the level at which two (or more) functions control the same hazard at this level, being however completely independent."

Potential common cause failures have to be carefully taken into consideration.

The standard [5] explains in Chapter 10 in more detail how to apportion requirements and how to allocate SILs. The transformation of functional safety requirements to an adequate system architecture is given, as well as the assignment of SIL to functions on subsystem level and applying safety measures.

It should be noted that components performing tasks for several operational functions, e.g., controllers, have to fulfill the requirements allocated from the most demanding function but no more. Thus, the requirements resulting from other functions performed by this component are considered as being covered.

It is clearly stated that "SIL shall not be used for non-functional safety. SIL assigned to mechanical or electro-mechanical equipment shall be considered not applicable and in the demonstration process this shall be stated, as this standard introduces only measures and techniques for electronic systems" [5].

4.6 Misuse of SIL

Railway engineers are sometimes faced with requirements as, e.g., "the wind-screen wiper system has to be carried out as an SIL 2 system." As outlined earlier, such requirements are not meaningful. Moreover, it is not possible to fulfill it. Apparently, the requirement has been meant as being the THR related to SIL 2. But THR is not the same as SIL. If one looks at the integrity measures related to SIL 2, it becomes clear that they are not written for and not applicable to windscreen wipers, but only to electronic equipment.

If requirements are wrongly requested and expressed as earlier, the problem needs to be addressed and clarified between customer and supplier at an early stage of the project or better at the tender phase. Thus, very often, the problem is related to the contract and not to the standard.

An SIL is an attribute of functional safety. It cannot be assigned to nonfunctional requirements, e.g., "cutting fingers at sharp edges." Such requirements usually need to be covered by "state-of-the-art" quality standards or mutual agreements.

In EN 50126-2:2017 [5], it is pointed out that SILs should not be used for describing systems attributes. E.g., a computer should not be called "an SIL 4 computer." The correct wording would be "this is a computer capable of performing the specified SIL 4 safety-related functions, provided that all Safety-related Application Conditions (SRAC) associated to this computer are applied."

4.7 Conclusions

The integrity concept, especially if applied to safety, is a very important concept, which requests concentrating not only on requirements for random failures but also on systematic aspects as well. The concept can be applied to any technology, whereas the measures dedicated to the classes of integrity are only commonly agreed and standardized for safety-related electronics, called SILs.

SIL cannot be applied to nonfunctional requirements but only to functional requirements.

Safety integrity requirements often have to be apportioned from the higher level to the related functions at the lower subsystems. This has to be done very carefully, and the respective rules must be followed. Those rules are laid down in EN 50126-2 [5]. This standard is valid for all application areas of railways, which are signaling, rolling stock, and fixed installations.

In addition, following these rules ensures correctly applying the concept and does prevent any misuse.

Abbreviations

E/E/PE	electrical/electronic/programmable electronic
SIL	safety integrity level
TFFR	tolerable functional unsafe failure rate
THR	tolerable hazard rate

References

1. International Electrotechnical Commission, IEC 61508: *Functional Safety of Electrical/Electronic/ Programmable Electronic Safety-Related Systems (IEC 61508 Series)*, IEC, Geneva.
2. CEN, EN 50128:2011: *Railway Applications—Communication, Signalling and Processing Systems— Software for Railway Control and Protection Systems*, CEN, Brussels.
3. CEN, EN 50129:2003: *Railway Applications—Communication, Signalling and Processing Systems— Safety Related Electronic Systems for Signalling*, CEN, Brussels.
4. CEN, EN 50657:2017: *Railway Applications—Rolling Stock Applications—Software Onboard Rolling Stock*, CEN, Brussels.
5. CEN, EN 50126-2:2017: *Railway Applications—The Specification and Demonstration of Reliability, Availability, Maintainability and Safety (RAMS): Part 2: System Approach to Safety*, CEN, Brussels, 2018.

5

SIL Apportionment and SIL Allocation

Hendrik Schäbe

CONTENTS

5.1 Introduction

Technical systems become more and more complex. An increasing number of technical systems contains electronics and software, and therefore, functional safety has an increasing importance. The safety integrity level (SIL) is a discrete number that defines a set of measures against random and systematic failures depending on the requirements for risk reduction. The concept of SILs has been developed within different systems of standards.

When discussing the safety architecture of a system, a main question arises: How can components or subsystems of a lower SIL be combined to give a system with a higher SIL? The answer to this question would allow the use of already existing and certified components to build up a system with a required SIL, perhaps also with a higher SIL than that of the components; see e.g., Schäbe (2004, 2017) and Schäbe and Jansen (2010).

The concept of SILs is defined and used in different standards as IEC 61508, DEF-STAN-0056 (Ministry of Defence 1996), EN 50126 (CEN 1999), EN 50128 (CEN 2011), EN 50129 (CEN 2003), and many more; see, e.g., the study by Gräfling and Schäbe (2012). It is common practice within these standards to define four different SILs.

The SIL consists of two main aspects:

1. A target failure rate, which is a maximal rate of dangerous failures for the system safety function that must not be exceeded in order to cope with random failures
2. A set of measures that is dedicated to cope with systematic failures

Note that for software, only systematic failures are considered, and no target failure rate is given. This is caused by the fact that normal software does not have random failures.

5.2 SILs

Table 5.1 gives the four SILs and their tolerable hazard rates (THRs) as defined in the three standards IEC 61508, EN 50129, and DEF-STAN-00-56.

The THR is the maximal rate of dangerous failures of the technical equipment that is allowed for a certain SIL within a defined standard. First of all, we note that the THR values are identical for IEC 61508 and EN 50129 and different for DEF-STAN-00-56. Therefore, the SILS are not comparable. Even though the THR values for IEC 61508 and EN 50129 are the same, the measures against systematic failures are different so that the SILs are not identical.

The standards EN 50126 and EN 50128 do not provide target failure rates. EN 50126 requires only the existence of SILs. EN 50128 is dedicated to software and software SILs without numeric values for THRs.

DEF-STAN-00-56 implicitly gives the target rates by stating verbal equivalents and presenting numbers for those in another place.

5.3 Combining SILs

In this section, we will describe the rules of combinations of SILs as they are used in different standards.

5.3.1 DEF-STAN-0056

The rules are provided in clause 7.4.4, Table 5.8, of DEF-STAN-00-56. The reader is warned not to mix up these SIL combination rules (DEF-STAN-00-56) with the SILs for EN 50129, since these are different characteristics.

TABLE 5.1

THR Values for Different SILs and Standards

SIL	IEC 61508/EN 50129	DEF-STAN-00-56
1	$10^{-6}/h \leq THR < 10^{-5}/h$	Frequent $\approx 10^{-2}/h$
2	$10^{-7}/h \leq THR < 10^{-6}/h$	Probable $\approx 10^{-4}/h$
3	$10^{-8}/h \leq THR\ 10^{-7}/h$	Occasional $\approx 10^{-6}/h$
4	$10^{-9}/h \leq THR < 10^{-8}/h$	Remote $\approx 10^{-8}/h$

- Combining two devices with an SIL 3 in parallel yields an SIL 4 system.
- Combining two devices with an SIL 2 in parallel yields an SIL 3 system.
- Combining two devices with an SIL 1 in parallel yields an SIL 2 system.
- Combining two devices with an SIL x and an SIL y in parallel yields an SIL max(x;y) system.

Note that "combining in parallel" means that the two devices or functions are combined in such a manner that a dangerous failure of both components/functions is necessary for the system to fail in a dangerous manner. From these rules, we see that a combination of two devices would lead in the best case to an SIL that is just once higher. Also, a system with a certain SIL cannot be built by combining devices or functions without an SIL—at least not using these general SIL combination rules.

5.3.2 Yellow Book

Another interesting source is the *Yellow Book*. The *Yellow Book* is a national UK regulation which has become obsolete with the coming of the common safety methods. Nevertheless, it provides interesting information. The SILs in the *Yellow Book* are the same as those in EN 50129, but it is quite different from the ones defined in DEF-STAN-00-56 (Table 5.2).

5.3.3 IEC 61508

IEC 61508 does not give such a nice combination rule but provides nevertheless a possibility to achieve an increased SIL via combination. The general rule is as follows (see IEC 61508-2, clause 7.4.4.2.4):

> Selecting the channel with the highest safety integrity level that has been achieved for the safety function under consideration and then adding N safety integrity levels to determine the maximum safety integrity level for the overall combination of the subsystem.

TABLE 5.2

SIL Combination Rules from *Yellow Book*

| Top-Level SIL | SIL of Lower-Level Function | | |
	Main	Other	Combinator (If Present)
SIL 4	SIL 4	None	None
	SIL 4	SIL 2	SIL 4
	SIL 3	SIL 3	SIL 4
SIL 3	SIL 3	None	None
	SIL 3	SIL 1	SIL 3
	SIL 2	SIL 2	SIL 3
SIL 2	SIL 2	None	None
	SIL 1	SIL 1	SIL 2
SIL 1	SIL 1	None	None

Source: *Engineering Safety Management* (*Yellow Book*), Vols 1 and 2: Fundamentals and Guidance.

Here N is the hardware fault tolerance of the combination of parallel elements, i.e., the number of dangerous faults that are tolerated by the system.

Note that to achieve a certain SIL, the hardware failure tolerance and safe failure fraction criteria for the system built by combination must be fulfilled according to the table from IEC 61508-2 in addition. Here, one has to distinguish type A and type B elements/systems (IEC 61508-2, chapter 7.4.4.1.2).

An element can be regarded as type A if, for the components required to achieve the safety function, the following apply:

1. The failure modes of all constituent components are well defined.
2. The behavior of the element under fault conditions can be completely determined.
3. There is sufficient dependable failure data to show that the claimed rates of failure for detected and undetected dangerous failures are met.

Other elements/systems fall into type B.

Both Tables 5.3 and 5.4 clearly show that the IEC 61508 does not give a simple rule for combining SILs. Not only a combination of systems and, therefore, the hardware fault tolerance decide about the SIL, but also the safe failure fraction. However, one might observe that with increasing the hardware failure tolerance by one and keeping the safe failure fraction and using subsystems of the same type (A or B), the SIL is incremented by one. That means, if

TABLE 5.3

Rules for Achievable SILS for Type A Systems

Safe Failure Fraction of an Element (%)	Hardware Fault Tolerance		
	0	1	2
<60	SIL 1	SIL 2	SIL 3
60–<90	SIL 2	SIL 3	SIL 4
90–<99	SIL 3	SIL 4	SIL 4
>99	SIL 3	SIL 4	SIL 4

Source: IEC, IEC 61508-2. *Functional Safety of Electrical/ Electronic/Programmable Electronic Safety-Related Systems–Part 2: Requirements for Electrical/Electronic/ Programmable Electronic Safety-Related Systems,* Table 5.2, IEC, Geneva, 2010.

TABLE 5.4

Rules for Achievable SILS for Type B Systems

Safe Failure Fraction of an Element (%)	Hardware Fault Tolerance		
	0	1	2
<60	Not allowed	SIL 1	SIL 2
60–<90	SIL 1	SIL 2	SIL 3
90–<99	SIL 2	SIL 3	SIL 4
>99	SIL 3	SIL 4	SIL 4

Source: IEC, IEC 61508-2. *Functional Safety of Electrical/ Electronic/Programmable Electronic Safety-Related Systems–Part 2: Requirements for Electrical/Electronic/ Programmable Electronic Safety-Related Systems,* Table 5.3, IEC, Geneva, 2010.

two subsystems have the same SIL and the same safe failure fraction, a combination would increase the SIL by one. Combining subsystems from different types and/or safe failure fractions might lead to other results, and one would need to consult the table anyhow.

5.3.4 SIRF 400

The Sicherheitsrichtlinie Fahrzeug (SIRF) (2011) is a German regulation for the safety of rail vehicles prepared together by German railway industry, German Rail, Society of German Railway operators, and the German Federal Railway Authority. In Germany, one could easily make reference to this document; however, outside Germany, acceptance is not automatically guaranteed.

The following principles are explained in the document.

By connecting two subsystems in series (e.g., using an OR gate in a fault tree), the smallest SIL would also be the resulting SIL of such a system.

For parallel combinations of subsystems, the following rules are provided:

1. An SIL > 0 must not be constructed from SIL 0 elements.
2. The SIL may be released only by one SIL within an AND gate.
3. Exclusion from 2: One branch completely takes over the safety function.
4. Exclusion from 2: A common cause failure analysis is carried out.
5. In case of 4, a suitable systematic method (failure mode effects analysis, hazard and operability study, etc.) has to be used down to the lowest level of the hazard tree to show that common cause/mode failures are excluded.

Note that the SIRF uses the term *SAS*, which is mainly equivalent to *SIL*, but not completely the same.

The allowed AND combinations are depicted in Figures 5.1 through 5.4. Here, light gray means that the combination is allowed; dark gray shows a forbidden combination; and white means that only based on a deep analysis showing independence of the subsystems,

00	01
10	11

FIGURE 5.1
Allowed combinations for SIL 1 taken from SIRF 400.

00	01	02
10	11	12
20	21	22

FIGURE 5.2
Allowed combinations for SIL 2 taken from SIRF 400.

00	01	02	03
10	11	12	13
20	21	22	23
30	31	32	33

FIGURE 5.3
Allowed combinations for SIL 3 taken from SIRF 400.

00	01	02	03	04
10	11	12	13	14
20	21	22	23	24
30	31	32	33	34
40	41	42	43	44

FIGURE 5.4
Allowed combinations for SIL 4 taken from SIRF 400.

the combination is allowed. Figures 5.1 through 5.4 show the combinations of SIL that are allowed according to SIRF 400.

By neglecting the combination of two independent SIL 2 subsystems for an SIL 4 system, we mainly see that a combination of two subsystems with the same SIL would yield a system that has an SIL one step higher.

5.3.5 Numerical Approach

In this subsection, we will carry out a calculation based only on the rates of dangerous failures, i.e., the THRs to be achieved by the subsystems. This does not take into account the measures against systematic failures.

We will use the following assumptions for the analysis:

1. A combinator is not necessary.
2. The inspection interval is t. At inspection, all failures and faults are detected and removed so that the subsystem will be as good as new.
3. The system is constructed of two subsystems that are connected in parallel and have the same SIL.
4. The system is intended to have an SIL which is one increment higher than those of the subsystems.

Roughly, the rate of dangerous failure of the combined system would be

$$\lambda = \lambda_1 \cdot \lambda_2 \cdot t,$$

where λ_1 is the rate of first system; λ_2 is the rate of second system; and t is the inspection interval.

Table 5.5 gives the result for a time interval of 10,000 hours, i.e., approximately one year.

TABLE 5.5

SILs and THRs of Subsystems and System for an Inspection Interval of 10,000 h

System		Subsystems		
SIL	THR	Required Rate	SIL	THR
4	10^{-8}/h	10^{-10}	3	10^{-7}/h
3	10^{-7}/h	10^{-8}	2	10^{-6}/h
2	10^{-6}/h	10^{-6}	1	10^{-5}/h

It can be seen that for all three cases (SILs 2, ..., 4 for the system), the subsystems would fulfill the requirement (required rate) if the subsystems have the same SIL, which is one step lower than the required SIL of the system.

However, this computation still needs to be augmented by a common cause failure analysis. For this sake, we use IEC 61508-6. Although, EN 50129 gives another approach in B3.2, we use IEC 61508 for the simple reason that it provides us with a figure. A worst-case beta factor would be 10%. This is the fraction of the failure rate that has to be used to describe common cause failures. Then, for the rate of dangerous failures of the combined system, the common cause failures would dominate. If now a subsystem has an SIL n, the rate of dangerous failures of the combined system would be

$$10\% \text{ of } 10^{-(n+4)}/h = 10^{-(n+5)}/h,$$

so that the combined system would have an SIL $n + 1$ in the best case. The conclusion would be that without further assumption on common cause failures, one would get an improvement of the SIL by one step, when combining two subsystems with the same SIL.

5.3.6 Short Summary on SIL Combination

Besides the target rates, design requirements have to be considered when subsystems of a lower SIL are combined with the intention to construct a system with a higher SIL.

DEF-STAN-00-56 requires in clauses 7.3.3 that "design rules and techniques appropriate to each safety integrity level . . . . shall be determined prior to implementation. . . ." No particular rules are given.

IEC 61508 (part 2, annex A3, annex B) and EN 50129 (annex E) give different design methods for different SILs. The most extensive set of methods are required for SIL 4.

The set of methods cannot be easily transferred and for all possible systems into a simple rule for combination of subsystems of a lower SIL to form a system with a higher SIL.

However, a general rule seems that the improvement of the SIL by one step is possible by combining two subsystems of the same SIL.

5.4 Examples

5.4.1 Example 1

The system consists of two subsystems and does not contain software. No comparator is necessary. Each subsystems checks the difference between the other subsystem and itself and switches the other subsystem off, in case there are differences. This means that switching off the entire system would be a safe situation. The block diagram in Figure 5.5 shows the example.

If both subsystems are in SIL 3 and they are independent, they could be combined to an SIL 4 system. Design rules are not very different for SILs 3 and 4.

If the system is required to have SIL 2, it could be combined from two SIL 1 subsystems.

If both subsystems have an SIL 2 and the system is required to have SIL 3, deeper investigation regarding the system is needed. Design rules required for SIL 3 (system) differ from those for SIL 2.

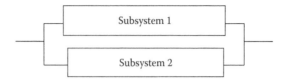

FIGURE 5.5
Block diagram for example 1.

5.4.2 Example 2

The system is mostly similar to example 1; however, the subsystems are operated by software, and the same software is used in both subsystems; see Figure 5.6.

Since the same software is used, it is depicted in the block diagram as being in series with the two hardware subsystems.

If the system has SIL 4, the software shall also have SIL 4. (The software SIL must be at least as good as the system SIL.) An SIL 2 system can be constructed from two parallel SIL 1 systems with an SIL 2 software. If the system is required to have SIL 3, the software must also have SIL 3. If the hardware is SIL 2, additional considerations have to be made as for the system in example 1.

5.4.3 Example 3

This system is again similar to the system in example 1; however, it contains diverse software. The block diagram is shown in Figure 5.7.

There are different and diverse types of software used in both subsystems. The same considerations as in example 1 apply regarding the SIL apportionment. An SIL 4 system can be constructed from two SIL 3 subsystems, each equipped with SIL 3 software. An SIL

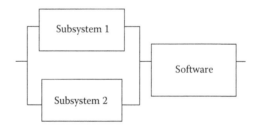

FIGURE 5.6
Block diagram for example 2.

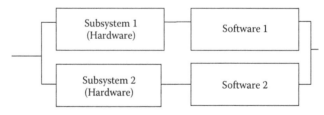

FIGURE 5.7
Block diagram for example 3.

2 system can be constructed from two SIL 1 subsystems. For constructing an SIL 3 system from two SIL 2 subsystems, additional considerations must take place.

5.4.4 Example 4

The system consists of one hardware channel but redundant software; see Figure 5.8. The software "redundancy" can come from two different software packages or from redundant programming techniques (diverse software). In any case, diversity of software must be ensured.

If the system is required to have an SIL 4, the hardware must have an SIL 4, and both software versions must be at least according to SIL 3. In addition, it must be proven that each failure of the hardware is detected by the software, i.e., by both software versions and that there are means to bring the system into a safe state.

If the system shall have SIL 2, the hardware has to have SIL 2 and two independent software versions with an SIL 1 each.

For an SIL 3 system, however, a detailed study is necessary if the hardware is SIL 3 and the software versions are SIL 2.

The question of the independence of two software versions running in the same hardware is not trivial. In any case, the software must be diverse.

5.4.5 Example 5

In this example, we study an electronic subsystem consisting of hardware and software and another hardware system acting in parallel (hardware bypass); see Figure 5.9.

If the "hardware bypass" has the same SIL as required for the system, hardware 1 and software 1 do not need to have any SIL. Also, the same logic as in example 1 can be applied: SIL 4 system can be constructed from SIL 3 subsystems (hardware 1 and software 1 on the one side and hardware bypass on the other side). "Software 1" must have the same SIL as "hardware 1" or better.

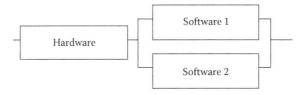

FIGURE 5.8
Block diagram for example 4.

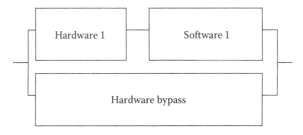

FIGURE 5.9
Block diagram for example 5.

5.5 Conclusion

A general rule for SIL apportionment as given in DEF-STAN-00-56, *Yellow Book*, or SIRF cannot be provided for all countries and all situations. Target failure rates and/or inspection intervals have to be taken into account. General rules can only be given for subsystems connected in parallel and for some SIL combinations (see, e.g., *Yellow Book* and SIRF). In any case, common cause failures need to be duly taken into account. A general rule of thumb might be to achieve an SIL one step higher by connecting two subsystems in parallel. Other system architectures have to be studied in detail. A good indication whether the chosen architecture would meet an SIL requirement is when the target failure rate of the system SIL is not exceeded by the rate of the system, computed from the rates of its subsystems.

Generally, a combination of subsystems in series gives a system that has an SIL that is the minimum of the SILs of the subsystems.

References

CEN (European Committee for Standardization). EN 50126 (1999) *Railway Applications—The Specification and Demonstration of Reliability, Availability, Maintainability and Safety (RAMS)*, CEN, Geneva.

CEN. EN 50129 (2003) *Railway Applications—Communication, Signalling and Processing Systems—Safety Related Electronic Systems for Signaling*, CEN, Geneva.

CEN. EN 50128 (2011) *Railway Applications—Communication, Signalling and Processing Systems—Software for Railway Control and Protection Systems*, CEN, Geneva.

Ministry of Defence. DEF-STAN-0056 (1996) *Safety Management Requirements for Defence Systems: Part 1: Requirements, Part 2: Guidance*, issue 2, February 13, 1996, Ministry of Defence, London.

Gräfling, S., and Schäbe, H. (2012), *The Agri-Motive Safety Performance Integrity Level—Or How Do You Call It? 11th International Probabilistic Safety Assessment and Management Conference & The Annual European Safety and Reliability Conference*, ESREL 2012/PSAM 11, Proceedings, paper 26 Fr2_1.

International Electrotechnical Commission. IEC 61508 (2010), *Functional Safety of Electrical/Electronic/Programmable Electronic Safety-Related Systems—Parts 17*, April 1, 2010, IEC, Geneva.

Schäbe, H. (2004) Definition of safety integrity levels and the influence of assumptions, methods and principles used, *PSAM 7/ESREL 2004 Proceedings: Probabilistic Safety Assessment and Management* (Eds C. Spitzer, U. Schmocker, and V. N. Dang), Taylor & Francis, Vol. 4, pp. 1020–1025.

Schäbe, H. (2017) SIL apportionment and SIL allocation, *ESREL 2017 Safety and Reliability—Theory and Applications* (Eds M. Cepin and R. Briš), Taylor & Francis Group, London, pp. 3393–3398.

Schäbe, H., and Jansen, H. (2010) Computer architectures and safety integrity level apportionment, *Safety and Security in Railway Engineering* (Ed. G. Sciutto), WIT Press, Ashurst, UK, pp. 19–28.

SIRF (2011), *Sicherheitsrichtlinie Fahrzeug Ausführungsbestimmungen*, Rev. 1, June 1, 2011.

Yellow Book (2007) *Engineering Safety Management (Yellow Book)*, Vols 1 and 2: Fundamentals and Guidance, Issue 4.

The *Yellow Book* is replaced by the guidance on the application of the Common Safety Method on Risk Evaluation and Assessment. However, the guidance does not contain the information, e.g., for combining SILs.

6

Prognostics and Health Management in Railways

Pierre Dersin, Allegra Alessi, Olga Fink, Benjamin Lamoureux, and Mehdi Brahimi

CONTENTS

6.1 Introduction and Historical Perspective

Industrial maintenance has evolved considerably over the last 70 years. Roughly speaking, one could say that the first generation, until approximately 1950, was characterized by a purely "corrective" perspective, i.e., failures led to repairs, and then came the second generation, characterized by scheduled overhauls and maintenance control and planning tools (roughly the period of 1950–1980). From the 1980s onward, the notion of condition-based maintenance (CBM) gained ground. With the turn of the twenty-first century, a great interest in "predictive maintenance" has emerged, along with the concept of prognostics and health management (PHM). A number of rail companies and original equipment manufacturers now have a PHM department.

Let us backtrack and take a closer look at those concepts and try to understand this evolution.

First of all, what is maintenance? The standard IEC 60300-3-14 defines maintenance as "the combination of all technical and administrative actions, including supervision actions, intended to retain an item in, or restore it to, a state in which it can perform a required function." It is divided into two broad categories: preventive, i.e., maintenance operations taking place before a failure, and corrective, after a failure, as illustrated in Figure 6.1.

As far as preventive maintenance is concerned, there are two main categories: scheduled (or systematic) maintenance, which takes place at fixed intervals (defined by time, distance, or other measures of usage), and CBM, which takes place when the condition of the asset warrants it. Maintenance costs (corrective and preventive) of complex systems such as railways represent a considerable percentage of asset life cycle cost (several times the initial investment), and, therefore, there is considerable economic pressure to reduce them. At the same time, the direct and indirect costs of a service-affecting failure (SAF) can be considerable (in terms of lost missions, punctuality, and service disruptions in the case of railways). Therefore, there is substantial incentive to reduce their frequency of occurrence.

The prevalent belief in the 1950s and 1960s was that every item on a piece of complex equipment had a right age at which complete overhaul was needed to ensure safety and reliability. This attitude had led to the conclusion that the new Boeing 747 would require 4 hours of maintenance for each hour of flight because it contained many more parts to be maintained than in previous models! To escape this conclusion, which defied common sense and would have meant that the new aircraft was economically not viable, a task force was set up in 1960 by the US Federal Aviation Administration. The task force discovered that scheduled overhauls have little effect on the overall reliability of a complex item unless the item has a dominant failure mode and that there are many items for which there is no effective form of scheduled maintenance. This effort led to the Maintenance Steering Group (MSG), which was able at the same time to reduce maintenance costs of

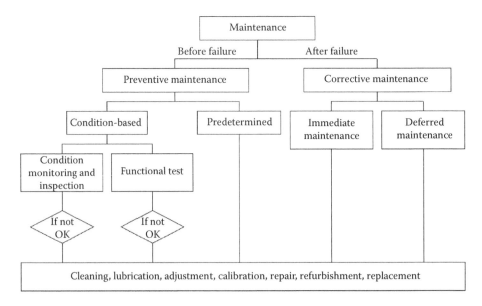

FIGURE 6.1
Classification of maintenance types according to IEC 60300-3-12. (Courtesy of International Electrotechnical Commission, Geneva, Switzerland, 2004.)

civilian aircraft and increase their reliability. Eventually, in 1978, the success of the MSG led to the "reliability-centered maintenance" (RCM) methodology, pioneered by the US Department of Defense. RCM is a process used to determine the maintenance requirements of an asset in its operating context (see standard IEC 60300-3-11). It is based on functional failure modes and effects analyses (FMEAs) and systematic cost-performance trade-off analyses, which lead to the definition of the appropriate maintenance policy. A key concept part of RCM is the "P–F curve," where P denotes a point in time when a progressive deterioration in the maintained asset is detected and F denotes the point in time when failure takes place if no maintenance action is performed after that detection. CBM can be based on periodic inspections (whose periodicity must be adapted to the speed of evolution of the failure mechanisms) or on continuous monitoring.

The notion of predictive maintenance is a special case of CBM: CBM infers from the condition of the asset that there is an impending failure, i.e., a failure will take place if no action is taken, and predictive maintenance intends to predict, in a probabilistic sense, after how much time the failure will take place if no action is taken. That time is called the remaining useful life (RUL). It is, of course, very much dependent on the assumed future load profile of the asset.

Since the first decade of the new millennium, several factors have favored the effective development and deployment of CBM and predictive maintenance: the appearance of low-cost wireless sensors and the expansion of the Internet of things, the significant progress of advanced multivariate statistics and machine intelligence methods, and the ability to manage and store very large data sets (big data) and to perform concurrent processing at affordable costs.

The expression "prognostics and health management" refers to the methods, means, and tools (both hardware and software) that are used to perform diagnostics, health assessment, and prognostics of industrial assets.

The digitalization of railways is a major trend (as, for instance, illustrated at InnoTrans 2016), and PHM both benefits from that context and influences it.

6.2 Links between PHM and Reliability, Availability, and Maintainability

The concepts of reliability engineering and PHM are closely related. However, they are often not considered jointly in practical applications. There is a great potential to improve the reliability and availability of the system and to reduce the maintenance costs by integrating both approaches already starting in the concept phase and to develop designs for reliability and for PHM and by continuously updating and improving the implemented sensors, models, algorithms, and the decision support systems during the operating phase.

Traditional reliability analysis typically relies on time-to-failure data to estimate the lifetime distributions and to evaluate the reliability of the "average" asset operated under "average" operating conditions. With the highly customized components and subsystems, and very different operating conditions of the components within the fleet, the traditional analysis provides a good estimate for the maintenance decisions on the aggregated level, but not on the level of an individual component. With the increased reliability of the critical systems, full time-to-failure trajectories are often available only for noncritical systems, as critical system failures are rare events. Additionally, the average behavior of the systems does not sufficiently decrease the temporal uncertainty of the required maintenance actions. With the increased availability of advanced sensors and measuring devices, communication networks, and computer processing power, PHM approaches enable the assessment of the reliability of a component or a system under its actual application conditions and its actual system health state.

The main difference between the reliability and the PHM approaches can be seen in the change of the perspective from a fleet or a set of components operated under diverse conditions to the perspective of an individual component with its unique characteristics and unique operating conditions. Both are complementary.

PHM can be considered as a holistic approach to an effective and efficient system health management of a component or an overall system. PHM integrates the detection of an incipient fault, its isolation, the identification of its root cause, and prediction of the RUL. The system health management goes one step beyond the predictions of failure times and supports optimal maintenance and logistic decisions by considering the available resources, the operating context, and the economic consequences of different faults.

While reliability considerations are important during design for taking the design decisions, during the operation, on the contrary, early fault detection, diagnostics, and the prediction of the RUL at the level of the individual component combined with an uncertainty in the prediction become more important. By aggregating the predicted RUL of the components within the fleet, the reliability can be considered by quantifying the mean residual life and using it for CBM decision making. Indeed, the mean residual life is the mathematical expectation of the RUL. It is derived from the reliability function or the failure rate.

While PHM methods are not able to improve the inherent reliability of a component, they are able to avoid operational consequences of a fault and thereby preserve or extend the operational reliability of the asset. PHM also influences the operational availability by reducing the frequency of occurring failures affecting the operational reliability and reducing the time for fault finding and improving the preparedness of the maintenance organization and by improving the maintainability of the system.

PHM can therefore be considered as an enabler for improved operational reliability and availability.

Conversely, reliability, availability, and maintainability (RAM) engineering is a powerful tool for PHM methodology: FMEAs, or their extension failure modes, mechanisms, and effects analysis, make it possible to pinpoint the key causes of SAFs and thereby determine some of the assets that stand most to benefit from PHM; and they constitute the first step toward identifying the physical variables to be monitored.

6.3 ISO 17359 and Other Standards

Given the impact of PHM technologies on any process, it is necessary to ensure the interoperability of various systems as well as a standardized approach to apply a PHM technology to any process. Having a standard which is adopted throughout the same field is beneficial, as it provides a generally accepted technical specification which enables different parties across different domains to interact and understand the practices adopted. For this reason, a dedicated safety unit already exists: the European Railway Agency, which monitors all rail safety in the European Union and develops common technical standards and approaches to safety, promoting interoperability. In the United Kingdom, the Rail Safety and Standard Board is an independent and nonprofit company which aims to improve the performance of the railway in Great Britain and defined the rules and standards for operating the network. It is also important to underline that as the number of standards for the same topic increases, their effectiveness and usefulness significantly diminish: a series of competing standards go against the very nature and aim of the standard itself.

While a PHM standard per se does not yet exist, the International Organization for Standardization (ISO) has developed several guidelines for condition monitoring and diagnostics of machinery, for which ISO 17359 (*Condition Monitoring and Diagnostics of Machines—General Guidelines*) is the parent document, containing references to prognostics as well. In the same manner, the Institute of Electrical and Electronics Engineers Standards Development Committee has written a PHM standard, P1856, which is undergoing approval process at the time of writing.

Both ISO 17359 and P1856 introduce a series of capabilities requirements that are outlined in the following phases:

- Data acquisition
- Data processing
- Detection
- Diagnostics
- Prognostics
- Advisory generation

Each of these phases is defined as necessary and fundamental for the overall application of the PHM process to a system. These steps pertain to the actual application of PHM to a system and do not delve on matters of feasibility, cost-effectiveness, and business scenarios for the implementation of PHM. While it is true that the standards are meant to cover thoroughly the PHM process, a consideration of such matters is strongly advised before diving into the technical aspects in a real application. These are briefly addressed in the informative section of the P1856 standard, as preliminary considerations.

The first step is data acquisition, which consists of identifying the right technology and methods to acquire the data from the system on which PHM will be applied. This step implicitly requires a previous evaluation of the failure modes of the system as a whole, to define the critical elements which will be the key to a successful PHM application. The critical elements are often those which fail more often, cause the most severe failures, cause the longest downtimes, or cost the most to replace. The decision of which parts of the system constitute the elements which require monitoring needs to be taken from a hybrid of RAM and safety approach and cost–benefit analysis. Once these key elements are identified, the parameters and data to be measured and recorded can be defined. The acquisition of these data is not necessarily straightforward, as the railway industry has static infrastructure (rails and signaling) as well as moving assets (rolling stock), and the acquisition of data from both needs to be coherent and coordinated into the same platform.

The second step is data processing, where the data acquired from the system are treated to extract the relevant information, pertinent to producing a PHM process. This is a step common to any use of data, as rarely is the raw signal acquired ready for the analysis.

The third step is detection, where signals are taken and analyzed to evaluate whether the behavior of the monitored system is indicating any incipient degradation mechanisms. If this is the case, then the PHM system detects that there is an anomaly in the functioning of the system and triggers the evaluation of the next steps.

The fourth step is diagnostics, where the anomaly which has been detected is analyzed, and its location, severity, and root cause are identified. This information is vital, as it differentiates a PHM process from a CBM process, which merely detects the condition of a system.

The fifth step is prognostics, which is the study and prediction of the future evolution of the health of the system being monitored, to estimate the RUL of the system. By calculating these values, it is possible to define the optimum time for maintenance intervention, to maximize the use of the assets and plan the maintenance actions at a time when the impact on productivity and availability is the lowest.

The sixth and final step is advisory generation, also called health management, where the data collected and evaluated in the previous phases are combined with the information of the system process and the results of the processing are readily available for use. This information can be sent to a visualization interface or integrated in the current maintenance planning platform.

While ISO 17359 and P1856 offer a good framework for reference on how to apply and work with PHM, the application to the railway sector still lacks a well-defined standard. Other industries, such as the aerospace industry, are regulated by the plethora of standards on maintenance and health management systems, which are drafted and released by the Society of Automotive Engineers (SAE) Aerospace Propulsion System Health Management Technical committee. In the coming years, because of the increased awareness toward PHM and its contributions, as well as the increase in importance of the railway sector, it is expected that standards well defining these practices will be released.

However, while a series of standards for PHM exists, there is still a lack of harmonization and coordination in the overall field, which often causes confusion. For this reason, it is imperative that when proceeding with the development of PHM technology, the user ensures to comply with the standard that they deem most appropriate.

6.4 Data Acquisition Challenge

Whatever "analytics" are used to process data collected from the field, the output will not be of great value if the input data are of poor quality.

The reliability of the entire acquisition chain, including sensors, data loggers, and communication channels, is of paramount importance, and adequate noise filtering must be performed. In addition, the frequency of acquisition (both how often data are acquired and with what frequency signals are sampled) must be tailored to the type of signal processing method, which the analyst intends to use and to the time constant of the degradation processes. Additional considerations are presented in Section 6.6 ("Integration and Human–Machine Interface").

6.5 Performance Indicators

In the industry, performance indicators or key performance indicators (KPIs) are nowadays the fundamental quantitative values on which managers rely to justify the success or failure of projects. In the past decade, the issue of defining KPIs for PHM has been raised in several papers. For example, the works of Saxena et al. [1], Roemer and Byington [2], and Dzakowic and Valentine [3] can be cited. For PHM, six types of KPIs are defined accordingly to the main PHM steps.

6.5.1 Data Acquisition KPIs

The data acquisition KPIs quantify how well the sensors, data acquisition board, wires, and network infrastructure perform in faithfully translating physical quantity into analog values, digitalizing them and sending them as fast as possible while minimizing the loss of information. Usually, engineers rely on values such as accuracy, sampling rate, resolution, and bandwidth to give a short list.

6.5.2 Data Processing KPIs

The data processing KPIs quantify how well the features represent the raw data. In the case of PHM, a set of good features must principally show (a) a very good sensitivity to degradations and (b) a minimum correlation between the components. To quantify the first aspect, many KPIs issued from the domain of sensitivity analysis can be used, for example, the Fisher criterion [4], the Sobol indices [5], or the Morris method [6]. The L1 term of the Lasso [7] can also be used for that matter. For the second aspect, the best indicator of the correlation between features is the covariance matrix.

6.5.3 Detection KPIs

The detection KPIs quantify how well the anomaly detection performs. The most common approach is to compute the confusion matrix, as explained by Wickens [8]. Usually, for industrial purposes, the KPIs of interest are the false alarm ratio and the good detection ratio. One can also use a receiver operating characteristics (ROC) curve, which can help in tuning the detection threshold in function of the true positives/false positives ratios targeted. The area under the curve, or similarly the Gini coefficient [9] of the ROC, can also be used as a direct KPI of the detection performance. In the field of big data, some other KPIs, derived from the confusion matrix, are also commonly used, for example, accuracy, prevalence, negative/positive likelihood ratios, and diagnostics odds ratio.

6.5.4 Diagnostics KPIs

The diagnostics KPIs quantify how well the classification of the degradation type performs. A common way of quantifying this type of performance is to compute the classification matrix for the degradation types. At this point, the reader must be warned that a necessary condition for that is to have labeled data. The classification matrix gives, for each set of degradation type label, the results of the classification process. A perfect classification would have all the values in the diagonal of the matrix. In the case of PHM, industry practitioners are usually interested not only in the type of degradation occurring in their system but also (and sometimes more) in its localization. Indeed, many systems are now modular, according to the principle of line replaceable unit (LRU), and identifying which LRU is degraded is the only necessary information needed for maintenance. In this case, a localization matrix can be used instead of (or complementarily to) the classification one [10].

6.5.5   Prognostics KPIs

The prognostics KPIs quantify how well the assessment of the RUL performs. Some authors have recently focused their research on this topic, such as Saxena et al., who gives in their study [11] a list of metrics that can be used to quantify both the online performance of prognostics and the uncertainty in predictions. For example, the online performance can be assessed from the RUL online precision index and the dynamic standard deviation that quantify, respectively, the precision of RUL distributions by quantifying the length of the 95% confidence bounds and the stability of predictions within a time window. The uncertainty can be evaluated by using the β-criterion that specifies the desired level of overlap between the predicted RUL probability density function and the acceptable error bounds.

6.5.6   Advisory Generation KPIs

The advisory generation KPIs quantify how well the predictive maintenance itself performs. Indeed, because it is the last layer of the PHM process, its requirements are directly given in terms of service reliability and availability, for example, the rate of SAFs, the service availability, the time to repair or, even better, if it is possible, the difference between these values taken without and with PHM with respect to the total cost of the PHM system.

6.6   Integration and Human–Machine Interface

Once the sensors and the PHM algorithms have been validated in laboratory conditions, the integration consists of pushing the hardware and software to a real-life field environment. Multiple questions are then to be answered.

6.6.1   Questions and Issues for Hardware Integration

Is the equipment compliant with railway standards and does it necessitate safety case reopening? Indeed, as soon as integration is addressed, the issue of compliance with the domain standards is raised. These standards can impose, for example, electromagnetic and vibration resistance for sensors or specific shielding for wires. Intrusivity is also a very important factor for integration: if the sensor is intrusive, it is very likely that a safety case will have to be reopened and significant additional costs are then to be expected.

Does the equipment necessitate dedicated processing/storage unit? Of course, as long as you stay in laboratory conditions, the processing and storage capabilities are not an issue, as they can be easily and quickly scaled to the need. However, in a field environment, these resources can be limited, especially in a location where the empty spaces are rare, such as in bogies. For some cases, the already installed processing unit may be enough to process the measurements, for example, in local control units for heating, ventilation, and air-conditioning, but most of the time, a dedicated unit will have to be designed and integrated in the train architecture.

Does the equipment necessitate dedicated communication network? The trains of the new generation are generally equipped with a remote diagnostics system that sends the train status variables (speed, Global Positioning System coordinates, etc.) and some discrete events (mainly fault codes) to a cloud server in which dynamic availability analysis are performed. However, this system uses 4G routers, which are often not designed to handle PHM-related data in addition to the events. In some cases, it can be necessary to design a dedicated communication system to download data from the train, for example, a Wi-Fi or Ethernet connection in the depot.

What is the reliability of the installed hardware? PHM hardware includes sensors, wires, and processing units which have their very own reliability properties. Thus, even if neither the safety nor the service reliability is impacted, a failure of one of the PHM data acquisition chain components will trigger some corrective maintenance that will necessitate workload and spare parts. Because this hardware is generally multiplied by the number of a given subsystem occurrence, a low reliability can have such a significant impact that operators prefer to shut down the system. Hence, performing reliability analysis before the integration of the hardware is of paramount importance.

6.6.2 Questions and Issues for Software Integration

Does the communication network constrain the bandwidth? In case that a dedicated communication system for PHM cannot be designed, and if the bandwidth of the existing one is limited, it is not an option to send sampled physical quantities from sensors. In this case, the main solution is to perform the data manipulation (features extraction) onboard and to send only the features to the PHM off-board server that runs the next processing steps. The issue of the onboard/off-board split is very important for PHM integration because it dimensions the onboard computation and storage needs. Usually, if the bandwidth permits, sending the raw sampled measurements to be able to modify the features extraction logic as often as needed is preferred.

Do the algorithms necessitate a specific safety integrity level (SIL) and do some parts of my process need to be recoded in another language? Whatever the onboard/off-board split, there will always be a part of the PHM algorithms onboard, at least to trigger the acquisition according to specific operating modes of the system (example: launch acquisition of the bogie acceleration if the speed of the train is above 100 km/hour) because storing all the measurements during the whole mission is not wanted for obvious volumetry issues. In some cases explained before, the feature extraction logic can also be performed onboard. For each of those embedded tasks, it has to be checked as early as possible in the design process if an SIL is required (in case that it interacts with a safety-related unit) or if a specific language is required.

What are the different machines/servers needs for the information technology (IT) architecture? For a fully instrumented fleet, the volume of collected data per day and the computational power needed to process them can be enormous. In the future, it will be more and more standard to use a cloud-based architecture. The key for success at this level is to ensure that the architecture is not only dimensioned for the current data but also scalable for future evolutions. Nowadays, several companies are providing cloud-based service, under the form of raw infrastructure (infrastructure as a service [IaaS]), platform (platform as a service [PaaS]), or high-level software (software as a service [SaaS]). The PaaS is a compromise between the high complexity (coding and networking) of the IaaS and the low flexibility of the SaaS.

What is the global availability of the IT architecture? Similarly to the hardware reliability, a study of the IT architecture availability has to be performed to assess the service level agreement (SLA) that can be sold to the end user. Most of the time, this SLA is in the scope of the platform provider.

Another aspect of the integration is the interface with the end user, the so-called human–machine interface. Beyond the purely network and coding language issues, the human–machine interface is the emerging part of the iceberg and has to be (a) compliant with the client requirements, i.e., functionalities are tuned with respect to the needs; (b) user friendly and aesthetic; and (c) scalable and highly reliable. According to ISO 13374, the minimal information that a PHM human–machine interface shall contain is the current value of the state (healthy on degraded), the degradation type identified, the health index, and the RUL, and this for each asset of the fleet. Ideally, the possibility to visualize the history of these values should be given to the end user.

6.7 Importance of PHM in the Railway Industry

Investments and projects based on monitoring assets and their functionalities have significantly increased in numbers in the past years in the railway industry. This is due both to the significant cost of scheduled maintenance on railway assets and infrastructure and to the impact of a failure on the overall functioning of the railway industry. Failures of components of the railway industry, in terms of both infrastructure (e.g., deformed rail, failure to move a turnout, signaling failure) and assets (e.g., traction faults, bogie failure), cause delays, which are often costly and can cause severe accidents.

According to the European Commission, passenger and freight transportation in Europe is expected to, respectively, double and triple in 2020 [12]. This increase in demand and, consequently, in service frequency, greater axle loads, and an overall increased attention to the railway industry cannot be simply met by building new infrastructure, which is often expensive and not time effective. To meet this challenge, it is pivotal to ensure the safety and reliability of the current network, which will become of critical importance in the coming years. This, together with the high costs of scheduled maintenance and planned overhauls, which often turn out to be not necessary, has pushed the industry to seek more advanced methods to monitor the real-time state of the assets.

Implementing a monitoring strategy on assets allows for early identification of failures, thereby increasing railway availability and safety while reducing maintenance costs.

6.8 Case Studies

6.8.1 Railway Turnouts

Railway turnouts are devices which enable trains to be directed from one track to another, by mechanically moving a section of railroad track. They are composed of a motor which generates power, which is used in turn to move sections of track to the desired position. Their functioning is central to allowing a functional infrastructure and an undisturbed flow of traffic. When a turnout is faulty and therefore unable to move to the desired position, it forces the network operator to have to reroute the trains when possible, or halt the traffic and send a team to repair the turnout as rapidly as possible to recommence the normal flow of operations.

While turnouts play a simple role in the greater complexity of the railway network, they have a great impact on the train circulation and overall delay times. In the United Kingdom, Network Rail indicated that the total delay of passenger strains amounts to 433,400 minutes a year in 2014, due to turnout failures [13], while Die Bahn in Germany indicated that turnout failures are responsible for a total of 19% of all minutes of delay, approximately 28 million minutes in 2010 [14]. While these figures significantly impact the scheduling of traffic and the customer's opinion of the network operator, they also carry a significant economic burden. Other than the official refunds that are often dispatched to passengers of delayed trains, in some cases, there are governmental fines if the target for delays set by the office of regulation is not met. For example, by 2019 Network Rail is expected to ensure that 9 out of 10 trains run on time and fewer than 3 out of 100 trains on the Virgin route and about 4 in 100 on the East Coast are severely delayed or cancelled [15]. In 2015, Network Rail was fined £53.1 million for missing nearly all the performance targets of the fourth control period and was required to invest an additional £25 million to improve the resilience of the network. Hence, an improvement in the reliability and availability of turnouts is a direct earning for the network operator.

Most maintenance operations on turnouts are aimed to improve their operation, as the problems associated with turnouts involve wear, misalignments of components, and lack of lubrication. The maintenance actions aim at regulating and adjusting the various components to compensate for these issues [16]. Maintenance actions for turnouts are a significant cost of the overall railway infrastructure cost. For example, in the United Kingdom, £3.4 million are spent every year for the maintenance of point machines on 1000 km of railways [17]. These maintenance operations are lengthy and costly, although vital, as poor maintenance can result in deadly accidents, such as the Potters Bar crash on May 10, 2001, where poor maintenance of the turnout resulted in a deadly derailment [18]. For these reasons, the application of a PHM approach in the field or railway turnouts has been deemed necessary. This reflects the current increasing trend of publications and case studies being provided on the topic.

The main focus of the current research on railway turnouts relies on the correct detection of incipient faults and degradation mechanisms in the turnout, as this is the first step in a PHM process and the key to ensuring safety and reducing failures. In 2015, Alstom instrumented a turnout test bench and simulated different kinds of defects over thousands of maneuvers while monitoring several signals [19]. From these signals, the data from each maneuver are processed to extract only the information relevant to the functioning and the health of the machine, and context factors are taken into account. The extracted information is then processed through an artificial neural network (ANN), and the health of the machine during each maneuver is calculated. Moreover, the effect of the wear of the machine, simulated through a series of incremental tests with a hydraulic brake device, which generated a resistive force during the movement, can be seen on the calculated health of the machine. Through the observation of the machine due to its wear, it is possible to estimate the evolution of the health, to calculate the RUL. The degradation trend can be seen in Figure 6.2.

From these results, it becomes evident that it is possible not only to monitor the current health state of the turnouts, but also to predict their evolution over time and therefore assess the optimum time for a maintenance intervention, optimizing costs, and availability.

6.8.2 Bogies

Bogies are one of the main subsystems of a train: together with the traction unit, they provide the rolling capability. People usually see bogies as axles and wheels only, but they are actually

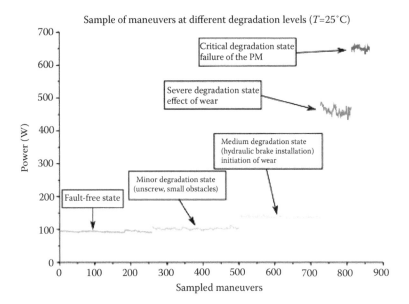

FIGURE 6.2
Evolution of the effects of simulated degradation on the power signal. PM: point machine. (Reproduced from C. Letot et al., *IFAC-PapersOnLine*, 48, 958–963, 2015. With permission. © 2015 International Federation of Automatic Control.)

much more complex: they also include primary and secondary suspension systems, some part of the traction chain, gearboxes, axle bearings, beacons, and many sort of embedded electronics. Bogies are certainly the train subsystem which has the highest impact in terms of costs. Indeed, not only are maintenance operations long and expensive, but also a main failure of a bogie obviously incapacitates the train. Thus, bogies are a very good candidate for PHM.

If one wants to develop a PHM system for bogies, the first step is to identify the targets. Three different types of targets can be identified:

- *Target SAFs*: To increase the service reliability, the objective is to decrease the occurrence of the most critical SAFs by providing predictive functions.

- *Target "cost-affecting maintenance actions"*: To decrease the maintenance costs, the objective is to optimize (space up, automate) the most costly maintenance actions by providing some monitoring functions.

- *Target "availability-affecting maintenance actions"*: To increase the availability, the objective is to optimize the most time-demanding maintenance actions by providing monitoring functions.

For the SAF targets, the first step consists of an analysis of the different events defined in the existing failure detection system and their occurrence. For example, according to railway standards, high-speed bogies must be equipped with an instability detection system which triggers alarms when the lateral acceleration is too important. This event is an SAF because the train is then not authorized to operate normally. Consequently, if it appears that the occurrence of this event on the field is not negligible, then it can be identified as a target. Similarly, hot axle box alarms and nonrotating axle alarms can be SAF targets.

For the cost-affecting maintenance actions targets and availability-affecting maintenance actions targets, the life cycle cost analysis and the maintenance plan are the main reference documents. For the bogies, the maintenance plan is divided into two types of operations:

- *Examinations*: done at a high frequency and composed of online light maintenance tasks such as visual checks, inspections, tank/reservoir refilling, or nondestructive examinations that do not necessitate any unmounting
- *Overhauls*: done at a low frequency and composed of heavy off-line maintenance tasks such as bogie disassembling, unmounting of bearings from axle, complete washing, or component replacement

Then, the idea behind the targeting is the following: (a) to decrease the maintenance costs, the most costly maintenance operations are targeted, and (b) to increase the availability, the online maintenance operations that have the highest duration are targeted. For example, for the bogies, the price of the trailer and motor wheels overhaul makes it a good target for cost reduction and the duration of the online frame and wheels visual check makes it a good candidate for availability improvement. Finally, PHM requirements can be derived from the targets, such as the following, for example:

- *Requirement 1*: The PHM system shall improve service reliability by predicting and avoiding the occurrence of hot axle box alarm at least one operation day before the event with at least 90% certainty of the event occurrence.
- *Requirement 2*: The PHM system shall reduce the maintenance cost by extending the periodicity for the motor and trailer wheels overhaul from 1 to 2 million km by detecting 90% of the depart from the specified performances with at least 90% certainty.
- *Requirement 3*: The PHM system shall improve the availability by reducing the duration of the visual check of frame and wheels at E12 by 90% by automating the detection of the failure mechanisms.

Then, the next step is to perform a physics-based analysis of the main failure mechanisms (the use of the term *mechanism* and not *mode* is significant here) associated with the requirements and the symptom from which detection should have been made possible. In our example, the main failure mechanisms and symptoms can be as follows:

- For requirement 1
  - Failure mechanisms: fatigue, wear, cracking, plastic deformation, corrosion, electrical erosion of the axle bearings [20], temperature aging or contamination of the axle bearings grease, and drift/offset of the phonic wheel
  - Symptoms: abnormal vibrations on the axle box, increase in temperature, abnormal noise for the bearings, presence of particles/water, abnormal viscosity for grease, and aberrant value of the phonic wheel measure
- For requirement 2
  - Failure mechanisms: shelling, spalling scaling, surface/subsurface fatigue crack propagation, and wheel–rail contact wear [21,22]

- Symptoms: presence of cracks on the surface/under the surface of the wheel, abnormal wheel surface condition, and abnormal vibrations on the wheels
- For requirement 3
  - Failure mechanisms: fatigue crack propagation on the frame/axle/wheels, impacts, atmospheric/chemical corrosion, and material deformation
  - Symptoms: presence of cracks on the surface of the frame/axle/wheels, presence of impacts traces, abnormal surface condition, and abnormal vibration on the bogie and/or carbody (dynamic characteristics)

Eventually, the list of physical quantities that have to be measured to provide the symptoms observation capability is established. This list of so-called relevant physical quantities will then be communicated to engineers in charge of the data capture to study the feasibility, maturity, and cost of existing sensor technologies that could answer the need. For example, common sensors are temperature probes, accelerometers, displacement sensors, viscosity sensors, strain gauges, and piezoelectric sensors.

6.8.3 Infrastructure

Railway infrastructure maintenance is a vast field, which addresses a large number of assets geographically distributed over large distances, such as signaling, electrification system, track and ballast, and civil works. Moreover, it is labor intensive and has to be coordinated with train operations and traffic, which represents a challenging optimization task [23,24].

In this context, PHM can be a key enabler for effective and operational infrastructure maintenance to achieve objectives such as high availability and safety and lower maintenance costs. Most of the existing PHM solutions are adapted to complex industrial mechanical systems or mechanical components which are not compatible with the large-scale distributed systems that make up the railway infrastructure. Consequently, as the development of a PHM solution must be adapted to the railway infrastructure, this has an impact on the PHM architecture, the monitoring systems, the diagnostic techniques, and the prognostic algorithm development [25,26].

Current research on PHM for the railway infrastructure is mostly focused on monitoring systems and diagnostic techniques for the railway track and the overhead contact system (OCS).

In the context of track health management, two main groups of track defects are considered: (a) the geometrical defects, which are associated with low vibrations and have a wavelength longer than 3 m and are related to defects affecting track geometry, such as longitudinal leveling, alignment, gauge, cross level, and twist, and (b) the short length track irregularities, which have a wavelength shorter than 3 m and are generally related to the rail degradation mechanism, including defects such as squats, corrugation, insulated joints, large head spalls, internal corrosion, and surface wear.

Today, measurement cars equipped with health-monitoring systems provide principal track geometrical data (track gauge, longitudinal level, cross level, alignment, twist), which are computed to assess the geometrical state of the track and calculate limit values defined by standards [27]. Guler [28] proposed an ANN model based on different data collected during 2 years (track structure, traffic characteristics, track layout, environmental factors, track geometry, and maintenance and renewal data) to predict the railway track geometry deterioration.

For short length defects, 12 types of rail defects (squat, corrugation, etc.) [29] are considered. These defects are monitored and diagnosed usually using ultrasonic inspection cars (UICs) or axle box accelerations (ABAs). In general, the UIC allows the detection and identification of rail breakage and internal cracks [30], while ABA is used to assess surface defects such as squat or corrugation [31]. In addition, several prediction models for rail failure have been developed, in particular, physics models based on crack growth [32]. However, those models have to be improved by using measurement data, environmental conditions, and operational parameters.

For the OCS, we distinguish three main categories of monitoring data: geometrical, mechanical, and electrical data. The geometrical data comprise the information about system dimensions (contact wire height, stagger, and wear of the contact wire). The mechanical measurements characterize the dynamical behavior of the pantograph–catenary interaction; the data measured are, for example, the contact force between the catenary and the pantograph, uplift of steady arms, and contact wire elasticity. The main measurement in electrical data type is the arcing rate, which is used to assess the current collection quality; in fact, the loss of contact between the pantograph and the catenary implies under certain conditions an arcing between pantograph strips and the contact wire. Additionally, vision techniques are often used for automatic inspection of the OCS. Based on the data measured, several methodologies were developed to achieve diagnostics and prognostics. For example, high-speed cameras are used by Petitjean et al. [33] to detect and count the droppers; the method was applied to a simple catenary with a precision of 98%. The prediction of contact wire wear based on mechanical measurements was studied by Bucca and Collina [34], and a heuristic model of wear is established. Moreover, a data-driven approach was proposed by Shing [35] for contact wire wear prediction; an ANN model is proposed to achieve prediction of contact wire wear based on operating data, wear measurement, geometrical data, and mechanical measurement. There are other techniques based on arc measurement or contact fore in the literature to achieve diagnostics. Combining all these techniques can lead to robust wire wear prediction models and diagnostic methods for the OCS.

For both track systems and OCS, many approaches have been developed to achieve diagnostics and prognostics of failures. For a railway infrastructure asset, the lack of run-to-failure data due to system size and monitoring techniques and the evolving environment of the system can constitute an obstacle for the development of data-driven approaches. However, the data collected using inspection trains can lead to understanding the evolution of failure mechanisms regarding system geometry, environmental conditions, operating history, external conditions, etc., which is difficult to model with a physics-based model in view of the significant number of parameters. Therefore, the development of PHM for railway infrastructure requires the development of sophisticated hybrid approaches. Consequently, integrating measurement systems on service trains, streaming data from those trains to clusters, and deploying data analytic solutions based on both models and data can be a key enabler for PHM deployment.

6.9 IT Challenges

In addition to the hardware architecture aspects described in Section 6.6, software questions lie at the heart of PHM algorithm design and implementation. The scope of this handbook does not permit an exhaustive treatment, but suffice it to say that at least the following aspects deserve close attention, some of which are not specific to PHM:

- Seeking parallel processing in algorithm implementation
- Optimizing database structure to permit storage and repeated access to raw data
- Studying the trade-offs between decentralized and centralized architectures
- Ensuring scalability, i.e., easy adaptation of data platform to increasing load and expanding system size
- Deciding on implementation on the cloud versus on premises
- Selecting either proprietary or open software
- Optimizing the software architecture to permit processing at various speeds (such as with the "lambda architecture")
- Last but not least, assuring cybersecurity in presence of various threats

6.10 Challenges and Future Perspectives

The area of PHM currently undergoes a very fast evolution, and the scope of applications increases daily. Some approaches are primarily data driven: they consider the asset of interest as a black box, without taking into account any expert knowledge; others are rather based on physics of failures or expert knowledge; and still others rely to some extent on the classical methods of reliability engineering, i.e., a statistical population-based view. Increasingly, progress is being made with hybrid methods, which combine several viewpoints, such as methods that are data driven and physics of failures based.

A key challenge is to develop and validate, for a given application area, algorithms that rely on the best combination of those approaches to provide the most efficient detection, diagnostic, or prognostic capabilities (as measured by the KPIs described in Section 6.5).

Another challenge resides in decentralization, i.e., achieving distributed intelligence, close to the assets, which would avoid sending huge volumes of raw data to a central processing station.

And, finally, as the ultimate goal of PHM is to provide decision support to maintenance, one must consider the challenge of moving from an asset-based PHM to a fleet-based vision, i.e., integrating the health indicator, diagnostic, and prognostic information arising from various distributed individual assets to make global maintenance and operation decisions at fleet level: this is a nontrivial optimization problem.

It is the authors' conviction that links with other disciplines provide a strong potential for progress in PHM. For instance, progress in data science increasingly permits the complementary use of unsupervised methods (such as for clustering) with supervised methods (for classification); "deep learning" (relying on multilevel ANNs) is beginning to be applied to automatic feature generation or automatic pattern recognition. Some applications of those techniques to PHM are now reported in the literature.

Also, a closer link with traditional reliability engineering methods and tools would be fruitful. For instance, the Cox proportional hazard model [36] is a convenient way to model reliability dependence on stresses, some of which can be monitored. A stronger link with traditional maintenance engineering (such as the characterization of maintenance efficiency) seems desirable as well.

Ultimately, through PHM, the following vision may become reality:

- Near-zero SAF rate
- Drastically reduced maintenance costs
- Dynamic balancing of load in depots

All of these will open a new era in railway system reliability and availability and life cycle cost optimization.

6.11   Conclusions

The maturity level of PHM applications in railways has significantly increased over the last years, and several new applications have been introduced both in railway rolling stock and in infrastructure systems. Pilot projects and field tests have been increasingly turned into successful applications applied either network-wide or fleet-wide. New players have emerged in the field of railway PHM applications, introducing entire or only parts of PHM systems.

New sensor technology has enabled better understanding of system conditions. Also, existing condition monitoring data have been used more proactively. Wayside monitoring devices, for example, are progressively used not only for safety-related monitoring tasks, but also for monitoring the evolution of the system condition in time. Not only are dedicated measurement trains able to measure the infrastructure system condition, but also information captured by regular revenue service trains can be used either to complement or to replace some of the dedicated measurements or even to provide new information on system condition that was not available previously. One of the open issues, which is also an opportunity, is how to combine information from different data sources, including information collected from nondedicated measurement devices.

In most of the new calls for tenders and new development projects, requirements on PHM are included. However, the design-to-PHM mind-set has not yet been widely implemented. Also, the integration of reliability and PHM considerations, particularly in the early design stages, needs to be further incorporated in the engineering processes. Successful PHM implementations also require adaptations in maintenance organization and scheduling, making it more flexible and dynamic.

With the increased availability of condition monitoring data, new challenges are faced by railway operators and maintainers. In many cases, the decisions based on PHM are affecting only service reliability and availability. However, numerous PHM decisions are also aiming at substituting the traditional scheduled inspections and preventive maintenance tasks of safety critical components. In this case, new regulatory frameworks and guidelines need to be developed for adjusting the maintenance regimes not only of existing but also of newly developed systems.

There are several research challenges to be addressed to support a more pervasive implementation of PHM systems in railways, including a combination of physical and data-driven models, development of effective canary devices, further development of fleet

PHM to enable transfer of fault patterns between systems operated under different conditions, and implementation of control systems for the RUL. The proactive control of RUL will not be limited to solely predicting the evolution of RUL, but will enable to influence the operating parameters proactively to prolong the RUL, taking the operational requirements and resource constraints into consideration.

From the regulatory stakeholders, guidelines, standardization, and support are required to enable industry-wide standardized interfaces and supplier-independent decision support systems.

Acknowledgment

The authors thank the International Electrotechnical Commission (IEC) for permission to reproduce information from its international standards. All such extracts are copyright of IEC, Geneva, Switzerland. All rights reserved. Further information on IEC is available on www.iec.ch. IEC has no responsibility for the placement and context in which the extracts and contents are reproduced by the author nor is IEC in any way responsible for other content or accuracy therein.

References

1. A. Saxena, J. Celaya, E. Balaban, K. Goebel, B. Saha, S. Saha, and M. Schwabacher, Metrics for evaluating performance of prognostic techniques, *PHM 2008: International Conference on Prognostics and Health Management*, 2008.
2. M. J. Roemer and C. S. Byington, Prognostics and health management software for gas turbine engine bearings, *ASME Turbo Expo 2007: Power for Land, Sea, and Air*, 2007.
3. J. E. Dzakowic and G. S. Valentine, Advanced techniques for the verification and validation of prognostics and health management capabilities, *Machinery Failure Prevention Technologies* (MFPT 60), pp. 1–11, 2007.
4. M. Dash and H. Liu, Feature selection for classification, *Intelligent Data Analysis*, vol. 1, pp. 131–156, 1997.
5. I. M. Sobol, Sensitivity estimates for nonlinear mathematical models, *Mathematical Modelling and Computational Experiments*, vol. 1, pp. 407–414, 1993.
6. M. D. Morris, Factorial sampling plans for preliminary computational experiments, *Technometrics*, vol. 33, pp. 161–174, 1991.
7. R. Tibshirani, Regression shrinkage and selection via the lasso, *Journal of the Royal Statistical Society. Series B (Methodological)*, pp. 267–288, 1996.
8. T. D. Wickens, *Elementary Signal Detection Theory*, Oxford University Press, Oxford, 2002.
9. C. Gini, Concentration and dependency ratios, *Rivista di Politica Economica*, vol. 87, pp. 769–792, 1997.
10. B. Lamoureux, *Development of an Integrated Approach for PHM—Prognostics and Health Management: Application to a Turbofan Fuel System*, 2014.
11. A. Saxena, S. Sankararaman, and K. Goebel, Performance evaluation for fleet-based and unit-based prognostic methods, *Second European Conference of the Prognostics and Health Management Society*, 2014.

12. T. Asada and C. Roberts, Improving the dependability of DC point machines with a novel condition monitoring system, *Proceedings of the Institution of Mechanical Engineers, Part F: Journal of Rail and Rapid Transit*, vol. 227, pp. 322–332, 2013.

13. Network Rail, *Annual Return 2014*, https://www.networkrail.co.uk/wp-content/uploads/2016/11/annual-return-2014.pdf, 2014.

14. T. Böhm, Accuracy improvement of condition diagnosis of railway switches via external data integration, *The Sixth European Workshop on Structural Health Monitoring*, Deutsche Gesellschaft für Zerstörungsfreie Prüfung, Dresden, 2012.

15. Network Rail, *Network Rail Monitor—Quarter 4 of Year 5 of CP4*, http://orr.gov.uk/__data/assets/pdf_file/0004/13792/network-rail-monitor-2013-14-q4.pdf, 2014.

16. V. Atamuradov, F. Camci, S. Baskan, and M. Sevkli, Failure diagnostics for railway point machines using expert systems, *SDEMPED 2009: IEEE International Symposium on Diagnostics for Electric Machines*, 2009.

17. P. Bagwell, The sad state of British railways: The rise and fall of Railtrack, 1992–2002, *The Journal of Transport History*, vol. 25, pp. 111–124, 2004.

18. F. P. G. Márquez, D. J. P. Tercero, and F. Schmid, Unobserved component models applied to the assessment of wear in railway points: A case study, *European Journal of Operational Research*, vol. 176, pp. 1703–1712, 2007.

19. C. Letot, P. Dersin, M. Pugnaloni, P. Dehombreux, G. Fleurquin, C. Douziech, and P. La-Cascia, A data driven degradation-based model for the maintenance of turnouts: A case study, *IFAC-PapersOnLine*, vol. 48, pp. 958–963, 2015.

20. B. N. Dhameliya and D. K. Dave, Causes and failure patterns of bearings in railway bogies and their remedies, *International Journal of Application or Innovation in Engineering & Management*, vol. 2, 2013.

21. CEN (European Committee for Standardization), EN 15313: *In-Service Wheelset Operational Requirements*, CEN, Brussels, 2010.

22. CEN, EN 13749: *Method of Specifying the Structural Requirements of Bogie Frames*, CEN, Brussels, 2011.

23. T. Lidén, Railway infrastructure maintenance-a survey of planning problems and conducted research, *Transportation Research Procedia*, vol. 10, pp. 574–583, 2015.

24. E. Zio, M. Marella, and L. Podofillini, Importance measures-based prioritization for improving the performance of multi-state systems: Application to the railway industry, *Reliability Engineering & System Safety*, vol. 92, pp. 1303–1314, 2007.

25. M. Brahimi, K. Medjaher, M. Leouatni, and N. Zerhouni, Development of a prognostics and health management system for the railway infrastructure—Review and methodology, *Prognostics and System Health Management Conference (PHM-Chengdu)*, 2016.

26. E. Fumeo, L. Oneto, and D. Anguita, Condition based maintenance in railway transportation systems based on big data streaming analysis, *Procedia Computer Science*, vol. 53, pp. 437–446, 2015.

27. CEN, EN 13848-1: *Railway Applications—Track—Track Geometry Quality—Part 1: Characterisation of Track*, CEN, Brussels, 2008.

28. H. Guler, Prediction of railway track geometry deterioration using artificial neural networks: A case study for Turkish state railways, *Structure and Infrastructure Engineering*, vol. 10, pp. 614–626, 2014.

29. S. Kumar, *A Study of the Rail Degradation Process to Predict Rail Breaks*, 2006.

30. L. Podofillini, E. Zio, and J. Vatn, Risk-informed optimisation of railway tracks inspection and maintenance procedures, *Reliability Engineering & System Safety*, vol. 91, pp. 20–35, 2006.

31. M. Molodova, Z. Li, A. Núñez, and R. Dollevoet, Automatic detection of squats in railway infrastructure, *IEEE Transactions on Intelligent Transportation Systems*, vol. 15, pp. 1980–1990, 2014.

32. R. Enblom, Deterioration mechanisms in the wheel–rail interface with focus on wear prediction: A literature review, *Vehicle System Dynamics*, vol. 47, pp. 661–700, 2009.

33. C. Petitjean, L. Heutte, V. Delcourt, and R. Kouadio, Extraction automatique de pendules dans des images de caténaire, *XXIIe Colloque GRETSI (traitement du signal et des images)*, September 8–11, Dijon, France, 2009.
34. G. Bucca and A. Collina, Electromechanical interaction between carbon-based pantograph strip and copper contact wire: A heuristic wear model, *Tribology International*, vol. 92, pp. 47–56, 2015.
35. W. C. Shing, *A Survey of Contact Wire Wear Parameters and the Development of a Model to Predict Wire Wear by Using the Artificial Neural Network*, 2011.
36. D. R. Cox, Models and life-tables regression, *Journal of the Royal Statistical Society*, vol. 34, Series B, pp. 187–220, 1972.

7

Human Factors and Their Application in Railways

Birgit Milius

CONTENTS

7.1 What Are Rail Human Factors?

Engineers have to know what, how, and why aspects influence a system and the people working or using it to develop a successful system.

The main aim of a railway company is providing safe, reliable, and economic transport. The railway system is a typical sociotechnical system (Wilson 2007):

- It is a purposeful system that is open to influences from and, in turn influences, the environment (technical, social, economic, demographic, political, legal, etc.).

- The people within it must collaborate to make it work properly.

- Success in implementation of change and in its operation depends upon as near as possible jointly optimizing its technical, social, and economic factors.

The main aspects of the sociotechnical system are also mirrored by the definition of human factors (HFs). In the study by the Health and Safety Executive (1999), it is defined that "Human factors refer to environmental, organizational and job factors, and human and individual characteristics which influence behavior at work in a way which can affect health and safety. A simple way to view HFs is to think about three aspects: the job, the individual and the organization and how they impact people's health and safety-related behavior." The International Ergonomics Association (2016) defines HF as follows:

> Ergonomics (or human factors) is the scientific discipline concerned with the understanding of the interactions among humans and other elements of a system and the profession that applies theory, principles, data and methods to design in order to optimise human well-being and system performance.*

In the Transport for New South Wales's guide (TfNSW 2014), three general categories for HFs are distinguished:

- Physical—This is concerned with human anatomical, anthropometric, physiology, and biomechanical characteristics as they relate to physical activity within the workplace.
- Cognitive—This is concerned with mental processes such as perception, memory, reasoning, mental workload, and decision making, as related to the nonphysical aspects of a job or specific task.
- Organizational—This is concerned with the organization of sociotechnical systems, including organization structures, policies and processes, and teamwork.

HFs for railways is often called rail HFs and incorporates, e.g., different approaches, processes, and methods. What makes rail HFs special is the focus on people working and interacting with and in a railway system. By this, the environment the people are working in as well as the people and tasks are very well defined. This allows identifying, discussing, assessing, and adjusting the (rail) HFs exactly to the tasks at hand.

Figure 7.1 gives an overview of common HF topics (taken from TfNSW [2014]).

Discussing HFs is often difficult as, for example, people with different knowledge backgrounds (e.g., psychologists, engineers, computer scientists, mechanical engineers) and practitioners (e.g., operators, train drivers) and theorists (e.g., from universities) have to find a common understanding. Also, the awareness of the importance of HFs widely varies worldwide, with a huge gap even in Europe. Whereas primarily, in the United Kingdom and Scandinavia, HF involvement in railway projects is normal, Germany only started considering such aspects. This chapter will be of interest to people who are rather new at HFs. It tries to give an overview of the topic. It is not a detailed guide on how to apply HFs in actual project. To learn more about HFs in general, we suggest looking at material published by the British Rail Safety and Standards Board (RSSB). Regarding human reliability in detail, we suggest referring to Chapter 11 of this book.

---

* Literature varies in its use of HFs and ergonomics. For this chapter, we assume that both terms refer to the same area of research. We will use the term *human factors*.

machine interface
Level of automation
Users with disabilities
anthropometric data
Human-computer and human Workstation design
Operating and maintenance manuals
Workload and job design
Alarms and alerts System design Error and violation
Training

FIGURE 7.1
Common topics of Rail HFs. (From TfNSW, *AEO Guide to Human Factors Integration*, Version 2.0, TfNSW, Sydney, 2014.)

The chapter is structured as follows:

First, we look at HF research in general and discuss shortly why railways should look at the importance of humans in their system. As the book addresses reliability, availability, maintainability, and safety (RAMS) engineers, the phases of the system life cycle are presented, looking at what EN 50126 (European Committee for Electrotechnical Standardization 1999)—probably one of the most known and most important RAMS standards in Europe—requires regarding HFs and discussing some ideas why and how HFs should be taken into account. We will present the concept of HF integration. To give the reader a better idea about the practical application of HFs, we present three interesting topic: performance shaping factors (PSFs), human reliability, and usability.

We do not look at the psychological basis of the HF research, as this would lead too far from the focus of the book.

7.2 Human Factors Research

Once a human started using technology, HFs played a role. Early hand axes could be too big, too small, too slippery. So one would expect that serious HF research has been done for a long time. However, this is not the case. Hollnagel (2012) distinguishes three ages of HF research:

- The first is the bottleneck age, where limitations in the human capacity were seen as reasons why system performance were not as, e.g., productive, fast, precise as technology would allow. Training, better design, and some automation were seen as concepts to overcome these limitations.
- The second age of HF research began with the accident in the Three Mile Island power plant. It became obvious from very early on that the operator actions played a significant role. Research in the following years was predominantly aiming at reducing and eliminating errors as well as containing the consequences of possible errors. During these years, humans were mainly seen as liability.

This view started to change in the last years and with it the third age of human factors has started. All sociotechnical, complex systems are underspecified, which means that details might be missing or that not all processes and dependencies are completely understood (see also Perrow [1984] and Leveson [2011]). At this point, humans suddenly become necessary, as they are adept at finding effective ways of overcoming problems. Hollnagel states that "the mission is to understand the interdependences and to develop ways to sustain or strengthen the individual and collective performance variability. . . ."

7.3   History of Railway Developments

Traditionally, developments in railways were triggered by accidents which often lead to changes in railway operations or railway technology, trying to eliminate, e.g., human errors or technical problems. With the introduction of computers, changes often included automation as it is/was supposed to make everything safer. Operators, who used to sit where they more or less could see the controlled trains, can now sit hundreds of kilometers away, in large operating centers. Instead of handling every single train, they are often passive, only ensuring that the technology is working correctly. Also, train drivers were affected by changes. Modern trains can drive more or less automatically with the train driver just monitoring the systems and the route ahead. Experience shows that even though the human has less chances to make mistakes, when he/she has to take action, errors might more easily happen. Operators are no longer "part of the loop," which might influence their situational awareness (Bainbridge 1983). They have less experience and can more easily be overwhelmed. Also, humans are rather bad at passive tasks. This means that the systems have to be designed exceptionally well so that, e.g., operators and train drivers do work well despite their limitations. Railway system development is still strongly influenced by the so-called second age of HF research. It does not yet see the human in the system as a positive influence. Therefore, HF research in the railway domain is necessary for best results. For these results to be as theoretical as necessary and as applicable as possible, we need to better understand where and how the human is part of the railway system and where and how to take him/her into account when changes occur. This needs to be done in a systematic manner and become part of the relevant standards.

7.4   Human Factors in the Life Cycle Process

As previously said, applying rail HF research can take very different forms and will influence all involved parties. To give RAMS practitioners with little to no experience regarding HFs in railways an idea about the importance of HF research, we take a look at the probably most relevant RAMS for railways standard, EN 50126 (50126:1999 [CEN 1999]). The standard EN 50126, *Railway Applications—Demonstration of Reliability, Availability, Maintainability and Safety (RAMS)*, describes in part 1 the basic requirements and the generic process of RAMS demonstration. As such, HFs as RAMS-influencing factors should be part of the standard. The standard focuses on the description of the life cycle process (Figure 7.2).

The life cycle process gives an overview of all activities involved when a new system is developed or an existing system is majorly changed and put in operation. As the phases consider the tasks of the railway organization as well as those of the manufacturer, it can be used as guidance for a general overview of where the human impact could and should be considered and by whom. The last phase of the life cycle is decommissioning. We will not look at this phase as the effects of rail HFs are limited.

The information given regarding HFs in the standard is not coherent. A whole subclause is discussing RAMS influencing factors and focuses on HFs. However, already in the first paragraph, the effect of HFs is limited to the requirements phase:

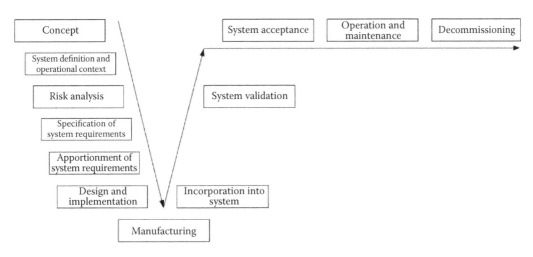

FIGURE 7.2
Life cycle process according to EN 50126.

> This subclause introduces and defines a process to support the identification of factors which influence the RAMS of railway systems, with particular consideration given to the influence of human factors. These factors, and their effects, are an *input to the specification of RAMS requirements* for systems.

Later on, it was said that "an analysis of human, factors, with respect to their effect on system RAMS, is inherent within the 'systems approach' required by this standard."

This again would foster the idea of taking HFs into account in all phases. This is further highlighted in Section 4.4.2.6: "This analysis should include the potential impact of human factors on railway RAMS within the design and development phases of the system." The standard proposes the use of a given list to derive detailed influencing factors. The nonconclusive list gives 37 factors which should be used during the assessment: "This evaluation shall include a consideration of the effect of each factor at each phase of the lifecycle and shall be at a level which is appropriate to the system under consideration. The evaluation shall address the interaction of associated influencing factors. For human factors, the evaluation shall also consider the effect of each factor in relation to each." Even though the idea in general is good, it cannot be applied as expected by the standard. The complexity of considering each effect in combination with another at every phase of the life cycle is generating lots of effort and might have the effect that the important issues are overlooked.

After making a point regarding the importance of HFs, one would expect that in the following chapters, more information about the use of HFs during the life cycle RAMS assessments is given.

The idea of a life cycle is not unique to EN 50126. A combination of HFs and life cycle already exists in IEC 62508. It is an interesting, comprehensive introduction to human aspects of dependability. However, neither is it focusing on railway aspects, nor is the life cycle identical to the one used in 50126. Integrating new aspects must be based on existing and used standards, and as such, HF integration has to be based on EN 50126 and not the other way around. Therefore, we will not focus on IEC 62508 in the following chapters.

7.4.1   Phase 1, Concept, and Phase 2, System Definition and Operational Context

Phases 1 and 2, the concept and the definition of a new or changed system, respectively, are usually done by the railway organization. The standard expects the definition of the mission profile, the boundaries, the application conditions, etc. In the whole section with detailed information, there is just one mentioning of human interaction, that is, that as part of the boundary definition, the interface to the human needs to be described.

The little information about the human leave it up to the railway organization to determine which information is necessary for the system development. Sometimes it is said that it is supposedly best to just exclude the human from the system definition and focusing on the technical part only, as by this, many uncertainties can be avoided. However, leaving the human out of the definition will quite probably lead to the effect that the technical system is not very well integrated in the existing railway system.

For the future, the interactions between persons and a technical system should be described in detail. Just as operational requirements, the requirements a person has on the system, and vice versa, should be noted. As guidance, a defined set of PSFs could be developed which can be used for systematically evaluating the expected influences on persons and the effects on the developing system. This set can be based on the list given in Section 4.4.2.11 of EN 50126. Expected difficulties by using the system or effects of changed properties on the operation of railways should be collected. It might become more common that in this phase of the life cycle, usability analyses are done to get a better idea of how the new or changed system will work.

7.4.2   Phase 3: Risk Assessment

The aim of the risk assessment is the derivation of safety requirements. These requirements can be qualitative or quantitative. Over the last years, the process of risk analysis was in the center of different research and standardization efforts. Therefore, lots of literature exists. Based on the results of phases 1 and 2, the hazards need to be derived. For technical system, guidewords are suggested be used to systematically look at typical failures of functions. This does not work properly for humans as the variance of human action is much bigger than the variance of possible failures. Especially for humans, it is necessary to look at hazards which arise due to the fact that a human has done too much or something unexpected, but necessarily a mistake.

Once the hazards are all known, a risk assessment is done which derives how safe a certain function needs to be. For this step, human reliability data are often used. Especially when an explicit risk assessment is done, the reliability of humans needs to be taken into account. Several categorizations exist, which often have significant shortcomings. Later in this chapter, we take a more detailed look at reliability data in railways. Methods which are typically used today for risk assessments are only partially usable when the consequences of human action need to be assessed again, because of the bigger variance in possible actions and reactions.

7.4.3   Phase 4: Specification of System Requirements

Part of the system requirements is the quantitative data from the risk assessment. But more than this, lots of qualitative data of, e.g., the overall RAM program and the maintenance, are needed. Information about human-related data is scarce in the standard. But it is clear that it should definitely play a part as, e.g., times, tasks, and ergonomics need to take the human interaction into account. Unfortunately, especially when replacing an existing system with a new one, the railway organizations often expects the new system to look and interact just as the old one, often not realizing the advantages of modern technology. In some European countries, it is common to include usability data in the systems requirements. These might not only come from the old systems, but can also be the result from specifically done usability analyses.

7.4.4 Phase 5, Apportionment of System Requirements, and Phase 6, Design and Implementation

At phase 5, the job of the manufacturer starts. The manufacturer has to make sure that all process steps are safely done and according to the rules, so that the final system is correct. As these tasks are done by humans, HFs play a huge part in providing the best results. However, these HFs issues are not railway specific and, as such, are not part of this chapter.

At a first glance, the manufacturer is not concerned with the human in the railway system as he/she gets all the requirements from the railway organization. However, this is not completely true as, e.g., usability is an important issue not only for, e.g., operators and train drivers, but also for maintenance. A system that has shown to be especially usable and, as such, can potentially reduce the error probability of persons, has an edge against competitors. Also, often the railway organization does not give enough detailed information but leaves room for the manufacturer to make its own decisions which can be influenced by HFs. Furthermore, it can be necessary for the manufacturer to keep the HFs in mind when reusing an existing design for another railway. Often, even though the general functionality seems to be the same, underlying factors, e.g., from operations or safety culture, might mean that adaptations are necessary.

7.4.5 Phase 7, Manufacturing, and Phase 8, Installation

For the manufacturing and installation phases, rail HFs are less important as most design decisions are finished. However, influences are possible during tests. Here, it can be sensible to choose test conditions, especially for hardware tests, taking HFs into account as they can significantly influence the performance and can even show limitations of a design. As part of the installation phase, the maintenance personnel is educated. Here, the manufacturer has to make sure that the processes and scenarios are well prepared and take into account, e.g., the experiences of the people and the environments they have to work in. Usability tests as well as interviews should be used to show any problem areas which might arise in maintenance.

7.4.6 Phase 9, System Validation, and Phase 10, System Acceptance

As part of system validation, the end user gets to know their new completed system. Even though the system should be final by now, any complaints should be taken seriously and be evaluated. Just as often new systems are only slowly accepted, sometimes important (informal) requirements are not known and only become obvious once the final system is installed. HFs can help guide users in a new system and, by this, retract important information.

7.4.7 Phase 11, Operation and Maintenance

While in operation, dealing with HFs remains an important issue. The performance of all people involved should be monitored to learn if all processes work as planned. When problems arise, a thorough investigation should determine the causes. Usability remains an issue to learn more about the behavior of the humans in the system. A close cooperation between railway organization and manufacturer can help adapt and change systems once more data are available.

7.4.8 Are There Enough Human Factors in the RAMS Life Cycle Process?

As the short overview earlier has shown, there are very many possibilities where during the life cycle of a system, the human should be taken into account. It is obvious that HFs are a topic for about every involved partner in a railway development project. Secondly, the information given in the general RAMS standard is not enough to allow a systematic handling of HF issues. It is not understandable that in the general sections of the standard, an inclusion of HFs is described as necessity, but in the final description of the phases, hardly any information is given.

The described scenarios for considering HFs are just examples. In reality, people associated with a project have to carefully evaluate how the human might be impacted. As said earlier, taking the easy route and leaving human interaction outside of the scope might be easy to begin with but will lead to more problems afterward. Having a closer look at the human is not the job of, e.g., psychologists or persons working in ergonomics alone. Experience has shown that the input of people with a thorough knowledge with the system at hand is crucial for successful results. Also, involving the people who actually have to work with the system is not necessarily easy in the beginning, but will lead to better and more accepted results in the end (Heape 2014).

7.5 Human Factors Integration

Considering where the RAMS standards come from, it will be no surprise that detailed information about HFs is scarce. Especially not only in the UK railway sector, but also, e.g., in Australia, more information dealing with HFs issues during the development process exist. The concept is called "human factors integration/human system integration" (HFI) and exists since the 1980s, first introduced by the American military (Heape and Lowe 2012). It is defined as the "systematic approach to the identification, tracking and resolution of human-system issues in order to ensure the balanced development of both the technological and human aspects of operational capability" (ISO/TS18152 [ISO 2010a]). Especially for railways, in TfNSW (2014), a standard from the New South Wales railways, it is pointed out that the goals of HFI are the following:

- There is structured, rigorous consideration of human factor (HF) issues from the beginning of the asset life cycle (feasibility) and continuing throughout the life cycle to disposal.
- There is systematic treatment of HF issues throughout projects and the asset life cycle.

- The project has a human focus and uses iterative design.
- HF principles, good practice, and appropriate techniques, tools, methods, and data are applied.
- The design reflects the concept of operation, meets end user needs, and matches end user characteristics and operator and maintainer organizational requirements.
- The design adopts a multidisciplinary approach.
- End users are involved in system design and evaluation.

As Bentley (2013) points out, the financial benefits of HFI can be huge. His examples are from military and aviation, but it can be expected that similar effects—the savings are several times higher than the cost of HFI—are possible. Besides important cost benefits, several projects mentioned by him have said that the safety of the system also improved. In the study by Railway Advisory Committee (2016), the benefits of a successful HFI are listed as follows:

- Improved system performance through better consideration of user capabilities and limitations
- Reduction in system whole life cycle costs by identifying the correct balance of investment in people and complex automated systems
- Reductions in procurement cost and risk by identifying potential usability, labor, and training issues earlier in the development process
- Improved recruitment and retention by improving the working environment and fitting the task to the user
- Removing health hazards and thereby reducing liabilities for both the employer and employee
- Reduced likelihood and severity of accidents

In the study by Heape and Lowe (2012), it was shown that the V-model of the EN 50126 life cycle can also be used to display (some) tasks for HFI (Figure 7.3). It is obvious that the work packages of this model fit to the early process steps of the life cycle in EN 50126. However, the model stops at the commissioning phase. It should be clear that HFI should be extended to the operational phase as well.

The approach shown in Figure 7.3 will especially appeal to persons with a background in HFs as it is very detailed and already looks at specific tasks. There exists another, less detailed but more complete example for a combination of the EN 50126 life cycle and the correlating HF tasks. It is from Intergo (2009) and shown in Figure 7.4. It will appeal to people with less experience in HFs as it only shows the starting points for the corresponding HF tasks and does not go into detail. An advantage of the concept is that it distinguishes between customer, supplier, and assessor. By this, it becomes obvious that good HFI is always a collaboration between different shareholders.

The given examples show that HFI is obviously an important concept but that lots of research is still necessary so that it can be applied more often. The approaches so far show that a combination with the EN 50126 life cycle is possible and can lead to an adapted life cycle.

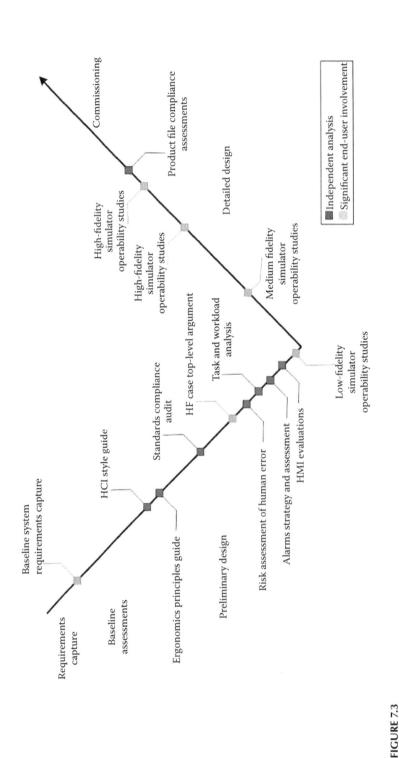

FIGURE 7.3

HF life cycle V-process (HCI, human–computer interface; HF, human factor). (Based on S. Heape, and C. Lowe, *Effective Human Factors Integration in the Design of a Signalling and Train Control System for the Metro Rail Industry*, https://www.researchgate.net/publication/265010612_Effective_human_factors_integration_in_the_design_of_a_signalling_and_train_control_system_for_the_metro_rail_industry, 2012.)

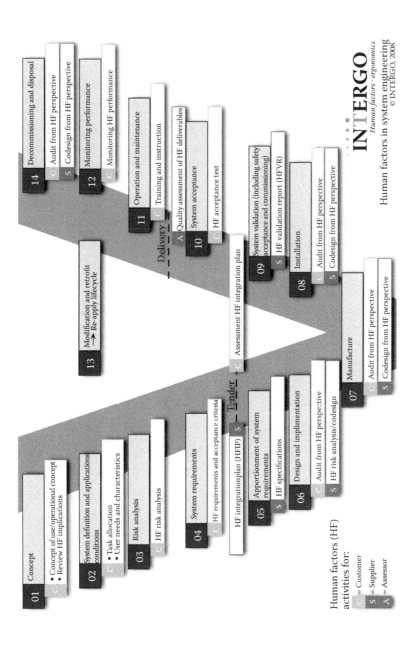

FIGURE 7.4
Adapted life cycle process. (From Intergo, *Human Factors and Human Error in the Railway Industry,* http://www.intergo.nl/public/Downloads/79/bestand/RSRA.pdf, Intergo, Utrecht, 2009.)

7.6 Concepts of Human Factors Research

The first sections of this chapters looked at the life cycle of new and changed systems and how to integrate HFs into it. For the remaining pages of this chapter, we have a closer look at some interesting concepts which can help to better understand and apply rail HFs.

It is not possible to discuss all methods and areas of application in a chapter of a book. However, looking at what defines the human action and human interaction, one will arrive at the conclusion that in almost all aspects, factors are needed which describe influences on human performance. These are called PSFs, sometimes also called "performance influencing factors" (SHERPA method) or "error-producing conditions" (HEART method).

Following an introduction to PSFs, we will have a short look at human reliability data as needed by, e.g., risk assessments, and will finish the chapter by looking at the concept of usability, as usability analysis can be helpful for railway organizations as well as manufacturers. It can be done at different times of the system life cycle with different goals and, as such, is an important concept.

7.6.1 Performance Shaping Factors

In the study by Boring, Griffith, and Joe (2007), PSFs are defined as "... influences that enhance or degrade human performance and provide basis for considering potential influences on human performance and systematically considering them in quantification of human error probabilities (HEPs)." PSFs are not only most often thought of as a negative influence on human behavior (decreasing the operator's ability) but can also have a beneficial effect on human reliability. PSFs can be used for a variety of purposes during the system development process, e.g., defining the environment (phase 1), deriving reliability data (phase 3), and discussions with practitioners regarding the old and the new system (phases 1, 2, 8, 10, and 11).

The general idea of how PSFs have an effect on task solving is shown in Figure 7.5, which was taken from Milius (2013).

It shows in a very simplified form how a task is translated into actions. It is obvious that the success in solving the task very much depends on the cognitive couplings and PSFs working together. As the cognitive effects are very difficult to influence, it is more sensible to look at PSFs to understand and influence human behavior. Even when the effects of a PSF on task solving in general are well understood, we have to keep in mind that the effect that

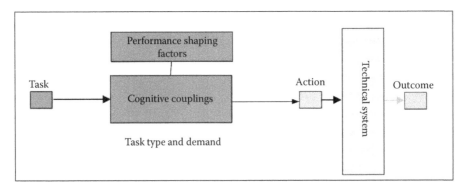

FIGURE 7.5
Model describing the effect of PSFs on the decision making process. (From B. Milius, *Rail Human Factors: Supporting Reliability, Safety and Cost Reduction*, CRC Press, Boca Raton, FL, 2013, pp. 692–700.)

a PSF has on a concrete person cannot be easily calculated as it might widely vary depending on, e.g., personal experiences and interdependencies with other PSFs occurring. This has to be kept in mind when PSFs are used within a system development process.

PSFs can take very different forms, and brainstorming might lead to a huge amount of different effects. A more structured approach is to analyze actual accident and incident data. This was done in a Germany, funded by the German Research Foundation project (Schwencke 2014; Lindner et al. 2013). A similar study was done by Kyriakidis (2012) for the United Kingdom. Both studies show a wide variety of different aspects which have played a role when incidents have occurred. It can easily be seen (Table 7.1) that the PSFs do not all match, but that similarities exist.

It is also obvious that room for discussion exists. This is a good example why it is difficult to understand and compare PSF as well as their relationships. The level of detail varies, and interdependencies are not or differently taken care of. A very good example for these are PSF "irregularities." This PSF does probably also relate to PSFs such as distraction, time pressure, or workload. All this might lead to stress. But looking at Table 7.1, all the effects are listed as separate PSF. One can expect that this has effect when PSF are used during the system lifecycle. Even worse, when quantifying PSF, this can significantly influence the results.

PSF taxonomies exist, but there are several, and they are not necessarily compatible. A first step when developing a taxonomy is the categorization of the PSF. They can be distinguished in, e.g., internal (specific to a person, e.g., mood, fitness, stress level) and external (describing a situation or system-dependent and environmental (temperature, noise, safety culture). Another classification (Lindner et al. 2013) can be direct (the effect can be observed directly) and indirect (cannot be observed directly). Also, different categorizations can be used on different levels. However, Galyean (2007) suggest that three categories

TABLE 7.1

Railway-Related PSF

PSF for SPADS[a]	PSF for Railways[b]
Irregularities	Distraction—concentration
Distraction	Experience/familiarity
Comprehension (positive)	Expectation—routine
Visibility conditions	Communication
Fatigue	Safety culture
Job experience	Training—competence
Just culture	Perception
Signal location	Quality of procedures
Stress/strain	Weather conditions
Signal design	Human–machine interface quality
	Fatigue (sleep lack, shift pattern)
	Time pressure
	Fit to work—health
	Workload
	Visibility
	Coordination of work—supervision
	Stress
	Risk awareness

Source: [a]T. Lindner et al., *Proceedings of European Safety and Reliability*, 2013, pp. 435–442; [b]M. Kyriakidis, *Journal of the Transportation Research Board*, 2289, 2012, pp. 145–153.

TABLE 7.2

Structure of Influencing Factors as Given by Hammerl and Vanderhaegen (2009)

	Anthropometry	Basic Layout of Working Environment
Physical factors	Working conditions	Physical conditions, e.g., temperature, humidity
	Design of HMI	Position and layout of HMI
		Usability
		Quality of feedback
Personal factors	Individual factors	Health, age, gender
		Emotional tension
	Dependent factors	Tiredness
		Skills, experience
		Motivation, safety awareness
Organizational factors	Employee related	Roster planning
		Leadership
		Education, training
		Social aspects, safety culture
	Standard factors	Standards
		Rules and guidelines
		Task design

Source: Given by M. Hammerl, M. and F. Vanderhaegen, *3rd International Rail Human Factors Conference*, Lille, 2009.

Note: HMI, human–machine interface.

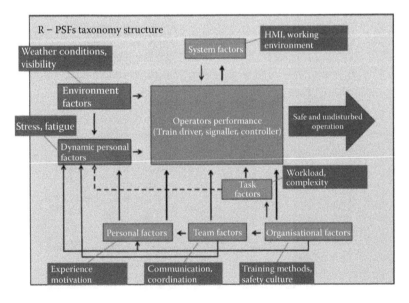

FIGURE 7.6
Rail-PSF structure and examples for PSF. (From M. Kyriakidis et al., *Safety Science*, 78, 60–76, 2015.)

are enough: individual, organizational, and environment. Hammerl and Vanderhaegen (2009) provide a more detailed classification based on the evaluation of HF analysis methods (Table 7.2).

Another example is the structure given in Figure 7.6, which was developed by Kyriakidis (2012) especially for railways.

The number of PSFs needed to describe human performance for a certain function or system is disputed and, of course, depends on the used definitions and taxonomy. In the study by Boring (2010), it is shown that, usually, three categories are too little, but the exact number varies depending on the task at hand. In general, too many PSFs are distracting and not helpful.

To use PSF throughout the life cycle of a system in a structured manner, a taxonomy needs to be agreed on so that railway organizations as well as manufacturers have the same understanding. Using such a taxonomy a guidance for statistical data collection and evaluation can lead to a sound basis for future projects.

7.6.2 Basic Idea of Human Reliability Assessment

There are very different methodologies to be used for human reliability assessment. In this section, we focus on just two of them, aiming at giving an idea how reliability data can be used. As both approaches include some type of PSF (here called conditions), a connection with the preceding paragraph is given. A more detailed discussion of human reliability assessment can be found in Chapter 11.

As shown earlier, when doing a risk assessment, a qualitative or quantitative assessment of the human reliability is necessary. In Germany, a so-called approach after Hinzen (1993) is taken. It is based on the Rasmussen model (e.g., Rasmussen [1979]) and classifies tasks as rule based, knowledge based, and skill based. These categories are then combined with the aspects of stress and general environmental conditions. The quantification of this model was developed by Hinzen based on accident and incident data from the 1960s and 1970s nuclear power plants in the United States. Experience has shown that the model is easy to apply but leads to conservative results. It is known that the method has flaws, e.g., due to the basis not being railway related and it not being adapted to newer research. However, to this date, it can still be applied to German risk assessments and was the basis for newer risk assessment methods (0831-103 2014). The Hinzen model is shown in Table 7.3.

A newer method which is a significantly a more detailed version of the Hinzen concept and much more elaborated is the Railway Action Reliability Assessment (Gibson et al. 2013). It is also based on the Rasmussen model. However, it not only allows to include a large number of so-called error-producing condition (which are actually much like PSF),

TABLE 7.3

Table for Human Error Probability

	Favorable Environmental Conditions			Adverse Environmental Conditions		
	Stress due to Too Little Demands	Optimal Stress Level	Stress due to Excessive Demands	Stress due to Too Little Demands	Optimal Stress Level	Stress due to Excessive Demands
Skill-based	2×10^{-3}	1×10^{-3}	2×10^{-3}	1×10^{-2}	5×10^{-3}	1×10^{-2}
Rule-based	2×10^{-2}	1×10^{-2}	2×10^{-2}	1×10^{-1}	5×10^{-2}	1×10^{-1}
Knowledge-based	2×10^{-1}	1×10^{-1}	5×10^{-1}	1	5×10^{-1}	1

Source: A. Hinzen, *Der Einfluss des menschlichen Fehlers auf die Sicherheit der Eisenbahn*, Doctoral thesis, RWTH Aachen University, Aachen, 1993.

but it also gives more so-called generic task types within the Rasmussen framework to even better choose the ideal benchmark task. All aspects of this model are quantified. However, it is not known to the author where exactly the numbers come from and if they are directly related to the British railway system. Furthermore, an application for the German railway system or any other railway system would only be possible when a thorough comparison of both systems shows that they are reasonably alike. Due to effects of, e.g., safety culture, work force, training, and supervision, the reliability data could widely vary between different environments. This is a good example of how difficult it is to transfer quantitative data between railway companies of different countries.

Reliability data are relevant to railway organizations and manufacturers alike. As shown earlier, there are railway-related tables, which are rather easy to use. However, only a very limited number of scenarios is usually given so choosing one that fits the task is not simple. In general, it is assumed that today's methods lead to conservative results and, as such, are safe to use. For the future, research should aim to develop a more rigid method for deriving reliability data. As a basis for this, more real-life data of not only accidents, but also incidents including the relevant PSF have to be gathered.

7.6.3 Usability and User Experience

Usability is a powerful tool to gather information during several phases of the life cycle.

Usability is defined by ISO 9241 [ISO 2010b] as the "extent to which a system, product, or service can be used by specified users to achieve specified goals with effectiveness, efficiency, and satisfaction in a specified context of use" with

- Effectiveness being the accuracy and completeness with which specified users can achieve specified goals in particular environments
- Efficiency being the resources expended in relation to the accuracy and completeness of goals achieved
- Satisfaction being the comfort and acceptability of the work system to its users and other people affected by its use

In the book by Rubin and Chisnell (2010), the goals of usability are listed as follows:

- Informing design (e.g., designing useful, valued products that are easy to learn and help people to be effective and efficient)
- Eliminating design problems and frustration
- Improving profitability (e.g., minimizing cost of maintenance, minimizing risks, and creating a usability log to use as benchmarks)

The list shows that just as with the product life cycle, usability is a topic which is relevant for the railway organization as well as the manufacturer. Both sides have an incentive to try to make new or changed products as usable as possible.

Systems, which have to interact with humans, have to be designed to take the human into account. Therefore, before a system development starts, usability analysis should be performed to identify which requirements are necessary and will therefore deliver information for the first and second phases of the life cycle. During the development process, more analysis is necessary to confirm that the developed product meets the requirements. This delivers at least information for the phases design, incorporation, and system validation.

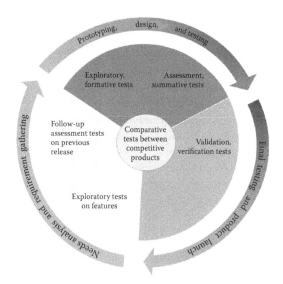

FIGURE 7.7
Usability tests during a system life cycle. (Based on J. Rubin and D. Chisnell: *Handbook of Usability Testing.* 2008. Copyright Wiley-VCH Verlag GmbH & Co. KGaA. With permission.)

When a system is in use, another type of analysis looks at if and how users use the new system (Figure 7.7).

In general, usability is known for decades but only became more relevant for railways in the last years. A usable system will have an overall positive effect on people's work and will lead to less errors and a better understanding of the process. The usability concept focuses on people getting a job done well. This approach was complemented in research by the concept of user experience (for definition of user experience, see, e.g., the study by allaboutUX [2016]). To get best results, a system should not only help a person to fulfill a task, but also if the person is enjoying was he/she is doing, then the results of the tasks are even better and the person works more reliably.

Usability testing can be as simple as having a questionnaire and can be as complicated as building a mock-up and having a simulation study. The right degree of usability and user experience research is determined, e.g., by the goals of the development, the nature of the changes, and the impact the new system will have. In every case, usability and user experience should always be researched with railway engineers and HF specialists closely interacting.

7.7 Conclusion

The chapter gave an overview of HFs. It showed that they can have an influence in about any phase of the system life cycle and are in the responsibility of railway organizations and manufacturers alike. Even though in some countries HF research is established, in many others, it is a new topic. Nevertheless, using the (existing) results of HF research from other countries or other domains should be carefully done, as the results very much depend on formal and (more important) informal aspects of the individuals, the organizations, and the environment, taking into account, e.g., long-standing traditions, values, and beliefs.

To allow for a better incorporation of HFs into the system development process, not only should be RAMS engineers be aware of the implications, but the relevant RAMS standards should also give more and detailed guidance. Until then, RAMS engineers should rely on existing standards which deal with HF in a general manner and interpret these. A very extensive collection of information can be found in the standards of ISO 9241 [ISO 2010b], a multipart standard covering ergonomics of human interaction. Also, ISO 62508 can be used. It is rather short and gives a first impression of relevant aspects. Last but not least, even though results of HF research should not be transferred from one country to another, it can be helpful to look for reports of what other institutions, e.g., RSSB, have done.

For the future, another area of HF research will be emerging. It will not be possible to deal with information technology security issues without learning and understanding potential attackers better. Here, more and completely new HF research will also be necessary.

References

allaboutUX. 2016. *User Experience Definitions.* http://www.allaboutux.org/ux-definitions (accessed October 31, 2016).

Bainbridge, L. 1983. Ironies of automation. *Automatica,* Vol. 19, No. 6, pp. 775–779.

Bentley, T. 2013. *Integrating Human Factors into a System's Lifecycle.* http://www.slideshare.net/informaoz /dr-todd-bentley.

Boring, R. L. 2010. *How Many Performance Shaping Factors are Necessary for Human Reliability Analysis?.* Preprint PSAM 2010, https://inldigitallibrary.inl.gov/sti/4814133.pdf.

Boring, R. L., Griffith, C. D., and Joe, J. C. 2007. The measure of human error: Direct and indirect performance shaping factors. In *Joint 8th IEEE Conference on Human Factors and Power Plants/13th Conference on Human Performance, Root Cause and Trending (IEEE HFPP & HPRCT),* Idaho National Laboratory, Idaho Falls, ID, pp. 170–176.

European Committee for Electrotechnical Standardization. 1999. EN 50126-1:1999: *Railway Applications: The Specification and Demonstration of Reliability, Availability, Maintainability and Safety (RAMS): Basic Requirements and Generic Process.* European Committee for Electrotechnical Standardization, Brussels.

Galyean, J. 2006. Orthogonal PSF taxonomy for human reliability analysis. In *Proceedings of the 8th International Conference on Probabilistic Safety Assessment and Management,* American Society of Mechanical Engineers, New Orleans, LA.

Gibson, W. H., Mills, A. Smith, S., and Kirwan, B. 2013. Railway action reliability assessment, a railway-specific approach to human error quantification. In: *Rail Human Factors: Supporting Reliability, Safety and Cost Reduction,* eds: Dadashi, N., Scott, A., Wilson, J. R., and Mills, A. CRC Press, London.

Hammerl, M., and Vanderhaegen, F. 2009. Human factors in the railway system safety analysis process. *3rd International Rail Human Factors Conference,* Lille.

Health and Safety Executive. 1999. *Reducing Error and Influencing Behavior.* Guidebook. http://www .hse.gov.uk/pubns/books/hsg48.htm (accessed October 19, 2016).

Heape, S. 2014. *Importance of Identifying and then Including Representative Users Throughout Design.* Keynote at the First German Workshop on Rail Human Factors, Braunschweig.

Heape, S., and Lowe, C. 2012. *Effective Human Factors Integration in the Design of a Signalling and Train Control System for the Metro Rail Industry.* https://www.researchgate.net/publication/265010612 _Effective_human_factors_integration_in_the_design_of_a_signalling_and_train_control _system_for_the_metro_rail_industry.

Hinzen, A. 1993. *Der Einfluss des menschlichen Fehlers auf die Sicherheit der Eisenbahn*, Doctoral thesis, RWTH Aachen University, Aachen.

Hollnagel, E. 2012. The third age of human factors: From independence to interdependence. In *Rail Human Factors around the World: Impacts on and of People for Successful Rail Operations*, eds: Wilson J. R., Mills, A., Clarke, T., Rajan, J., and Dadashi, N., CRC Press, Boca Raton, FL, pp. 1–8.

Intergo. 2009. *Human Factors and Human Error in the Railway Industry*. http://www.intergo.nl/public /Downloads/79/bestand/RSRA.pdf. Presented at Rail Safety & Risk Assessment, Amsterdam, June 4 and 5, 2009.

International Ergonomics Association. 2016. *Definition and Domains of Ergonomics*. http://www.iea .cc/whats/ (accessed October 19, 2016)

ISO (International Organization for Standardization) 2010a. ISO/TS18152:2010: *Ergonomics of Human– System Interaction—Specification for the Process Assessment of Human–System Issues*. ISO, Geneva.

ISO. 2010b. ISO 9241-210:2010: *Ergonomics of Human–System Interaction—Part 210: Human-Centered Design for Interactive Systems*. ISO, Geneva.

Kyriakidis, M. 2012. The development and assessment of a performance shaping factors for railway operations. *Journal of the Transportation Research Board*, Vol. 2289, pp. 145–153.

Kyriakidis, M., Majumdar, A., and Ochieng, W. Y. 2015. Data based framework to identify the most significant performance shaping factors in railway operations. *Safety Science*, Vol. 78: pp. 60–76.

Leveson, N. 2011. *Engineering a Safer World*. MIT Press, Cambridge, MA.

Lindner, T., Milius, B. Schwencke, D., and Lemmer, K. 2013. Influential Factors on Human Performance in Railways and their Interrelations. In *Safety, Reliability and Risk Analysis: Beyond the Horizon (Proc. ESREL 2013)*, pp. 435–442. CRC Press/Balkema. The Annual European Safety and Reliability Conference (ESREL) 2013, 30.9.-2.10.2013, Amsterdam, Niederlande. ISBN 978-1-138-00123-7.

Milius, B. 2013. A new approach for the assessment of human reliability. In *Rail Human Factors: Supporting Reliability, Safety and Cost Reduction*, eds: Dadashi, N., Scott, A., Wilson, J. R., and Mills, A. CRC Press, Boca Raton. FL, pp. 692–700.

Perrow, C. 1984. *Normal Accidents: Living with High-Risk Technologies*. Basic Books, New York.

Railway Advisory Committee. 2016. *Human Factors Integration*. http://orr.gov.uk/__data/assets/pdf_file /0003/5592/hf-integration.pdf.

Rasmussen, J. 1979. Skills, rules, and knowledge: Signals, signs, and symbols, and other distinctions in human performance models. *IEEE Transactions on Systems, Man, and Cybernetics*, Vol. 13, No. 3, pp. 257–266.

Rubin, J., and Chisnell, D. 2008. *Handbook of Usability Testing*, Second Edition, Wiley, Hoboken, NJ.

Schwencke, D. 2014. Menschliche Zuverlässigkeit im Bahnbereich—Einflussfaktoren und Bewertungsmethoden. In *Eisenbahn Ingenieur Kalender 2015 EIK—Eisenbahn Ingenieur Kalender*. DVV Media Group, Hamburg, pp. 251–261.

TfNSW (Transport for New South Wales). 2014. *AEO Guide to Human Factors Integration*. Version 2.0. TfNSW, Sydney.

Wilson, J. R., Farrington-Darby, T., Cox, G., Bye R., and Hockey, G. R. J. 2007. The railway as a socio-technical system: Human factors at the heart of successful rail engineering. *Proceedings of the Institution of Mechanical Engineers: Part F: J. Rail and Rapid Transit*, Vol. 221, pp. 101–115.

8

Individual Risk, Collective Risk, and F–N *Curves for Railway Risk Acceptance*

Jens Braband and Hendrik Schäbe

CONTENTS

8.1 Introduction

The exposition of an individual or a group of persons to a hazard is important, when a risk analysis is carried out to derive a safety integrity level (SIL). The same holds true when the achieved risk of a system is computed, e.g., for a safety case. Questions such as individual risk, collective risk, and their relation are discussed in Sections 8.2.1 and 8.2.2. The exposure time is mentioned in several standards when the SIL is derived. There, the exposition is used as a parameter, e.g., in IEC 61508 as "exposure in the hazardous zone."

Risks are sometimes given per calendar year, sometimes per hour of use. In other cases, risk is presented per person kilometers for a traffic system. The level of risk might differ, depending on the time unit which is used for the indication of the risk. This will be the subject of Section 8.2.3. In Section 8.2.4, we provide examples of risk and exposure time.

In Section 8.3, we propose statistical tests to judge collective risk and its changes based on statistical values. The tests are based on the fact that the accident data that have to be collected can be described as a compound Poisson process (CPP). Often, such processes are assumed to follow a homogeneous compound Poisson process (HCPP), but in practice, seasonal variations occur, e.g., due to different weather conditions during a year. A CPP can be decomposed into parts: the counting process itself and the distribution height of the jumps describing the size of the accidents. Note that this testing approach differs from the one proposed in the commission decision (European Parliament and Council of the European Union 2009) of the Electronic Retailing Association Europe (ERA) since this is not efficient as shown in Braband and Schäbe (2012, 2013).

It is also essential to have an idea about the distribution function of the accident severities. This will be derived from an *F–N* curve, and a statistical test will be given to test hypotheses whether accident severities, e.g., in different countries or in different years, have the same distribution. This is the subject of Section 8.4.

8.2 *F–N* Curve

8.2.1 First Considerations

The exposition of an individual or a group of persons plays an important role, if risk analysis (e.g., the analysis to determine an SIL) is carried out or if the residual risk is computed that is achieved after having implemented risk reduction measures. The latter is important for a safety case.

A simple example can illustrate this. A rare event, e.g., bungee jumping is considered. There is a certain probability that despite all measures, a technical failure occurs, e.g., the rope breaks, leading to the death of the jumper. Since bungee jumping has become a sport, there are less than 10 cases known where the rope has broken (see Wikipedia 2010). Taking into account unreported failures and taking into account that millions of jumps have been carried out, the risk of dying during bungee jumping due to technical defects might be 1:10,000 or even less. However, if the risk of an individual per year is considered, then one might ask how many jumps per year are carried out. The exposition time per jump, normally several seconds, can be neglected. The example shows that it makes no sense to compute a risk per hour of exposition, e.g., a hypothetical hour of uninterrupted bungee jumping. Similar examples can be constructed for technical systems, e.g., the exposition of a vehicle driver on a railway level crossing or the exposition of a worker to an automatically closing door.

The exposition time is used in different standards in a different manner; see Table 8.1.

Studying the use of exposition time for SIL determination in these standards leads to the question how exposition time has to be understood to be computed, and to be taken into account. In this chapter, we consider this problem. A comparison between collective and individual risks is given in Sections 8.2.2 and 8.2.3. In Section 8.2.4, we provide examples of risk and exposition time.

TABLE 8.1

Use of Exposition Time in Different Standards

Standard	Use of Exposition Time
IEC 61508	IEC 61508-5, Annex E, Risk graph, parameter F
EN 50126	None
EN 50129	None, but CENELEC (1999) report uses parameters N_i (number of persons) and E_i (exposure)
IEC 62061	Parameter F
ISO 25119	ISO 25119-2, 6.2.5 parameter E (exposition)
ISO 26262	ISO DIS 26262-3, clause 7.4.5.3, parameter E
IEC 61511	IEC 61511-3, annex D.2 parameter F (exposition)

8.2.2 Collective Risk

EN 50126 defines risk as "the probable rate of occurrence of a hazard causing harm and the degree of severity of the harm." EN 50126 means that risk is a combination of the severity of damage caused by the hazard (not of the hazard itself) and the occurrence probability per time unit. The risk arises from a scenario of events that consists of the hazard, an initiating event, and a chain of consequence events leading to an accident with the damage mentioned earlier.

Collective risk is a risk related to damage to a group of persons (risk group). Typical groups of persons are passengers; railway personnel (e.g., train staff and trackworkers); and third persons, e.g., users of level crossings.

In order to simplify the considerations, we will study only accidents with fatalities. In this context, we note that severe injuries and slight injuries can be taken into account via equivalent fatalities; see *Yellow Book* (Rail Safety and Standards Board 2007). For the computation of equivalent fatalities, 10 severe injuries are considered as one fatality, and 10 (sometimes 20) slightly injured persons, as one severely injured person.

For each risk group, there exists a probability $P(k)$ that an accident with k fatalities can occur. This probability depends on two main factors:

1. The probability that a certain accident scenario occurs
2. The (generally conditional) probability that during this accident, a certain number of persons are killed (or injured)

Assuming that the accidents occur statistically independently, for each type of accident, this can be described as a CPP. The accidents occur with a rate of occurrence of $\lambda(t)$, and for each accident type, there exists a certain "jump height," e.g., the number of fatalities (or equivalent fatalities), described by realizations of an independent random variable Y_T, where T describes the accident type. For a fixed interval $[0,t]$, the process X_t describes the cumulated number of fatalities and N_t describes the accumulated number of accidents; see Figure 8.1.

This CPP can be used to derive characteristics for collective risk. Frequently, collective risk is represented in the form of so-called $F–N$ curves (Modarres 2006). These curves express the probability of an accident (F) as a function of the minimal number of persons (N) killed during an accident. The probability is normally related to a fixed time interval, mostly 1 year. Then, $F(N)$ is the probability of accidents with N or more fatalities that occur during 1 year. This can be described with the help of the CPP as follows. If until time t a

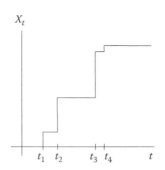

FIGURE 8.1
Compound Poisson process.

number of N_t accidents have occurred with damage $Y_1, Y_2, \ldots, Y_{Nt}$, then $F(N)$ is the probability that max $(Y_1, Y_2, \ldots, Y_{Nt})$ is larger than N. Therefore, $F(N)$ can be derived with the help of so-called extreme value distributions; see Leadbetter et al. (1986).

The group the collective risk is related to is an abstract group. This becomes clear if one considers the number of passengers. As an example, consider railway passengers. Normally, not the overall number of passengers is counted, but the number of voyages. A person who has traveled k times during 1 year by train will be counted k times, i.e., in the same manner as k different passengers. Note that a passenger is not permanently subjected to the railway system and its risks. He/she is only subject to these risks during his/her journey. In the remaining time, he/she is subjected to the influence of other, sometimes more dangerous systems. Therefore, it makes sense to represent the exposition of a passenger to the railway system by means of the number of voyages and their corresponding duration (parameters N_i and E_i according to (1999); for an extended discussion, see Bepperling [2009]). For onboard train personnel, the exposition time per year is mainly fixed. Note that they are not permanently onboard the train or under the influence of the railway system, since this time is limited by law, regulations, and trade union agreements.

One easily sees that a collective risk is frequently given per calendar year; the risk groups, however, are subject to the risks according to different exposition time. This exposition time then needs to be specified.

8.2.3 Individual Risk

The individual risk is a quantity defined for an arbitrary, but fixed individual. This individual can be seen as a fictive average person. The individual risk can be given as a probability that the individual is killed or injured during a certain time (e.g., 1 hour or 1 year). Here, the risk can also be given per calendar time. Then, assumptions have to be made about the exposure of the person to the risk, e.g., the number of train journeys and their duration. In CENELEC (1999), the figures $N_i = 500$ and $E_i = ½$ h are used.

Now the fatality rate of the individual can be expressed as the probability (per unit time) that an accident occurs with k fatalities, combined with the probability that the individual is among the fatalities.

The latter probability is derived from the fact how often and how long all members of the risk group are subjected to the risk of the system and from the coinciding figures for the individual.

Now it is worthwhile to introduce the exposition time, i.e., the time an individual is exposed to a risk or the system. Analogously, one can compute the accumulated risk exposition for a group of persons.

For a train voyager, one would have an exposition time of $N_i E_i = 250\,\mathrm{h} = (500 \times 1/2\,\mathrm{h})$ for the individual per year, i.e., the product

$$E = \text{number of journeys} \times \text{average duration of a journey}. \tag{8.1}$$

Now it is also possible to convert a collective risk into an individual one.

If a collective risk is given per calendar time, the individual risk can be obtained by

$$K_r \times E_1 / E, \tag{8.2}$$

where K_r is the collective risk, E_1 is the exposition time of the ith individual per calendar year, and E is the cumulated exposition time of all persons of the risk group per calendar year.

This means that the individual shares such a part of the overall risk that is proportional to its individual part in the exposition time of the group.

8.2.4 Examples of Risk

Many sources provide risk target values. In this chapter, we will show how risk values depend on relating quantities as, e.g., exposition time. For the correctness of the values cited, no guarantee can and will be given. The intention is not to draw conclusions from the numerical values regarding different safety levels of different means of transport.

In the literature, risk values are partially given per calendar hour, partially per hour of exposition. E.g., Melchers (1987) provides the Table 8.2.

TABLE 8.2

Some Societal Risks as Given by Melchers (1987)

Activity	Approximate Death Rate ($\times 10^{-9}$ Deaths/h of Exposure)	Typical Exposure (h/Year)	Typical Risk of Death ($\times 10^{-6}$/Year)
Alpine climbing	30,000–40,000	50	1500–2000
Boating	1,500	80	120
Swimming	3,500	50	170
Cigarette smoking	2,500	400	1000
Air travel	1,200	20	24
Car travel	700	300	200
Train travel	80	200	15
Coal mining (UK)	210	1,500	300
Construction work	70–200	2,200	150–440
Manufacturing	20	2,000	40
Building fires	1–3	8,000	8–24
Structural failures	0.02	6,000	0.1

Source: R. E. Melchers: *Structural Reliability, Analysis and Prediction.* 1987. Copyright Wiley-VCH Verlag GmbH & Co. KGaA. Reproduced with permission.

One can clearly see how risk is given per hour of exposition and how exposition time per year and the fatality risk per calendar year are given. Also, Kuhlmann (1986) and Kafka (1999) give the risk per hour of exposition. On the contrary, Proske (2008) provides in a table risk values for different causes of fatality in "mortality per year"; see Tables 8.3 and 8.4. For various areas of occupational life, the data are given in these tables.

Surely, one needs to assume a typical working time between 1600 and 1900 hours per year to derive the risk per hour of exposition for Table 8.3. The risk per hour of exposition would then be larger by a factor of [8760/1900, …, 8760/1600], i.e., [4.61, …, 5.48].

The interpretation of the figures in Table 8.4 from the same source is much more complicated.

Here the question for the exposition time becomes very important. One has to ask how long the soldier has been in Iraq, how many space flights an astronaut participates in during 1 year, how many hours a person is onboard a train (Germany and the United States). The question is most interesting when considering accidents at home: are we talking about a housewife who works 90 hours per year or the lazy chauvinist who is injured when he/she falls from the sofa, where he/she was sleeping? Generally, risk data without a reference to exposition times are questionable.

For the astronauts, one might learn from Kafka (1999) that for Crew Transfer Vehicles, one out of 500 missions is unsuccessful. Reference is made to a work by Preyssl from 1996. One gets the impression that the probability of 2×10^{-3} for loss of spacecraft for one flight is converted to a probability per year, assuming one flight per year but not indicating this

TABLE 8.3

Risk Values per Year according to Proske (2008)

Risk Group	Mortality per Year
Artist on the trapeze (US)	5×10^{-3}
Worker in agriculture	7.9×10^{-5}
Heavy industry worker (UK, 1990)	1.8×10^{-3}
Deep sea fishing	1.7×10^{-3}
Offshore platform worker (UK, 1990)	1.3×10^{-3}
Oil and gas industry worker	10^{-3}
Mining (mine; US, 1970)	8.4×10^{-4}

Source: With kind permission from Springer Science+Business Media: *Catalogue of Risks*, 2008, D. Proske.

TABLE 8.4

Risk Values according to Proske (2008)

Risk Group	Mortality per Year
US soldiers during second Iraq war in 2003	3×10^{-3}
Parachuting (US)	2×10^{-3}
Astronauts (European Space Agency crew recovery vehicle)	2×10^{-3}
Accidents at home	10^{-4}
Traffic accidents (UK)	9.1×10^{-5}
Drowning (US, 1967)	2.9×10^{-5}
Railway	5.1×10^{-7} to 4.4×10^{-6}

Source: With kind permission from Springer Science+Business Media: *Catalogue of Risks*, 2008, D. Proske.

conversion factor. Other sources provide the exposition time explicitly or implicitly as well, when risk per year is given.

Assuming a certain exposition time, the data of Proske can be converted to mortality per hour of exposition; see Table 8.5. One can easily convert the risk data given per calendar time into data per exposition times. The conversion factor is given in the fourth column; the result of the conversion is presented in the third column. One can clearly see the differences that arise from taking into account the exposition time. Assuming different exposition times during one year, differences of about one order of magnitude can easily arise. Moreover, the transport capacity of different transport means per hour can significantly differ; see Table 8.6.

Note that the data for traffic accidents are quite small. Possibly this is caused of a small underlying traffic density. Unfortunately, details about this are not known.

We have seen that a risk is presented best for a general unit of exposition time, e.g., per hour. This holds true if exposition time cannot be neglected as, e.g., for bungee jumping. If risk is given per calendar time, the assumptions about exposition need to be considered in the form of exposition time per event and the number of events or the cumulative exposition time in order to avoid misunderstanding. When talking about transport systems, the risk per unit of transport capacity, i.e., person kilometer, is also used. This makes sense for collective risks. If exposition time is incorrectly taken into account, one might easily have differences of one order of magnitude or even more.

TABLE 8.5

Risk Values per Year with Exposition Time

Risk Group	Mortality Risk per Year	Mortality Risk per Hour of Exposition	Assumption
Aviation (10,000 mi/year; regular flights)	6.7×10^{-5}	2.4979×10^{-6}	Travel velocity of 600 km/h,
Railway (200 h of exposition per year)	1.5×10^{-5}	7.5×10^{-8}	
Traffic accidents (10,000 mi/year, careful driver)	8×10^{-6}	2.4×10^{-8}	Travel velocity 30 mph; 1 mi = 1609.344 m

Source: With kind permission from Springer Science+Business Media: *Catalogue of Risks*, 2008, D. Proske.

TABLE 8.6

Mortality Risk in the European Union (2003) according to European Transport Safety Council (2003)

Transport Means	Mortality Risk (per 10^9 Person Kilometers)
Road traffic (sum)	9.5
Motorbike	138
Pedestrian	64
Bicycle	54
Automobile	7
Bus	0.7
Ferry	2.5
Civil aviation	0.35
Railway	0.35

Source: European Transport Safety Council, *Transport Safety Performance in the EU Statistical Overview*, European Transport Safety Council, Brussels, 2003.

8.3 *F–N* Curves and Statistical Tests for Comparing Data

8.3.1 *F–N* Curves

Assume the collective risk for a certain group of persons is given with the help of an *F–N* curve. Then, conversion of risk is somewhat more difficult.

An *F–N* curve is usually given in the form

$$K(N) = C/N^{\alpha}. \tag{8.3}$$

Here, $K(N)$ is the number of accidents with N or more fatalities. Note that $K(N)$ is proportional to $F(N)$. C denotes a constant. The exponent α usually takes values between 1 and 2, but is always larger than 1, which means risk aversion. In the general case, a value of 1.5 is a good guess.

The overall number of fatalities according to Equation 8.3 is then

$$\int_0^{\infty} N \, dK(N) = \int_0^{\infty} K(N) \, dN = \int_1^{\infty} CN^{-\alpha} \, dN = \frac{C}{\alpha - 1}. \tag{8.4}$$

The individual risk *IR* is computed as

$$IR = \int_0^{\infty} N \, dK(N) \frac{E_1}{E} = \frac{CE_1}{(\alpha - 1)E}, \tag{8.5}$$

where E_1 is the exposition time of a typical individual in hours (e.g., during 1 year) and E is the cumulated exposure time of all affected persons in hours (e.g., during 1 year).

8.3.2 Statistics

In practice, one does not observe an *F–N* curve. Rather, one can observe numbers of accidents with certain severities, i.e., the process X_t that is described earlier. If one now wants to compare the statistics of accidents of different countries, companies, or years, it is not sufficient to draw the *F–N* curve but to carry out a statistical analysis.

Let us now first describe the CPP X_t in more detail.

Often, such processes are assumed to follow an HCPP, but in practice, seasonal variations occur, e.g., due to different weather conditions during a year. A CPP can be decomposed into parts: the counting process itself and the distribution height of the jumps describing the size of the accidents.

We first give a generic definition of the problem we want to discuss in this section: we assume an HCPP with N_t depicting the number of accidents until time t and $S_1, S_2, \ldots,$ an independent, identically distributed (i.i.d.) sequence of nonnegative random variables describing the severity of the accidents. The sum of all accident severities until time t $X(t)$ is then distributed as

$$\sum_{i=1}^{N_t} S_i. \tag{8.6}$$

For a HCPP, we have

$$E\big(X(t)\big)=\lambda t E(S_1)\text{ and }V\big(X(t)\big)=\lambda t E\big(S_1^2\big), \tag{8.7}$$

where λ denotes the accident rate per year, and we assume that the moments exist. From the book by Birolini (2007), we also know the exact distribution, if the jumps are independent and i.i.d.

$$P(X(t)\le x)=1-F_1(t)+\sum_{n=1}^{\infty}G_n(x)\big(F_n(t)-F_{n+1}(t)\big), \tag{8.8}$$

where F_i and G_i denote the *i*-fold convolution of the distribution functions of the time between events and the severity, respectively. Given a sufficient number of accidents per year, limit theorems for the distribution of $X(t)$ may apply, e.g., the central limit theorem.

In the following sections, we will propose tests for H_0, both on a random walk approach and by a classic decomposition approach from statistical test theory.

These tests will form an alternative to the comparison procedures for national reference values (NRVs) as required by the ERA; see the directive by the European Parliament and Council of the European Union (2009).

In order to compare different periods, we assume that $X_1(t)$ and $X_2(t)$ are CPP over the same time span, e.g., a year. For an example, see Figure 8.4. Under H_0, "Safety performance has not changed," $X_1(t)$ and $X_2(t)$ are i.i.d. It immediately follows that the difference $X_1(t)-X_2(t)$ is a one-dimensional random walk with expectation 0, which may lead to distribution-free test statistics for maximum deviations, at least asymptotically. By convolution of the densities given by Equation 8.8, the distribution of the difference is analytically known, but may be hard to explicitly evaluate. Also, particular distribution assumptions might be hard to justify. Other interesting test statistics would include the area between the differences.

Generally, the maximum difference or absolute maximum difference of $X_1(t)-X_2(t)$ is distributed under H_0 as a sum of independent random variables with random signs

$$Z=\sum_{i=1}^{N_1+N_2}(-1)^{I_i}S_i, \tag{8.9}$$

where I_i represents i.i.d. Bernoulli random variables with success probability of ½ and N_i represents the sample sizes. Note that distribution depends, in the general case of CPP, not explicitly on the jump times, but only on the ordering of the jump times of the two samples.

It is immediately clear that $E(Z)=0$ and $\mathrm{Var}(Z)=2E(N)(\mu^2+\sigma^2)$, assuming the moments μ and σ of S exist. If F_S has a continuous distribution, it follows that the randomly signed random variable has the density

$$f(x)=\frac{1}{2}f_S\big(|x|\big). \tag{8.10}$$

Figure 8.2 shows a sample plot of Z for a sample size of 50 $N(5,1)$ distributed random variables.

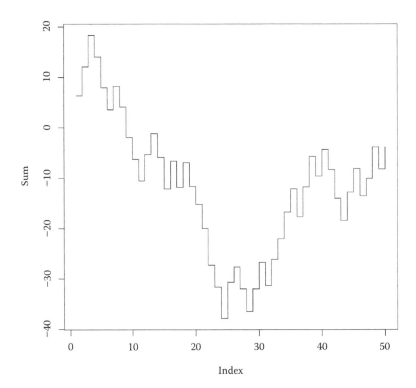

FIGURE 8.2
Sample path plot of Z.

For large numbers of observations, the distribution of Z will become normally distributed due to the central limit theorem. Alternatively, bounds based on Chebychev inequality could be applied to determine test statistics. However, such tests would depend only on deviations of the complete sum and not on the maximum deviation over time as in the Kolmogorov–Smirnov-type tests.

8.3.3 Testing with the Random Walk

The most intuitive test would be a kind of Kolmogorov–Smirnov-type test, i.e., to reject H_0 if the maximum observed difference between $X_1(t)$ and $X_2(t)$ exceeds a threshold. If the jump distribution is identical to 1, then for a particular realization with n_1 and n_2 jumps, respectively, $F_i(t) = X_i(t)/n_i$ has similar properties as a distribution function, and the maximum difference of $F_1(t) - F_2(t)$ follows the same distribution as the test statistics $D_{1,2}$ of the two-sample Kolmogorov–Smirnov test. As in the Kolmogorov–Smirnov test, the distribution does not depend on the particular values in the sample, but only on the relative order of the sample values.

For generally distributed jumps S, the distribution of Z could be exactly calculated by Equation 8.8; however, this would become quite complex. If S follows a discrete distribution, $D_{1,2}$ should be distributed as Kolmogorov–Smirnov statistics in the presence of ties. In the general case, we can apply limit theorems such as Erdös and Kac (1946), which are of the type

$$\lim_{n \to \infty} P\left(\max \left(X_1(t) - X_2(t) \right) < \zeta \sigma' \sqrt{n} \right) = 2\Phi(\zeta) - 1, \tag{8.11}$$

where σ denotes the standard deviation of the signed jump distribution and Φ is the distribution function of the standard normal distribution. This is an interesting result as it shows that the limiting distribution is indeed distribution free and depends only on the standard deviation. This result also holds, in particular, for inhomogeneous CPP. In practice, however, we would have to estimate σ by an estimate s based on the sample, so that Equation 8.11 is not exact, but for large sample sizes, the deviation should be negligible. Assuming a desired probability of the first kind, an error of 10% would result in a threshold of approximately $1.64\,\sigma\sqrt{n}$. If the deviation is larger, H_0 is then rejected.

For the absolute maximum, Erdös and Kac (1946) also give a limiting distribution, but this is more complex

$$\lim_{n\to\infty} P\left(\max\left|X_1(t)-X_2(t)\right| < \zeta\sigma'\sqrt{n}\right)$$

$$= \frac{4}{\pi} \sum_{m=0}^{\infty} \frac{(-1)^m}{2m+1} \exp\left(-\frac{(2m+1)^2\pi^2}{8\zeta^2}\right). \tag{8.12}$$

Figure 8.3 shows a Monte Carlo simulation for sample size 50 for the distribution of the absolute maximum deviation of Z for $N(0,1)$ random variables smoothed by a kernel estimator. The quantiles for $p = 0.1$, 0.05, and 0.01 are 13.4, 15.4, and 19.3, respectively. In order to apply this to the sample from Figure 8.1, we would have to stretch the quantiles by a factor of $\sigma' = 5.09$, so that the sample from Figure 8.1 would not lead to a rejection of H_0.

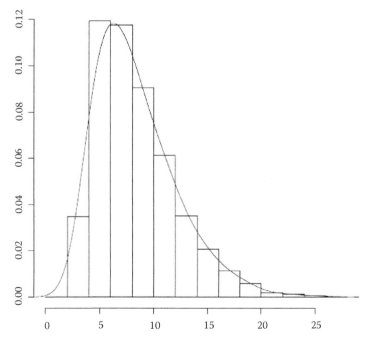

FIGURE 8.3
Sample distribution for the absolute maximum deviation of Z.

8.3.4 Testing with a Decomposition Approach

As the jumps and the occurrence times are independent, we could factor the likelihood function and apply optimal tests to the separated datasets. At times T_i, we have observed jumps heights S_j. The jump heights have the distribution function $G(x)$ and the jump times T_i distributions F_i. We have observed n sample pairs (t_i, s_i) of jump times and jump heights. Observed data are denoted by lowercase characters.

Then, the likelihood function can be written as

$$L(t_i, s_i) = \prod f_i(t_i) g(s_i). \tag{8.13}$$

Where appropriate, indices ranging from $i = 1$ to n have been omitted. The likelihood function can be expressed as two factors:

$$L(t_i, s_i) = L(t_i) L(s_i) = \prod f_i(t_i) \prod g(s_i). \tag{8.14}$$

Then, the Neyman–Pearson lemma (see Schmetterer [1966]), for testing the H_0 hypothesis "The times and jumps have probability densities $f_i(t)$ and $g(x)$" against the alternative hypothesis H_A "The times and jumps have probability densities $h_i(t)$ and $k(x)$," is a test that rejects H_0 for

$$\prod f_i(t_i) \prod g(s_i) > M \prod h_i(t_i) \prod k(s_i), \tag{8.15}$$

where M is a properly chosen constant so that the failure of first kind is α. Now, the test can be decomposed into two independent tests:

- Test A: Reject H_0 if

$$\prod f_i(t_i) > M_1 \prod h_i(t_i), \tag{8.16}$$

  where M_1 is chosen so that the failure of the first kind of this partial test is α_1.
- Test B: Reject H_0 if

$$\prod g(s_i) > M_2 \prod k(s_i), \tag{8.17}$$

  where M_2 is chosen so that the failure of the first kind of this partial test is α_2. The hypothesis H_0 is rejected whenever at least one of the tests rejects the hypothesis. Then, the failure of the first kind of the combined test is

$$\alpha = 1 - (1 - \alpha_1)(1 - \alpha_2). \tag{8.18}$$

It must be noted that $M = M_1 M_2$ does not necessarily need to hold. This is caused by the fact that one of the tests might reject H_0, but nevertheless, Equation 8.15 can hold.

Assuming a particular model for the distribution functions, we can now construct-tests. Because the general test has been decomposed into two tests; the test for the jump heights (severities), i.e., the distribution function $G(s)$, and the tests for the distribution functions $F_i(t)$ for the jump times can be discussed separately. The most interesting situation, however, is a two-sample test, i.e., to test whether two realizations $\{(t_i, s_i), i = 1, ..., n\}$ and

$\left\{\left(t'_j, s'_j\right), j = 1, \ldots, m\right\}$ come from the same distribution. Now, we have the following simple and efficient two-sample test at our disposal.

8.3.4.1 Test for Jump Times

The simplest assumption is a homogeneous Poisson process for the times t. Then, F_i are Erlang distributions of order i and intensity parameter λ. Sufficient statistics are given by the pairs (t_n, n) and (t'_m, m) with $2nt_n$ and $2mt'_m$ being chi-square distributed with $2n$ or $2m$ degrees of freedom, respectively (Johnson, Kotz, and Balakrishnan 1994; Mann, Schafer, and Singpurwalla 1974). An efficient two-sample parametric test for equality of the parameter λ, i.e., intensity of the homogeneous Poisson process, is an F-test based on the fact that nt_n / mt'_m has an F distribution with $(2n; 2m)$ degrees of freedom. The test would be two sided, rejecting the H_0 hypothesis for large or small values using the quantiles $F_{\alpha 1/2}(2n; 2m)$ and $F_{1-\alpha 1/2}(2n; 2m)$. Since this test is constructed with the help of the Neyman–Pearson lemma, it is efficient.

8.3.4.2 Test for Jump Heights

For the jump heights, several assumptions can be made. The observed samples are samples $(s_i; i = 1, \ldots, n)$ and $(s'_i; i = 1, \ldots, m)$. One assumption could be a common normal distribution with mean μ and standard deviation σ. Using the sufficient statistics sample mean and sample standard deviation, we can construct the test (Zacks 1971). First, an F-test is carried out to test the equality of standard deviations using the fact that $(n-1)s^2 / (m-1)s'^2$ has F distribution with $((2n-2);(2m-2))$ degrees of freedom. Here, s^2 and s'^2 denote the sample variances of the two samples, respectively. Assume that the failure of the first kind is α_{21}. The second part consists of a two-sample t-test using the fact that

$$T = \sqrt{\frac{nm}{n+m}} \left| \frac{h' - h}{\omega} \right|$$

has a t distribution with $n + m - 2$ degrees of freedom (Bickel and Doksum 1977). Here, h and h' denote the sample means of both samples, and ω is $n(s^2 + ms'^2)/(n + m - 2)$. The hypothesis H_0 is rejected if T is larger than the quantile $t_{n+m-2}(1 - \alpha_{22}/2)$ of the t distribution with α_{22} being the error of the first kind. Note that $1 - \alpha_2 = (1- \alpha_{21})(1 - \alpha_{22})$, since the hypothesis H_0 is rejected as soon as one of the tests (F-test or t-test) rejects the hypothesis.

The jump heights may also be exponentially distributed. This is motivated by the fact that in many cases, the probability of the occurrence of accidents with a number of fatalities decreases with the number of accidents by a certain factor. In that case, the sample means h' and h have the property that $2nh/\mu$ and $2mh'/\mu$ have chi-square distributions with $2n$ or $2m$ degrees of freedom, respectively. Here again, μ denotes the mean of the distribution of $(s_i; i = 1, \ldots, n)$. Then, $nh/(mh')$ has an F distribution with $(2n; 2m)$ degrees of freedom. Using the quantiles

$$F_{\alpha 2/2}(2n; 2m) \text{ and } F_{1-\alpha 2/2}(2n; 2m),$$

a two-sided test can be constructed for the jump heights in case of exponentially distributed jump heights. We may also use a rank test for testing the hypothesis of the equality of jump height distribution. An important-test is the Wilcoxon test (Hajek and Sidak 1967), which is well known to be the optimal rank test for logistic distributions.

All the test procedures mentioned earlier are optimal since they are directly derived from the Neyman–Pearson lemma, provided the assumptions made on the parametric models hold true. A popular assumption is a homogeneous Poisson process on the times and an exponential distribution on the jump heights. Any such test must be more powerful than the statistical test proposed by the ERA in the commission decision (European Parliament and Council of the European Union 2009) and analyzed by Braband and Schäbe (2012), since the test provided by the ERA is less efficient, compared with an optimal test.

8.3.5 Testing: An Example

We consider the simulated data shown in Figures 8.4 and 8.5. All simulations and computations have been carried out with R (R Project 2013). The true underlying distributions

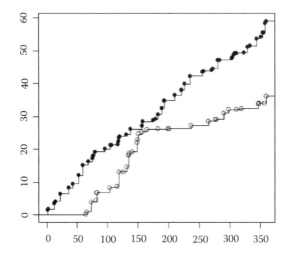

FIGURE 8.4
Simulated example data—first set.

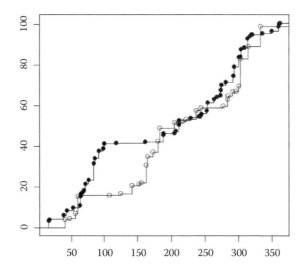

FIGURE 8.5
Simulated example data—second set.

have been chosen as homogeneous Poisson processes with intensity 30/365 and 50/365 and exponential jump height distributions. In the first case, we assume equal jump height with parameter 1; in the second case, with parameters 0.3 and 0.5. Thus, in the first case, the expected NRVs are different, but in the second case, they are the same. Note that a classic comparison of the two NRVs according to the rules of the ERA see commission decision by the European Parliament and Council of the European Union (2009), is not possible, since there would be only two values NRV_1 and NRV_2 to compare.

The data set for case 1 contains 36 accidents in year 1 and gives the accumulated sum of severities $NRV_1 = 36.2$, while the second year has had 55 accidents and yields $NRV_2 = 59.1$. In case 2, the NRVs are almost equal, but the number of accidents and the respective accident severities differ. But is there sufficient evidence to conclude that safety performance in year 2 has become worse? For the following statistical tests, we have chosen the error of the first kind α as 0.1, giving $\alpha_1 = \alpha_2 = 0.052$.

8.3.5.1 Tests for Jump Times

The two-sided Kolmogorov–Smirnov test accepts the equality of the distributions in both cases. The F-test for the equality of the interarrival times of the accidents rejects the hypothesis of equality with p values of 0.028 and 0.003, respectively. This test exponentially assumes distributed times and is therefore more powerful than the Kolmogorov–Smirnov test but at the cost of making a particular distribution assumption.

8.3.5.2 Tests for Jump Heights

For the jump heights (FWSI = fatal and weighted serious injury values of the accident), the Wilcoxon test accepts the hypothesis of the equality of the distributions in both cases. The F-test for the equality of the jump heights rejects the hypothesis of equality with a p value of 0.013 in case 1, but accepts case 2 with a p value of 0.479. Again, this test assumes exponential distributions of the jump heights.

8.3.5.3 Random Walk Tests

We apply the test according to Equation 8.11. In case 1, the combined sample size is 91, the combined sample mean is 1.05 and sample variance is 0.79, and the maximum difference in the sample is 22.9. By Equation 8.11, we calculate that the threshold is 21.5, so that we can reject H_0. In case 2, the combined sample size is 83, the combined sample mean is 2.4 and the sample variance is 5.94, and we can calculate the threshold as 51.1, which is clearly not exceeded by the sample.

8.3.5.4 Comparison

It is interesting to notice that in both cases, the nonparametric tests for jump times and jump heights fail to detect the differences in the sample distributions, but the nonparametric random walk test detects them in case 1, but not in case 2. For the parametric tests, the differences are quite obvious, but they are tailored to these distributions, except in case 2 where the difference in jump height parameters is not detected. So, if distribution assumptions can be justified, the optimal parametric tests should be chosen. However, without particular distribution assumptions, the random walk test would be a reasonable alternative.

So we have proposed possible tests as alternatives to the test procedure defined by the ERA in the commission decision (European Parliament and Council of the European Union 2009). All tests have been easily compiled from standard statistical literature, and the parametric tests are well known to be efficient under regularity assumptions and certain parametric hypotheses that have been mentioned in this chapter. In addition, the tests use so-called sufficient statistics, which, in fact, are the cause for their optimality. Sufficient statistics contain the same amount of Fisher information as the original sample, provided the parametric model is true. The NRVs are not sufficient statistics; they are just weighted sums of FWSIs accumulated over 1 year. Therefore, we propose to use classic statistical tests based on the statistics of accident times and coinciding FWSIs rather than based on the NRVs.

If no particular distribution assumptions are to be made or can be justified, then test statistics based on random walk properties seem a good alternative, as the test distributions can be given for a large class of CPP. The only assumptions are that the second moment of jump distribution exists and that the sample size is sufficiently large (similar to the central limit theorem) (Braband and Schäbe 2014).

8.4 Derivation of the Distribution of Accident Severity

8.4.1 *F–N* Curve and the Distribution of Severities

Individual risk is simply the predicted frequency of the undesired outcome. Societal risk takes account of both the severity of the range of possible outcomes and the frequency at which they each are predicted to occur. Now, from Equation 8.1, one can derive the probability that an accident occurs with more than s fatalities. It is just

$$\Pr(S > s) = A \times s^{a} / A \times 0.1^{a} = B \times s^{a}, \text{ with } s \geq 0.1, \tag{8.19}$$

where B is a new constant. The value 0.1 is chosen since only serious injuries, i.e., 0.1 fatalities, are accounted for, but no light injuries. Equation 8.19 presents the survival function of the distribution function of accident severities with severities $S > s$ of at least 0.1. Therefore, Equation 8.19 is a conditional distribution function.

Obviously, the probability density of this survival function reads

$$f(s) = a \times B \times s^{a-1} = a \times s^{a-1} / 0.1^{a} \text{ for } s \geq 0.1 \text{ and zero otherwise.} \tag{8.20}$$

This is the Pareto distribution, which can be found in standard textbooks, such as in the book by Johnson, Kotz, and Balakrishnan (1994). This distribution is typical for the distribution of income.

8.4.2 Estimation and Testing

When considering the distribution function (Equation 8.3), it is obvious that only one parameter needs to be estimated. The second parameter B is fixed by the requirement that only accidents with at least 0.1 FWSI will be regarded. The estimator of a is given by Johnson, Kotz, and Balakrishnan (1994) as

$$\hat{a} = \log(G/0.1) - 1, \tag{8.21}$$

where G is the geometric mean of the sample

$$G = (\Pi S_i)^{1/n}, \tag{8.22}$$

where $\{S_i, i = 1, \ldots, n\}$ is a sample of observed accident severity values. The product Π runs over i from 1 to n.

This result can now be used to estimate the parameters of the F–N curve. Note that by the approach described earlier, one can estimate the parameter a of the F–N curve, $F(N) = C/N^a$. Parameter C can easily be estimated by setting N to the smallest value used in the statistics, say, N_0, be that 1 for fatalities or 0.1 for severe injuries and just counting all accidents with N_0 or more accidents. Count M would then be equal to C/N_0^a. Therefore,

$$\hat{C} = M \times N_0 \hat{a}. \tag{8.23}$$

Johnson, Kotz, and Balakrishnan (1994) mention that $2na/\hat{a}$ has a chi-squared distribution with $2(n - 1)$ degrees of freedom. This property can be used to test whether the sample comes from a certain population with given value a of the shape parameter. As soon as the inequalities

$$\chi^2\left(2(n-1); \ 1-\alpha/2\right) < 2na/\hat{a} < \chi^2\left(2(n-1); \ \alpha/2\right) \tag{8.24}$$

are fulfilled, one would accept the hypothesis that the sample comes from a population with shape parameter a. Here $\chi^2(2(n - 1);1 - \alpha/2)$ denotes a standard chi-square quantile from the table of the chi-square distribution. With α, we denote an error of the first kind of the test.

The result (Equation 8.24) can also be used to provide confidence intervals on a. So the parameter a would lie with probability $1 - \alpha$ within a two-sided confidence interval given by

$$\left[\hat{a}\chi^2\left(2(n-1); \ 1-\alpha/2\right)/(2n); \ \hat{a}\chi^2\left(2(n-1); \ \alpha/2\right)/(2n)\right]. \tag{8.25}$$

Based on this property, it is also possible to construct a two-sample test. Assume that $\hat{a}_1$ and $\hat{a}_2$ are estimators from two samples of size n_1 and n_2, respectively. Then,

$$T = (n_1\hat{a}_2)/(n_2\hat{a}_1) \tag{8.26}$$

has an F distribution with $(2n_1;2n_2)$ degrees of freedom. So, the test would accept the hypothesis of equality of the parameter a if

$$F(2n_1;2n_2;\alpha/2) < T < F(2n_1;2n_2;1-\alpha/2). \tag{8.27}$$

8.4.3 Example

Accident data from Switzerland for Swiss railway accidents have been used. The data have been collected over the last decade. Using Equation 8.21 for accidents with an FWSI value of 0.1 or more, the shape parameter a has been estimated to be 1.8. This is in the region of 1.5–2, which is frequently used, and to the value of two, used by Rijkswaterstaat (2002) for providing target data for group risks.

By computing the confidence interval for 95% coverage, i.e., for $\alpha = 1\%$, we arrive with the help of Equation 8.25 at the interval [1.60; 2.03].

A statistical test carried out for the value 2 would therefore accept the hypothesis. This is due to the fact that for one sample test, confidence intervals for a parameter can be used for testing, the computations being the same as those for obtaining critical values for the test statistic.

We test whether the Pareto distribution is a good model. This is done as follows. We just compare the counts of accidents in the following four classes:

$$[0.1;0,2], \ [0,2;1.5], \ [1.5,2.5], \ \text{and} \ [2.5,\infty]$$

with the computed values for these intervals. This is done with the help of the chi-square statistic

$$H = \sum (O_k - E_k)^2 / E_k. \tag{8.28}$$

Here, O_k and E_k denote the observed and expected numbers of events in the kth class. H is known to have a chi-square distribution with $r - 1$ degrees of freedom. With r, the number of calluses is denoted, and since one parameter is estimated from the same sample, the number of degrees of freedom is reduced. The computed test statistics is 97, which largely exceeds the critical value of 11.3 for a 1% error of the first kind. This is caused by the fact that in the smallest class, a smaller number of events has been counted that is expected according to the parametric model.

8.4.4 Outlook

A closer look at the data reveals, however, that while the general decaying shape suggests a Pareto distribution, there might be a better fit when approximating the distribution by a weighted sum of discrete random variables. To intensify this affect, we regard fatalities and weighted injuries (FWI), which are often defined by

FWI = <number of fatalities> + <number of serious injuries>/10
+ <number of light injuries>/100.

Figure 8.6 shows an example based on a simulation, where the number of fatalities, serious injuries, and slight injuries have been drawn for each accident as realizations of independent Poisson random numbers. The dominating number of accidents with FWI zero has been truncated.

Like in real accident data, the typical distribution shape emerges with a kind of step-like structure beginning at 0, 0.1, 0.2, and 1.0. This is due to the fact that the majority of accidents affects only a small number of people, often only one. In practice, however, the number of fatalities, serious injuries, and slight injuries would be dependent.

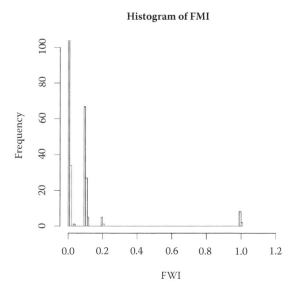

FIGURE 8.6
Typical sample FWI distribution.

8.5 Conclusions

We have seen that a risk is presented best for a general unit of exposition time, e.g., per hour. This holds true if the exposition time cannot be neglected as, e.g., for bungee jumping. If risk is given per calendar time, the assumptions about exposition need to be considered in the form of exposition time per event and the number of events, or the cumulative exposition time, in order to avoid misunderstanding. When talking about transport systems, also the risk per unit of transport capacity, i.e., person kilometer is used. This makes sense for collective risks. If exposition time is incorrectly taken into account, one might easily have differences of one order of magnitude or even more.

We have proposed possible tests as alternatives to the test procedure defined by the ERA in the commission decision (European Parliament and Council of the European Union 2009). All tests have been easily compiled from standard statistical literature, and the parametric tests are well known to be efficient under regularity assumptions and certain parametric hypotheses that have been mentioned in this chapter. In addition, the tests use so-called sufficient statistics, which, in fact, are the cause for their optimality. Sufficient statistics contain the same amount of Fisher information as the original sample, provided the parametric model is true. We propose to use these classic statistical tests based on the statistics of accident times and coinciding FWSIs rather than based on the NRVs.

Furthermore, we have given a theoretically well-founded derivation of the distribution for the jump height of the CPP, i.e., the accident severity, which can be used to describe counts of railway accidents. The distribution has been derived from the well-known F–N curve and turns out to be the Pareto distribution. We have used standard statistical textbooks to provide estimation and testing methods. In a practical example, we have derived the shape parameter that falls into an interval for parameters usually used for F–N curves. The distribution type, however, may in fact be more complex due to discretization effects due the definition of FWI.

References

Bepperling, S. (2009) *Validierung eines semi-quantitativen Ansatzes zur Risikobeurteilung in der Eisenbahntechnik* (*Validation of a Semi-quantitative Approach to Risk Assessment*), Dissertation, Braunschweig University of Technology, Braunschweig.

Bickel, P. J., and Doksum, K. A. (1977) *Mathematical Statistics*, Holden–Day, San Francisco, CA.

Birolini, A. (2007) *Reliability Engineering*, 5th edition, Springer, Berlin.

Braband, J., and Schäbe, H. (2011) Statistical analysis of railway safety performance, in the European Union, *ESREL 2011: Proceedings, Advances in Safety, Reliability and Risk Management*, CRC Press/Balkena, Leiden, pp. 1783–1787.

Braband, J., and Schäbe, H. (2012) Assessment of national reference values for railway safety—A statistical treatment, *ESREL 2012/PSAM 11*, IAPSAM & ESRA.

Braband, J., and Schäbe, H. (2013) Assessment of national reference values for railway safety: A statistical treatment, *Journal of Risk and Reliability* 227(4), 405–410.

Braband, J., and Schäbe, H., (2014) Comparison of compound Poisson processes as a general approach towards efficient evaluation of railway safety, *Proceedings of the ESREL 2013: Safety, Reliability and Risk Analysis: Beyond the Horizon*, Steenbergen et al. (Eds), Taylor & Francis Group, London, pp. 1297–1302.

CENELEC (European Committee for Electrotechnical Standardization) (1999) *Railway Applications—Systematic Allocation of Safety Integrity Requirements*, Report R009-004, CENELEC, Brussels.

European Parliament and Council of the European Union (2009) Commission Decision of June 5, 2009 on the adoption of a common safety method for assessment of achievement of safety targets, as referred to in Article 6 of Directive 2004/49/EC of the European Parliament and the Council; 2009/460/EC.

European Transport Safety Council (2003) *Transport Safety Performance in the EU Statistical Overview*, European Transport Safety Council, Brussels.

Erdös, P., and Kac, M. (1946) On certain limit theorems of the theory of probability, *Bulletin of the American Mathematical Society* 52, 292–302.

Hajek, J., and Sidak, Z. (1967) *Theory of Rank Tests*, Academia, Prague.

Johnson, N. L., Kotz, S., and Balakrishnan, N. (1994) *Continuous Univariate Distributions*, Vol. 1, 2nd edition, John Wiley & Sons, Hoboken, NJ.

Kafka, P. (1999) How safe is safe enough, *Proceedings of European Safety and Reliability Conference 1999* 1, 385–390.

Leadbetter, M., Lingren G., and Rootzen H. (1986) *Extremes and Related Properties of Random Sequences and Processes*, Springer, New York.

Kuhlmann, A. (1986) *Introduction to Safety Science*, Springer, New York.

Mann, N. R., Schafer, R. E., and Singpurwalla, N. D. (1974) *Methods for Statistical Analysis of Reliability and Life Data*, John Wiley & Sons, Hoboken, NJ.

Melchers, R. E. (1987) *Structural Reliability, Analysis and Prediction*, John Wiley & Sons, Hoboken, NJ.

Modarres, M. (2006) *Risk Analysis in Engineering: Techniques, Trends, and Tools*, CRC Press, Boca Raton, FL.

Proske, D. (2008) *Catalogue of Risks*, Springer, New York.

R Project (2013) *The R Project for Statistical Computing*, http://www.r-project.org/.

Rail Safety and Standards Board (2007) *Yellow Book: Engineering Safety Management Volumes 1 and 2, Fundamentals and Guidance*, Issue 4, Rail Safety and Standards Board, London.

Rijkswaterstaat (2002) *Normdocument veiligheid Lightrail*, v. 5.0, Rijkswaterstaat, Brussels.

Schmetterer, L. (1966) *Einführung in die Mathematische Statistik*, Springer-Verlag, Vienna.

Wikipedia (2010) Bungeespringen (bungee jumping), http://de.wikipedia.org/wiki/bungee-Springen, last access June 29, 2010, Wikipedia, St. Petersburg, FL.

Zacks, S. (1971) *The Theory of Statistical Inference*, John Wiley & Sons, Hoboken NJ.

9

Practical Demonstrations of Reliability Growth in Railways

Hendrik Schäbe

CONTENTS

9.1 Introduction

In this chapter, we provide methods to investigate trends in reliability. Analysis of trends is important since in a growing number of cases, debugging and fault correction are done in a test period or a qualification period. In all these cases, the questions what will be the result of these improvement actions and which level of reliability will be reached arise. This process is called reliability growth. This is the subject of the present chapter.

We briefly present the methods to be applied and the models that are useful. This is supported by an example.

The next section deals with technical measures for reliability improvement. Section 9.3 presents the mathematical model.

9.2 Technical Measures

Reliability improvement is only possible if measures are taken to improve the reliability. That means that failures and faults must be detected; the causes, analyzed; and measures, implemented to remove the causes for the faults to prevent further occurrence of these faults.

The draft standard EN 61014 (European Committee for Electrotechnical Standardization 2014) describes which steps need to be taken to ensure that reliability is improved as a result of this process. It is important to establish a process of reliability improvement management that systematically removes the causes of faults to ensure improvement. More detailed information on technical issues can be found in Chapters 4 and 5 of EN 61014 and in IEC 60706-6: Part 6. This includes establishment of a FRACAS (Failures Reporting analysis and Corrective action) process including at least the following:

- Reliability growth that needs to be planned
- Reliability that needs to be observed and indicators that need to be computed
- Reliability management that must be installed especially with the aspect of improvement

Reliability management must include the following:

- Ensure registration and detection of all faults and near fault situation.
- Include in-depth analysis of the faults, including root cause analysis. During root cause analysis, the entire life cycle proves needs to be analyzed to reveal flaws in the life cycle, including design analyses, production process, and testing. So, improvement must always include an improvement of the processes and not only reveal and remove the fault itself and its causes.
- Ensure that no new deficiencies are introduced. A main element is to have a functioning change management system, including impact analysis and regression testing. Each fault cause removed must be treated as a change and its impact must be thoroughly analyzed before being implemented. The impact analysis must also indicate which tests must be repeated and which additional new tests need to be carried out.

9.3 Nonhomogeneous Poisson Process

As a result of the recoded failure or fault times, statistical analysis needs to be carried out to derive a reliability prediction. In this chapter, we will describe the mathematical basis for the modeling. This information is also part of Chapter 8.5.2.1 of EN 61014.

Usually, lifetime is statistically modeled by a random variable, say, T, which is called the lifetime variable. It is described by its distribution, say,

$$F(t) = P(T \leq t),\qquad(9.1)$$

which satisfies the condition

$$F(0) = 0,\qquad(9.2)$$

i.e., $T > 0$ with probability one. Normally, it is assumed that $F(t)$ is absolutely continuous, and the derivative of $F(t)$ is $f(t)$, which is called the probability density. Moreover, the cumulative hazard function $\Lambda(t)$ is defined as

$$\Lambda(t) = -\ln\left(1 - F(t)\right).\qquad(9.3)$$

The most simple model for a distribution function is the exponential distribution:

$$F(t) = 1 - \exp(-\lambda t). \tag{9.4}$$

In this model, it is assumed that the system is not repairable, i.e., it is put into exploitation at time zero and fails at time T, and the exploitation of the system stops.

Obviously, this model is not applicable to a situation where the reliability improves. In that situation, the system generates a flow of faults, and it is repaired after each fault, and the fault cause is removed from the system so that the behavior of the system should improve over time. An alternative model is needed for this case. A frequently used model is the nonhomogeneous Poisson process (HNPP).

The HNPP is chosen is for the following reasons: The homogeneous Poisson process (HPP) is used to model repairable systems with exponentially distributed times between the faults. A HNPP is a simple generalization of the HPP, as is an arbitrary distribution function for the lifetime of a nonrepairable system.

Let the NHPP be denoted by $N(t)$, where $N(t)$ is the number of faults observed within the interval $[0,t]$ with

$$N(0) = 0, T_i = \inf\{t: N(t) \geq i\}. \tag{9.5}$$

Hence, T_i is the time of occurrence of the ith fault.

The expectation function of the NHPP is denoted by $m(t) = E\{N(t)\}$ and is assumed to be continuous and differentiable with first derivative $v(t)$.

The NHPP is defined by the following properties; see, e.g., Engelhardt and Bain (1986):

1. $m(t)$ is strictly increasing and approaches infinity as $t \to \infty$.
2. The number of events within the interval $(t,s]$, $t < s$ $N(t) - N(s)$ is a random variable with Poisson distribution having parameter $m(t) - m(s)$.
3. For nonoverlapping time intervals $(t_1,s_1]$ and $(t_2,s_2]$ with $t_1 < s_1$ and $t_2 < s_2$, $N(t_1)$ $N(s_1)$ and $N(t_2) - N(s_2)$ are statistically independent random variables.
4. There are no simultaneous events.

A link between the two models, i.e., the NHPP and the lifetime variable, can be formally established as follows. Let us consider the distribution of T_1, the first event of the NHPP. It has distribution

$$P(T_1 < t) = 1 - \exp\{-m(t)\}. \tag{9.6}$$

The derivative of $m(t)$, i.e., dm(t)/dt, gives the Rate of Occurrence of Failures (ROCOF); see e.g., Ascher and Feingold (1984).

By equating the mean value function $m(t)$ and the cumulative hazard function $\Lambda(t)$ by

$$m(t) = \Lambda(t), \tag{9.7}$$

we have established a formal link between both models. The meaning of the link between both models can be explained using the minimal repair model; see Barlow and Proschan

(1983). If upon failure a system having lifetime T with distribution $F(t)$ undergoes minimal repair, i.e., it is restored into the state just prior to failure, the failure times form a NHPP with mean value function

$$m(t) = \Lambda(t). \tag{9.8}$$

For an analysis of reliability growth, the Weibull process is used, also well known as the Duane model. This process is the counterpart of the Weibull distribution for nonrepairable models, but the Weibull process must not be confused with the Weibull distribution. It must be noted here that there is a principal difference between a model for a nonrepairable system, characterized by a lifetime distribution and the model for a nonhomogenous counting process characterized by its ROCOF; see Ascher and Feingold (1984).

The best overview of the Duane model can be found in the study of Crow (2011); more material is provided in Finkelstein (1976), Bain (1978), Bain and Engelhardt (1980), Lawless (1987), and Härtler and Schäbe (1991).

This process is obtained from an HPP by a time transformation

$$t' = t^{\beta}/\Theta. \tag{9.9}$$

Therefore, the expected and cumulated number of events up to time t is

$$m(t) = t^{\beta}/\Theta. \tag{9.10}$$

If now β is greater than 1, the cumulated number of faults increases faster than linearly; for $\beta = 1$, it increases linearly; and for $\beta < 1$, more slowly than linearly. For $b = 1$, the process would be a HPP, where the times between the failures is exponentially distribution and, therefore, the system reliability does not change.

For $\beta < 1$, faults become more and more seldom; therefore, we can establish an improvement; for $\beta > 1$, there is deterioration, i.e., there are more faults per time interval if the system becomes older.

So, the power law model or Weibull process or Duane model is as flexible as the Weibull distributions for nonrepairable systems. Moreover, from Equations 9.3 and 9.4, we see that the time to first fault follows a Weibull distribution.

The parameter β can be estimated from the data of the fault times t_i, $i = 1, \ldots, n$, by

$$\beta_s = \sum \ln(t_i/T), \tag{9.11}$$

where the sum Σ is over all i from 1 to n and T is the time the process has been observed. The t_i are the times of occurrence failures.

Now $2n\beta_s/\beta$ has a chi-square distribution with $2(n-1)$ degrees of freedom so that a confidence interval for β is given by *MIL HDBK 189C* (Department of Defense 2011):

$$\left[2n\beta_s/\chi^2_{2(n-1)}(\alpha/2); \ 2n\beta_s/\chi^2_{2(n-1)}(1-\alpha/2) \right]. \tag{9.12}$$

These values can also be used as critical values to test the hypothesis $\beta > 0$ (i.e., the number of failures is increasing which is not desirable) or $\beta < 0$ (the number of failures is decreasing).

9.4 Example

In our example, we have simulated data following a Weibull process to demonstrate the method. That means, it is assumed that the system has been monitored and that all faults have been registered.

Figure 9.1 shows the cumulated number of failures and the model fitted to these data. In the example, we have computed the following for $n = 158$ events:

- $\beta_s = 0.496$
- $\Theta_s = 0.00319$,

where the index s denotes that the values are computed from data and are therefore a statistical estimate.

For the shape parameter β with 90% coverage, the following is computed:

$$[0.4396, \ 0.5718].$$

Consequently, for a statistical test with the hypothesis H_0 "$\beta \geq 0$" with 95% coverage, the hypothesis would be rejected since the upper confidence bound is smaller than 1. So, there is reliability improvement for this process, i.e., the number of failures is decreasing over time in the statistical average. It is important to carry out a statistical test for this hypothesis and not only use the point estimator; the point estimator might, e.g., be smaller than 1, suggesting reliability improvement. If, however, the upper confidence bound is larger than 1, then the statistical test would not support the hypothesis on reliability improvement.

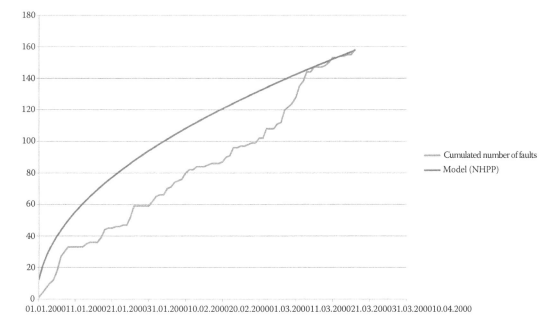

FIGURE 9.1
Cumulated number of faults and model.

Figure 9.2 shows the intensity of the process and the model.

The example shows how the model of the Weibull process can be applied.

How can the mean time between failures (MTBF) be predicted now? Assume that the tendency of decreasing intensity is continued since the process is stable regarding the technical side, i.e., finding faults and their causes and removing them.

By using the parameters calculated, we can now predict the intensity of the fault process after 1 year, i.e., 365 days (December 31, 2000). An MTBF in days is then just the inverse of the rate.

To be on the safe side, the pessimistic variant for the prediction of the MTBF should be used, for which the upper confidence bound for beta is used.

The values given in Table 9.1 assume that the process of fault detection and removal is carried out until December 31, 2000, i.e., for 365 days and would then be stopped. Note that without having a stable process of finding and removing faults, the statistical prediction is worthless.

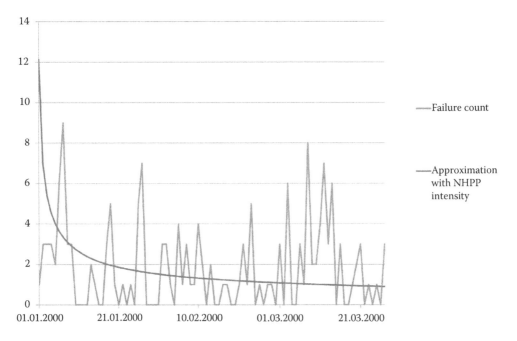

FIGURE 9.2
Number of failures and intensity of HNPP.

TABLE 9.1

Predicted Values

Variant	Rate in Events per Day	MTBF in Days
Using lower confidence band for beta (optimistic variant)	0.227	4.403
Using point estimate	0.437	2.291
Using upper confidence band for beta (pessimistic)	1.059	0.944

9.5 Conclusions

We have provided methods to investigate trends in reliability. The methods presented here can be applied for any system, where reliability growth plays a role. This can be transport systems, plants, communications systems, and, of course, railway systems, since the methods are general mathematical methods. A main precondition is that not only a statistics is elaborated, but also a stable technical process of fault detection and removal is in place. Also, for the prediction of failure intensities of MTBF values, the upper confidence limit of the shape parameter beta should be used.

References

Ascher, H., and Feingold, M. (1984) *Repairable Systems Reliability: Modelling, Inference, Misconceptions and Their Causes*. Marcel Dekker, New York.

Bain, L. J. (1978) *Statistical Analysis of Reliability and Life Testing Models*. Marcel Dekker, New York.

Bain, L. J., and Engelhardt, M. (1980) Inferences on the parameters and current system reliability for a time truncated Weibull process. *Technometrics* 22: 421–426.

Barlow R., and Proschan F. (1975) *Statistical Theory of Reliability*. Holt, Rinehart and Winston, New York.

Crow, L. H. (2011) Reliability growth planning, analysis and management. *Reliability and Maintainability Symposium*, January 2011.

Department of Defense (2011) MIL-HDBK-189C: *Handbook Reliability Growth Management*. June 14, 2011. Department of Defense, Arlington, VA.

Engelhardt, M., and Bain, L. J. (1986) On the mean time between failures for repairable systems. *IEEE Transactions on Reliability* 35: 419–423.

European Committee for Electrotechnical Standardization (2014) EN 61014, *Programmes for Reliability Growth*, European Committee for Electrotechnical Standardization, Brussels.

Finkelstein, J. M. (1976) Confidence bounds on the parameters of the Weibull process. *Technometrics* 18: 15–117.

Härtler, G., and Schäbe, H. (1991) Algorithms for parameter estimation of nonhomogeneous Poisson processes. *Computational Statistics Quarterly* 2: 163–179.

International Electrotechnical Commission (1994) IEC 60706-6: Part 6—Section 9: Statistical Methods in Maintainability Evaluation.

Lawless, F. (1987) Regression method for Poisson process data. *Journal of the American Statistical Association* 82: 808–815.

Schäbe, H. (2017) Trapped with availability, *ESREL 2017: Safety and Reliability—Theory and Applications*. Cepin, M., and Briš, R. (Eds). Taylor & Francis, London, pp. 575–580.

10

Methods for RAM Demonstration in Railway Projects

Pierre Dersin and Cristian Maiorano

CONTENTS

10.1 Introduction to RAM Demonstration in Railway Projects

Railway systems design activities are carried out to deliver a system which complies with the applicable requirements. In this framework, reliability, maintainability, and availability (RAM) analyses are carried out to influence design choices and to increase the confidence that the system will perform as predicted. Despite these efforts to identify and quantify the credible failure modes and their potential effects on the system and mission, the in-field experience has to verify the completeness and accuracy of design phase studies up to system acceptance.

This chapter is structured on different levels to overall present the methods used for successful RAM demonstration in the field. Primarily, the principles of data collection (failure reporting, analysis, and corrective action system [FRACAS]) and processing are presented. The parameters of interest relating to infrastructure and trains, data collection means, failure data, and root cause assessment are investigated.

Later, the methods used to demonstrate compliance with the RAM requirements and the underlying theory are illustrated.

Main reliability demonstration techniques are explained. Methods for availability demonstration tests are discussed and illustrated, and practical examples are proposed. The maintainability demonstration topic is addressed as well, including sample sizing and selection criteria.

10.2 Data Collection Principles

An effective data collection process, often referred to as RAM monitoring, is the necessary condition to conduct the RAM demonstration activities described in this chapter. Despite the apparent simplicity of this task, it may quickly become very hard to put it in place when multiple stakeholders have to contribute. This is what often happens in a complex railway project where many actors work together.

However, it is not the aim of this chapter to illustrate the whole data collection process and the share of responsibilities among the various contributors. Rather, we want to illustrate what the main steps are, from the data collection framework definition to the data validation, prior to commencement of data processing and analysis.

10.2.1 FRACAS

FRACAS is essential to the achievement of RAM performance of railway systems, equipment, and associated software. This process aims to report both software and hardware failures. Collected data are then used to identify, implement, and verify corrective actions to prevent or reduce the probability of further recurrence of these failures.

Corrective action options and flexibility are greatest during design, when even major changes can be considered to eradicate or significantly reduce vulnerability to known failure causes. These options and flexibility become more limited and expensive as the product or project moves along its life cycle. Nonetheless, time and budget constraints are variables to be considered and weighted against the benefits of a corrective action, in order to substantiate decisions.

A comprehensive FRACAS process should be instantiated to encompass the following steps:

- Event reporting: It includes an exhaustive description of event symptoms and effects, failure data and time, failure location, and all the data needed to allow investigation and traceability. The affected item (system, subsystem, module, and component) has to be clearly identified. Item part number is habitually used for this purpose. Serial number may be needed as well in order to keep the history of the events, identify recurrent failures, and make the link with repair data (in case of repairable items).

- Problem investigation: Each failure event is the subject of an investigation at both software and hardware levels, aimed at identifying failure causes and mechanisms. Immediate failure cause is studied in the first place; it is based on the failure diagnosis provided at repair level and the intervention information provided by the intervention team. However, it must be understood that the immediate failure cause is not the root cause. It helps with determining the root cause which can be identified by an in-depth analysis and investigation.

- Corrective action identification: Once the failure cause is known, an action plan should be defined to eradicate or reduce the failure mode probability. Corrective action could also aim at mitigating the effect of such a failure mode. This could happen when the corrective action is applied to products/systems already in operation. At this stage, it is sometimes impossible or too expensive to modify the design. A palliative measure could be introduced instead to make the failure effects acceptable. Once the failure analysis and the resulting corrective actions are put in place, the effectiveness of the corrective action should be demonstrated.

- Problem closure: Once the corrective action is in place, its effectiveness is validated after a given observation period.

A FRACAS process has to be initiated in order to implement (at least) the following:

- Data collection: The data framework is to be defined according to the scope.
- Workflow management: This is done to ensure that all the steps are correctly implemented and to ensure data traceability.

This can be achieved either by the use of commercial tools or in-house developed tools or by a structured data validation process.

10.2.2 Parameters of Interest

The choice of the parameters of interest, when performing failure data collection, depends on many factors. On the one hand, the information required has to be kept at a minimum, as far as reasonable, in order to avoid the opposite result: bad quality and missing data. On the other hand, some details are necessary to allow failure investigation and correct data processing. The following questions, for instance, should be answered to decide whether a parameter is necessary or not:

- Are all the items under the same operational conditions?
- Do we need to characterize the observed item depending on the application conditions?

- Does the failure responsibility affect the failure classification?
- Is a failure relevant only when it is demonstrated to be caused by a hardware fault?

The same parameter, for instance, could be considered not relevant when the object monitored is a single product undergoing a factory test, but could instead become crucial when monitoring a complex railway system which enters revenue service, with multiple products working each in different application conditions.

Consider monitoring the trackside signaling equipment (i.e., point machines, track circuits, etc.) of a heavy metro system during operations. Environment may not be relevant when the scope is to demonstrate compliance with system reliability requirements, which have been taken into account to define the system design in those application conditions. The latter become suddenly vital if it is desired to demonstrate the achievement of reliability figures of point machines installed in tunnel. The point machines installed at open-air line section will, on average, exhibit poorer reliability performance due to the effects of the sun exposure and consequent higher working temperature. Incomplete data may then lead to inaccurate conclusions.

Parameters of interest can be split in three main categories:

1. Operational data: These are the data independent of the failure events observed. They rather define the application conditions of the sample items.
2. Event data: These represent the data which are actually collected when a failure event occurs.
3. Repair data: The information which enables investigation about the nature of the failure (failure classification) and the failure root cause.
    - The parameters of interest belonging to each set of data listed earlier are to be defined according to the needs.
    - Operational data may include those shown in Table 10.1.

Event data for atypical railway project may include those shown in Table 10.2.

Finally, repair data which help identify the immediate cause and the root cause of a failure may include those shown in Table 10.3.

TABLE 10.1

Parameters of Interest—Operational Data Example

Parameter	Description	Scope
Working ratio	The time ratio during which the item under test is powered on. This is necessary to define the working hours of the sample.	Reliability indicators
Temperature control	Is the item installed in air-conditioned environment? Is there any forced air ventilation system?	Traceability and investigation
Weather protection	Is the item exposed to sunbeams or other weather agents?	Traceability and investigation
Humidity	What is the humidity percentage?	Traceability and investigation
Mechanical stress	What is the intensity of mechanical stress level the item under test is exposed to?	Traceability and investigation

TABLE 10.2

Parameters of Interest—Event Data Example

Parameter	Description	Scope
Event date	Date of the event	RAM indicators
Project phase	Phase of the project during which the event occurred	Investigation and traceability
Event description	Circumstances and consequences of the detected anomaly	Investigation and traceability
Event location	Station ID (sector) or train set ID	Investigation and traceability
Symptoms	What the user sees: message from interface, symptoms, etc.	Investigation and traceability
Delay at event	Delay due to the event; may be linked with a contractual definition (at terminus, first impacted train, and so on).	RAM indicators
Emergency brake	If emergency brake is applied	Investigation and traceability

TABLE 10.3

Parameters of Interest—Repair Data Example

Parameter	Description	Scope
Repair details	Any particular comment on repair activity in manufacturing unit	Investigation and traceability
Repair diagnosis	Defines the failure cause for classification of the failure	RAM indicators
Repair action	Describes particular actions for corrective maintenance in manufacturing unit	Investigation and traceability
To be repaired	Informs whether repair is scheduled or not	Investigation and traceability
Repair unit	Manufacturing unit responsible for repairs of the failed item	Investigation and traceability

10.2.3 Failure Review Board

The roles of the failure review board (FRB) are the following:

1. Review failure sentencing
2. Arbitrate any dispute over failure responsibility
3. Supervise the RAM performance monitoring process
4. Make decisions relative to in-service reliability and maintainability demonstration

FRB may take place at different levels and not only between the final client and the main supplier. Subcontractors may attend FRB meetings on occasions, when assessment of failures involving their respective scopes of supply warrants it. Additional FRB may be set up (recommended) between main supplier and subcontractors.

The FRB is generally cochaired by the client's and the supplier's representatives and meets up regularly to review failure data. The FRB must be prepared before the actual meeting. Indeed, both parties must come with a clear idea about the chargeability of the failure and defend it by providing the needed information. The FRB will decide on the priority of RAM improvement actions according to the trends evidenced by failure analysis reports. It will also assess the effectiveness of already undertaken improvement actions.

Figure 10.1 provides an example of the FRB workflow.

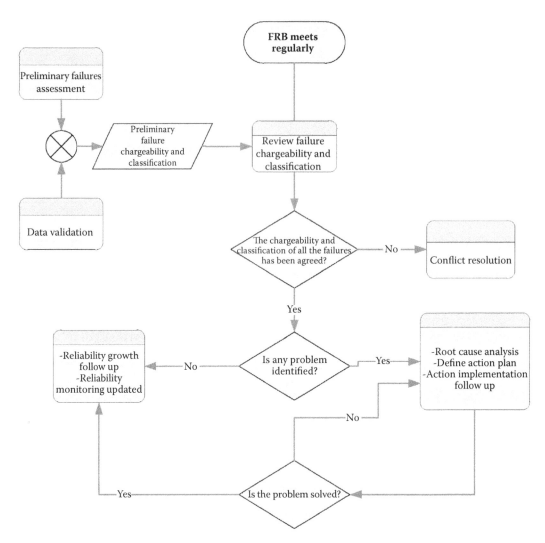

FIGURE 10.1
FRB workflow.

10.2.4 Failure Data and Root Cause Analysis

Failure data analysis aims not only at defining the nature of the failure and its chargeability (especially when dealing with contractual obligations) but also at identifying the root cause of a failure and, based on this, at conducting the appropriate corrective or mitigating actions. As already mentioned, this process mainly consists of three steps: collection, analysis, and solution.

This section focuses on solution, which assumes that a proper data collection process is already in place, and therefore, all the necessary information to understand and analyze the nature of a problem are available.

Despite the fact that the root cause is what we are finally looking for, its assessment goes through two intermediate steps. The first one is the knowledge of the immediate cause or, in other words, the event which has, at the end, determined the failure. The information provided by the maintenance team which operates on the field, and the repair data provided by the manufacturer, are essential to determine the immediate cause. This will basically provide us with the last link of the root cause–effect chain. The second step is to analyze and isolate all the factors which may have contributed to the occurrence of the immediate cause. The first chain of the root cause chain is likely to be among those.

Once all the possible causes are identified, several techniques, such as the fishbone diagram, the way chart or fault tree, may apply to break down all the possible links between each cause and the final effect. One by one, the possible causes are evaluated and discarded, starting from the least likely to occur. This should finally bring to the identification of the root cause. Further tests aimed at reproducing the failure may be conducted to exclude root cause candidates.

The final step is to identify a possible solution to the problem. The aim is to break the chain going from the root cause to the failure.

10.3 Reliability Demonstration Tests

10.3.1 Introduction

In most railway projects, it is necessary to demonstrate the achievement of reliability targets either during development or in operations.

A sample of items is therefore tested, their failures are observed, and collected data are finally analyzed to estimate and demonstrate the reliability of the item under test.

Early deployment of such tests provides the greatest benefits at the minimum cost. Nonetheless, RAM demonstration is often required once a railway system is put in operations.

The aim of this section is to present the most common methods used for estimation and demonstration of a constant failure rate λ (in that case, mean time between fault [MTBF] $= 1/\lambda$, and it is equivalent to demonstrate an MTBF value). Two types of reliability demonstration test plans will be described: the classical, time-terminated test, where test duration is fixed in advance, and the sequential test plans. It is also assumed that as the units under test fail, they are replaced.

Other types of tests, such as "failure terminated" (i.e., the test ends when a predetermined number of failures has been recorded), or test units not replaced, are presented in known standards and handbooks such as IEC 61124 [1] and MIL-HDBK-781A [2].

The more general variable failure rate case is presented in Chapter 12.

10.3.2 Fixed Duration Tests

It is desired to statistically prove that the observed MTBF exceeds a specified target.

For that purpose, the "design MTBF," denoted MTBF_0, is considered.

It is assumed that the producer takes a margin, i.e., will have designed the system for $\text{MTBF}_0 > \text{MTBF}_1$, where MTBF_1 is the minimum acceptable MTBF value from the consumer's perspective.

The ratio $\text{MTBF}_0/\text{MTBF}_1$ (greater than 1) is called the discrimination ratio; the higher that ratio, the higher will be the margin and, as will be seen, the shorter the test period for given risk levels.

In what follows, we denote MTBF_1 by θ_1 and MTBF_0 by θ_0 for conciseness.

Hypothesis H_0 (null) is defined as $\text{MTBF} > \text{MTBF}_0$.
Hypothesis H_1 (alternative) is defined as $\text{MTBF} < \text{MTBF}_1$.

A procedure will now be given for accepting or rejecting the reliability of a product, i.e., testing hypothesis H_0 against hypothesis H_1.

10.3.2.1 Acceptance Rule

The rule outlined in the following is based on a fixed duration test, i.e., the duration of the test is predefined. It is also assumed that the failed units are replaced (other types of tests exist which make different assumptions).

The acceptance rule is constructed as follows: a number c is defined such that if the number of failures observed during the test period T exceeds c, then hypothesis H_0 is rejected, i.e., the test is failed, and if the number of failures observed is less than or equal to c, then hypothesis H_0 is accepted or the test is passed.

Thus, test rules are the following:

- *Passed* if $k \le c$
- *Failed* if $k > c$

where k is the number of failures observed during period T.

10.3.2.2 Type I and Type II Errors (Producer's and Consumer's Risks)

The maximum acceptable producer's risk (type I error) is denoted as α; and the maximum acceptable consumer's risk (type II error), β.

The consumer's risk is, by definition, the probability that the test is passed (no more than c failures occur), when in fact MTBF < $MTBF_1$. The producer's risk is, by definition, the probability that the test is failed (more than c failures occur), when, in fact, true MTBF is greater than $MTBF_0$.

A more detailed discussion of those two types of risk is included in Chapter 12.

10.3.2.3 Determination of Test Parameters

Given values for θ_0, θ_1, α, and β, it is desired to determine the minimal possible test length T and the failure threshold c.

The process is repeated until satisfactory values of α and β are found. It is not always possible to have α and β equal, which is generally the desired scenario to ensure that both consumer and producer share the same risk level.

The results are tabulated, as shown, for instance, in Table 10.4, which is excerpted from IEC [2]. The inputs to Table 10.4 are the following:

- The maximum acceptable consumer risk β
- The maximum acceptable producer risk α (in the examples of the table, the two are taken equal)
- The specified and design MTBF values, $MTBF_1$, and $MTBF_0$

Outputs from the table are the following:

- The test rejection threshold c
- The ratio of the minimal test duration T to $MTBF_0$

TABLE 10.4

Test Duration T and Rejection Failure Threshold c as Functions of Risk and Discrimination Ratio D

Risk $\alpha = \beta$	$D = \dfrac{\text{MTBF}_0}{\text{MTBF}_1} = 1.5$	$D = \dfrac{\text{MTBF}_0}{\text{MTBF}_1} = 2$	$D = \dfrac{\text{MTBF}_0}{\text{MTBF}_1} = 3$
0.1	$C = 39$	$C = 13$	$C = 5$
	$\dfrac{T}{\text{MTBF}_0} = 32.14$	$\dfrac{T}{\text{MTBF}_0} = 9.47$	$\dfrac{T}{\text{MTBF}_0} = 3.1$
0.2	$C = 17$	$C = 5$	$C = 2$
	$\dfrac{T}{\text{MTBF}_0} = 14.3$	$\dfrac{T}{\text{MTBF}_0} = 3.93$	$\dfrac{T}{\text{MTBF}_0} = 1.47$
0.3	$C = 6$	$C = 2$	$C = 1$
	$\dfrac{T}{\text{MTBF}_0} = 5.41$	$\dfrac{T}{\text{MTBF}_0} = 1.85$	$\dfrac{T}{\text{MTBF}_0} = 0.92$

Source: IEC, IEC 61124. *Reliability Testing—Compliance Tests for Constant Failure Rate and Constant Failure Intensity*, 2nd edition, IEC, Geneva, 2006. © 2006 IEC, Geneva, Switerland.

The iterative procedure provided earlier does not always permit reaching the exact values of the maximum allowable risks that have been specified. Then, the true risks achieved by the test can be obtained.

In Table 10.5 (excerpted from IEC [2]), the true values of risks corresponding to a number of time-terminated test plans, next to the nominal, or target risks, are shown.

TABLE 10.5

Nominal Risks and True Risks

Test Plan No.	Characteristics of the Plan			Test Time for Termination	Acceptable Numbers of Failure	True Risk for	
	Nominal Risk		Discrimination Ratio			MTBF = θ_0	MTBF = θ_1
	α (%)	β (%)	D	T/θ_0	c	α' (%)	β' (%)
B.1	5	5	1.5	54.1	66	4.96	4.84
B.2	5	5	2	15.71	22	4.97	4.99
B.3	5	5	3	4.76	8	5.35	5.4
B.4	5	5	5	1.88	4	4.25	4.29
B.5	10	10	1.5	32.14	39	10	10.2
B.6	10	10	2	9.47	13	10	10.07
B.7	10	10	3	3.1	5	9.4	9.9
B.8	10	10	5	1.08	2	9.96	9.48
B.9	20	20	1.5	14.3	17	19.49	19.94
B.10	20	20	2	3.93	5	20.4	20.44
B.11	20	20	3	1.47	2	18.37	18.4
B.12	30	30	1.5	5.41	6	29.99	29.95
B.13	30	30	2	1.85	2	28.28	28.54

Source: IEC, IEC 61124. *Reliability Testing—Compliance Tests for Constant Failure Rate and Constant Failure Intensity*, 2nd edition, IEC, Geneva, 2006. © 1997 IEC, Geneva, Switzerland.

10.3.3 Sequential Testing

The motivation behind sequential test plans (first introduced by Wald [3]) is to shorten the duration of a test.

For instance, if in a fixed duration test, the hypothesis is rejected when $k > c$ (k is the number of failures observed), and $c + 1$ failures are observed much before the end of the theoretical duration T of the test, it seems that the decision can be achieved earlier. Thus, sequential tests aim at making decisions earlier. Decision rules for a sequential test can be analytically derived and finally plotted on a failure–time coordinate system in the form of two parallel lines:

$$\text{the accept line: } y_1(x) = ax + b \tag{10.1}$$

$$\text{and the reject line: } y_2(x) = ax + c. \tag{10.2}$$

As soon as during the test one of the lines is crossed, the test is terminated and a decision is made accordingly. Even though on average, the sequential test is of shorter duration, the duration of this test cannot be predicted in advance: the r value could remain indefinitely between the Accept and reject lines, so that no decision could ever be made. In order to remedy this situation, a truncated sequential test is usually adopted. The method for truncating a sequential test was developed by Epstein and Sobel [4].

10.3.3.1 Sequential Test Design

Let us consider the null hypothesis H_0: MTBF $> \theta_0$ and the alternative hypothesis H_1: MTBF $< \theta_1$. The truncated sequential test plan may be analytically generated for any given θ_0, θ_1, α, and β.

We want to determine the discrimination ratio, the accept–reject criteria, the truncation points, and the slope and ordinate intercepts of the test plan curves. Finally, we want to plot the test plan. The solution proceeds as follows:

1. The discrimination ratio is determined as

$$D = \frac{\text{MTBF}_0}{\text{MTBF}_1}. \tag{10.3}$$

2. Two constants A and B are calculated as they allow finding the reject and acceptance lines:

$$A = \frac{(1-\beta)(D+1)}{2\alpha D}, \tag{10. 4}$$

$$B = \frac{\beta}{1-\alpha}. \tag{10.5}$$

3. Determine the slope and ordinate intercepts of the reject and accept lines:

$$a = \frac{\left(\dfrac{1}{\theta_1} - \dfrac{1}{\theta_0}\right) - 1}{\ln\left(\dfrac{\theta_0}{\theta_1}\right)}, \tag{10.6}$$

$$b = \frac{\ln(B)}{\ln\left(\dfrac{\theta_0}{\theta_1}\right)}, \tag{10.7}$$

$$b = \frac{\ln(A)}{\ln\left(\dfrac{\theta_0}{\theta_1}\right)}. \tag{10.8}$$

4. Compute the points of truncation as follows: Search the chi-square tables until a point is reached at which

$$\frac{\chi^2_{(1-\alpha;2r)}}{\chi^2_{(\beta;2r)}} \geq \frac{\theta_1}{\theta_0}. \tag{10.9}$$

When this point is found, the corresponding number of degrees of freedom is set equal to $2k_0$. The value of k_0 is always rounded to the next higher integer. From this, the maximum time T_0 can be found as follows:

$$T_0 = \frac{\theta_0 \chi^2_{(1-\alpha;2r_0)}}{2}. \tag{10.10}$$

At this point, the test plan can be plotted on a failure–time coordinate system (Figure 10.2). The test should not last longer than k_0 failures or T_0 hours. In this case,

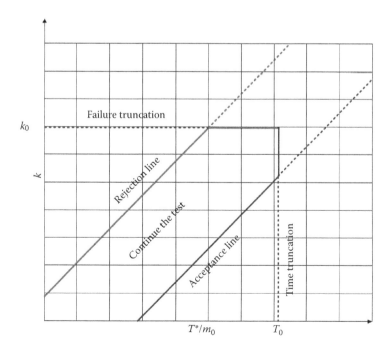

FIGURE 10.2
Truncated sequential test plan. (Courtesy of International Electrotechnical Commission, Geneva, Switzerland, 2004.)

however, the supplier and client risks are no longer those initially specified. In order to modify the risks only slightly, the truncation point should be far enough (the rule of thumb is three or four times the expected testing time when MTBF = $MTBF_0$ or $MTBF_1$).

10.3.4 Test Method Based on the Chi-Square Confidence Interval

Given the relationship between the Poisson and chi-square random variables, it is known that for a constant failure rate (with replacement of failed equipment), a confidence interval for the MTTF can be derived from failure observations of a sample over a total test time of *T*.

The lower bound of the confidence interval, for a confidence level of $1 - \delta$, is given by

$$\text{MTTF}_{\text{low}} = \frac{2T}{\chi^2_{(2n+1,1-\delta/2)}}, \tag{10.11}$$

where *n* is the number of failures observed during the operating time *T* and $\chi^2_{(2n+1,1-\delta/2)}$ is the $1 - \delta/2$ quantile of the chi-square random variable with $(2n + 2)$ degrees of freedom, i.e., the probability that that random variable takes a value less than the quantile is equal to $1 - \delta/2$.

Some railway clients are not comfortable or not familiar with the notion of discrimination ratio. As a result, they propose a demonstration test based on the confidence interval.

With that method, the hypothesis H_0: MTTF > $MTTF_{\text{target}}$ is accepted if and only if the confidence interval lies entirely above the target MTTF, denoted by $MTTF_{\text{target}}$, which is equivalent to the lower bound of the interval being greater than the target.

Thus, in this approach, the test is passed if and only if

$$\text{MTTF}_{\text{low}} > \text{MTTF}_{\text{target}}, \text{ with MTTF}_{\text{low}} \text{ given by Equation 10.11}.$$

Thus, given the confidence level $1 - \delta$ and the target $MTTF_{\text{target}}$, the corresponding test procedure is as follows:

> Over the operating time period *T*, count the number of failures *n*, and calculate $MTTF_{\text{low}}$ by Equation 10.11 as a function of δ and *n*. Then, accept the hypothesis that MTTF > $MTTF_{\text{target}}$ if and only if $MTTF_{\text{low}} > MTTF_{\text{target}}$.

At first glance, this method presents the following apparent advantages:

- There is no need to select a discrimination ratio.
- The notion of producer's risk and consumer's risk is not invoked, only the more familiar notion of confidence level.

However, even if the notion of risk does not explicitly appear, it, of course, exists: there is a risk that the producer fails the test despite the true MTTF exceeding the $MTTF_{\text{target}}$. The evaluation of that risk can be obtained from the operating characteristic curve (OC) [1].

It can be shown [5] that in reality, that method leads to asymmetric risks and most of the time to a producer risk much greater than the consumer risk.

The apparent simplicity of this method is therefore quite deceptive: in fact, it can be shown [5] that such a test is equivalent to that of IEC 61124 [1] but with a much higher target MTTF than the declared one. Therefore, this type of test, with implicit risk, can be very dangerous for the supplier and should be avoided.

10.4 Availability Estimation and Demonstration

When the designer of a complex system, such as is found in electric power networks and automated manufacturing or in the railways and aerospace industries, must make contractual commitments to an availability level; two needs naturally arise during the ascending branch of the V cycle: (a) estimating availability on the basis of operational data and (b) demonstrating to the client that the target has been reached; both despite a limited data sample.

Compared with the standard statistical tests for reliability and maintainability estimation and demonstration, existing methods for estimating or demonstrating the availability of a repairable system in continuous operation are still somewhat incomplete.

In the first part, a method based on the Fisher distribution is presented for availability demonstration, in the case of constant failure and repair rates, or its immediate generalization, in the case of Erlang-distributed times to failure and repair times. In the second part, the Chebyshev inequality is shown to lead to a confidence interval for availability estimation without any assumption on the failure and repair events except for their independence (failure and repair process is modeled as an alternating renewal process). These results have been excerpted from the study by Dersin and Birolini [6], where their proof can be found. They are also described in the book by Birolini [7].

10.4.1 Availability Demonstration Test (Constant Failure Rate or Erlangen) Distributions

10.4.1.1 Demonstration of Availability

In the context of acceptance testing, the demonstration of the asymptotic and steady-state point and average availability $PA = AA$ is often required, not merely its estimation. For practical applications, it is useful to work with the unavailability $\overline{PA} = 1 - PA$. The main concern is to test a zero hypothesis H_0: $\overline{PA} < \overline{PA_0}$ against an alternative hypothesis H_1: $\overline{PA} > \overline{PA_1}$ on the basis of the following agreement between client and supplier.

The item should be accepted with a probability nearly equal to (but not less than) $1 - \alpha$ if the true (unknown) unavailability $\overline{PA}$ is lower than $\overline{PA_0}$ but rejected with a probability nearly equal to (but not less than) $1 - \beta$ if $\overline{PA}$ is greater than $\overline{PA_1}$ $(\overline{PA_0}, \overline{PA_1} > \overline{PA_0}, \ 0 < \alpha < 1 - \beta < 1$ are given [fixed] values).

$\overline{PA_0}$ is the specified unavailability and $\overline{PA_1}$ is the maximum acceptable unavailability. α is the allowed supplier's risk (type I error), i.e., the probability of rejecting a true hypothesis H_0: $\overline{PA} < \overline{PA_0}$. B is the allowed client's risk (type II error), i.e., the probability of accepting the hypothesis H_0: $\overline{PA} < \overline{PA_0}$ when the alternative hypothesis H_1: $\overline{PA} > \overline{PA_1}$ is true. Verification of the agreement stated earlier is a problem of statistical hypothesis testing, and different approaches are possible.

Since constant failure and repair rates are assumed ($\lambda(x) = \lambda$, $\mu(x) = \mu$), the procedure is as follows:

1. For given (fixed) $\overline{PA_0}$, $\overline{PA_1}$, α, and β, find the smallest integer n (1, 2, ...) which satisfies

$$F_{2n,2n,1-\alpha} \cdot F_{2n,2n,1-\beta} = \frac{\overline{PA_1}}{\overline{PA_0}} \cdot \frac{PA_0}{PA_1} = \frac{(1 - PA_1)PA_0}{(1 - PA_0)PA_1}, \tag{10.12}$$

where $F_{2n, 2n, 1-\alpha}$ and $F_{2n, 2n, 1-\beta}$ are the $1 - \alpha$ and $1 - \beta$ quantiles of the F distribution and compute the limiting value

$$\delta = F_{2n,2n,1-\alpha} \cdot \overline{PA_0}/PA_0 = F_{2n,2n,1-\alpha} \cdot (1 - PA_0)/PA_0. \tag{10.13}$$

2. Observe n failure-free times $t_1 + \ldots + t_n$ and the corresponding repair times $t'_1 + \ldots + t'_n$ and

$$\text{Reject } H_0: \overline{PA} < \overline{PA_0} \text{ if } \delta \leq \frac{t'_1 + \ldots + t'_n}{t_1 + \ldots + t_n}$$

$$\text{Accept } H_0: \overline{PA} < \overline{PA_0} \text{ if } \delta > \frac{t'_1 + \ldots + t'_n}{t_1 + \ldots + t_n}. \tag{10.14}$$

Table 10.6 gives n and δ for some values of $\overline{PA_1}/\overline{PA_0}$ used in practical applications. It must be noted that the test duration is not fixed in advance which may represent a major problem in a real railway application.

10.4.1.2 Generalization to Erlangian Distributions

If failure-free and/or repair times are Erlang distributed, i.e., the failure-free times are sums of β_λ exponential times, and the repair times are sums of $\beta\mu$ exponential times, then $F_{2k,2k,1-\beta_2}$, and $F_{2k,2k,1-\beta_1}$, have to be replaced by $F_{2k\beta_\lambda,2k\beta_\mu,1-\beta_2}$ and $F_{2k\beta_\lambda,2k\beta_\mu,1-\beta_1}$, respectively (unchanged MTTF* and MTTR†).

10.4.2 Confidence Intervals for Availability

While the preceding methods are extremely useful, they are restricted to the assumption of exponential or Erlangian failure-free times and repair times.

It is worth recalling that a general expression exists for the cumulative distribution function of down time over a time interval t, under the only assumption that failure-free times are independent and identically distributed and so are repair times (as well as being independent of failure times), but without restriction on their probability distribution. This expression results from the assumption that the stochastic process of failures and repairs is an alternating renewal process. Then, the first two moments of DT/T are given by

$$E(DT/T) = \text{MTTR}/(\text{MTTF} + \text{MTTR}), \tag{10.15}$$

---

* Mean time to failure.
† Mean time to restore.

TABLE 10.6

Values for n and δ to Demonstrate Unavailability

	$\overline{PA_1}/\overline{PA_0} = 2$	$\overline{PA_1}/\overline{PA_0} = 4$	$\overline{PA_1}/\overline{PA_0} = 6$
$\alpha \approx \beta \lesssim 0.1$	$n = 28$	$n = 8$	$n = 5$
	$\delta = 1.41 \dfrac{\overline{PA_0}}{PA_0}$	$\delta = 1.93 \dfrac{\overline{PA_0}}{PA_0}$	$\delta = 2.32 \dfrac{\overline{PA_0}}{PA_0}$
$\alpha \approx \beta \lesssim 0.2$	$n = 13$	$n = 4$	$n = 3$
	$\delta = 1.39 \dfrac{\overline{PA_0}}{PA_0}$	$\delta = 1.86 \dfrac{\overline{PA_0}}{PA_0}$	$\delta = 2.06 \dfrac{\overline{PA_0}}{PA_0}$

$$\mathrm{Var}\left(DT/T\right) = (\mathrm{MTTR}_2 \cdot V_1 + \mathrm{MTTF}_2 \cdot V_2)/T(\mathrm{MTTF} + \mathrm{MTTR})^3, \tag{10.16}$$

where V_1 and V_2 denote, respectively, the variance of failure-free time and repair time.

One can then use the Chebyshev inequality, which states that for any random variable X and any positive number ε,

$$\Pr\left\{\left|X - E(X)\right| > \varepsilon\right\} < \mathrm{Var}(X)/\varepsilon^2. \tag{10.17}$$

When Equation 10.17 is applied to $X = DT/T$ (for T given), there follows, in view of Equation 10.16

$$\Pr\left\{\left|DT/T - \frac{\mathrm{MTTR}}{(\mathrm{MTTF} + \mathrm{MTTR})}\right| > \varepsilon\right\} < \frac{(\mathrm{MTTR}_2 \cdot V_1 + \mathrm{MTTF}_2 \cdot V_2)}{\varepsilon^2 T(\mathrm{MTTF} + \mathrm{MTTR})^3}. \tag{10.18}$$

From this inequality, a confidence interval can be obtained for the unavailability DT/T over an observation period T.

In the case of constant failure and repair rates (λ, μ), Equations 10.15 and 10.16 become

$$E\left(DT/T\right) = \frac{\lambda}{\lambda + \mu}, \; \mathrm{Var}\left(DT/T\right) = \frac{2\lambda\mu}{T \cdot (\lambda + \mu)^3}, \tag{10.19}$$

or, assuming $\lambda \ll \mu$,

$$E\left(DT/T\right) = \frac{\lambda}{\mu}, \; \mathrm{Var}\left(DT/T\right) = \frac{2\lambda}{T\mu^2}. \tag{10.20}$$

Then a confidence interval can be obtained from the following form of the Chebyshev inequality:

$$\Pr\left\{PA = \mathrm{AA} < 1 - \varepsilon - \hat{\lambda}/\hat{\mu}\right\} \le 2\hat{\lambda}/(T\hat{\mu}^2\varepsilon^2), \tag{10.21}$$

where the true values for λ and μ have been replaced by their estimates $\hat{\lambda}$ and $\hat{\mu}$. Given a confidence level, one determines ε from Equation 10.21. With a probability at least γ,

$$PA > 1 - \hat{\lambda}/\hat{\mu} - \left[2\hat{\lambda}/\left(T\hat{\mu}^2(1 - \gamma)\right)\right]^{1/2}. \tag{10.22}$$

Thus, Equation 10.22 yields a one-sided confidence interval with confidence level equal to γ.

10.5 Maintainability Demonstration Tests

The objective of a maintainability demonstration test is twofold:

- To demonstrate contractual target achievement, such as a maximum value for MTTR, MACMT* or ACMT-p[†]
- To verify qualitative aspects such as adequacy of test support documentation and ease of maintenance [8]

  The maintainability demonstration test considers both corrective and preventive maintenance tasks. Maintenance tasks in general include not only those performed by maintenance personnel but also those which may be performed by operations personnel, such as train drivers.

An important question is who should be conducting the maintainability tests.

The Union of Rail Equipment Manufacturers in Europe argues that they should be carried out by supplier personnel in order to not mix two factors: the intrinsic maintainability of the equipment and the effectiveness of training. However, attention should be paid to selecting well-trained personnel to perform the tests.

10.5.1 Lognormal Distribution Hypothesis

Typically, it is desired to test hypothesis H_0 against the alternative hypothesis H_1:

$$H_0: \text{MACMT} = \mu_0,$$
$$H_1: \text{MACMT} = \mu_1 > \mu_0. \tag{10.23}$$

There are a number of tests (such as those presented in MIL-HDBK-470 A [8] or IEC 60706-3 [9]) to test such a hypothesis. The one proposed here is designated as "Test method 1—Test A" in MIL-HDBK-470 A [8] and as "B.1—Test Method 1" in IEC 60706-3 [9]).

For this test, it is assumed that the probability distribution of active corrective maintenance times follows a lognormal law, and furthermore, an estimate σ^2 of the variance of the logarithm of maintenance times is assumed to be known.

NOTE: The lognormal law is such, by definition, that its logarithm follows a normal distribution; here σ^2 is the variance of that normal distribution.

10.5.2 Maintainability Test Method

Due to the difference in the quantity of equipment, the method to be used should not be necessarily the same for each railway application.

---

* Mean active corrective maintenance time.
[†] Mean active corrective maintenance time percentile.

It is possible to perform either random sampling, as there will be a statistically signifi-cant failure sample during the demonstration period or failures simulations.

When the system to be tested consists of few elements, it is necessary to simulate fail-ures. The same statistical methods then apply as for the random sampling case.

10.5.3 Test Design

Under those assumptions, the test procedure is as follows:

1. Determine the minimum sample size n:

$$n = \frac{(Z_\beta \mu_1 + Z_\alpha \mu_0)^2}{(\mu_1 - \mu_0)^2} (e^{\sigma^2} - 1), \tag{10.24}$$

where α and β are the producer's risk and the consumer's risk, respectively and Z_α (resp. β) is the standardized normal deviate exceeded with probability α (resp. β). In other words,

$$P\{Z > Z_\alpha\} = \alpha, \tag{10.25}$$

where Z is the normal random variable with mean 0 and variance 1.

In addition, the sample size n must be at least equal to 30 (this is to ensure that the central limit theorem, leading to a normal distribution, can be applied).

2. Make n observations of the ACMT* (TTR†), denoted by x_i ($i = 1, 2, \ldots, n$).
3. Calculate the sample mean:

$$\bar{x} = 1 / n \sum_{i=1}^{i=n} x_i. \tag{10.26}$$

4. Calculate the sample variance:

$$\langle d^2 \rangle = 1/(n-1) \left(\sum_{i=1}^{i=n} x_i^2 - n\bar{x}^2 \right). \tag{10.27}$$

5. The acceptance rule is as follows:
   - Accept H_0 if

$$\bar{x} \leq \mu_0 + Z_\alpha \frac{\langle d \rangle}{\sqrt{n}}; \tag{10.28}$$

   - otherwise, reject H_0.

Under the lognormal assumption, the variance d^2 of $\bar{x}$ can be written as

$$d^2 = \mu^2 \left(e^{\sigma^2} - 1 \right). \tag{10.29}$$

---

* Active corrective maintenance time.
† Time to repair.

TABLE 10.7

Number of Maintainability Tests for Each LRU

LRU	Predicted λ_i (h^{-1})	n_i	Contribution in Global λ	
Item 1	1.80×10^{-4}	355	35.59%	15
Item 2	1.85×10^{-4}	183	18.86%	8
Item 3	3.95×10^{-5}	538	11.84%	5
Item 4	3.63×10^{-5}	538	10.88%	5
Item 5	1.42×10^{-5}	858	6.79%	3
Item 6	8.60×10^{-6}	858	4.11%	2
Item 7	1.25×10^{-5}	538	3.75%	2
Item 8	2.05×10^{-5}	286	3.27%	1
Item 9	2.85×10^{-5}	178	2.83%	1
Item 10	3.50×10^{-6}	1076	2.10%	1

10.5.4 Sample Selection

Consider that a maintainability test is to be conducted to demonstrate that a given subsystem, composed of several LRUs,* achieves the contractual MACMT requirement. Let us consider the case where all the LRUs of the subsystem under consideration are characterized by a lognormal distributed ACMT with the same variance.

According to the method described in the previous section, we are able to determine the sample size (the number of tests to be conducted to conclude about the conformity of the subsystem to the MACMT requirement) and the acceptance rule of the test. We now come to the problem of how we sample the LRUs to be tested among all the LRUs composing the subsystem under analysis. One solution consists into breaking down the sample among the LRUs according to their respective failure rates. The items most influencing the subsystem MACMT are tested more times since, in theory, they would fail more often.

Let us consider a subsystem composed by 10 different LRUs and that the sample size defined according to the method mentioned earlier is 43. The LRUs of the subsystem are listed in decreasing order of contribution to the subsystem failure rate. The latter is used to determine how many of the 43 tests are to be conducted on each LRU. An example is provided in Table 10.7, where λ_i is the failure rate of the ith LRU of the subsystem and n_i is the quantity of the ith LRU in the subsystem.

10.5.5 Example

Estimate the acceptance conditions and the (OC for the demonstration of MACMT = MACMT$_0$ = μ_0 = 1.75 hours against MACMT = MACMT$_1$ = μ_1 2.25 hours with $\alpha = \beta = 0.2$ and $\sigma^2 = 0.5$.

Solution:

For $\alpha = \beta = 0.2$, it follows that $Z_\alpha = Z_\beta \approx 0.84$. From Equation 10.24, it follows that $n = 30$. The acceptance condition is then given by Equations 10.26, 10.28, and 10.29:

$$\frac{\sum_{i=1}^{i=30} x_i}{n} \leq 1.75 \times \left(1 + 0.84\sqrt{\frac{e^{0.5} - 1}{30}}\right) = 1.97 \text{ h.}$$

---

* Line replaceable unit.

10.6 Conclusions

Every RAM demonstration test includes a notion of risk, which must not be neglected if it is desired to avoid incurring unfair and costly contractual clauses. The naive method, which consists of comparing a point estimate, to the target does not allow for assessing those risks.

Demonstration tests have to be designed taking into account the time and cost constraints, by using the adequate methodology in order, on the one hand, to minimize the length of the testing period, therefore, the cost, and, on the other hand, to minimize the risk of a wrong decision.

The main concern when approaching the design of a RAM test is, "how long will it take?" Railway projects always follow tight program schedules, and demonstration periods are usually short (from a few months to a couple of years). Time issue is often blocking especially when the testing program has to be made compatible with a tight development or acceptance schedule. On the other hand, with highly reliable products, a long time is typically required to witness enough failures to achieve sufficient confidence that the reliability targets are met. This situation explains that more often than not, incorrect testing methods are proposed (such as the one described in Section 10.3.4). However, some alternative approaches may be adopted in order to reduce the testing time. Beside the use of sequential test instead of failure/time terminated test, which helps reduce the testing time considerably, a different testing method may be preferred: accelerated life test (ALT). In typical reliability tests, life data from a sampling of units operating under normal conditions are used. If one is interested in obtaining reliability results more quickly than is possible with data obtained under normal operational conditions encountered in the field, quantitative accelerated life tests can be put in place.

ALTs permit to increase the damage factor (function of different stresses applied to the product under test) of a product and, thus, to decrease its time to failure.

Acknowledgment

The authors thank the International Electrotechnical Commission (IEC) for permission to reproduce information from its international standards. All such extracts are copyright of IEC, Geneva, Switzerland. All rights reserved. Further information on IEC is available on www.iec.ch. IEC has no responsibility for the placement and context in which the extracts and contents are reproduced by the author nor is IEC in any way responsible for other content or accuracy therein.

References

1. IEC (International Electrotechnical Commission), IEC 61124: *Reliability Testing—Compliance Tests for Constant Failure Rate and constant Failure Intensity*, 2nd Edition, IEC, Geneva, 2006.
2. US Department of Defense, MIL-HDBK-781A: *Handbook for Reliability Test Methods, Plans and Environments for Engineering, Development Qualification, and Production*, US Department of Defense, Arlington, VA, 1996.

3. Wald, A., *Sequential Analysis*, Chapter 1, pp. 5–20; Chapter 2, pp. 2–32, Dover, Mineola, NY, 1943; reprinted, 1973.
4. Epstein, B., and Sobel, M., 1953, Life testing, *J. Amer. Stat. Association*, Vol. 48, pp. 486–502, 1953.
5. Dersin, P., Implicit risks associated with the decision rule in some reliability demonstration tests, *Proceedings of the Lambda-Mu 19 Symposium*, 2014.
6. Dersin, P., and Birolini, A., Statistical estimation and demonstration of complex system availability, *Proceedings of the Lambda-Mu 15 Symposium*, 2006.
7. Birolini, A., *Reliability Engineering—Theory and Practice*, Chapter 7, pp. 323–327, 8th Edition, Springer, New York, 2017.
8. US Department of Defense, MIL-HDBK-470 A: *Designing and Developing maintainable Products and Systems*, US Department of Defense, Arlington, VA, 1997.
9. IEC, IEC 60706-3: *Maintainability of Equipment—Part 3: Verification and Collection, Analysis and Presentation of Data*, 2006.

11

Guide for Preparing Comprehensive and Complete Case for Safety for Complex Railway Products and Projects

Ruben Zocco

CONTENTS

11.1 Safety Case: Definition and Purpose

Before discussing how to prepare a safety case, one has to agree on what the safety case actually is. The EN 50129 standard [1] defines it as "the documented demonstration that the product complies with the specified safety requirements."* This definition traces back to the original purpose of EN 50129 in addressing the acceptance and approval requirements of safety-related electronic systems in the railway signaling field. In the course of the years, the application of safety cases has been extended to a variety of other systems, such as traction power supply, civil infrastructure, and up to the whole railway system. One of the most reasonable alternative definitions to the CENELEC standards has been formulated by the British Ministry of Defence [5]. They define a safety case as "a structured argument, supported by a body of evidence that provides a compelling, comprehensible and valid case that a system is safe for a given application in a given environment." This definition identifies all the key characteristics that a safety case shall display to be

---

* The CENELEC has rolled out a 2015 draft update of the EN 50126 part 1 [2] and part 2 [3]; however, at the time of writing, these documents have yet to become European Standards. While this chapter refers to the currently valid EN 50126 [4], it shall be noted that the new drafts do not introduce significant changes to the approach identified by the EN 50129 [1] with regards to the safety case. On the contrary, the new EN 50126 part 1 dedicates a new section to the safety case that is in line with that defined in the current valid EN 50129.

successful in demonstrating safety. It also emphasizes that the clarity of the information provided is as important as the factual evidence supporting safety. Although the British Ministry of Defence does not explicitly mention it in their definition, absolute safety is unachievable. Safety cases exist to convince someone that the system is safe "enough" based on the acceptance criteria defined by the local regulatory authorities.

Further to the preceding definition and in consideration of the current industry development, three key properties can be identified:

- *Structured*—The safety case exists to communicate an argument, based on factual evidence. Given the complexity of the task, the safety argumentation shall be structured in such a clear and convincing way that someone can reasonably conclude that a system is safe from the body of evidence available.
- *Valid*—The system for which safety is argued has to be accurately represented in its design intent and its physical, operational, and managerial aspects. Safety cannot be argued in a context-free fashion: the safety case applies to a specific system configuration and shall be maintained as the system evolves. Additionally, the information presented in the case shall be consistent and correct.
- *Incremental*—The system development moves along a life cycle and so does the safety management. The safety case should not be considered only a summary of the safety activities performed across the life cycle, to be released before the system handover; on the contrary, it should be seen as integral part of the system life cycle and a supporting document for the achievement of key milestones.

The following sections discuss the preceding properties and elaborate on safety case typical failures in displaying them.

11.2 Structured

Although in some instances the safety case is referred to as a logical concept, in most of the railway domain applications the safety case is considered a physical artifact. Almost all its definitions, including the more abstract ones, can be associated with the documentation that presents the safety case. The EN 50129 standard goes in this direction by defining that the safety case is a report and what its contents shall be, starting from the structure defined by Figure 11.1.

By doing this, the CENELEC implicitly states that the safety case nature is to refer to, and pull together, many other pieces of information; thus, the importance of presenting the case information in a structured manner. However, this proves to be challenging when we consider that the amount of "pieces" the safety case has to pull together increases with the complexity of the system and that increased complexity means increased difficulty in conveying a clear and convincing argument to the stakeholders. Hence, the importance of developing structured safety arguments.

11.2.1 Development of Structured Safety Arguments

EN 50129 very clearly defines what the objectives of a safety case are, what supporting evidence shall be produced, and where the latter shall be placed within the document. One may

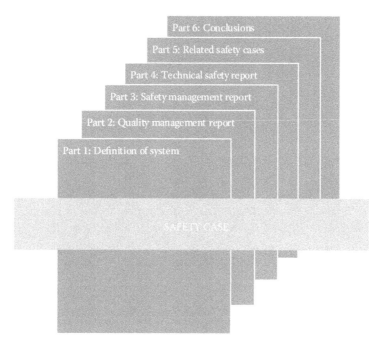

FIGURE 11.1
EN 50129 safety case structure.

think that a safety case that has scrupulously adhered to the standard requirements will automatically reach its main objective as defined in Section 11.1. However, based on the author's personal experience as safety assessor, this often is not the case because the safety case may still fail in delivering a convincing safety argument. The most common issue is that a huge amount of supporting evidence is presented (e.g., failure mode and effect analyses, functional safety analyses), but little attention is paid in explaining how these evidences link to the objectives; i.e., how one can reasonably conclude that the objectives are achieved based on the available evidence. This link, often neglected, is the "safety argument," as shown in Figure 11.2.

It shall be noted that both evidence and arguments are essential in establishing a convincing case: evidence without arguments leads to an unclear, often overwhelmingly complicated case, while arguments without evidence lead to a baseless case and thus to an unfounded safety claim.

In the typical safety case, arguments are expressed as free text, and the only structure available is given by the document table of contents. For example, in an EN 50129 compliant safety case, the physical independence of the components involved in a given safety function would be dealt within the Technical Safety Report part, section 3, "Effect of Faults." Figure 11.3 shows a possible safety argument related to physical independence.

While this approach can be considered a well-structured way of expressing a single argument, two common problems affect it. Firstly, the argument is only as good as the text that describes it: the intelligibility of the arguments and, consequently, of the overall case can be undermined by poor writing skills despite excellent engineering practices. Secondly, the table of contents alone does not provide a structure flexible enough to efficiently depict a multiobjective, structured safety claim where hundreds of arguments about the physical and functional properties of the system, subsystems, and components are tied up with, possibly, thousands of other documental references.

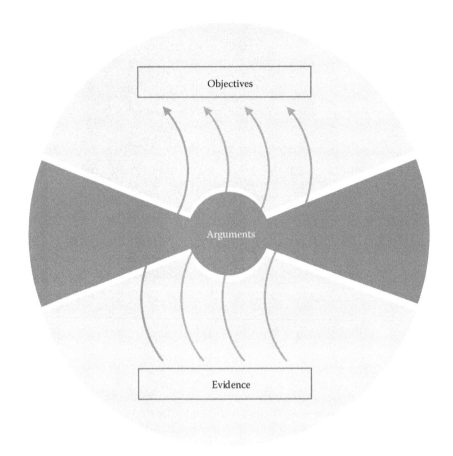

FIGURE 11.2
Safety argumentation.

The physical internal independence of the microswitch couples P1, P2 and P3, P4 is achieved by complete separation of the detection circuitry and redundancy of the position measurement rods, as it can be seen in the design documents [1] and [2]. Electromagnetic coupling is avoided bt using double-shielded control cables as required by the electromagnetic study (see section X); evidence can be found in [3].

FIGURE 11.3
Example of a textual argument.

As is often true in many domains, the graphical representation of complex concepts, in this case arguments, is found to ease and quicken the understanding of the argumentation strategy by the stakeholders. In addition, by graphically depicting the relationships between evidences and arguments, the information retrieval is strongly supported. This concept is illustrated by Figure 11.4.

The notation used in the preceding representation is the goal structuring notation (GSN) [6], which is a structured technique specifically developed to address the problems of

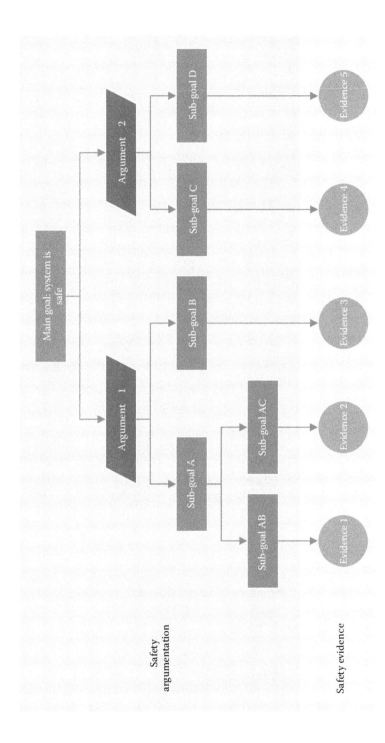

FIGURE 11.4
Graphical representation of safety.

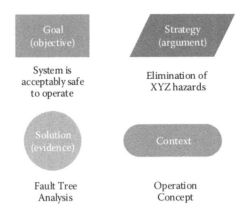

FIGURE 11.5
Key elements of the GSN.

clearly expressing and presenting safety arguments. Through the GSN, the individual elements of any case (objectives or goals, strategies or arguments, evidences or solutions, and context) are represented along with the relationships existing between them. Figure 11.5 illustrates the key elements used to build a GSN.

This notation supports the decomposition of goals (or claims about the system) into subgoals to a point where claims can be directly supported by referencing the available evidence (solutions).

Coming back to the textual argument example of Figure 11.3, one of its possible representations, GSN-like graphical argumentation, is given in Figure 11.6.

The original example has been inserted into a simplified wider argumentation for a point machine to show the benefits introduced by using a graphical notation and a goal-oriented decomposition.

As mentioned earlier, safety can be argued only within a specific context, i.e., different safety cases need to be produced for different installations of the same system. Considering that the amount of information to be presented in the case increases with the complexity of the system, a successful approach to the safety case development should also encompass the reuse of safety arguments.

11.2.2 Reuse of Safety Arguments

A key element in controlling the development costs is the support for modular certification, i.e., a strategy based on the reuse of arguments that have already been certified, expressed within well-defined modules. EN 50129 takes modularity into account when it defines three different types of safety case:

- The *generic product safety case*, which is focused on arguing safety as a property of a product or set of products (e.g. a point machine)
- The *generic application safety case* that provides evidence that a product or a collection of products is safe in a given class of application (e.g., an interlocking system for driverless operations)
- The *specific application safety case*, which argues the safety of a given configuration and physical installation of a product or collection of products

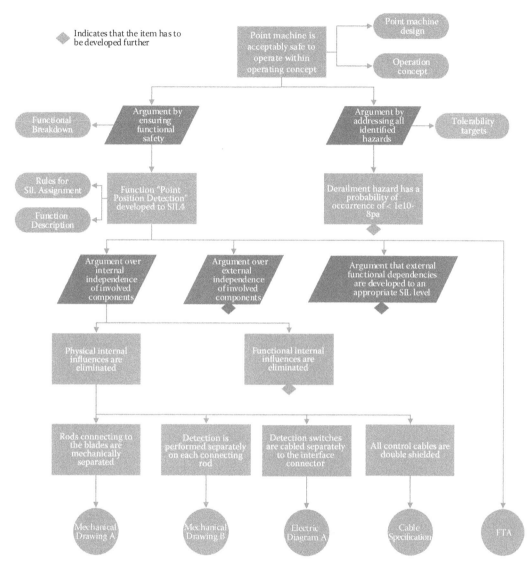

FIGURE 11.6
Example of GSN-like argumentation.

The underlying idea is that a generic product safety case will argue the intrinsic safety of a given product independently from its application, so that it can be deployed in a variety of different safety-related applications. The generic application safety case will present the case for an application by identifying and analyzing the generic properties that a product or a system (intended as set of products) should have. The specific application safety case will then contextualize a specific product into its target environment and argue the safety as a property of an integrated, operable, and maintainable system.

Theoretically, a generic application safety case does not need to identify the products to be used when arguing the safety of the application, as these can be specified through a well-defined set of properties (or requirements). In reality, the majority of the products are known in advance in most of the safety-critical applications due to the continuous

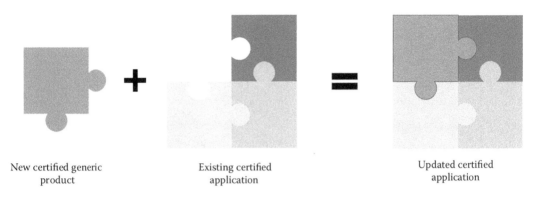

New certified generic   Existing certified   Updated certified
      product   application   application

FIGURE 11.7
Modular approach in certification.

platform development effort put in place by the manufacturers. In this context, the generic products are the "building blocks" of a generic application, as illustrated by Figure 11.7.

Since safety is dependent on the context in which a product is used, both generic product and generic application cases have to specify the limitations to the type of reuse of the product or the system is subjected to. These limitations are conditions that must be fulfilled in any safety-related application, and at any times, so that the safety properties are unaffected. These limitations are what the EN 50129 [1] calls "safety related application conditions."

11.2.3 Project or System-Wide Safety Case

The purpose of a system wide safety case is to demonstrate that the rail system as a whole is safe and to substantiate dependencies and claims made on it by the individual safety cases (e.g., systems interfaces, common services, and emergency arrangements). This concept is illustrated in Figure 11.8. The project safety case should show that the individual safety cases are comprehensive, consistent, and adequately integrated.

In addition, the system-wide safety case should cover project level aspects such as safety policy, safety management, safety culture, and organizational capability. These topics should also be addressed, as appropriate, in the individual safety cases, but the system-wide case should be able to demonstrate that the overall project has adequate organizational structure and resources, safety policy, and safety management arrangements. In this context, particular focus should be placed on the governance function of the project level safety organization, and the associated verification and validation activities performed.

The project-wide safety case should enable the reader to understand the significance of key services, major hazards, and significant safety issues for the system as a whole. The reader should be able to understand the main arguments substantiating safety, how hazards are properly controlled, why the top risks are acceptably low, and the improvements necessary in the interest of safety.

11.3 Valid

The safety case exists to argue the safety of a given system, either for a generic application or for a specific one, as seen in Section 11.2.2. Logically, one would expect that this system

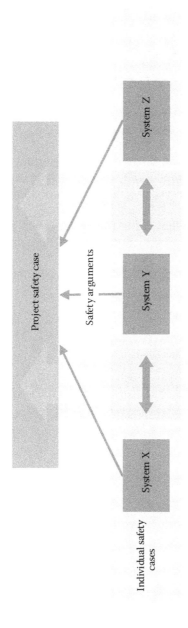

FIGURE 11.8
Relationship between individual and project-wide safety cases.

should be accurately defined by the safety case, and this is in fact what the EN 50129 requires when it dedicates one whole part of the safety case report to the description of the system. While the enumeration of the functionalities and properties of the system may add value to the overall description, it should also be pointed out here that the key elements defining the system are requirements, design documentation, and application documentation. Once again, the mere enumeration of the contractual document list is not enough to define a system. Requirements, design, and application documentation are interdependent items, and their consistency is vital in ensuring safety. Congruity and consistency of the system development are achieved through the application of configuration and change management practices. In particular, baselines and status of changes are the two key inputs that the safety case shall acknowledge. The former defines a consistent set of the three key elements at a given point in time (e.g., final design baseline for interlocking), and the latter defines if any modification is pending on the baseline (e.g., addition of turnout at location X) and what is its status. The relationship between the system definition and the configuration management is illustrated in Figure 11.9.

When baselines are not formally identified, the safety case shall explain in great detail how the consistency between requirements, design, and application documentation is ensured (e.g., via a design review process) and shall minutely detail the modification status of each element.

The validity of the safety case is not limited to the development life cycle of the system, but to its entire life span. Systems can be evolutionary as they are subject to perfective maintenance or technology obsolescence. Moreover, changes to the system might invalidate safety evidence or argument; the original claims might become unsupported because

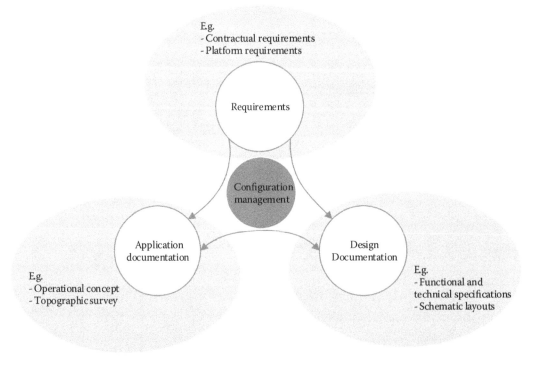

FIGURE 11.9
Key elements of the system definition.

they reflect old development artifacts or old assumptions that are no longer valid. Hence, it is almost inevitable that the safety case will require modifications throughout the lifetime of the system. However, safety case maintenance is often a painful exercise. Two common problems are the lack of traceability of safety case versus its system and inadequate documentation of the safety case dependencies. The adoption of a structured approach in the development of the safety case (e.g., the use of the graphical goal-oriented notation discussed in Section 11.2.1) helps in overcoming these issues by establishing a configuration model that can be used in evaluating the impact of a change.

A valid safety case is also a case that provides correct and complete information. Starting from its conception, the safety case should be prepared and updated by competent and experienced safety engineers. Although this may seem obvious, there are instances where the safety case is assigned to young professionals or is outsourced to consultants who may have enough theoretical knowledge but are likely to have little practical knowledge of designing, operating, and maintaining the system being considered by the case. One may argue that a review process will reduce the risk of development oversights, and indeed, it will provided that it is thorough. It shall be noted that since the safety case is a multidisciplinary document that draws on the outputs of several engineering processes, e.g., configuration, requirements, and quality, its review should also be multidisciplinary: the key engineering specialists involved in the various system development processes should be called to review the safety case qualities associated with their scope. Such a review may be performed in different fashions: via a safety case review panel, a specific interdisciplinary review process or as part of a project wide review process.

Further to a thorough documental review, another fundamental piece supporting the safety case validity is the safety verification. EN 50129 defines the verification as "the activity of determination, by analysis and test, at each phase of the life cycle, that the requirements of the phase under consideration meet the output of the previous phase and that the output of the phase under consideration fulfils its requirements." It can be inferred that the safety verification specifically concerns the safety life cycle activities, a fact that is further supported by the number of safety verification tasks mandated by the CENELEC standards. Because the safety case is built on the outputs of the safety life cycle, it is easy to appreciate the importance of verifying that for each of the life cycle phases, inputs, outputs, and processes in between are aligned. One of the characteristics of a successful safety case is to be incremental (discussed in Section 11.4), and this extends to the verification arguments. However, based on experience, a typical safety case failing is the lack thereof. This may be due to the safety verification activities being performed but not adequately documented or to the misconception that safety verification means verification of the system safety requirements. The lack of safety verification arguments undermines the credibility and robustness of the safety case and shall be regarded as threat to the safety of the system.

11.4 Incremental

It has been increasingly recognized by the industry that the safety case preparation is not an activity to be left at the end of the safety life cycle. In fact, safety standards and local regulatory authorities often require the incremental development and presentation of the safety case at each major Project milestone, until the system handover to the operation and maintenance authorities. Although variations exist on the safety life cycle requirements between different regulatory domains, the most common safety cases are the following:

- *Preliminary safety case*—after the completion of the requirements specification phase
- *Design safety case*—also called "intermediate safety case," after the completion of the design and initial validation activities
- *Final safety case*—also called "operation safety case," after the completion of the validation activities and before entering into service

The relationship between the traditional development life cycle and the production of these safety cases is shown in Figure 11.10.

Interestingly, one might question the value of preparing a preliminary safety case since the latter is developed prior to the detailed design specifications being available, and thus, any detailed safety analysis is not yet possible. However, the preliminary safety case advantage is to provide a skeleton for the safety argument that can be used to manage the evidence to be gathered at later stages and to enable early discussion with the regulatory authorities and other stakeholders. This may also list already existing constraints and assumptions relevant in situations such as an operational railway that is being upgraded.

Other than supporting the achievement of certain project stages, the safety case has to be forward looking, i.e., contain enough details to give confidence that the safety objectives will be achieved in the subsequent stages and identify any constraints imposed on them. This is of particular significance for the stages belonging to the right side of the development life cycle: the transitions from installation to static and dynamic testing and from dynamic testing to trail running may in fact hide insidious challenges. Firstly, the Health Safety Environment site management alone may not be enough to guarantee an adequate safety level during testing activities, and secondly, increased focus on delivery and compressed site activities schedules may shift the attention away from the project safety objectives. In the context of a traditional rail system development life cycle, the following safety cases may be produced:

- *Commissioning safety case*—Delivered after the completion of the site installation; it demonstrates that static and dynamic tests can safely commence.
- *Trial operation safety case*—Produced after the completion of dynamic testing; it argues that trial operation may safely commence.

Characteristics and objectives of the five staged releases of the safety case are shown in Table 11.1, which is an adaptation of the UK Office for Nuclear Regulation guidelines [7] to the rail industry.

FIGURE 11.10
Safety case life cycle integration.

TABLE 11.1

Principal Stages of a Rail System Development Life Cycle and Associated Safety Cases

Major Stage	Associated Safety Case	Key Objectives
Preliminary design	Preliminary safety case	• To make a statement of design philosophy, the consideration of design options, and a description of the resultant conceptual design sufficient to allow identification of main railway safety hazards, control measures, and protection systems • To provide a description of the process being adopted to demonstrate compliance with the legal duty to reduce risks to workers and the public • To provide an overview statement of the approach, scope, criteria, and output of the deterministic and probabilistic safety analyses • To define the application and environment conditions in which the system will be operated • To provide explicit references to standards and design codes used justification of their applicability and a broad demonstration that they have been met (or exceptions justified) • To provide information on the quality management arrangements for the design, including design controls, control of standards, verification and validation, and the interface between design and safety • To give a statement giving details of the safety case development process, including peer review arrangements, and how this gives assurance that risks are identified and managed • To provide information on the quality management system for the safety case production • To identify and explain any novel features[a] including their importance to safety • To identify and explain any deviations from modern international good practices • To provide, where appropriate, information about all the assessments completed or to be performed by the safety regulatory authority or other third parties • To identify outstanding information that remains to be developed and its significance

Final design	Design safety case	• To explain how the decisions regarding the achievement of safety functions ensure that the overall risk to workers and public will be reduced to an acceptable level
		• To provide sufficient information to substantiate the claims made in the preliminary safety case
		• To demonstrate that the detailed design proposal will meet the safety objectives before construction or installation commences and that sufficient analysis and engineering substantiation has been performed to prove that the plant will be safe
		• To provide detailed descriptions of system architectures, their safety functions, and reliability and availability requirements
		• To confirm and justify the design codes and standards that have been used and where they have been applied, noncompliances, and their justification
		• To identify the initial safety-related application conditions that shall be respected to maintain the desired safety integrity
		• To confirm
		  • Which aspects of the design and its supporting documentation are complete
		  • Which aspects are still under development and identification of outstanding confirmatory work that will be addressed

(Continued)

TABLE 11.1 (CONTINUED)

Principal Stages of a Rail System Development Life Cycle and Associated Safety Cases

Major Stage	Associated Safety Case	Key Objectives

Installation, testing, and commissioning	Commissioning safety case	• To demonstrate that the system, as manufactured and installed, meets relevant safety criteria and is capable of begin safe static and dynamic testing
		• To enable the production of a program of safety commissioning activities that will
		  • Demonstrate as far as practicable the safe functioning of all systems and equipment
		  • Prove as far as practicable all safety claims
		  • Confirm as far as practicable all safety assumptions
		  • Confirm as far as practicable the effectiveness of all safety-related procedures
		• To demonstrate that the commissioning activities can and will be carried out safely and that the operating procedures for commissioning are supported by the safety case
		• To demonstrate that there are no aspects of safety that remain to be demonstrated after completion of commissioning activities

Trial operation	Trial operation safety case	• To demonstrate that the system (as built and commissioned) meets the safety standards and criteria set down in the precommissioning safety case • To demonstrate that detailed analysis and testing have been undertaken to prove that the system operation will be safe • To capture any relevant design change that occurred during the commissioning activities and to analyze and justify its effect on the safety of the system • To identify the safety-related application conditions that shall be respected during trial operations to maintain the desired system safety integrity, inclusive of any temporary mitigations, and restrictions
Passenger operation	Final safety case	• To demonstrate that the system (as tested in operation-like conditions) meets the safety standards and criteria set down in the precommissioning safety case and it is ready to commence passenger operation • To identify the final safety-related application conditions that shall be respected during passenger operation to maintain the desired safety integrity • To demonstrate compliance with the legal requirements to reduce risks to workers and the public to an acceptable level

[a] A novel feature can be defined as any major system, structure, or component not previously licensed in a railway system anywhere in the world.

While it is clear that for each key life cycle stage, there exist a supporting safety case, it is less clear when it is the right time to release or update a particular stage safety case. In complex projects, where the scope of supply is decomposed in a number of work packages and each of them progresses at different paces, it might prove challenging to determine the best delivery timeline. A common misconception behind this is the belief that a safety case should provide a comprehensive and conclusive argument since its initial release. As mentioned in the beginning of this section, the safety case is incrementally developed; thus, it is perfectly normal to release a safety case before the evidence required is fully available, provided that any missing information or open issues are clearly identified. In some cases, the five stages identified may not be sufficient and subdivisions would be useful or beneficial. For example, a safety case for construction may need to be divided into civil construction and civil systems installation stages. Similarly, a safety case for commissioning may need to be divided into more substage-related to the different technologies, such as power supply and signaling.

Typically, the initial version of any given safety case is released to discuss the approach and the structure of the safety demonstration with the project stakeholders and to provide a safety justification and demonstration covering the most advanced work packages.

11.5 Safety Case Shortcomings and Traps

Despite the effort and expertise put into the development and review of the safety cases, there are still many things that can go wrong. The Nimrod Review [8], an independent review into the loss of the RAF Nimrod XV230 over Afghanistan in 2006, includes a comprehensive dissection of the problems and shortcomings of the safety case for the Nimrod aircraft. The more generic types of these issues are found to bear relevance to all domains where safety cases are used and have been in fact reproduced by the UK Office for Nuclear Regulation in their guidelines [7] and further reported in the following. This encompasses work by Dr. Tim Kelly of the University of York [9] and endorsed by Charles Haddon-Cave, the author of the Nimrod Review.

11.5.1 Shortcomings

1. Bureaucratic length:

   Safety cases and reports are too long, bureaucratic, and repetitive and comprise impenetrable detail and documentation. This is often for "invoice justification" and to give safety case reports a "thud factor."

2. Obscure language:

   Safety case language is obscure, inaccessible, and difficult to understand.

3. Wood-for-the-trees:

   Safety cases do not see the wood for the trees, giving equal attention and treatment to minor irrelevant hazards as to major catastrophic hazards and failing to highlight and concentrate on the principal hazards.

4. Archeology:

   Safety cases for "legacy" platform often comprise no more than elaborate archeological exercises of design and compliance documentation from decades past.

5. Routine outsourcing:

   Safety cases are routinely outsourced to outside consultants who have little practical knowledge of operating or maintaining the platform, who may never even have visited or examined the platform type in question and who churn out voluminous quantities of safety case paperwork in back offices for which the clients are charged large sums of money.

6. Lack of vital operator input:

   Safety cases lack any, or any sufficient, input from operators and maintainers who have the most knowledge and experience about the platform.

7. Disproportionate:

   Safety cases are drawn up at a cost which is simply out of proportion to the issues, risks, or modifications with which they are dealing.

8. Ignoring age issues:

   Safety cases for "legacy" systems are drawn up on an "as-designed" basis, ignoring the real safety, deterioration, maintenance, and other issues inherent in their age.

9. Compliance only:

   Safety cases are drawn up for compliance reasons only and tend to follow the same, repetitive, mechanical format which amounts to no more than a secretarial exercise (and, in some cases, have actually been prepared by secretaries in outside consultant firms). Such safety cases also tend to give the answer which the customer or designer wants, i.e., that the platform is safe.

10. Audits:

    Safety case audits tend to look at the process rather than the substance of safety cases.

11. Self-fulfilling prophecies:

    Safety cases argue that a platform is "safe" rather than examining why hazards might render a platform unsafe and tend to be no more than self-fulfilling prophecies.

12. Not living documents:

    Safety cases languish on shelves once drawn up and are in no real sense "living" documents or a tool for keeping abreast of hazards. This is particularly true of safety cases that are stored in places or databases which are not readily accessible to those on front line who might usefully benefit from access to them.

11.5.2 Traps

1. The "apologetic safety case":

   Safety cases which avoid uncomfortable truths about the safety and certifiability of systems in production so that developers do not have to face the (often economically and politically unacceptable) option of redesign ("X doesn't quite work as intended, but it's OK because . . . .").

2. The document-centric view:

   Safety cases which have as their aim to produce a document. The goal of safety cases should not simply be the production of a document; it should be to produce a compelling safety argument. There was a danger of "spending a lot of money to produce a document" of no safety benefit.

3. The approximation to the truth:

   Safety cases which ignore some of the rough edges that exist. For example, safety cases which claim in a goal-structured notation diagram that "all identified hazards have been acceptably mitigated" and direct the reader to the hazard log when, in reality, the mitigation argument is not so straightforward.

4. Prescriptive safety cases:

Safety cases which have become run of the mill or routine or simply comprise a parade of detail that may seem superficially compelling but fails to amount to a compelling safety argument.

5. Safety case shelfware:

Safety cases which are consigned to a shelf, never again to be touched. The safety case has failed in its purpose if it is "so inaccessible or unapproachable that we are happy never to refer to it again."

6. Imbalance of skills:

The skills are required of both someone to develop the safety case and someone to challenge and critique the assumptions made. Too often, the latter skills are missing.

7. The illusion of pictures:

People are "dazzled" by complex, colored, hyperlinked graphic illustrations which gives both the makers and viewers a warm sense of overconfidence. The quality of the argument cannot be judged by the node count on such documents or number of colors used.

Abbreviations*

CENELEC	European Committee for Electrotechnical Standardization (*Comité Européen de Normalisation Electrotechnique*)
EN	European Norm
RAF	Royal Air Force
RAMS	Reliability, Availability, Maintainability, and Safety
SRAC	Safety-Related Application Condition

References

1. CENELEC (2003) EN 50129: *Railway Applications—Communications, Signalling and Processing Systems—Safety Related Electronic Systems for Signalling*, CENELEC, Brussels.
2. CENELEC (European Committee for Electrotechnical Standardization) (2015) *Draft prEN 50126 Railway Applications—The Specification and Demonstration of Reliability, Availability, Maintainability and Safety (RAMS)—Part 1: Generic RAMS Process*, CENELEC, Brussels.
3. CENELEC (2015) *Draft prEN 50126 Railway Applications—The Specification and Demonstration of Reliability, Availability, Maintainability and Safety (RAMS)—Part 2: Systems Approach to Safety*, CENELEC, Brussels.
4. CENELEC (1999) EN 50126: *Railway Applications—The Specification and Demonstration of Reliability, Availability, Maintainability and Safety (RAMS)*, CENELEC, Brussels.

---

* These can be considered universal abbreviations in the industry.

5. Ministry of Defence (2007). *Safety Management Requirements for Defence Systems*, Defence Standard 00-56 (Issue 4), UK Ministry of Defence, London.
6. Origin Consulting (York) Limited (2011) *GSN Community Standard Version 1*, Origin Consulting, (York) Limited, York.
7. Office for Nuclear Regulation (2016) *The Purpose, Scope, and Content of Safety Cases*, NS-TAST-GD-051 Revision 4, July 2016, Office for Nuclear Regulation, Bootle, UK.
8. Haddon-Cave, Charles (2009) *The Nimrod Review: An Independent Review into the Loss of the RAF Nimrod MR2: Aircraft XV230 in Afghanistan in 2006*, The Stationary Office, London, October 2009.
9. Kelly, Tim (2008) Are 'safety cases' working, *Safety Critical Systems Club Newsletter*, Volume 17, No. 2, January 2008, pages 31–3.

12

Reliability Demonstration Tests: Decision Rules and Associated Risks

Pierre Dersin and Cristian Maiorano

CONTENTS

12.1 Introduction and Historical Perspective

The purpose of this chapter is to show on a theoretical basis and to illustrate by practical examples that reliability demonstration tests necessarily include an element of risk and to explain how they can be organized in order to minimize that risk.

The underlying theory is that of hypothesis testing. Type I and type II errors (or, in this context, producer's and consumer's risk, respectively) are complementary. Good test procedures aim at making those two risks acceptable for both parties while fulfilling the time and cost constraints. Reducing the risks usually entails a higher test cost.

Fixed duration tests as well as sequential tests are reviewed and compared. Examples of erroneous methods are also given, and the reason for their fallacy is explained.

Unfortunately, such methods are sometimes found in calls for tenders, which is not in the interest of either consumers or producers.

12.2 Notion of Discrimination Ratio

The discrimination ratio* is an index characterizing a test plan by its ability to distinguish between an acceptable and an unacceptable dependability measure. An example of reliability measure is the probability of success over a given period.

In the case of a reliability test, the null hypothesis H_0: $R > R_0$ is tested against the alternative hypothesis H_1: $R < R_1$ (with $R_1 < R_0$).

R_0 is commonly defined as the design target, while R_1 is the minimum acceptable reliability. R_0 is commonly defined according to the system specification and is a nonflexible input to the test design.

If we denote by p_0 and p_1, the quantities $1 - R_0$ and $1 - R_1$, the discrimination ratio is

$$D = \frac{p_1}{p_0}. \tag{12.1}$$

The discrimination ratio is a figure of merit for the test plan.

12.3 Implicit Risks Associated with Decision Rules in Some Tests (the Danger of Some "Simple" Method)

Both the consumer and the producer must be aware that any decision rule based on statistical observations entails a risk. The acceptable levels of risk α and β have to be negotiated between consumer and producer.

Unfortunately, quite a number of consumers and producers still do not realize this and propose contractual clauses of the type: MTTF, as defined by T/n (total observation time divided by number of failures), must be at least equal to 100,000 hours, over a period of 1 year.

The risk arises from the fact that the estimate of the mean time to failure (MTTF) obtained as the ratio T/n may be far from the true value and is the more likely to be so, the shorter the observation period. The contractual requirement must refer to the true value.

Furthermore, the use of an MTTF metric should be avoided, when possible, to improve clear understandable communication about reliability. The estimate of MTTF ignores the changing nature of a failure rate, which provides much more information useful for the understanding of test planning and results.

The notion of risk is fundamental to any reliability, maintainability, and availability (RAM) demonstration test. Let us take the example of a reliability demonstration test; the idea is similar for the other tests. On the basis of a limited set of observations—failure data in the case of a reliability test—it is desired to decide whether the true reliability value is greater than a certain value R_0. The decision must be "Yes" or "No."

In fact, this decision may be wrong: it may happen, for instance, that during the limited duration of the test, many failures will occur (more than would, on average, on a very long period). On the basis of these observations, it would be decided to reject the null

---

* IEC 61124, second edition.

TABLE 12.1

Hypothesis Testing Decision Risk

	H_0 Is False	H_0 Is True
Accept H_0	Type II error (consumer's risk β)	Correct decision
Reject H_0	Correct decision	Type I error (producer's risk α)

hypothesis, when, in fact, that hypothesis would be true. This is the risk incurred by the producer: his/her product is rejected despite being reliable enough. Conversely, it may happen that during the limited duration of the test, very few failures occur (fewer than would, on average, on a very long period). Then, on the basis of these observations, the consumer will be led into believing that the product is reliable enough, when, in fact, it is not. This is the risk incurred by the consumer: he/she accepts the producer's product despite this not being reliable enough. These two risks are a special case of the risks involved in any hypothesis testing.

Table 12.1 shows how a test outcome may truly or falsely reflect the truth and result into the previously described risks.

The only way to reach a zero risk would be to use an infinitely long observation period (or equivalently an infinitely big sample size).

Thus, there is an education process that must take place in order to create awareness of that risk.

Clauses such as the preceding ones should not be accepted as stated but must be reformulated in a RAM demonstration test plan.

We shall now analytically express the constraints that the consumer's risk should not exceed β and that the producer's risk should not exceed α.

Expressions for the two risks will be derived in terms of R_0, R_1, the sample size n, and the acceptance threshold c.

The consumer's risk is, by definition, the probability that the test is passed, when in fact $R < R_1$.

Otherwise stated, it is the probability that $k \leq c$ given $R < R_1$. The producer's risk is, by definition, the probability that the test fails, when, in fact, $R > R_0$. Otherwise stated, it is the probability that $k > c$ given $R > R_0$.

12.4 Demonstrating Reliability

12.4.1 Introduction

Although several reliability demonstration test methods exist when the assumption of exponential distribution for the time to failure is made (which entails a constant failure rate λ), only few approaches are studied which apply to the more generic (and realistic) case of variable failure rate. The aim of this section is to present a possible approach to designing a generic demonstration test plan built around reliability (instead of using a specific metric). The proposed method is based on the binomial equation, which is nevertheless independent of the reliability distribution deemed to be applicable to the reliability of the equipment under test.

The methodology allows for determining the test sample size n, provided that the acceptable failure quantity is fixed or by fixing the sample size, for calculating the acceptable number of failures. The proposed plans are time terminated, and therefore, the time is supposed to be fixed.

Let us consider a reliability test run on n samples. The test for each sample can end up with an unknown reliability value R.

We wish to test the hypothesis $H_0: R > R_0$ against the hypothesis $H_1: R < R_1$. The producer's risk should be smaller than α for $R = R_0$. The consumer's risk should be smaller than β for $R = R_1$.

12.4.2 Two-Sided Sampling Plan

Let us consider that the following parameters are given:

- The design reliability value: R_0
- The minimum acceptable reliability: R_1
- The producer's risk: α
- The consumer's risk: β

Assuming H_0 is true, the constraint that the producer's risk should not exceed α, can be written as

$$P\left[k > c \,|\, R > R_0\right] = \sum_{i=c+1}^{n} \frac{n!}{i!(n-i)!} \cdot (1-R)^i \cdot R^{(n-i)} \geq 1-\alpha. \tag{12.2}$$

Assuming H_0 is false, the constraint that consumer's risk should not exceed β, can be written as

$$P\left[k \leq c \,|\, R < R_1\right] = \sum_{i=0}^{c} \frac{n!}{i!(n-i)!} \cdot (1-R)^i \cdot R^{(n-i)} < \beta. \tag{12.3}$$

We want to compute the smallest values of c (acceptable failures threshold) and n (sample size) able to satisfy the preceding written inequalities.

12.4.3 One-Sided Sampling Plan

Sometimes, only one reliability value and one error are specified, around which to build a demonstration test. Therefore, the null and alternative hypotheses become as follows:

- $H_0: R > R_0$
- $H_1: R < R_0$

Alternatively, the value R_1, together with the type II error, may be specified, leading to the consequent hypothesis definitions.

In this case, only one of inequalities in Equations 12.2 and 12.3 can be used.

By this approach, different combinations of c and n can be found, able to satisfy the inequality considered. According to the value selected for c and the one calculated for n, different values of the type II error β can be obtained.

Depending on the approach used, advantage can be given to either the consumer or the producer.

12.4.4 Weibull Distributed Reliability Case

For the purpose of this example, a Weibull reliability distribution is assumed, which is one of the most commonly used for modeling reliability behavior of electronic, mechanical, and electromechanical equipment.

$$\sum_{i=0}^{r} \frac{n!}{i!(n-i)!} \cdot (1-\hat{R})^i \cdot \hat{R}^{(n-i)} = 1 - \text{CL},\qquad(12.4)$$

where CL is the desired confidence level; r is the maximum desired number of failures; n is the quantity of items under test; $\hat{R}$ is the reliability calculated at the time $\hat{t}$, which represents the duration of the test.

$1 - \text{CL}$ is the probability of passing the demonstration test. Depending on the value of R used in Equation 12.4, the probability of passing the test would be either the type II error or the value of type I error.

Under the assumption of a Weibull reliability distribution, the parameter $\hat{R}$ is a function of the test time $\hat{t}$, the shape parameter β, and the scale parameter η $\left(\hat{R} = f(\hat{t}, \beta, \eta)\right)$.

The Weibull reliability function is (assuming location parameter $\gamma = 0$)

$$R(t) = e^{-\left(\frac{t}{\eta}\right)_\beta}.\qquad(12.5)$$

The application of such a method assumes an a priori knowledge of the reliability model and therefore of the shape parameter β.

In order to determine the scale parameter, we can solve the Equation 12.5 for η, where R is equal to R_{req} (the reliability requirement), t is the time at which the desired reliability R_{req} is defined, and β is the preliminary known shape parameter.

$$h = \frac{t}{\left(-\ln(R_{req})\right)^{\frac{1}{\beta}}}.\qquad(12.6)$$

Once all the parameters are known, we can calculate the effective reliability value $\hat{R}$ to be included in the test, by applying Equation 12.5 with the known values of β, η, and $\hat{t}$ (effective test duration).

The last step is to substitute those values into Equation 12.4 and to solve for n, to find the number of items to be put under test.

If instead it is desired to calculate the necessary test duration $\hat{t}$ by a priori knowledge of the test sample size n, we can inject Equation 12.5 into Equation 12.4, which leads to

$$1 - CL = \sum_{i=0}^{r} \frac{n!}{i!(n-i)!} \cdot \left(1 - e^{-\left(\frac{\hat{t}}{\eta}\right)_\beta}\right) \cdot \left(e^{-\left(\frac{\hat{t}}{\eta}\right)_\beta}\right)^{(n-i)}. \tag{12.7}$$

Once the scale parameter η is calculated as explained before, Equation 12.7 can be solved for $\hat{t}$.

The proposed method can also be used to determine the necessary sample size n or the test duration $\hat{t}$ to demonstrate a given MTTF requirement at a given time t. The approach is the same, but the scale parameter is calculated by using

$$MTTF = \eta \cdot \Gamma\left(1 + \frac{1}{\beta}\right) \tag{12.8}$$

and solving for η,

$$\eta = \frac{MTTF}{\Gamma\left(1 + \frac{1}{\beta}\right)}. \tag{12.9}$$

The remaining steps are the same as those mentioned earlier.

12.4.5 Example

A test must be designed to demonstrate MTTF = 75 hours with a 95% confidence if during the test no failure occurs (therefore, the acceptable failures threshold is fixed at 0). Let us assume a Weibull distribution with a shape parameter $\beta = 1.5$. We want to determine the test sample size to be used, knowing that the test duration is 60 hours.

The first step is to evaluate the scale parameter η. Applying Equation 12.9 yields

$$\eta = \frac{75}{\Gamma\left(1 + \frac{1}{1.5}\right)} \approx \frac{75}{0.90} = 83.1. \tag{12.10}$$

The next step is to calculate the value $\hat{R}$ by using Equation 12.5:

$$\hat{R} = e^{-\left(\frac{60}{83.1}\right)_{1.5}} = 0.541. \tag{12.11}$$

By solving Equation 12.7 for n, we found that the required sample size is ≈ 4.88 units, which, rounded to the next higher integer, gives as result five units to be tested.

If during the test a failure occurs, the test is failed. At this time, a new test can be initiated or the time extension necessary to demonstrate the required reliability can be calculated by assuming the sample size previously calculated and one acceptable failure.

12.5 Constant Failure Rate Demonstration

Although this approach is applicable to a limited number of cases, the constant (in time) failure rate assumption remains one of the most widely used, especially in industry.

Indeed, when mainly electronics is concerned, the time-independent failure rate assumption is a well-working approximation.

Under the constant failure rate assumption, if $MTTF = MTTF_1$, the probability distribution of the number of failures during a period of length T is a Poisson variable of mean T/θ_1, and therefore the conditional probability of at most c failures over a period of length T given that $MTTF \le MTTF_1$ is

$$P\left[k \le c | MTTF \le MTTF_1\right] \le P\left[k \le c | MTTF = MTTF_1\right]$$
$$= \sum_{k=0}^{c} \frac{\left(T/MTTF_1\right)^k \cdot e^{-T/MTTF_1}}{k!}. \tag{12.12}$$

The constraint that the consumer's risk should not exceed β is therefore expressed by the following inequality:

$$\sum_{k=0}^{c} \frac{\left(T/MTTF_1\right)^k \cdot e^{-T/MTTF_1}}{k!} \le \beta. \tag{12.13}$$

The producer's risk is, by definition, the probability that the test is failed, when, in fact, true MTTF is greater than $MTTF_0$.

Otherwise stated, it is the probability of more than c failures over period T given that true MTTF is greater than $MTTF_0$.

The constraint that the producer's risk should not exceed α is equivalent to the constraint that the conditional probability of more than c failures over a period of length T given that $MTTF \ge MTTF_0$ should not exceed α, or

$$P\left[k > c | MTTF \ge MTTF_0\right] \le \alpha \tag{12.14}$$

or, equivalently,

$$P\left[k \le c | MTTF \ge MTTF_0\right] \ge 1-\alpha. \tag{12.15}$$

Since

$$P\left[k \le c | MTTF \ge MTTF_0\right] \ge P\left[k \le c | MTTF = MTTF_0\right], \tag{12.16}$$

it follows that Equation 12.4 will be achieved if

$$P\left[k \le c | MTTF = MTTF_0\right] \ge 1-\alpha. \tag{12.17}$$

Under the constant failure rate assumption, if MTTF = MTTF_0, then the number of failures during a period of length T is a Poisson variable of mean T/θ_0. Therefore, Equation 12.6 is equivalent to

$$\sum_{k=0}^{c} \frac{\left(T/\text{MTTF}_0\right)^k \cdot e^{-T/\text{MTTF}_0}}{k!} \geq 1-\alpha. \tag{12.18}$$

12.5.1 Example

Let us evaluate the level t of this risk in a real case.

Consider that the following requirement applies into the framework of a railway project:

- The MTTF of the system must be at least equal to 100,000 hours. The producer will demonstrate the achievement of the MTTF target during the 1-year defect liability period (DLP). All the events leading to a train delay greater than 10 minutes are to be taken into account for system MTTF indicator.

In many cases, especially when both parties are not aware of the risk hidden behind such requirement, a demonstration method would be put in place according to the following rule:

- During the DLP, all the failures under producer's responsibility will be considered to estimate the system MTTF as follows:

$$\text{MTTF}_{\text{System}} = T/k,$$

where T is the total operating time from the start of demonstration period (for more samples under test, T increases additively) and k is the observed number of failures during the elapsed time.

The requirement is considered as achieved if at the end of the DLP, the observed MTTF is greater than the target or if 0 (zero) failures are observed.

It has to be noted that in this particular case, MTTF_0 is equal to MTTF_1. The direct consequence is that the consumer's risk and the producer's risk are complementary.

The producer is, in other words, asked to design a system unlikely to fail during the 1-year DLP. With such a statement, there is a probability of about 92% that the consumer will finally accept the system provided that the true MTTF is equal to the required one.

Let us denote by D the event "max 0 failures occur during the DLP." The probability for the consumer to accept a system compliant to the stated target is, in this case, the probability that event D happens provided that the true MTTF is the required one ($\text{MTTF}_1 = \text{MTTF}_0 = 100{,}000$ hours). When the failure times follow an exponential distribution, the number of failures in the time interval T follows a Poisson distribution with associated parameter T/MTTF. The relationship is given by Equation 12.13.

Hence, in this case, $MTTF_1 = 100,000$ hours (under the hypothesis of λ constant) and c is equal to 0.

$$P(D) = e^{-\left(\frac{8760}{100,000}\right)} \frac{\left(\frac{8760}{100,000}\right)^0}{0!} \approx 0.92. \tag{12.19}$$

On the other hand, there is only 8% risk (in this case producer's risk is complementary to consumer's risk calculated in Equation 12.13) that the system, designed by the producer to achieve a 100,000 hours MTTF, fails at least once during the DLP (system is rejected). Therefore, there is only 8% risk that the system is not accepted by the consumer if it is true that MTTF achieves the requirements.

The producer could take advantage of this opportunity and decide to underdesign the system, if he/she is willing to increase the risk that the system is finally not accepted by the consumer.

Let us consider that the producer is willing to accept a risk of 20% that the system provided is not accepted by the consumer. In other words, the producer decides to design the system in such a way that the probability that at least one failure occurs during the DLP is equal to or lower than 20%.

By applying Equation 12.17, it can be found that the minimum true MTTF which entails a risk lower/equal to 20% for producer is about 40,000 hours. As a result, the producer, by accepting a small risk of 20%, has the opportunity to sell a system with an MTTF lower by a factor of 2.5 than the required MTTF, with considerable money savings.

Said otherwise, there is a trade-off between risk and cost: the risk the producer is willing to take versus the cost he/she is willing to incur to improve the design.

12.6 Notion of Operating Characteristic Curve

The operating characteristic curve is the curve that plots the probability of acceptance (or nonrejection) of the null hypothesis H_0 as a function of the true value of the dependability measure. For instance, it gives the probability of accepting the hypothesis "MTTF > $MTTF_0$," as a function of the true value of MTTF. The operating characteristic curve can be plotted from the "probability of acceptance" formula (Equation 12.2).

Examples of operating characteristic curves are given in Figure 12.1 (excerpted from IEC 61124), for several test plans.

It plots the acceptance probability as a function of the ratio $MTTF/MTTF_0$ (denoted m/m_0).

As expected, this probability goes to 1 as m/m_0 becomes large. It can be seen that test plan B.5 has the best operating curve, i.e., the probability of acceptance remains close to zero as long as the true MTTF does not approach the target. In contrast, B.8 gives a higher probability of acceptance for low values of the true MTTF. This is consistent with B.8 having a higher discrimination ratio (comparing m_0 with $5 \times m_0$ instead of with $1.5 \times m_0$). But B.5, the more desirable test plan, implies a much greater duration.

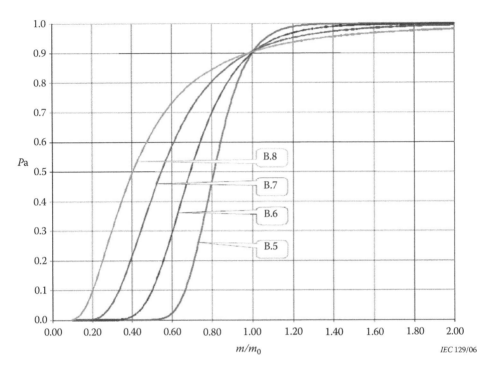

FIGURE 12.1
Operating characteristic curves. (Extracted from IEC [International Electrotechnical Commission], IEC 61124. *Reliability Testing—Compliance Tests for Constant Failure Rate and Constant Failure Intensity*, © 2006 IEC, Geneva, Switzerland.)

12.7 Comparison between Time/Failure-Terminated and Truncated Sequential Test Plans

A truncated sequential test can be described by the following:

- During the test, the test items are continuously monitored or at short intervals, and the accumulated relevant test time and the number of relevant failures are compared with the criteria given to determine whether to accept, reject, or continue testing.

Similarly, a time/failure-terminated test can be described by the following:

- During the test, the test items are continuously monitored or at short intervals, and the relevant test time is accumulated until either a predetermined amount of relevant test time has been exceeded (accept) or a predetermined number of relevant failures has occurred (reject).

Figure 12.2, excerpted from the IEC 60300-3-5 standard, illustrates the comparison between time-terminated or failure-terminated tests and the truncated sequential test, both with the same risk.

It can be seen from Figure 12.2 that the maximum duration of a truncated sequential test exceeds the duration of a fixed-duration test for the same parameters (risks and discrimination ratio).

On the other hand, the average duration of a sequential test is shorter, as again illustrated in Figure 12.3, also excerpted from IEC 60300-3-5.

Consequently, the comparison in Table 12.2 (also extracted from IEC 60300-3-5) between fixed and sequential test plans can be drawn.

In conclusion, the sequential test leads on average to an earlier decision, particularly for very reliable or very unreliable items. However, its duration is not certain, and if it has to be continued until truncation, it could be longer than the fixed-duration test. The duration

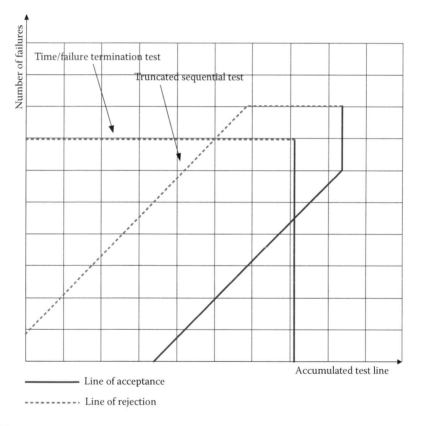

FIGURE 12.2
Comparison between classical tests and truncated sequential test. (Extracted from IEC, IEC 60300-3-5. *Dependability Management—Part 3–5: Application Guide—Reliability Test Conditions and Statistical Test Principles,* © 2001 IEC, Geneva, Switzerland.)

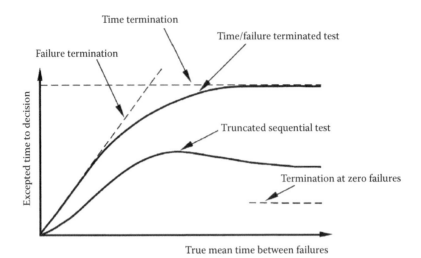

FIGURE 12.3
Expected test time to decision as a function of true MTTF. (Extracted from IEC, IEC 60300-3-5. *Dependability Management—Part 3–5: Application Guide—Reliability Test Conditions and Statistical Test Principles*, © 2001 IEC, Geneva, Switzerland, 2012.)

TABLE 12.2

Comparison of Statistical Test Plans

Type of Statistical Test Plan	Advantages	Disadvantages
Time/failure-terminated test	Maximum accumulated test time is fixed; therefore, maximum requirements for test equipment and labor are fixed before testing begins.	On average, the number of failures and the accumulated test time will exceed those of a similar truncated sequential test.
	Maximum number of failures is fixed prior to testing; therefore, the maximum number of test items can be determined in case of testing without repair or replacement.	Very good items or very bad items need to experience the maximum accumulated test time or number of failures to make a decision, which can be made sooner with a similar truncated sequential test.
	Maximum accumulated test time is shorter than that for a similar truncated sequential test.	
Truncated sequential test	Overall, the number of failures to a decision is lower than for the corresponding fixed time/failure-terminated test.	The number of failures and therefore the test item costs will vary in a broader range than for similar time/failure-terminated tests.
	The test has fixed maxima with respect to accumulated test time and number of failures.	Maximum accumulated test time and number of failures could exceed those for the equivalent time/failure-terminated test.
	Overall, the accumulated test time to a decision is a minimum.	

Source: IEC, IEC 60300-3-5. *Dependability Management—Part 3–5: Application Guide—Reliability Test Conditions and Statistical Test Principles*, © IEC, Geneva, Switzerland.

(even until truncation) is unfortunately not known in advance since it depends on the (unknown) value of the true MTTF.

When an accurate prediction of the MTTF is available, from theoretical prediction or in the best-case historical data analysis, however, the test plan can be selected by computing the maximum and average durations of the tests that would obtain if the true MTTF value were the predicted one.

Acknowledgment

The authors thank the International Electrotechnical Commission (IEC) for permission to reproduce information from its international standards. All such extracts are copyright of IEC, Geneva, Switzerland. All rights reserved. Further information on IEC is available on www.iec.ch. IEC has no responsibility for the placement and context in which the extracts and contents are reproduced by the author nor is IEC in any way responsible for other content or accuracy therein.

13

Hazard Log: Structure and Management in Complex Projects

Rohan Sharma

CONTENTS

13.1 Introduction and Definition

A hazard log is the most important component of the safety management system of a railway project. It is a "document in which all safety management activities, hazards identified, decisions made and solutions adopted are recorded or referenced" [1]. This chapter aims to explain about the structure of an "ideal" hazard log document but does not go into the details of the process of hazard management.

13.2 How Many Hazard Logs Should a Project Have?

This is one of the key decisions that needs to be taken early on in a rail project and the answer to this varies from project to project, depending on the scope split, organizational structure, contractual obligations, etc. Without going to different types of contractual arrangements, let us consider a simple scenario where a rail project is to be executed by multiple organizations of the following types:

1. The entity which owns (or which will ultimately own) the railway assets is the "owner." Usually, the owner does not create a hazard log; rather it gets hazard log(s) from the project developer before handover and/or at completion. Thereafter, the

owner will be responsible for maintaining and updating the hazard logs during operations and for the entire railway system life span.

2. The entity that is responsible for system-level project development is the "developer." Sometimes the owner is also the developer or is usually appointed by the owner to undertake all system level activities. In large, complex, multiorganizational projects, this level is usually a consortium composed of members that will undertake further subsystem level activities. This entity shall be responsible for a system or consortium hazard log and shall undertake a system hazards analysis.

3. The entity(s) responsible for subsystem level design is the "infrastructure designer" or the "operating system designer," and so on. Each of these shall be responsible for the subsystem-level hazard logs of their corresponding work scope.

4. The entity responsible for construction is a "main contractor." Usually a building or construction company maintains a health and safety hazard log and risk assessments (which is out of scope of this chapter) but not a project design hazard log. The main contractor relies on the design supplied by the designers and the developer. It should be the responsibility of the designers and the developer to produce and maintain robust and comprehensive hazard logs, which the main contractor can refer to. However, the main contractor has a greater role to play during hazard verification and validation, where it assists the developer in "closing" out hazards. The ultimate responsibility of hazard verification and validation and closure still lies with the developer.

5. The entity responsible for part of the construction or for supplying and installing particular subsystem or components is a "subcontractor" or a "supplier." This entity should usually supply, install, or build based on design and specifications provided by the designers and the developer. However, such a supplier should also provide details of product specifications that also include hazards and risks associated with each component, along with mitigations. Such equipment can also be "commercial off-the-shelf" items which follow design standards, test, and installation procedures. As such, component level hazards are usually considered during design development and procurement. However, if they introduce or change hazard risks assessments, then these shall also be part of the associated subsystem-level hazard log.

6. The entity responsible for operations and maintenance (O&M) during trial and revenue service of the rail system will also develop its O&M hazard log. This does not mean the developer does not contribute to the O&M hazards analysis, which is a key requirement for the developer, whose scope is to deliver a system that can be operated safely while the O&M organization's scope is to perform safe operation and maintenance. The O&M organization also has to analyze the hazards transferred* from the developer.

13.3 Responsibility, Accountability, and Ownership

A hazard log is a live document throughout the project life cycle. The ultimate accountability and ownership of a hazard log lies with the project safety assurance manager or

---

* Refer to Section 13.8, where hazard transfers between project organizations are discussed.

equivalent, appointed by the developer organization for the relevant scope of works. The responsibility for updating hazard logs also lies with other members of the organization, who from time to time, identify new hazards or update the verification or validation status. The ownership of a hazard log document should not be confused with the ownership of a hazard, which is usually by other members* of the project organization.

A hazard owner is someone who is responsible for the management of a hazard right from the time it is identified until it is closed. This typically coincides with the hazard belonging to the owner's scope of work. The exception to this is when the hazard is supposed to be transferred to another organization or another team within the same organization, and then the hazard ownership can also be transferred, in accordance with the formal hazard transfer process[†].

13.4 Hazard Log Structure

After safety management planning, hazards analysis formally commences with the preliminary hazard analysis (PHA) and hazard identification (HAZID).

Setting up a clear, structured, hierarchical, and yet flexible hazard log is necessary to support hazard analysis throughout the project life cycle. Table 13.1 lists all fields that a well-developed hazard log should have along with sample values for each field. These field entries are considered important for a hazard log to be deemed acceptable by the safety organization within the project or by external assessors. It is recommended that all field entries be used for completeness and consistency. This will also satisfy the objectives and criteria set out for hazard and risk analysis in Chapter 7 of IEC 61508 [2].

13.5 Hazard Log Management Tools: Excel, DOORS, and ComplyPro

Hazard logs are live documents and need control and regular updates throughout the project life cycle. Therefore, it is essential that hazard logs are formally maintained, whether using Microsoft Excel or a more specifically developed database management tools such as IBM Rational DOORS® or ComplyPro®. These two tools are widely used for requirements management processes as well.

Each has its advantages and disadvantages, and depending on the project complexity, one tool may be a better choice than the others. Eventually, it is up to the project owner and/or developer to decide which software the project should employ. Irrespective of the tool used, remember, each hazard requires a database entry and hazards should not be combined, merged, or split across multiple entries.

Table 13.2 gives a brief comparison[‡] of these three software tools.

---

* These are usually members of the teams directly responsible for design work and processes or relevant individuals, including stakeholders or third parties. The process of hazard identification is covered in Chapter 8.
† Hazard transfer process is discussed in greater length in Section 13.8.
‡ These are experience-based comparisons only. Further choices with these tools may be possible by updated versions of software or by more experienced users. The suppliers of these software tools should be able to provide greater details on functionality and flexibility for use on projects.

TABLE 13.1

Hazard Log Fields, Description, and Example Values

Field	Field Description	Example Values
Hazard ID	A unique ID for each hazard	H-SYS-0123
System (or subsystem) hazard ID	Reference to system (or subsystem) level hazard depending on the location of the hazard log in the project hierarchy (could be from/to external organization)	SSHA-SIG-0123
Hazard source	Origin of generic hazard defined during safety planning or by HAZID process	MODSafe; HAZID *reference*; *third party*
Text type	Determines if it is a hazard or otherwise	Title; hazard
Date identified	The date the hazard log entry was created	*Date format*
Date last updated	The last date that the hazard status was revised	*Date format*
Hazard owner organization	The organization responsible for the management of the hazard	*Organization code for* operator; rolling stock supplier; signaling system
Hazard leader	The safety manager responsible for the management of the hazard	*Personnel name*
Mode	Mode of operation in which the hazard is expected to occur	Normal; degraded; emergency; maintenance
Hazard category	High-level hazard or accident type defined during safety planning	Derailment; collision; fire; impact; entrapment; electrocution; environmental
System	Discipline that is primarily related to the hazard cause by design or mitigation	Rolling stock; signaling; track; PSD; system level
Subsystem/team	Subsystem directly responsible for technical hazard assessment	*Rolling stock subsystems*—car body; braking; traction power
Population at risk	Group of people potentially affected in case of hazard occurrence	Passenger; staff; stakeholders
Hazard description	Brief description of hazard scenario	*Free text*
Hazard location	Physical location(s) of hazard occurrence within the railway system scope	Tunnel; station; depot; viaduct; all
Interface with	Interface with other organization(s) in relation to hazard cause, effect and/or mitigations	*Organization code for* operator; rolling stock supplier; signaling system
Cause	Reasons contributing to hazard scenario and aggravation	*Free text*
Effect	Effects in case of hazard occurrence on the population at risk	*Free text*
Existing mitigation/ controls	Hazard control or mitigation measures already identified or applied	*Free text*
Initial frequency	Existing likelihood of hazard occurrence in accordance with risk matrix, as defined during safety planning	(F1) Incredible; (F2) remote; (F3) unlikely; (F4) occasional; (F5) frequent
Initial severity	Existing severity in case of hazard occurrence in accordance with risk matrix, as defined during safety planning	(S1) Insignificant; (S2) marginal; (S3) critical; (S4) catastrophic
Initial risk	Assessment of risk based on current levels of hazard frequency and severity and already existing control measures	(R1) Intolerable; (R2) undesirable; (R3) tolerable; (R4) negligible

(Continued)

TABLE 13.1 (CONTINUED)

Hazard Log Fields, Description, and Example Values

Field	Field Description	Example Values
Additional mitigation/controls	Further hazard control or mitigation measures applied to reduce risks to acceptable levels	*Free text*
Residual frequency	Residual likelihood of hazard occurrence in accordance with risk matrix, as defined during safety planning	(F1) Incredible; (F2) remote; (F3) unlikely; (F4) occasional; (F5) frequent
Residual severity	Residual severity in case of hazard occurrence in accordance with risk matrix, as defined during safety planning	(S1) Insignificant; (S2) marginal; (S3) critical; (S4) catastrophic
Residual risk	Assessment of risk based on estimated levels of hazard frequency and severity after additional control measures have been applied	(R1) Intolerable; (R2) undesirable; (R3) tolerable; (R4) negligible
Mitigation type	Type of mitigation measure applied that eliminates or minimizes the hazard	Elimination; reduction; procedure implementation
Mitigation status	Status of mitigation measure applied that eliminates or minimizes the hazard	Defined; resolved; implemented; transferred
Mitigation evidence	Reference to document(s) that indicate how the hazard has been mitigated by design, e.g., by design specification and drawing	Document title, document number, revision
Verified by	Organization and/or person who has verified the status of the mitigation	*Organization code/personnel name*
Interface or transfer reference	Reference to interface control document, external organization hazard transfer reference and/or related documents	ICD no., transfer reference, etc.
Risk control owner	The organization responsible for mitigation of the hazard, esp. if different from hazard owner organization	*Organization code for* operator; rolling stock supplier; signaling system
Safety function	The safety function which is affected or not achieved because of the hazard	Braking; train identification
Component	The component to which the safety function related to the hazard applies	Brakes; train identification
Safety requirement reference	Requirement ID reference that relates to the hazard	REQ-SYS-0123
Risk acceptance principle	The method of approach to hazard assessment and risk acceptance criteria that have been established during safety planning	Adherence to standard/code of practice; reference system cross acceptance; quantitative risk assessment; common safety method risk assessment
ALARP argument	Explanation for the residual risks to be accepted by the project	*Free text*
SRAC	Description of SRAC	*Free text*
Hazard status	The status of hazard item at the time of issue of the hazard log	Identified; resolved; closed; transferred; obsolete; deleted
Comments	Internal comments within or external to organization to monitor hazard assessment progress and modifications	*Free text*

TABLE 13.2

Comparison between Hazard Log Management Tools

Criterion	MS Excel	IBM Rational DOORS	ComplyPro (ComplyServe)
Size of project	Good for smaller and simpler projects undertaken by one or two organizations	Good for complex projects with multiple subsystems and multiple organizations	Good for complex projects with multiple subsystems and multiple organizations
Number of hazards	Suitable for less number of hazards, e.g., up to 100–200 hazards	Suitable for large number of hazards, e.g., 500+ hazards	Suitable for large number of hazards, e.g., 500+ hazards
Ease of use	As easy as using a spreadsheet	Requires major training for safety managers and end users	Requires major training for safety managers but minor training for end users, fairly easy user interface
Cost	Cost to buy MS Excel package	Cost to buy user licenses, possible to transfer license to another user	Cost to buy user licenses, possible to transfer license to another user
Number of users	As many as required	Suitable for up to 5–10 trained users, limited by number of user licenses bought	Suitable for several trained users, limited by number of user licenses bought
Database control	Owned and maintained by information and technology of the project team	Owned and maintained by tool provider	Owned and maintained by tool provider
Change management	Requires process and control of master version; no history can be saved	Baselines can be frozen by the tool at planned milestones; full history for each hazard entry can be generated	Baselines can be frozen by the tool at planned milestones; full history for each hazard entry can be generated
Traceability	Depends on manual entry of references to other spreadsheets and row IDs	Top–down or sideways linkages can be made between modules, e.g., system–subsystem and to other databases such as interfaces	Top–down or sideways linkages can be made between modules, e.g., system–subsystem and to other databases such as interfaces
Time consumption	Time required for day to day use is dependent on manual entry of data and Excel functions required	Fast for day-to-day usage for less number of users; multiple linkage formations can be made quickly	Needs more time for day to day usage as system requires to update itself frequently, especially with multiple users at the same time; multiple linkage formation requires time
Statistics	User-defined statistics can be generated	User-defined statistics can be generated	User-defined statistics can be generated
Compatibility and integration	Suitable to use with other MS packages, DOORS, or ComplyPro	Database design is by tool provider; import/export from/to other software such as Excel is possible	Can be easily tailor-made as per end user requirements; import/export from/to other software such as Excel is possible

13.6 System-Level Hazard Log Feeding

So far, this chapter has discussed how many hazard logs are required on a complex rail project. It is equally important that all these hazard logs are linked, both across the disciplines (interfaces) and across system and subsystems (hierarchy levels). Figure 13.1 depicts hazard logs at system and subsystem levels and illustrates how hazard log levels and links can be established. The system hazard log determines which hazards belong to which subsystem(s). For example, a "train collision"–related hazard can be considered a system level hazard, but it has multiple causes. This hazard is therefore dependent on multiple subsystem hazards such as "failure of signaling" or "failure of braking system."

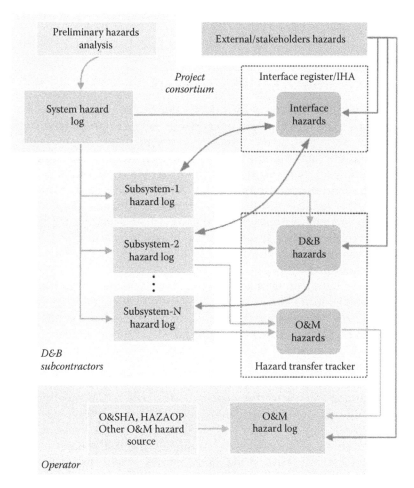

FIGURE 13.1
Hazard log hierarchy and interactions.

A system-level hazard analysis should be considered as complete only when all associated subsystem level hazards analyses have been concluded. When a hazard requires analysis by more than one subsystem, it becomes an interface hazard (refer to Section 13.9).

Figure 13.1 depicts a single system hazard log which is linked "downward" to multiple subsystem-level hazard logs. The interface hazards block consists of a list of interface hazards linked to interfacing subsystems. Note that this is also linked to external interface hazards. The design and build (D&B) hazards and the O&M hazards blocks are parts of the hazard transfer tracker where hazards between project entities can be transferred. D&B hazards usually need to be transferred during design and construction stages and that need to be mitigated before handover. O&M hazards need to be transferred to the (future) operator. Note that the operator maintains its own hazard log, which will incorporate hazards transferred to the operator.

13.7 Mitigations versus Safety Requirements

Safety requirements and mitigations for safety hazards are closely linked but are not the same thing and are often erroneously interpreted. Safety requirements form an important link between safety management/hazard assessment, and the overall requirements management but fall under the requirements management process. Safety requirements specify the minimum safety measures that the project is supposed to implement. These come from various sources including the employer's specifications, stakeholders, and regulations. On the other hand, mitigations to hazards are independently identified control measures and actions that need to be applied to eliminate or reduce the risk posed by the hazard. A safety requirement on itself is not a mitigation measure. The "means of adherence" to a safety requirement that inherently eliminates/reduces a hazard risk is essentially a mitigation measure.

This distinction is better understood when a project decides to implement additional mitigations. Additional mitigations are ideally identified during design development and retrospectively during later stages when existing measures are insufficient and associated risks need to be further reduced or eliminated. Additional measures are therefore not directly derived as means of adherence to particular safety requirements. They are standalone mitigations.

During the safety planning phase, it is crucial to establish clear definitions and processes to manage application of mitigations to hazards while meeting project safety requirements. While the requirement management process should focus on ensuring that requirements are met by the project, hazard logs should focus on independently determining or deriving from safety requirements, the mitigations measures toward hazard resolution. Thus, the hazard log also contains a field for referencing the safety requirements.

Another important aspect related to safety requirements and mitigations is the safety-related application conditions (SRACs). Although SRACs are not in the scope of this chapter, it is worth mentioning about them. SRACs come into picture when hazards and safety requirements lead to conditions or restrictions that should be implemented (in addition to mitigations) for ensuring safety of the system. SRACs are very important for the operator and should therefore be clearly identified in hazard logs.

13.8 Management of Transfer of Hazard and Mitigation Responsibilities

Hazards and risk transfer is an important aspect of the safety management system of a project and directly related to its system or subsystem hazard logs. This project-wide process must be clearly described during the safety planning phase. The hazard transfer process includes cases where ownership of hazards, risks, or additional mitigation actions gets transferred to and from any party in the project organization as well to and from third parties. Hazard transfers here refer to all possible types, including by the owner to the designer or by the designers to the developer, to the operator.

A hazard transfer "list" or "tracker" must be set up to ensure documentary evidence is maintained when hazards are transferred. A hazard transfer tracker should ensure that information and risks associated with hazards are not lost during the transfer and are clearly identified, transferred, accepted, and closed out. A typical tracker should have entries for the data fields as shown in Table 13.3.

TABLE 13.3

Hazard Transfer Tracker Fields, Description, and Example Values

Field	Field Description	Example Values
Hazard origin	The organization which identified the hazard	*Organization code for* operator, rolling stock supplier; signaling system
Hazard origin reference	The unique ID for the hazard from the origin hazard log	H-SYS-0123
Date of transfer	The date the hazard origin organization initiated the transfer	*Date format*
Hazard destination	The organization identified by the origin organization to which the hazard is being transferred to	*Organization code for* operator, rolling stock supplier, signaling system
Hazard description	Brief description of hazard scenario	*Free text (same as in origin hazard log)*
Initial risk	Assessment of risk based on current levels of hazard frequency and severity and existing control measures applied by the origin organization	(R1) Intolerable; (R2) undesirable; (R3) tolerable; (R4) negligible
Transfer type	To determine whether full hazard ownership is being transferred or only mitigation action	Hazard risk transfer; mitigation required
Transfer acceptance	Whether the destination organization accepts or rejects the hazard	Accepted; rejected; partly accepted
Comments by destination organization	Explanation in case hazard transfer is being rejected or partly accepted, additional notes about the hazard, etc.	*Free text*
Reference document	A document that discusses hazard ownership details, scope split or prior agreements, etc.	Document title, document number, revision
Destination reference	The unique ID for the hazard from the destination hazard log	H-O&M-0123
Residual risk	Assessment of risk based on estimated levels of hazard frequency and severity after additional control measures have been applied by the destination organization	(R1) Intolerable; (R2) undesirable; (R3) tolerable; (R4) negligible
Hazard status	The status of hazard item as determined by the destination organization	Identified; resolved; closed; transferred; obsolete; deleted

The hazard transfer tracker itself should be owned and maintained by one person, normally the project or consortium safety assurance manager, while other members of the project safety management system team will contribute to it. Only one master project hazard transfer list or tracker is recommended even if each entity may have its own copy for internal management. It is also important to identify other personnel who will be responsible for initiating or addressing hazard transfers at various levels within the project organization. It is crucial to link the hazard transfer tracker to the hazard log for a seamless information exchange.

Hazard transfers may involve not only just the members of the project organization but also external organizations, mainly direct stakeholders such as the local fire and rescue department. Ideally the goal must be to eliminate or minimize risks according to the risk acceptance principle applied by the project, however, in cases where there is a need to transfer risks to external organizations, the process has to be clearly explained to them and workshops may be required to resolve issues related to mitigations and risk ownership.

In case of disputes, a neutral party must be appointed. In case of hazard transfers within the project organization, typically the consortium-level safety assurance manager may assume the responsibility to resolve disputes. In case of hazard transfers to/from external bodies, a safety committee of key personnel, who also usually undertake other crucial roles in decision making for safety-related issues, may be constituted. On smaller projects, an interface hazard may sometimes be listed within the hazard transfer tracker but is then jointly addressed. However, this method to prevent disputes on ownerships is not advocated.

13.9 Management of Interface Hazards

Interface hazards require more attention as they involve two or more parties responsible for hazard ownership and mitigation. The organization that identifies the interface should not "transfer" an interface hazard but indicate the interfacing party and what part of the hazard is to be addressed by others. It is important that the overall interface management process of the project is followed for management of safety interfaces as well. As such, identified interface hazards should form a traceable chain to entries in the interface register and specific interface control documents (ICDs)* of the project. This can be done either by directly analyzing those hazards that are also interfaces or by separately listing interface hazards and performing an interface hazards analysis (IHA).

13.10 Verification and Validation Methodology for the Mitigations

Formal definitions of verification and validation as per EN 50126 [1] are provided in the following:

- Verification—"confirmation by examination and provision of objective evidence that the specified requirements have been fulfilled"
- Validation—"confirmation by examination and provision of objective evidence that particular requirements for a specified intended use have been fulfilled"

---

* Interface management process and related documentation, such as ICD, are beyond the scope of this chapter.

In other words, the objective of verification in a project stage is to demonstrate that the outcome of each project stage (e.g., traction power design) meets all requirements of that stage (e.g., traction power requirements). The objective of validation is to demonstrate that the system under consideration (e.g., the physical traction power conductor rail), when implemented, meets its requirements in all respects (e.g., traction power requirements). Basically, verification is to check that the system is being designed and implemented rightly, while validation is to check that the right system has been built. Validation takes place during the right hand side of the V-life cycle stages, and majority of it is performed during testing and commissioning. However, it can also take place during early stages of project life cycle for complex systems, such as a test bench validation during a factory acceptance test (FAT) or even earlier, during the design, via analysis and calculations.

However, this is slightly different from "verification of safety" in the context of a hazard log. A safety verification activity of the hazard log would be the safety manager checking a validated requirement (e.g., by design/engineering) for the validity of the mitigation evidence. Another example of safety verification would be checking traceability between hazard log entries and the hazard analyses.

The goal is to provide mitigation measures that can be practically designed and then implemented. By the end of the design stage, the project should be able to demonstrate that all hazard mitigations belonging to the design have been implemented. By the end of testing and commissioning stage, the project developer should be able to demonstrate that all hazard mitigations have been validated or transferred to the appropriate organization (and accepted). The hazard log must include all details of the evidence used to demonstrate performance of verification and validation activities. Examples of evidence documents toward verification and validation are listed in Table 13.4.

As it happens on many complex projects, when interface hazards require verification or validation inputs by multiple parties during a project stage, internal or external references must be referenced.

Each hazard is unique. It is also possible that the type of objective validation for one hazard is not entirely possible or practicable, e.g., an earthquake simulation may not be conceivable in order to validate associated hazard mitigations. Such hazards must also be identified in the hazard log and can still be validated by selecting an appropriate validation method. Therefore, a validation method must also be identified for each hazard.

TABLE 13.4

Verification and Validation Evidence Types

Verification	Validation
• Design drawings	• As-built or shop drawings
• Design specification report	• Material submittals or material inspection reports
• Requirements management verification	• Test bench simulation
• Product data sheets	• FAT reports
• Test plan and procedures	• Post installation checkout reports
• O&M manuals for products and equipment	• Dynamic test reports
	• Fire evacuation and rescue drill reports
	• O&M manuals for products and equipment

13.11 Hazard Log Statistics, Safety Progress, and Mitigation Summary

One of the key subjects of safety progress meetings and external assessments in a project will be the progress of hazard risk assessment and their status. Comprehensive hazard log statistics and results will assure the safety regulatory authority and safety auditors that a robust process has been followed and hazard management is under control. The hazard status summary should include detailed statistics so that a clear snapshot at any given time during the project life cycle can be obtained. The following is a list of the key information to be provided:

1. Number of hazards per discipline or subsystem and their status:
   a. Hazards internal to the organization
   b. Interface hazards
   c. Transferred hazards
2. Number of undesirable or intolerable residual hazards and "as low as reasonably practicable" (ALARP) arguments
3. Number of obsolete and deleted hazards
4. Number of mitigations and their status
5. Any hazards that was not successfully verified at any stage and the reason why

These figures can be presented in tabular or graphical formats as illustrated in Figure 13.2.

It is a good idea to focus on statistics for interface hazards separately. Table 13.5 shows a possible way of presenting interface hazard statistics across multiple organizations.

Lastly, a hazard shall be "closed" only if its mitigation(s) have been both verified and validated and its residual risk have been assessed to be acceptable. In case of transferred hazard, the new owner must close the hazard.

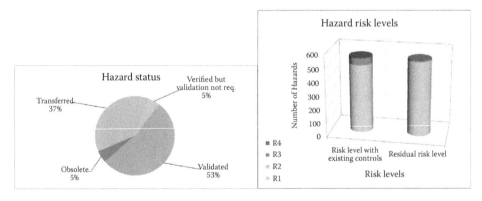

FIGURE 13.2
Simple graphical representation of hazard status and risk levels.

TABLE 13.5

Sample Summary of Status of Interface Hazards

Total Number of Hazards	Identified	Verified	Transferred	Validated
Interface with systems	188	44	136	39
Interface with contractor	38	32	2	31
External interfaces (with stakeholders)	25	10	15	10
Interface with operator	60	15	42	12
No interfaces (only civils)	243	220	11	204
Total	*554*	*321*	*206*	*295*

Abbreviations

ALARP	as low as reasonable practicable
CENELEC	European Committee for Electrotechnical Standardization (Comité Européen de Normalisation Électrotechnique)
D&B	design and build
DOORS	dynamic object-oriented requirements system
EN	European norm
FAT	factory acceptance test
HAZID	hazard identification
HAZOP	hazard and operability study
ICD	interface control document
IHA	interface hazard analysis
IEC	International Electrotechnical Commission
O&SHA	operating and support hazard analysis
PHA	preliminary hazard analysis
RAMS	reliability, availability, maintainability, and safety
SRAC	safety-related application condition
SSHA	subsystem hazard analysis

References

1. CENELEC (Comité Européen de Normalisation Électrotechnique). 1999. EN 50126-1:1999: *Railway Applications—The Specification and Demonstration of Reliability, Availability, Maintainability and Safety (RAMS)—Part 1: Basic Requirements and Generic Process*. CENELEC, Brussels.
2. International Electrotechnical Commission. 2010. IEC 61508-1:2010: *Functional Safety of Electrical/Electronic/Programmable Electronic Safety-Related Systems—Part 1: General Requirements*. International Electrotechnical Commission, Geneva.

14

Life Cycle Cost in Railway Asset Management with Example Applications

Simone Finkeldei and Thomas Grossenbacher

CONTENTS

14.1 Introduction

Over the past decades life cycle costs (LCCs) and their analysis are of growing importance. Rolling stock operators and infrastructure manager intend to know and to reduce the LCC of their assets. Knowledge of the LCC of the asset is the basis for the financial planning and can be used for decision making in case of various alternatives. The reduction of LCC is required due to the increased cost pressure on public undertakings and the grown competition based on the liberalization of the rail sector. The Council directive 91/400/EEC

(Council of the European Union 1991) requests that rolling stock operator and infrastructure are separate legal entities and that no rolling stock operator may be discriminated against in terms of track access.

LCCs are the sum of the costs that occur over the life span of a product, whereby revenues and depreciation of the assets is not considered in the LCC (standard EN 60300-3-3:2005-03). LCC analysis, respectively, life cycle costing, is the economic method of evaluating the costs over the life cycle of the product or a portion of the life span.

In this chapter, the life cycle of the product is based on standard EN 60300-3-3 (EN 60300-3-3:2005-03) divided in the following seven phases:

1. Concept and definition
2. Design and development
3. Manufacturing and construction
4. Assembly and integration
5. Operation
6. Maintenance
7. Disposal

However, depending on the application of the LCC analysis, it is recommended to define the phases according to its scope.

For phase 1, typically the rolling stock operator, respectively, infrastructure manager, is responsible for the conception and definition of the product. During phases 2, 3, and 4, the responsibility for the product usually lies in the hand of the railway industry to a large portion, whereas during phases 5, 6, and 7, the infrastructure manager or rolling stock operator is usually responsible. In the last years, maintenance contracts become more important for the maintenance of rolling stock. The manufacturer or a third-party contractor specializing on maintenance performs the maintenance for a defined price per performance unit of the train. The handover between phases 4 and 5 is accompanied with the payment of the acquisition price for the product, which means that the LCC arising in phases 1, 2, 3, and 4 is indirectly paid by infrastructure manager, respectively, rolling stock operator through the acquisition costs. Costs that occur in phases 5–7 are known as ownership costs. Operation and maintenance are taking place during the same time period of the products life cycle. As during those two phases, a huge portion of the LCC are arising; these phases are treated separately in this chapter.

Processes and methodologies for LCC analyses, the integration of LCC analyses in the management of railway assets, the application of LCC analyses in the different phases of the life cycle, and the integration of the LCC process in the quality process are described in this chapter. Furthermore, the objectives and benefits of LCC analyses are discussed. The chapter identifies cost elements of the LCCs of railway assets. Exemplarily influencing factors for maintenance production costs are named, and it is described how railway industries and operator can influence them. On the basis of two examples, the application of LCC analysis is illustrated. To complete the chapter, opportunities of LCC analyses for railway assets are discussed.

14.2 Process and Methodology of LCC Analyses

14.2.1 Process according to European Standard

The application guide life cycle costing IEC 60300-3-3 (International Electrotechnical Commission 2004) defines the steps of a LCC analysis, which are shown in Table 14.1.

These steps can be taken as a basis to create the LCC flow chart in a project, to define the LCC business planning processes, or to align the asset management process according the current life cycle.

14.2.2 Management Application of LCC Process

LCC analysis should be part of the business planning in the strategic and financial planning process of the enterprise. Essential for LCC analysis is a cyclic procedure. A mature asset management planning is the basis to determine new life cycle information of a product in order to prepare the information for the next deadline of life cycle planning within the entire asset portfolio.

As part of the management processes *planning of the analysis* and *definition of the analysis approach* (see Figure 14.1) can be generally defined in a management system. The management system should provide general information, e.g., forms, templates, and other auxiliaries, to *perform the analysis*. Additionally, the management system should define starting conditions and criteria to *perform the analysis*, e.g., on an annual basis or case related.

The definition of the management processes and the responsibilities for the LCC process in the organization supports to focus on LCC. The granularity must depend on the maturity of the organization and on the competencies regarding LCC process and methodology of the staff.

TABLE 14.1

LCCs that Can Be Influenced Depending on the Current Life Cycle Phase

		Influenceable Costs					
		Concept and Definition	Design and Development	Manufactoring, Construction	Assembly, Integration	Operation and Maintenance	Disposal
Current Life Cycle Phase	Concept and Definition						
	Design and Development					FUTURE	
	Manufactoring, Construction			PRESENCE			
	Assembly, Integration						
	Operation and Maintenance		PAST				
	Disposal						

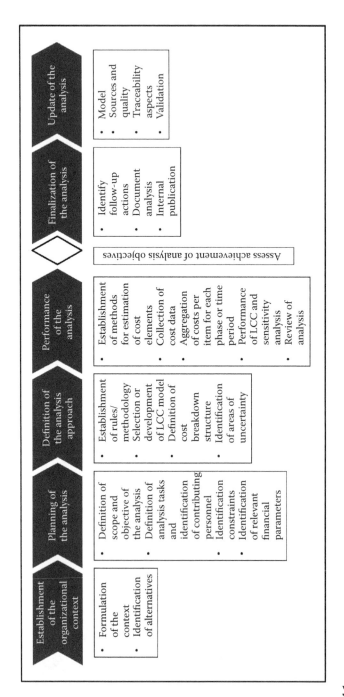

FIGURE 14.1
Steps of the LCC analysis process.

14.2.3 Application of LCC Analyses based on the Life Cycle Phase

Apart from the integration of the LCC process in the management system of the enterprise, it is important to define the specific application of the LCC analysis for the life cycle phases of the product.

As shown in Table 14.1, the LCC cover the following costs:

- Costs that occurred in the past (black fields in Table 14.1): These costs have typically been already paid and are well known and proved, but cannot be influenced anymore.

- Current costs (gray fields in Table 14.1): Those costs are mainly determined and occur for activities of the current life cycle phase. Their cost elements are typically well known. They can be influenced by challenging and optimizing process costs and purchasing costs for material. Inherent costs, depending on system properties that have been defined in previous life cycle, can be influenced only to a small portion.

- Future costs (white fields in Table 14.1): These can be influenced or changed by current activities and today's decisions. However, those costs are typically not known. They must be estimated or determined in advance.

Several organizations are processing specific LCC analyses, depending on the current phase of the life cycle. Thus, those organizations have specific management processes defined that are applicable according to the current life cycle phase. Hence, the process and the realization of the different steps (see Figure 14.1), e.g., the definition of analysis approach, the performance of the analysis, the finalizing of the analysis, and the update of the analysis, must be adapted based on the current life cycle phase and will in the following be described for phases 1–6.

14.2.3.1 LCC Process as Part of the Concept and Definition Phase

In the early stages of the life cycle of the product, the product, either a new system or a rework of an existing system, is just a documented idea. In the concept and definition phase, there is low or little experience available regarding the new product. However, reliability, availability, and maintainability (RAM) and LCC targets must be defined. Depending on the roles and responsibilities along the value chain, the transparency of cost and expenditures in the life cycle phases is not given for all actors. As a common scheme in railway applications, the operator draws up the *concept and definition* of a product, followed by *design and development, manufacturing or construction*, and *assembly and integration* of the product by a contractor. Then, the operator performs *operation and maintenance* or commissions the manufacturer, respectively, constructor or another contractor, to execute those activities. In that case, the designated operator or maintainer needs to have market information available for every activity to be commissioned. Data elaboration can be supported by analogies or with a parametrization methodology and the use of existing product information. The defined LCC targets must be documented and, in case of commissioning operation or maintenance, contractually agreed.

14.2.3.2 LCC Process as Part of the Design and Development Phase

The LCC process shall be aligned with the design and development processes. LCC targets must be apportioned to the systems and subsystems and the achievement of the targets shall be controlled as part of the design review of every design stage. It is recommendable that the operator (provider of *operation and maintenance*) participates in the LCC review. A feedback loop within the design phase allows compliance with the agreed values. Consequently, the LCC analysis must be updated. Hereinafter, the estimated values of the following life cycle phases have to be checked for plausibility and prepared as input for the following life cycle phase.

14.2.3.3 LCC Process as Part of the Manufacturing and Construction Phase

During the manufacturing and construction phase, the all actors of the value chain have defined roles and tasks. The objectives of LCC process are the following:

- Verify LCC based on RAM characteristics of the components and subsystems
- Gain information from the production regarding the components and their verification tests, e.g., of failure rates, energy consumption, or life span

Consequently, the LCC analysis must be updated. The estimated values for the following life cycle phases have to be checked for plausibility and prepared as input for the following life cycle phase.

14.2.3.4 LCC Process as Part of the Assembly and Integration Phase

During the assembly and integration phase, all actors of the value chain have defined roles and tasks. The objectives of LCC process are the following.

- Verify RAMS and LCC characteristics of the entire system
- Gain information regarding the integrated system and its verification tests, e.g., of failure rates, energy consumption or life span
- Elaborate accordance of agreed values between contractors or determination deviations

Consequently, the LCC analysis must be updated. The estimated values for the following life cycle phases must be clarified and completed. In that phase, the targets for operation and maintenance cost can be finally determined.

14.2.3.5 LCC Process as Part of the Operation and Maintenance Phase

In the operation and maintenance phase, the operating experience of the system within the operational environment must be collected. The LCC analysis must be updated cyclic or at special occasion. A cyclic (e.g., annual) update is helpful to support the business planning process in an optimal way. A specific update in advance of a major revision or renovation supports the selection of the right tasks for the planned activity. Additionally, the determined data can be used as input for new projects and concepts.

14.2.4 Methodology of LCC Analyses

Typically, railway undertakings and infrastructure managers as well as operation or maintenance providers are handling a wide portfolio of different products. Their assets can be structured according to the following categories:

- Fleets or vehicle families
- Components or subsystems of every fleet with similar properties
- Lines
- Type of equipment of every line with similar properties
- Customers

An overall LCC analysis including all assets can be performed if an evaluation is available for every specific asset.

The cost determination is a key issue to process LCC analyses. Applied product costs data should be up to date, and information upon the data validity should be available. Cost evaluation is often based on the operational accounting combined with a product cost analysis. Different methods for cost evaluation are detailed in Table 14.2.

The methods "engineering practice", "analogy", and "parameterization" are described in the application guide IEC 60300-3-3 (International Electrotechnical Commission 2014). Knowing that collecting input data can be the weak point of every LCC analysis, it may be helpful to take redundant sources into consideration or to use different methods to determine especially the most significant input parameters.

14.2.5 Quality Management for LCC Analyses

International standards, especially ISO 9001 and ISO 55000 (International Organization for Standardization 2014), define requirements regarding a management system. Section 14.2 explains how to integrate the LCC process in the management system and how to link it to other processes. Nevertheless, the LCC process follows an applied plan–do–check–act logic itself.

A few aspects of quality assurance should be taken in account for LCC processes:

- Knowledge and skill management (see EN ISO 9001:2015, Chapter 7.1.6)
- Competence and its management (see EN ISO 9001:2015, Chapter 7.2)
- Communication and publication, e.g., to contractors and suppliers (see EN ISO 9001:2015, Chapter 7.4)

TABLE 14.2

Methods for Cost Evaluation

	Methods for Cost Evaluation			
	Experience	**Engineering Practice**	**Analogy**	**Parameterization**
Origin of Data	Based on known product costs from the operational accounting, and/or service costs (updated if necessary)	Based on business product cost analysis	Based on known product costs from the operational accounting of a similar or scalable product or service	Costs are determined by cost estimating relationships
Requirements	Knowledge about validity of the information and knowledge about how to transform the figures to the present (or future)	Knowledge about the process, the materials, and their costs	Knowledge about the based product and about the effects of differences to the costs	Cost elements and product structure

- Steering of documented information (see EN ISO 9001:2015, Chapter 7.5.3)
- Secrecy and integrity (see EN ISO 9001:2015, Chapter 7.5.3)

Hence, every LCC analysis including the input data should be finally reviewed. All review results and the analysis must be documented.

14.3 Objectives and Benefits of LCC Analyses

Methods and processes described in Section 14.2 are the basis for the application of LCC analyses with the following objectives:

- Budget planning
- Project planning
- Cost reduction
- Decision making

14.3.1 Budget and Project Planning

LCCs are one share of the costs that have to be planned in the budget of rolling stock operator and infrastructure manager. They are relevant for both short- and long-term financial planning. Rolling stock operators and infrastructure managers perform LCC analyses for projects to estimate the required budget, its profitability, and the feasibility from the financial point of view.

For budget and project planning, it is required to estimate the complete LCC of all products. On the contrary, the usage of LCC analyses for decision making is only necessary to focus on the differences between the alternatives.

14.3.2 Cost Reduction

Due to the cost pressure on rolling stock operator and infrastructure manager and in order to be competitive, the reduction of costs is an important target. For this purpose LCC analyses are the basis with the following entry points:

- Identification of the cost driver of a product
- Seeing the big picture comprising acquisition and ownership costs

The identification of the cost drivers enables to recognize the cost elements or influencing factors that have a huge impact on the LCC and, hence, define the fields to focus on in order to optimize the costs.

The ownership costs (see Section 14.1) are a huge portion of the whole LCC, due to the long lifetime of railway assets. In the past, the procurement focus was set on acquisition costs, and the ownership costs were of minor relevance. Taking a big-picture approach supports to find an economic equilibrium between acquisition costs and ownership costs (see Figure 14.2) and to optimize the LCC of a product to gain an economic advantage. The target is to reduce the LCC and, in best case, to achieve the minimum. Generally spoken,

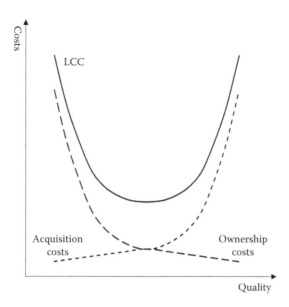

FIGURE 14.2
Trade-off between acquisition and ownership costs depending on quality.

the improvement of the product, respectively, product quality, leads to increasing acquisition costs and to reduced ownership costs.

Examples of influencing factors for acquisition and ownership costs are maintainability and reliability. To set a focus on the improvement of maintainability or reliability of a product during the design and development phase requires the utilization of engineering hours and can demand for additional requirements regarding the product, e.g., the selection of material of higher quality.

The focus on LCC which means to concentrate on the addition of acquisition and ownership costs instead of only on the acquisition costs leads to a product which is optimized with respect to the LCC. Therefore, LCC became more important in invitations to tender, and often, LCC target values or parameters that influence LCC are defined in contracts between the railway industry and rolling stock operator.

14.3.3 Decision Making

For the economic assessment of different alternatives, a LCC analysis can be performed to find the most cost-efficient alternative. For example, LCC can support choosing between the following:

- Similar products of different suppliers
- Different technologies
- New product acquisition or refit of the current product
- Different design solutions
- Different maintenance or operating strategies

In the case of decision making, only costs that relate to the differences between the alternatives need to be considered, and it is not required to calculate all costs of the product, in order to set up a ranking between the different options. However, a prerequisite is that the alternatives fulfill all operational, technical, and safety requirements. In the case that the operational and technical requirements are partly may criteria and, therefore, the decision is not only dependent on LCC, the result of the LCC analysis can serve as input for a classical cost–benefit analysis or a utility analysis in order to make the decision.

14.4 Cost Breakdown Structure for Railway Assets

A cost breakdown structure is the basis for a LCC analysis. Table 14.3 shows the first two levels of a breakdown structure of LCC for railway assets based on the phases of the product life cycle.

TABLE 14.3

LCC Breakdown Structure

Level 1	Level 2
Concept, definition	Engineering costs
	Costs for contract negotiations
	Project management costs
Design and development	Homologation costs
	Engineering costs
	Verification and validation costs
	Project management costs
Manufacturing, construction	Production costs
	Costs for production infrastructure and equipment
	Quality costs
	Logistics costs
	Procurement costs
	Costs for planning and support
	Project management costs
Assembly, integration	Production costs
	Logistics costs
	Costs for planning and support
	Project management costs
Operation	Operation costs
	Costs due to nonavailability
	Logistics costs
	Quality costs
Maintenance	Cost for the performance of the maintenance tasks
	Costs for planning and support
	Quality costs
	Logistics costs
	Costs for production infrastructure and equipment
Disposal	Demounting/dismantling costs
	Residue disposal costs
	Logistics costs

The list is not exhaustive and needs to be adapted to the application of the LCC analysis. In this chapter, only direct costs are considered. Semi-indirect costs such as capital costs, overhead costs, costs for planning of operation and maintenance, taxes, costs for training of employees, certification, and accreditation and indirect costs such as taxes, maintenance of equipment, and costs for information system are either not part of the LCC or often cannot be dedicated to a specific product and often occur in all phases of the product life cycle and are therefore not considered.

Exemplarily, the next structural levels of the cost breakdown structure of the costs for the performance of the maintenance for railway rolling stock are shown in Table 14.4.

For rolling stock, the costs for the performance of the maintenance are initially defined by the railway industries to a large portion; however, the maintainers can partly influence the costs themselves. The interval of the preventive maintenance is defined by the rolling stock manufacturer. The longer an interval is, the less often the maintenance task will be performed during the lifetime of the product, which leads to a reduction of the LCC. Maintenance tasks with short intervals even with short task times have a huge impact on the LCC over the long lifetime of up to 50 years. Based on experience, appropriate safety analyses, and safety control measures and dependent on technical feasibility, the operator can extend the maintenance intervals.

The failure rate is depending on the reliability of the component, and the higher the failure rate, the more often the maintenance task has to be performed in a given time and, which leads to an increase of the LCC.

The minimal maintenance task duration is defined by the design of the product and therewith by the rolling stock manufacturer. The manufacturer and operator can influence the material costs. The manufacturer impacts them by the design. Both can influence the material costs by standardization and/or negotiations, respectively, contracts with the component suppliers. Additionally, the order size impacts the material price.

The hourly rate of the maintenance staff is defined by the wage level. As machinery and equipment are assets as well as rolling stock, LCC analyses are necessary to determine the hourly rate for the machinery and equipment and will not be detailed here.

All parameters have to be defined for every component of the product, and then, the total costs can be calculated.

No residual value must be considered for the LCC over the whole life cycle, as at the end of the life cycle, the product is disposed of. However, in case the operator sells the rolling

TABLE 14.4

LCC Breakdown Structure for Costs for the Performance of Maintenance

Level 1	Level 2	Level 3	Level 4
Maintenance	Cost for the performance of the maintenance tasks	Preventive maintenance costs	Interval of the maintenance task
			Maintenance task duration/man hours
			Material costs
			Hourly rate for the maintenance staff
			Hourly rate for machinery and equipment
		Corrective maintenance costs	Failure rate
			Maintenance task duration
			Material costs
			Hourly rate for the maintenance staff
			Hourly rate for machinery and equipment

stock to another operator, the residual value should be considered in the financial planning as only a span of the life cycle is observed.

14.5 Example Applications of LCC Analyses

14.5.1 Technology Change: Hydraulic Axle Guide Bearing

Wheel wear is one of the driving factors of total LCC of rolling stock. Natural wear is to a huge extent correlating to the forces that occur between wheel and rail. The axle guidance stiffness, in particular the stiffness of the axle guide bearing, influences the ability of the wheelset to steer in a curve and, hence, the forces between wheel and rail. In curves, a low longitudinal stiffness is preferable to reduce the wheel guidance forces, but on straight tracks, the stiffness needs to be higher to ensure running stability. Freudenberg Schwab Vibration Control AG has developed a hydraulic axle guide bearing which changes its characteristic depending on the moving frequency of the wheelset (see Cordts and Meier [2012]). In curves with small radiuses, its stiffness is reduced, and at higher speeds on straight tracks or in curves with large radiuses, it is stiffer. Conventional axle guide bearings only provide one stable stiffness.

Swiss Federal Railways, the national rolling stock operator in Switzerland, operates three different types of passenger coaches, for which it had been decided to equip with hydraulic axle guide bearing (see Grossenbacher, Noll, and Edmaier [2016]). An LCC analysis has been performed to serve as basis for the decision.

To define the impact on the LCC of this technology change, the areas that are affected need to be identified:

- Project management costs
- Homologation costs for the new hydraulic axle guide bearings
- Costs for the installation of the new technology
- Costs for the maintenance of the hydraulic axle guide bearing/conventional axle guide bearing
- Costs for the maintenance of the wheel
- Costs for energy consumption during operation of the train
- Costs for maintenance of the track

Hereafter for each of those areas, the effects of the new technology will be described.

- *Project management costs*
  The technology change from the conventional bearing to the hydraulic bearing needs to be managed, and hence, a budget needs to be planned for the project management.
- *Homologation costs for the new hydraulic axle guide bearings*
  The homologation is based on test runs, safety assessment, and the preparation of deliverables for the application at the authorities.

- *Costs for the installation of the new technology*

  The conventional axle guide bearing must be exchanged during the overhaul of the bogie as it then is supposed to reach its end of life. As the interfaces of the conventional bearing and of the hydraulic bearing to the bogie are equal, no changes on the vehicles had to be implemented to install the new technology. Hence, it was decided to perform the technology change during the overhaul, and the only difference for the installation of the new technology for those fleets is caused by the higher price of the new product.

- *Costs for the maintenance of the hydraulic axle guide bearing/conventional axle guide bearing*

  The interval and duration of the visual inspection is equal for the conventional and hydraulic bearing; a higher failure rate was not expected and the exchange interval of the bearing stays the same. Hence, for the maintenance of the hydraulic bearing compared to the conventional bearing, the higher price is the only impact on the LCC.

- *Costs for the maintenance of the wheel*

  The basic maintenance tasks for a monobloc wheel are the following:

  - Visual inspections
  - Measurement of wheel profile, back-to-back distance and diameter
  - Wheel turning
  - Wheel exchange

  Visual inspections are mainly performed to detect failures on the wheel, e.g., wheel flats and cracks. The wheel profile is measured to identify the wheel wear and to decide if the next wheel turning, respectively, wheel exchange, is required. Wheel turning is performed to correct the profile, and in case the actual diameter is close to the wear limit, a wheel exchange instead of wheel turning is necessary.

  The impact on the wheel maintenance due to the technology change is the extension of wheel turning and wheel exchange interval, leading to a reduction of the number of wheel turnings and exchanges over the lifetime of the rolling stock. Hence, the costs for those tasks over the residual lifetime of the fleets are reduced. Additionally, the number of conveyances of coaches to those tasks during the residual lifetime is reduced, leading to a further reduction of the LCC.

- *Costs for energy consumption during operation of the train*

  As wheel wear means that energy is consumed to remove material, it is obvious that the reduction of wheel wear also leads to a reduction of the energy consumption by the train during operation.

- *Cost for maintenance of the track*

  Forces acting between wheel and rail cause rail wear, as well as they cause wheel wear. Hence, the technology change and therewith the reduction of the forces between the wheel and rail leads to the reduction of the maintenance costs for the track and, therefore, a reduction of the LCC to be paid by the infrastructure manager.

  Not all LCC impacts that lead to a reduction of LCC have been considered for the decision, as some of the costs are not easy to determine, and especially when the cost reduction does not affect the responsible legal entity, the reduction can hardly be argued. For example, the reduction of required maintenance of the track does not lead to an advantage for the rolling stock operator, but only for the infrastructure manager. However, the rolling stock operator is paying for the technology change and does not benefit from the advantages that are generated for the infrastructure manager. In the last years, some infrastructure manager started to implement track access charges which are dependent on the wear that is applied to the track. This would allow the rolling stock operator to calculate the difference of the track access charges in their LCC analysis, and hence, the improvement for infrastructure can be considered.

The new technology also reduces the noise emission, which cannot typically be expressed in a cost reduction.

14.5.2 Change of Wheel Material Quality

As already stated in Section 14.5.1, wheel wear and the resulting wheel maintenance in form of wheel reprofiling, and wheel exchange are driving factors for the LCC of railway rolling stock. Besides the reduction of forces between wheel and rail (see Section 14.5.1), another way to reduce the wheel wear is to enhance the wheel material quality.

The very small physical interface between the two systems' infrastructure and rolling stock with its interaction is characteristic for the wheel–rail contact. To show the application of LCC analysis for this example, the process introduced in Figure 14.1 is followed.

- *Establishment of the organizational context*

  The wheel–rail contact is influenced by the following inherent parameters:

  - Material and the related heat treatment
  - Shape of the wheel and rail profile in new condition and in wear condition until achieving the intervention criteria for maintenance
  - Dynamic and static reaction behaviors of both partners, the suspension system of the bogie with its springs and dampers, and the stiffness of the track body
  - Geometric and kinematic parameters such as radii and gradient

  Additionally, environmental parameters such as operational conditions, speed, accelerations, breaking intensity, load factors, operation strategy, load history, and more are contributing to total LCC of the railway system.

  Figure 14.3 shows the influence of wear and degradation of wheel and rail material during operation on the LCC costs due to the necessary maintenance activities.

  The creation of an LCC model considering the relevant parameters of all subsystems varying in the following aspects is necessary to realize a reliable LCC analysis:

  - Time
  - Condition
  - Operation

  However, business reality after dividing the railway companies in railway undertakings and infrastructure managers leads to decisions based on business cases for the subsystem of the affected company, not considering effects on the total system. Constellations of independent railway untertaking, infrastructure manager, and vehicle manager are even more difficult to handle. Hence, the effects of changes on the railway system are often only analyzed for the associated subsystem of the

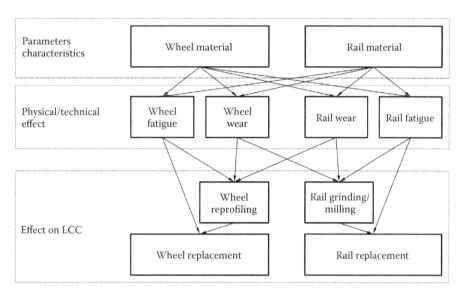

FIGURE 14.3
Modeling approach of the technical–economic relationships.

independent undertaking and not for all affected subsystems (see Figure 14.3). If there is an essential contribution on the LCC of other subsystems, the results typically contain unquantified effects on other railway subsystems. This mechanism and the resulting decisions harm the financial health of the railway system. Therefore, it is necessary to find criteria for the definition of boundary conditions for the systems, to be able to analyze the effects of simple modifications on one subsystem. As an objective, the change of wheel material quality must be rated as a preparation for the business decision.

- *Planning the analysis*

    To perform the analysis, knowledge about the reaction on both affected railway subsystems, rolling stock and track, is necessary. If data based on experience from former tests or former operational use of the new material are available, it must be checked in terms of validity for the planned change. If there is no database available, a pretest to gather data for the parameters would be an alternative and would have to be planned. The planning should also be scheduled in a common project management manor.

    Considering that the change of material in the ongoing life cycle must be rated, the analysis shall consider economic effects from this period up to the end of life of the affected rolling stock and track. Hence, the modeling and data collection must be performed based on this boundary condition.

- *Definition of the analysis approach*

    As the wheel and rail interface is an interactive system, the effects on both subsystems must be considered.

    1. *Process to reach a model simplification*

        The first step in considering the effects on to influenced railway subsystems (e.g., on the track body or on the vehicles bogies) is to build up a two-dimensional decision matrix (see Figure 14.4).

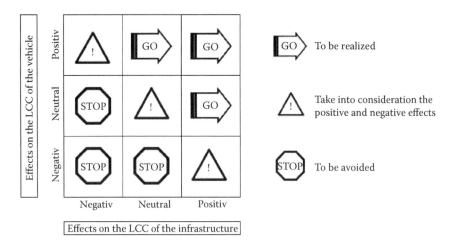

FIGURE 14.4
Two-dimensional decision matrix for subsystems vehicle and infrastructure.

A change or modification is analyzed in terms of negative, neutral, or positive effects on both subsystems. If the effect on both systems is positive, the implementation of change is strongly recommended. If the effect on one subsystem is positive and on the other it is neutral or unknown, the change is recommended, when the positive and negative effects on both rolling stock and infrastructure are considered in the model. If the effect on one subsystem is negative, or on both subsystems, is neutral, a modification should only be realized, if mandatory due to law and regulation, but not due to economical motivation.

Hence, the application of this simplified assessment matrix supports the decision if a change, e.g., of the wheel material quality shall be conducted. For example, the change is recommended to proceed in the following cases:

– The railway undertaking is responsible for the subsystem vehicle and the LCC analysis regarding the change and the effects on its subsystem has been performed. The result of the analysis is positive.

– The infrastructure manager points out that the influence of change of the wheel material quality on the tracks LCC is negligible.

2. *Application of an integrated LCC analysis*

The breakdown structure established in Section 14.4 can be adopted for each influenced subsystem and to gain transparently compared to each other (see Table 14.5).

Obviously, every application of an LCC model depends on the objectives to be achieved and on the area of application. In case of several modification variants, it is possible either to compare all variants with each other or to define a parametrized model. In the field of the wheel–rail interface, some steady variation of parameters lead to steady physical effects on the influenced subsystems. Hence, it can be expected that the effected costs react in a steady way.

However, discontinuities in maintenance planning and performing lead to discontinuous effects in the overall cost development.

So the extension of wheel life effects the LCC of the rolling stock if at least one wheel treatment will be omitted and no additional maintenance activities due to a reduced harmonization with other maintenance activities must be scheduled. If, for example, dismantling of the wheelset is still necessary for other maintenance activities, savings will be reduced. So, in this example, it is recommended to run an analysis for the starting point and for each new wheel material specification.

TABLE 14.5

LCC Breakdown Structure for a Wheel Material Quality Modification

Structural Level 1	Structural Level 2	Subsystem Vehicle	Subsystem Infrastructure
Concept, definition	Engineering costs	None	None
	Project management costs	Defining homologation concept	None
	Costs for contract negotiations	None	None
Design and development	Homologation	Tests to be performed: - homologation tests on rig - performance tests on rig - performance tests in operation	None
	Engineering costs	Documents to be delivered	None
	Project management costs	None	None
	Verification and validation	None	None
Manufacturing, construction	Project management costs	None	None
	Production costs	Cost difference due to modified material quality	None
	Costs for infrastructure	None	None
	Quality costs	None	None
	Logistics costs	None	None
	Procurement costs	None	None
	Costs for planning and support	None	None
Assembly, integration	Project management costs	None	None
	Production costs	None	None
	Logistics costs	None	None
	Costs for planning and support	None	None
Operation	Operation costs	None	None
	Costs due to non-availability	None	Cost difference due to more/less line availability (taking in consideration, that mostly the exchange of rails leads to unavailability of track)
	Logistics costs	None	None
	Quality costs	None	None

(continued)

TABLE 14.5 (CONTINUED)

LCC Breakdown Structure for a Wheel Material Quality Modification

Structural Level 1	Structural Level 2	Subsystem Vehicle	Subsystem Infrastructure
Maintenance	Production costs	Cost difference due to more/less lathe-treatments Cost difference due to more/less exchange of wheels (assembly/disassembly in maintenance)	Cost difference due tomore/less rail grinding/milling Cost difference due tomore/less rail exchange
	Costs for planning and support	None	None
	Quality costs	None	None
	Logistics costs	Cost difference due to more/less exchange of wheels (assembly/disassembly in maintenance)	None
Disposal	Demounting/Dismantling	None	None
	Residue disposal	None	None
	Logistics costs	None	None

- *Performance of the analysis*

    To perform the analysis, each cost element must be determined and filled in the model. Due to the integrated model of two subsystems the scaling unit must be defined at the beginning, e.g., the comparison of annual cost. Another approach is to find a specific indicator considering the compared subsystems performance like costs per gross mass and vehicle mileage (see Figure 14.5) or costs per gross mass and track length (see Figure 14.6).

    The plausibility can be checked by comparing the results of single cost elements with data of former years, e.g., of annual reports. Another possibility is to find

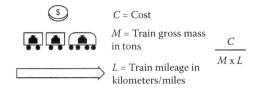

C = Cost

M = Train gross mass in tons

L = Train mileage in kilometers/miles

$$\frac{C}{M \times L}$$

FIGURE 14.5
Specific cost indicator for the subsystem vehicle.

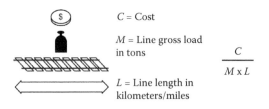

C = Cost

M = Line gross load in tons

L = Line length in kilometers/miles

$$\frac{C}{M \times L}$$

FIGURE 14.6
Specific cost indicator for the subsystem infrastructure.

benchmark partners, e.g., other railway undertakings or infrastructure managers, and to discuss and compare the results of the LCC analysis.

A sensitivity analysis should show the importance of considering the impact on infrastructure costs.

- *Finalization of the analysis*

    In the phase *performance of the analysis*, the focus is set on estimation of the cost element differences between the alternatives caused by different input values. The LCC analysis does not contain capital cost. Therefore it is necessary to consider annual interest rates to evaluate the net present value (for more information, see also EN 60300-3-3, Annex B).

- *Update of the analysis*

    An update of the analysis can be indicated when the following occurs:

    - A new technology in the investigated part of subsystems is coming up.
    - Input values significantly change, e.g., material prices.
    - Regulation leads to new money flows.

    In any case, an analysis should be frequently updated, if the results are part of frequent business planning processes. There is typically no need of updating if the analysis has been performed with the only objective to give a good basis on decision making for a new wheel material.

14.6 Discussion of the Opportunities of LCC Analyses for Railway Assets

Performance of LCC analyses is a complex and time-consuming task. Setting up an LCC breakdown structure and calculating the LCC theoretically allows reaching the minimal LCC of a product. This is not possible in practice due to the following facts:

- The LCC breakdown structure which covers all costs is highly complex; hence, a complete and totally consistent LCC analysis is not possible to reach. Double counting of costs for which it is not entirely distinct to which cost element they belong can occur. Additionally, it is possible to miss important cost elements, e.g., in the railway field, the focus is often set on calculating the "planned" costs not considering costs due to nonavailability or for time-consuming failure root cause analysis.

- Not all information is available, and therefore, some parameters need to be estimated, which stands in contradiction to a complete LCC without uncertainties.

- During the long lifetime of railway assets, input values for the LCC analysis and the whole market situation can drastically change, which cannot be foreseen over this long period, leading to inherent uncertainties of the LCC analysis.

- Different legal entities can influence the LCC of the product during the different phases of the lifetime. Those entities try to increase their own benefits and their targets can be in contradiction to the target to reduce the LCC of the product over the whole lifespan.

This knowledge and the uncertainties need to be considered when using LCC analyses and their results for financial planning and decision making. Furthermore, it has to be considered that LCCs are one input for the financial planning, but it is important to also incorporate revenues, residual values at disposal, capital expenditures, etc.

Especially for decision making, only the differences between the alternatives should be calculated in the LCC analysis to reduce the complexity. Besides that, it is possible, that the available fund is not sufficient to choose the alternative with the lowest LCC, but high acquisition costs. However, carrying out the LCC analysis enables to take a balanced decision considering LCC and the funding. A structured LCC breakdown structure supports a systematic analysis and helps reduce the uncertainties.

To improve LCC even though different legal entities are involved in the product life cycle, it is required to apply methods to challenge entities that are defining or impacting the level of the costs by rewarding them for determining low LCC. One method, for example, is to choose products with low LCC instead of low acquisition costs. Another example is wear-dependent track access charges. Those methods allow for the development of innovative products which are more expensive regarding the acquisition costs, but save expenses over the lifetime of the product.

References

Cordts, D., and Meier, B. 2012. Hydraulic axle guide bearings for railway vehicle applications. *Der Eisenbahningenieur* 3:69–73.

Council of the European Union. 1991. Council Directive 91/440/EEC: On the development of the Community's railways. *Official Journal of the European Communities* 25–28.

Grossenbacher, Th., Noll, U., and Edmaier, J. 2016. Effects of the current incentive systems in Switzerland on vehicle design—Findings and outlook. *ZEVrail* 140:173–183, Sonderheft Graz.

International Electrotechnical Commission. 2014. IEC 60300-3-3:2014-09: *Dependability Management—Part 3-3: Application Guide—Life Cycle Costing*. International Electrotechnical Commission, Geneva.

International Organization for Standardization. 2014. ISO 55000:2014: *Asset Management—Overview, Principles and Terminology*. International Organization for Standardization, Geneva.

15

System Assurance for Railway Mechanical Signaling Products

Kyoumars Bahrami

CONTENTS

15.1 Introduction

System assurance is a structured, systematic, and proactive approach based on adopting and implementing appropriate processes, procedures, tools, rules, and methodologies throughout the life cycle of products, processes, systems, and operations to ensure the requirements related to reliability, availability, maintainability, and safety (RAMS), i.e., gain confidence in integrity and correctness of safety and reliability.

In the case of railway mechanical signaling products, it is a framework to manage the risks and ensure the product has been designed and constructed considering all the RAMS critical factors.

As well as the moral obligation to avoid harming anyone, among the drivers for system assurance are compliance with the regulatory requirements (safety legislations) and the safety standards for avoiding accidents and managing safety.

Safety is defined as freedom from unacceptable risk of physical injury or of damage to the health of people, either directly or indirectly as a result of damage to property or to the environment. Functional safety is the part of the overall safety that depends on a system or equipment correctly operating in response to its inputs.

When the safety of a system relies on the control system, then the control system has a safety function (functional safety) and should fulfill the requirements of the relevant standard (e.g., ISO 13849).

Machinery/machine is defined in Section 3.1 of ISO 12100-1 as an "assembly, fitted with a drive system consisting of linked parts or components, at least one of which moves, and which are joined together for a specific application."

This chapter describes the railway mechanical signaling products and their differences with the electrical/electronic/programmable electronic (E/E/PE) products. It also covers the concept of performance level (PL) that can be used for the safety assurance of the railway mechanical signaling products.

15.2 Concept of Safety Integrity Level

The safety integrity level (SIL) concept was introduced in the UK Health and Safety Executive (HSE) programmable electronic system guidelines and subsequently extended in the development of IEC 61508. SILs are defined for E/E/PE systems only, or for the E/E/PE part of a mechatronic system.

If an SIL is assigned to a mechanic or mechatronic equipment, then this is to be considered not applicable (quote from CENELEC rail safety standard EN 50129).

The SIL table presented in IEC 61508/EN 50129 is the most commonly accepted for electronic systems and is provided for guidance. It is suggested that it may not be applicable to other systems, in particular, mechanical systems.

In such cases, other standards, the "proven in use" argument, or best practice could be of special value. Among other standards that can be used for mechanical systems is ISO 13849.

As such, the SIL concept can only be used for the E/E/PE safety-related technologies.

The rail derivative of IEC 61508 is EN 50129, *Railway Applications—Communication, Signalling and Processing Systems—Safety Related Electronic (E) Systems for Signalling*. Again, the SIL concept can be used only for electronic safety-related technologies (i.e., electronic systems that carry our safety functions).

The IEC 61508 and EN 50129 standards do not allow the allocation of SILs to non-E/E/PE systems or functions (e.g. mechanical, hydraulic, and pneumatics).

15.3 E/E/PE Safety-Related Systems

One mean of implementing a safety function is through electrical (e.g., electromechanical relays), electronic (e.g., nonprogrammable solid-state electronics), and programmable electronics (e.g., microprocessors).

Section 3.2.6 of IEC 61508-4 defines E/E/PE as based on electrical and/or electronic and/or programmable electronic technology.

The range of E/E/PE safety-related systems to which IEC 61508 can be applied includes the following:

- Emergency shut-down systems
- Fire and gas systems
- Turbine control
- Gas burner management
- Crane automatic safe-load indicators
- Guard interlocking and emergency stopping systems for machinery
- Medical devices
- Dynamic positioning (control of the movement of a ship when in proximity to an offshore installation)
- Railway signaling systems (including moving block train signaling)
- Variable speed motor drives used to restrict speed as a means of protection
- Remote monitoring, operation, or programming of a network-enabled process plant
- An information-based decision support tool where erroneous results affect safety

The dominant failure mode for an E/E/PE safety-related railway signaling system is characterized by a constant failure rate, i.e., random failures.

15.4 Non-E/E/PE Safety-Related Systems

Another mean of implementing a safety function is through non-E/E/PE safety-related systems.

Examples of non-E/E/PE systems in the rail industry are some mechanical, hydraulic, or pneumatic trackside products (mechanical points locks, mechanical point machines; air-operated point machines, hydraulic point machines, etc.), relays, contactors, pushbuttons, switching devices (e.g. limit switches), and guard interlocks.

Other examples of safety functions of mechanical products in machinery area are as follows:

- Emergency stop
- Manual reset
- Start/restart
- Control
- Muting (suspension of a safety function)
- Hold to run
- Activation function

- Mode selector for control or operating modes
- Prevention of unexpected start
- Isolating or energy dissipation

Mechanical products do not exhibit the constant failure rate. The dominant failure mode for a mechanical safety-related product is wear out.

15.4.1 Wear Out

The wear-out phase precedes the end of the product life. At this point, the probability of failure increases with time. While the parts have no memory of previous use during the random failure phase, yielding a constant failure rate, when they enter wear out, the cumulative effects of previous use are expressed as a continually increasing failure rate. The normal distribution is often used to model wear out. A Weibull distribution is more often used to approximate this and every other period. Scheduled preventative maintenance of replacing parts entering the wear-out phase can improve reliability of the overall system.

Failure intensity of a maintained system $\lambda(t)$ can be defined as the anticipated number of times an item will fail in a specified time period, given that it was as good as new (maintained) at time zero and is functioning at time t.

The failure rate of a component or an unmaintained system $r(t)$ is defined as the probability per unit time that the component or system experiences a failure at time t, given that the component or system was operating at time zero and has survived to time t.

Examples of failure mechanisms in wear-out phase are as follows:

- Fatigue—Constant cycle of stress wears out material
- Corrosion—Steady loss of material over time leads to failure
- Wear—Material loss and deformation, especially loss of protective coatings
- Thermal cycling—Stress
- Electromigration—Change in chemical properties, alloyed metals can migrate to grain boundaries, changing properties
- Radiation—Ultraviolet, X-ray, nuclear bombardment in environment changes molecular structure of materials

As components begin to fatigue or wear out, failures occur at increasing rates. For example, wear out in power supplies is usually caused by the breakdown of electrical components that are subject to physical wear and electrical and thermal stress. It is this period that the mean time between failures (MTBFs) or failures per million hours (FPMH) rates calculated in the useful life period no longer apply. The mean time to failure (MTTF) is the reciprocal of the failure rate if and only if the failure rate is constant in time.

A product with an MTBF of 10 years can still exhibit wear out in 2 years. No part count method can predict the time to wear out of components. Electronics in general are designed so that the useful life extends past the design life. This way, wear out should never occur during the useful life of a module. For instance, most electronics are obsolete within 20 years; as such, design a module with a useful life of 35 years or more. Note that power electronics may have a shorter useful life, depending on the stresses, and may therefore require preventive replacement.

15.4.2 Failure Rate Estimates for Mechanical Products

Weibull analysis can be used to make average failure rate estimates for items that do not exhibit a constant failure rate, such as mechanical components. Normally, failure occurs when stress exceeds strength, and therefore, predicting reliability amounts to predicting the probability of that event (i.e., the stress–strength approach).

Given a Weibull shape parameter (β), characteristic life (η), and a maintenance interval for item renewal T when the item is assumed to be restored to "as good as new," a tool (e.g., Excel spreadsheet) can be used to estimate an average item failure rate for the assumed maintenance concept.

The average failure rate is calculated using the following equation, where T is the maintenance interval for item renewal and $R(t)$ is the Weibull reliability function with the appropriate β and η parameters.

The characteristic life (η) is the point where 63.2% of the population will fail. This is because if $t = \eta$, then using Weibull reliability function, $R(t) = e^{-\left(\frac{t}{\eta}\right)^{\beta}}$ will result in $R(t) =$ exp(-1) = 37.8% and then $F(t) = 63.2\%$.

Example: A ball bearing has a Weibull shape parameter of 1.3 and characteristic life of 50,000 hours. What is the average failure rate for the bearing?

If the bearing is scheduled for preventive maintenance replacement every 10 years, the failure/hazard rate varies from a lower value of zero at "time zero" to an upper bound of 31 FPMH for bearings that survive to the 10-year point.

The average failure rate for a population of these bearings is estimated to be approximately 20 FPMH. Conversely, if the bearing is replaced every 2 years, the failure rate will vary between 0 at "time zero" to 19 FPMH at the 2-year point, with the average failure rate estimated to be 14 FPMH.

Because the average component failure rate is constant for a given maintenance renewal concept, an overall system failure rate can be estimated by summing the average failure rates of the components that make up a system.

15.4.3 MTBF, MTTF, and B10

MTBFs for constant failure rate products is a measure of time where the product probability of survival is 36.8% (there is 63.2% chance of component failure). MTBF is usually used for repairable systems and is also widely used for the case where the failure distribution is exponential. Whereas, MTTF, the mean time expected until the first failure, is a measure of reliability for nonrepairable products/systems.

For mechanical products, B10 reliability rating is calculated as a measure of time where the product probability of survival is 90% (there is 10% chance of component failure).

For mechanical products, it is assumed that MTBFd = 10 B10 (ISO 13849).

B10d stands for the number of cycles where 10% of the components fail to danger (wrong side failure/fail dangerously). T10d is the mean time until 10% of the tested samples fail to danger.

TABLE 15.1

Comparison of Characteristics of "E/E/PE products" versus "Non-E/E/PE Products"

Characteristics	E/E/PE Products	Non-E/E/PE Products (e.g., Mechanical/Pneumatic)
Reliability predictions	Lifetime calculation approaches (e.g., parts count or parts stress) based on the MTTF of each part (e.g., resistors and capacitors) are used to perform reliability prediction (e.g., MIL-HDBK-217F).	Reliability predictions are for a specific application based on the number of duty cycles/time interval for changing components subject to wear (e.g., $T_{10} = 90\%$ reliability).
MTBF	Environmental conditions such as excess temperature play an important role on MTTF.	MTBF depends on the number of cycles with which percentage of components fail (e.g., B_{10}), i.e., exceeded specified limits under defined conditions.
Failure mechanism	Failure mechanism is dependent on time.	Failure mechanism is dependent on the number and frequency of operations.
Lifetime	Product lasts for a long time (e.g., MTTF > 7500 years). Proof test is used to identify unrevealed failures.	Product lasts for a shorter time (e.g., MTBF ~ 75 years). Proof test is usually replacement (as good as new).
Failure mechanism	Product components fail randomly.	Wear out is the primary failure mode.
Failure rates	Product failure rates are constant with time (e.g., exponential distribution).	Product failure rates are proportional to a power of time (e.g., Weibull distribution).
Design	Design is complex.	Architectures are usually not complex.
Causes of failures	Causes of failures: • Environmental conditions such as excess temperature • Usage (time)	Causes of failures: • Natural end of life time due to mechanical wear, degradation or fatigue • Early end of lifetime due to improper application or insufficient maintenance (e.g., cracks, corrosion) • Environmental influences (e.g., thermal stress, sun, pollution, chemical degradation) • Defective components (e.g., broken pin and leakage)

15.5 Comparison of Characteristics of the E/E/PE and Non-E/E/PE Products

E/E/PE products behave fundamentally differently from non-E/E/PE "mechanical" products as seen in Table 15.1.

15.6 Key Requirements of ISO 13849

ISO 13849 provides instructions to developers to make products safe. It defines the complete safety life cycle, from risk assessment to modification of safety system.

The safety life cycle is the series of phases from initiation and specifications of safety requirements, covering design and development of safety features in a safety-critical system and ending in decommissioning of that system:

1. Quantifiable requirements
   - Reliability of the hardware ware (MTBFd)
   - Integrity of the measures to detect failures—Diagnostic coverage (DC)
   - Sufficient measures against common cause failures (CCF)
2. Qualitative (nonquantifiable) requirements
   - Sufficient safety integrity at failures according to the safety categories
   - (Deterministic approach of failures)
   - Application of quality management measures to avoid systematic failures
   - Verification and validation activities

15.6.1 What Is a Performance Level?

- The ability of safety-related parts of control systems (SRP/CS) to perform a safety function is allocated a PL
- A PL is established for the performance characteristic required for carrying out a safety function, i.e., the functions that are necessary to reduce each individual safety risk.
- PL is a measure of the reliability of a safety function, i.e., how well a safety system is able to perform a safety function under foreseeable conditions, i.e., risk reduction to a tolerable level.
- PL is proportional to the probability of a failure to danger over time (wrong-side failure rate).
- PL is divided into five levels (a–e). PL e gives the highest safety reliability and is equivalent to that required at the highest level of risk.

15.6.2 How to Assign a PL Rating

A PL rating should be assigned only after a risk analysis is performed. It is meaningless to assign a PL rating prior to completing such an analysis.

The following are the steps to be followed:

- Identify the hazard(s).
- Identify hazardous events (effects) that could be connected to each hazard.
- Determine whether a risk reduction is necessary or not.
- Determine how the required risk reduction shall be reached.
- Identify safety functions.
- Determine the PL for each safety function.

Determination of the required PL (risk graph):

S—Severity of injury
    S1: Slight (usually reversible) injury
    S2: Severe (usually irreversible) injury, including death
F—Frequency and/or duration of exposure to hazard
    F1: Rare to often and/or short exposure to hazard
    F2: Frequent to continuous and/or long exposure to hazard
P—Possibilities of hazardous prevention or damage limiting
    P1: Possible under certain conditions
    P2: Hardly possible

The risk graph method shown in Annex D of IEC 61508, Part 5, has the following definition for the "C" Consequence parameter:

- C1: Minor injury
- C2: … injury up to death of one person
- C3: Death to several people
- C4: Many people killed (catastrophic)

As shown earlier, both C2 and C3 (death to one and several people) can lead to SIL 3. C4 needs only SIL 4.

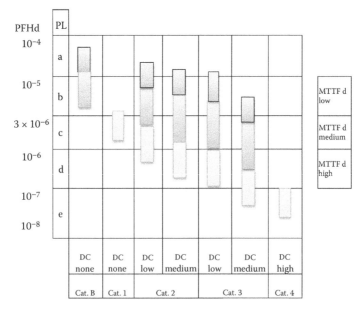

FIGURE 15.1
Risk graph for refers to severity level S2 that can lead to *death*.

15.6.3 Relationship between SIL and PL

SIL and PL can be compared based on the probability of random hardware failures. They are mutually transferable as shown in Table 15.2.

Based on the comparison table for SIL and PL, the highest equivalent integrity level that can be claimed for a mechanical product is SIL 3 (Figure 15.2). This is mostly because the consequence of a mechanical product failure is assumed never to be catastrophic (i.e., many people killed).

TABLE 15.2

Comparison of SIL ratings versus PL ratings

SIL	Probability of Dangerous Failures per Hour (h^{-1})	PL
–	$\geq 10^{-5}$ up to $<10^{-4}$	a
SIL 1	$\geq 3 \times 10^{-6}$ up to $<10^{-5}$	b
SIL 1	$\geq 10^{-6}$ up to $<3 \times 10^{-6}$	c
SIL 2	$\geq 10^{-7}$ up to $<10^{-6}$	d
SIL 3	$\geq 10^{-8}$ up to $<10^{-7}$	e

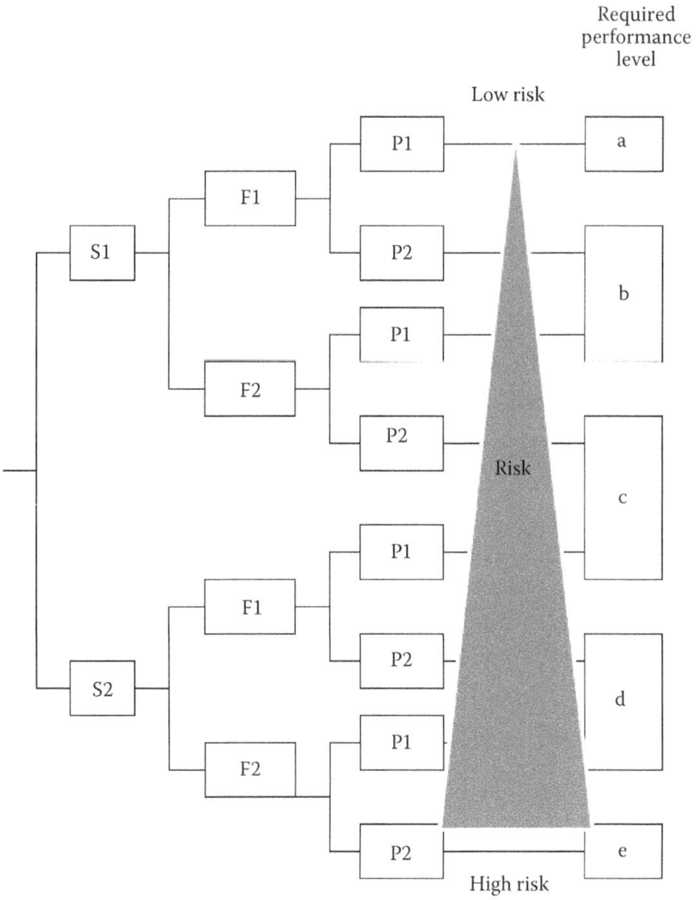

FIGURE 15.2
Risk graph method for determining performance level.

15.6.4 Validation of the PL Rating

PLs are defined in terms of *probability of dangerous failure per hour*, i.e., PFH_D (probability of dangerous failure per hour). The PFH_D of the safety function depends on the following:

- Architecture of the system (categories B and 1–4): How a safety-related control circuit must behave under fault conditions (identified through hardware failure modes and effects analysis)
- DC of the system: Extent of fault detection mechanisms, i.e., amount of fault monitoring
- Reliability of components: Mean time to dangerous failure of the components (MTTFd) that depends on the number of cycles with which 10% of components fail to danger (B10d)
- The system protection against a failure that knocks out both channels (i.e., CCF)— In case of redundancy
- The system protection against systematic errors (faults) built into the design

Other relevant issues are the following:

- Design process
- Operating stress
- Environmental conditions
- Operation procedures

15.7 Overview of EN ISO 13849

EN 13849 is a new standard that supports mechanical engineers. BS EN ISO 13849 (*Safety of Machinery—Safety-Related Parts of Control Systems*) can be used to cover the safety compliance of the mechanical products.

ISO 13849 applies to SRP/CS, and various kinds of products, regardless of the type of technology or energy source used (electrical, hydraulic, pneumatic, mechanical, etc.), for all kinds of machinery. It is aimed more at products developed under electromechanical technology rather than complex electronic and programmable electronic control systems.

ISO 13849 describes well-established strategies for designing *mechanical* products to minimize failure to danger situations as follows:

- Avoid faults (i.e., reliability)
- Detect faults (i.e., monitoring and testing)
- Tolerate faults (i.e., redundancy and diversity)

This standard provides safety requirements and guidance on the principles for the design, integration, and validation of SRP/CS. SRP/CS are parts of the product control system that are assigned to provide safety functions.

This standard has wide applicability, as it applies to all technologies, including electrical, hydraulic, pneumatic, and mechanical. The outputs of this standard to ensure the safety of machinery are PLs (PL a, b, c, d, or e).

15.8 Assurance Argument for Mechanical Products (Technical Documentation)

The objective of the technical document is to provide the necessary evidence, arguments, and information to demonstrate that the product under investigation will operate safely and reliably if applied in a way that meets the application of RAMS requirements and the conditions set within the technical document.

In doing so, the product technical document provides assurance that the product safety functions with PLx are in compliance with all relevant requirements of ISO 13849 PLx functions.

The requirement for technical documentation (in line with the safety case concept in EN 50129) from ISO 13849 is as follows.

The designer shall document at least the following information (ISO 13849—Requirement Section 10):

- Safety function(s) provided by the SRP/CS
- The characteristics of each safety function
- The exact points at which the safety-related part(s) start and end
- Environmental conditions
- The PL
- The category or categories selected
- The parameters relevant to the reliability (MTTFd, DC, CCF, and mission time)
- Measures against systematic failure
- All potential safety-relevant faults
- Justification for fault exclusions
- The design rationale (e.g., faults considered, faults excluded)
- Measures against reasonably foreseeable misuse

15.9 Conclusion

System assurance for railway signaling products is the application of management methods and analysis techniques to assure that the completed design meets the RAMS criteria requirements.

SIL ratings can be assigned to the safety functions of E/E/PE safety-related signaling products.

The SIL notion directly results from the IEC 61508 standard (i.e., applicable only to E/E/PE safety-related systems). As such, the railway (signaling) industry CENELEC safety standards EN 50129 and EN 50126 are derived from IEC 61508 and can be used only for the assurance of (E/E/PE) safety-related products and systems.

However, as mechanical products have dominant failure modes different from those of E/E/PE products (bathtub curve), the SIL concept is not directly applicable to them, and CENELEC railway signaling safety standards do not seem completely appropriate to be used for their assurance.

This chapter has described scenarios where the safety of a system relies on the mechanical control system, and the control system has a safety function (functional safety) and concluded that in this case, the functional safety should fulfill the requirements of the relevant standard (e.g., ISO 13849).

This chapter has also described mechanical signaling products and their differences with E/E/PE products. It has introduced a set of relatively new European (EN ISO) safety of machinery standards (EN ISO 13849-1 and 13849-2) for the assurance of mechanical products. It also covered the concept of PL that can be used for the safety assurance of mechanical products.

The chapter has concluded that *application of the new set of European safety of machinery standards (EN ISO 13849) will assist system assurance professionals to evaluate and assure the RAMS requirements for railway mechanical signaling products.*

15.10 Summary

- SIL terminology can be used for E/E/PE safety-related systems only.
- SIL terminology cannot be used for the safety functions performed by other technologies (e.g., mechanical systems—some of the railway signaling trackside products).
- For situations where the safety functions are performed by other technologies (e.g. mechanical products), EN ISO 13849 should be used.
- EN ISO 13849 uses PL. which is a technology-neutral concept, i.e., can be used for *mechanical signaling safety* systems.

References

CEN (European Committee for Standardization) (1999) EN 50126: *Railway Applications—The Specification and Demonstration of Reliability, Availability, Maintainability and Safety (RAMS).* CEN, Brussels.

IEC (International Electrotechnical Commission) (2000) IEC 61508: *Functional Safety of Electrical/Electronic/Programmable Electronic Safety-Related Systems.* IEC, Geneva.

CEN (2003) EN 50129: *Railway Applications—Communications, Signalling and Processing Systems—Safety Related Electronic Systems for Signalling.* CEN, Brussels.

ISO (International Organization for Standardization) (2006) ISO 13849-1:2006: *Safety of Machinery—Safety-Related Parts of Control Systems—Part 1: General Principles for Design.* ISO, Geneva.

ISO (2012) ISO 13849-2:2012: *Safety of Machinery Safety-Related Parts of Control Systems—Part 2: Validation.* ISO, Geneva

ISO (2010) ISO 12100: *Safety of Machinery—General Principles for Design—Risk Assessment and Risk Reduction.* ISO, Geneva.

US Military Handbook (1995): *MIL-HDBK-217F: The Military Handbook for "Reliability Prediction of Electronic Equipment"*

16

Software Reliability in RAMS Management

Vidhyashree Nagaraju and Lance Fiondella

CONTENTS

16.1 Summary and Purpose

This chapter provides an overview of software reliability and software reliability models. The chapter has been designed for individuals familiar with the basic concepts of reliability but who may lack knowledge of software reliability and could benefit from a refresher on reliability mathematics. To impart the practical skill of applying software reliability models to real failure data, we then discuss methods such as least squares and maximum likelihood estimation (MLE), which are techniques to identify numerical parameter values for a model to best fit a dataset. Given this background, we then introduce two of the most popular classes of software reliability models, including failure rate and failure-counting models, providing both the general theory as well as a specific model instance. Detailed numerical examples that apply these models to real datasets are also given as well as measures of goodness of fit to assess the relative utility of competing models. A tool to automatically apply software reliability models is also discussed and links to resources provided.

16.2 Introduction

Key to the success of software is its reliability [1], commonly defined as the probability of failure-free operation for a specified period of time in a specified environment. Unlike hardware, which fails due to physical deterioration, software fails due to design faults, operating environment, and inputs. In the context of railway systems [2,3], software reliability is important in critical applications such as dynamic control of safe separation between trains in "moving block" railway signaling, railway interlocking systems, monitoring and real time control software, and hardware control software.

Traditional software reliability growth models (SRGMs) employ techniques from probability and statistics on failure data acquired during software testing to predict how much additional testing is needed to reduce the failure rate to a desired level and to minimize the number of severe faults that elude discovery during testing [4]. These models can be used by organizations to establish confidence in the reliability of their software and improve the time to market or field a product.

The organization of this chapter is as follows. Section 16.3 reviews parameter estimation methods through a pair of motivating examples. Section 16.4 presents failure rate models, including the Jelinski–Moranda (JM) model. Section 16.5 covers nonhomogeneous Poisson process (NHPP) software reliability models, including the Goel–Okumoto (GO) model as well as goodness-of-fit measures. Section 16.6 summarizes with concluding remarks.

16.2.1 Mathematical Review

This section provides a review of basic mathematical identities used in subsequent sections such as differentiation, exponents, and logarithms.

Rules of differentiation:

$$\frac{d}{dx}c = 0, \tag{16.1}$$

where c is a constant;

$$\frac{d}{dx}cx = c \cdot dx, \tag{16.2}$$

$$\frac{d}{dx}x^n = nx^{n-1}\,dx, \tag{16.3}$$

$$\frac{\mathrm{d}}{\mathrm{d}x} a^x = a^x \ln a \cdot \mathrm{d}x, \tag{16.4}$$

$$\frac{\mathrm{d}}{\mathrm{d}x} \ln x = \frac{1}{x} \cdot \mathrm{d}x. \tag{16.5}$$

Properties of exponents:

$$a^x a^y = a^{x+y} \tag{16.6}$$

Equation 16.6 generalizes to

$$\prod_{i=1}^{n} a^{x_i} = a^{\sum_{i=1}^{n} x_i}, \tag{16.7}$$

$$\frac{a^x}{a^y} = a^{x-y}, \tag{16.8}$$

$$(a^x)^y = a^{xy}. \tag{16.9}$$

Properties of logarithms:

$$\log_b(xy) = \log_b x + \log_b y. \tag{16.10}$$

Similar to Equation 16.7, Equation 16.10 can be generalized to

$$\log_b \left(\prod_{i=1}^{n} x_i \right) = \sum_{i=1}^{n} \log_b x_i, \tag{16.11}$$

$$\log_b \left(\frac{x}{y} \right) = \log_b x - \log_b y, \tag{16.12}$$

$$\log_b(x^y) = y \log_b x. \tag{16.13}$$

16.3 Parameter Estimation Methods

Parameter estimation [5] is the process of finding numerical values of the parameters of a model that best fit the data. This section discusses two widely used estimation techniques:

- Least-squares estimation (LSE)
- Maximum likelihood estimation

To estimate the parameter of a model, first obtain the software fault detection time data, which consists of individual failure times $\mathbf{T} = \langle t_1, t_2,...,t_n \rangle$, where n is the number of faults observed. Alternatively, failure count data are represented as $\langle \mathbf{T}, \mathbf{K} \rangle = ((t_1, k_1), (t_1, k_1),..., (t_n, k_n))$, where t_i is the time at which the ith interval ended and k_i is the number of faults detected in interval i. These data should be collected during software testing and are dependent on the testing tools employed.

16.3.1 Least-Squares Estimation

Least squares estimation determines the parameters that best fit a dataset by minimizing the sum of squares of the vertical distances between the observed data points and the fitted curve. Given n-tuples (t_i, y_i) for failure time or failure count data, LSE minimizes

$$\sum_{i=1}^{n} \left(y_i - f\left(t_i | \Theta \right) \right), \tag{16.14}$$

where y_i is the cumulative number of faults detected and t_i is the ith failure time or time at the end of the ith interval. The function $f(t_i)$ is the model possessing the vector of parameters Θ to be estimated. Given a specific mean value function (MVF) $m(t)$ of software reliability, LSE minimizes

$$\text{LSE}(\Theta) = \sum_{i=1}^{n} \left(y_i - m(t_i; \Theta) \right)^2, \tag{16.15}$$

where $m(t_i; \Theta)$ is the number of faults predicted to occur by time t_i, given specific numerical values for Θ.

16.3.2 Maximum Likelihood Estimation

Maximum likelihood estimation [6,7] is a procedure to identify numerical values of model parameters that best fit the observed failure data. MLE maximizes the likelihood function, also known as the joint distribution of the failure data. Commonly, the logarithm of the likelihood function is maximized because this simplifies the mathematics, and the monotonicity of the logarithm ensures that the maximum of the log-likelihood function is equivalent to maximizing the likelihood function.

To obtain maximum likelihood estimates for the parameters of a model, first specify the likelihood function, which is the joint density function of the observed data:

$$L(\Theta; t_i, \ldots, t_n) = f(t_i, \ldots, t_n \mid \Theta) = \prod_{i=1}^{n} f(t_i \mid \Theta), \tag{16.16}$$

where $f(t)$ is the probability density function (PDF).

Consider the motivating example where an individual flips a coin five times and observes three heads and two tails. Based on these experiments, what is the most likely value of observing a head? Assuming the coin flips are independent and identically distributed, the probability may be characterized by a binomial distribution. Specifically, we can use our experimental results to write the probability of three heads and two tails as the following function of p:

$$L(p) = \binom{5}{3} p^3 (1-p)^2,$$

where $\binom{n}{k} = \dfrac{n!}{k!(n-k)!}$ is a binomial coefficient, p denotes a head, and $(1-p)$ denotes tails. Now, consider a discrete set of values for $p \in \left(\dfrac{0}{5}, \dfrac{1}{5}, \dfrac{2}{5}, \dfrac{3}{5}, \dfrac{4}{5}, \dfrac{5}{5} \right)$. Table 16.1 shows the probability of observing three heads and two tails for each value of p.

TABLE 16.1

Probability of Three Heads and Two Tails

Success Probability (p)	Probability
$\dfrac{0}{5}$	$\binom{5}{3}\left(\dfrac{0}{5}\right)^3\left(\dfrac{5}{5}\right)^2 = \dfrac{0}{625}$
$\dfrac{1}{5}$	$\binom{5}{3}\left(\dfrac{1}{5}\right)^3\left(\dfrac{4}{5}\right)^2 = \dfrac{32}{625}$
$\dfrac{2}{5}$	$\binom{5}{3}\left(\dfrac{2}{5}\right)^3\left(\dfrac{3}{5}\right)^2 = \dfrac{144}{625}$
$\dfrac{3}{5}$	$\binom{5}{3}\left(\dfrac{3}{5}\right)^3\left(\dfrac{2}{5}\right)^2 = \dfrac{216}{625}$
$\dfrac{4}{5}$	$\binom{5}{3}\left(\dfrac{4}{5}\right)^3\left(\dfrac{1}{5}\right)^2 = \dfrac{128}{625}$
$\dfrac{5}{5}$	$\binom{5}{3}\left(\dfrac{5}{5}\right)^3\left(\dfrac{0}{5}\right)^2 = \dfrac{0}{625}$

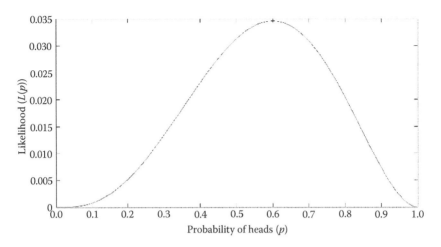

FIGURE 16.1
Likelihood function of motivating example.

Figure 16.1 shows a plot of the likelihood function for values of $p \in (0,1)$ for the motivating example. Figure 16.1 indicates that the value $p = \dfrac{3}{5}$ produces a higher probability than any of the other values considered, agreeing with the intuition that a reasonable estimate of p may be the number of heads divided by the total number of coin flips. Calculus-based techniques enable algebraic derivation of the MLE of the binomial distribution analytically. We demonstrate this next.

16.3.2.1 Binomial Distribution

The binomial distribution is built from a sequence of binomial random variables possessing probability mass function

$$P_x(X) = p^x(1-p)^{x-1},$$
(16.17)

where $0 \leq p \leq 1$ and $x = \{0, 1\}$ with $x = 1$ denoting success and $x = 0$ denoting failure. The joint probability is

$$\Pr(X_1 = x_1,\ X_2 = x_2,\dots,X_n = x_n) = \prod_{i=1}^{n} p^{x_i}(1-p)^{1-x_i} = p^{\sum x_i}(1-p)^{n-\sum x_i},$$
(16.18)

where the last equality follows from Equation 16.7. Therefore, the likelihood equation is

$$L(p) = p^{\sum x_i}(1-p)^{n-\sum x_i}.$$
(16.19)

Due to the monotonicity of the logarithm, one can maximize ln $L(p)$ in place of $L(p)$, which enables the simplification

$$\ln\left(L(p)\right) = \sum_{i=1}^{n} x_i \ln p + \left(n - \sum_{i=1}^{n} x_i\right) \ln\left(1-p\right), \tag{16.20}$$

which follows from Equation 16.13. Differentiating with respect to p,

$$\frac{\partial \ln\left(L(p)\right)}{\partial p} = \frac{1}{p}\sum_{i=1}^{n} x_i - \frac{1}{1-p}\left(n - \sum_{i=1}^{n} x_i\right), \tag{16.21}$$

which follows from Equation 16.5. Finally, setting the result equal to zero and solving for p,

$$\frac{n}{1-p} = \frac{1}{p}\sum_{i=1}^{n} x_i + \frac{1}{1-p}\sum_{i=1}^{n} x_i,$$

$$np = (1-p)\sum_{i=1}^{n} x_i + p\sum_{i=1}^{n} x_i,$$

$$\hat{p} = \frac{1}{n}\sum_{i=1}^{n} x_i, \tag{16.22}$$

where $\hat{p}$ is the maximum likelihood estimate of parameter p. Equation 16.22 is indeed the sum of the successes divided by the total number of experiments as conjectured from the numerical example with five coin flips introduced earlier.

16.3.2.2 Exponential Distribution

To illustrate the MLE in the context of a continuous distribution, the method is applied to the exponential distribution, which commonly occurs in reliability mathematics. The PDF of the exponential distribution is

$$f_x(x) = \lambda e^{-\lambda x}, \ x > 0.$$

Therefore, the likelihood equation is

$$L(\lambda) = \prod_{i=1}^{n} \lambda e^{-\lambda x_i} = \lambda^n e^{-\lambda \sum_{i=1}^{n} x_i}, \tag{16.23}$$

and the log-likelihood is

$$\ln\left(L(\lambda)\right) = n\ln\lambda - \lambda\sum_{i=1}^{n} x_i. \tag{16.24}$$

Next, differentiate with respect to λ:

$$\frac{\partial\ln\left(L(p)\right)}{\partial\lambda} = \frac{n}{\lambda} - \sum_{i=1}^{n} x_i. \tag{16.25}$$

Finally, set the result equal to zero and solve for λ:

$$\hat{\lambda} = \frac{n}{\sum_{i=1}^{n} x_i}. \tag{16.26}$$

Rearranging Equation 16.26,

$$\frac{1}{\hat{\lambda}} = \frac{\sum_{i=1}^{n} x_i}{n}. \tag{16.27}$$

When the x_i in Equation 16.27 are interpreted as failure times, this suggests that the MLE of the mean time to failure (MTTF) $\left(\frac{1}{\hat{\lambda}}\right)$ is the sum of the failure times divided by the number of items, which agrees with intuition as the simple average of the time to failure.

16.4 Failure Rate Models

Failure rate models characterize software reliability in terms of failure data collected during software testing. Some of the estimates enabled by failure rate models include the following:

- Number of initial faults
- Number of remaining faults
- MTTF of the $(n + 1)$st fault
- Testing time needed to remove the next k faults
- Probability that the software does not fail before a mission of fixed duration is complete

Intuitively, the failure rate decreases as the number of faults detected increases. Thus, additional testing time should lower the failure rate.

16.4.1 JM Model

The Jelinski-Moranda (JM) [8] model is one of the earliest failure rate models. It has also been extended by many subsequent studies. The assumptions of the JM model are as follows:

- The number of initial software faults $N_0 > 0$ is an unknown but fixed constant.
- Each fault contributes equally to software failure.
- Times between failures are independent and exponentially distributed with rate proportional to the number of remaining faults.
- A detected fault is removed immediately and no new faults are introduced during the fault removal process.

Let **T** be the vector of n observed failure times. The time between the $(i - 1)$st and ith failure is $t_i = (T_i - T_{i-1})$, which is exponentially distributed with rate

$$\lambda(i) = \phi\big(N_0 - (i-1)\big). \tag{16.28}$$

Here, $\phi > 0$ is a constant of proportionality representing the contribution of individual faults to the overall failure rate.

The probability distribution function of the ith failure is

$$
\begin{aligned}
f_i(t) &= \lambda(i)e^{-\int_0^t \lambda(i)dx} \\
&= \phi\big(N_0 - (i-1)\big)e^{-\int_0^t \phi(N_0-(i-1))dx} \\
&= \phi\big(N_0 - (i-1)\big)e^{-\phi(N_0-(i-1))t},
\end{aligned}
\tag{16.29}
$$

where the second equality is the hazard rate representation of the PDF.

The cumulative distribution function of the ith failure is

$$
\begin{aligned}
F_i(t) &= \int_0^t f_i(x)\,dx \\
&= \int_0^t \phi\big(N_0 - (i-1)\big)e^{-\phi(N_0-(i-1))x}\,dx \\
&= 1 - e^{-\phi(N_0-(i-1))t},
\end{aligned}
\tag{16.30}
$$

where the final equality follows from $\int_0^t \lambda e^{-\lambda x}\,dx = 1 - e^{-\lambda t}$.

The reliability of the software prior to the ith failure is therefore

$$R_i(t) = e^{-\phi(N_0-(i-1))t}, \tag{16.31}$$

and the mean time to the $(n + 1)$st failure is

$$\mathrm{MTTF}(n+1) = \int_0^\infty R_i\big(t|i=n+1\big)\mathrm{d}t_i$$

$$= e^{-\phi(N_0-n)t}\,\mathrm{d}t$$

$$= \frac{1}{\phi(N_0-n)},$$

where the final equality follows from $\int_0^\infty e^{-\lambda t} = \dfrac{1}{\lambda}$.

More generally, the amount of testing required to remove the next k faults is

$$\sum_{i=1}^{k} \mathrm{MTTF}(n+i) = \sum_{i=1}^{k} \frac{1}{\phi\big(N_0 - \big(n+(i-1)\big)\big)}. \tag{16.32}$$

The likelihood function of the JM model parameters is

$$L(t_i, t_2, \ldots, t_n;\ N_0, \phi) = \prod_{i=1}^{n} \phi\big(N_0-(i-1)\big)e^{-\phi(N_0-(i-1))t_i}$$

$$= \phi^n \left(\prod_{i=1}^{n} N_0 - (i-1) \right) e^{-\phi\sum_{i=1}^{n}(N_0-(i-1))t_i}. \tag{16.33}$$

The log-likelihood function is

$$\ln L = \ln\left(\phi^n \left(\prod_{i=1}^{n} N_0 - (i-1) \right) e^{-\phi\sum_{i=1}^{n}(N_0-(i-1))t_i} \right)$$

$$= \ln \phi^n + \ln\left(\prod_{i=1}^{n} N_0 - (i-1) \right) - \phi \sum_{i=1}^{n}\big(N_0-(i-1)\big)t_i \tag{16.34}$$

$$= n\ln\phi + \sum_{i=1}^{n} \ln\big(N_0-(i-1)\big) - \phi \sum_{i=1}^{n}\big(N_0-(i-1)\big)t_i.$$

The partial derivatives with respect to N_0 and ϕ are

$$\frac{\partial \ln L}{\partial N_0} = \sum_{i=1}^{n} \frac{1}{N_0-(i-1)} - \phi \sum_{i=1}^{n} t_i \tag{16.35}$$

and

$$\frac{\partial \ln L}{\partial \phi} = \frac{n}{\phi} - \sum_{i=1}^{n} N_0 - (i-1)t_i. \tag{16.36}$$

Equating Equation 16.35 to zero and solving for ϕ,

$$\hat{\phi} = \frac{\sum_{i=1}^{n} \frac{1}{N_0 - (i-1)}}{\sum_{i=1}^{n} t_i}. \tag{16.37}$$

Next, equating Equation 16.36 to zero and substituting Equation 16.37 for $\hat{\phi}$ reduces to the following fixed point problem with N_0 as the only unknown:

$$\frac{n \sum_{i=1}^{n} t_i}{\sum_{i=1}^{n} \frac{1}{\hat{N}_0 - (i-1)}} - \sum_{i=1}^{n} \left(\hat{N}_0 - (i-1) \right) t_i = 0. \tag{16.38}$$

After solving Equation 16.38 for $\hat{N}_0$, substitute the numerical result into Equation 16.37 to obtain the estimate $\hat{\phi}$.

Alternatively, one may equate Equation 16.36 to zero and solve for ϕ:

$$\hat{\phi} = \frac{n}{\sum_{i=1}^{n} \left(N_0 - (i-1) \right) t_i}. \tag{16.39}$$

Now, substitute Equation 16.39 into Equation 16.35 and equate to zero:

$$\sum_{i=1}^{n} \frac{1}{N_0 - (i-1)} - \frac{n}{\sum_{i=1}^{n} \left(N_0 - (i-1) \right) t_i} \sum_{i=1}^{n} t_i = 0. \tag{16.40}$$

Table 16.2 summarizes some additional failure rate models derived from the JM.

16.4.2 Illustrations of JM Model

This section illustrates the JM model for the SYS1 [9] failure dataset which consists of $n = 136$ faults. Figure 16.2 shows the plot of the log-likelihood function, which has been reduced to N_0 by substituting Equation 16.37 into Equation 16.34. Examination of Figure 16.2 suggests that the log-likelihood function is maximized at $N_0 \approx 142$, which can serve as the initial input to a numerical procedure to find the exact maximum.

TABLE 16.2

Additional Failure Rate Models

Model Name	Failure Rate $\lambda(i)$	Parameter Constraints
GO imperfect	$\phi\,(N_0 - p(i-1))$	$N_0, \phi > 0, 0 < p < 1$
Power-type function	$\phi\,(N_0 - (i-1))^\alpha$	$N_0, \phi, \alpha < 0$
Exponential-type function	$\phi\left(e^{-\beta N_0 - (i-1)} - 1\right)$	$N_0, \phi, \beta > 0$
Schick–Wolverton	$\phi\,(N_0 - (i-1))t_i$	$N_0, \phi > 0$
Modified Schick–Wolverton	$\phi\,(N_0 - n_{i-1})t_i$	$N_0, \phi > 0$
Lipow	$\phi\left(N_0 - (i-1)\right) \times \left(\dfrac{1}{2}t_i + \displaystyle\sum_{j=1}^{i-1} t_j\right)$	$N_0, \phi > 0$
Geometric model	ϕp^i	$\phi > 0, 0 < p < 1$
Fault exposure coefficient	$k(i)(N_0 - i),\ \ k(i) = k_0 + \dfrac{(k_f - k_0)i}{N_0}$	$N_0, k_0, k_f > 0$

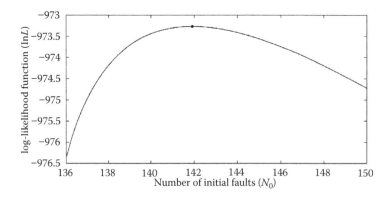

FIGURE 16.2
Log-likelihood function of JM model.

Figure 16.3 shows a plot of the MLE of N_0 specified by Equation 16.38. The value of N_0 that makes Equation 16.38 is the maximum likelihood estimate $\hat{N}_0$. This value is indicated with a dot. The grid lines suggest that this value is slightly less than 142.

The parameter estimates for the SYS1 dataset are as follows:

- Estimated number of initial faults: $\hat{N}_0 = 141.903$
- Failure constant: $\hat{\phi} = 0.0000349665$
- Estimated number of faults remaining: 5.902

$$(\hat{N}_0 - n = 141.903 - 136)$$

- Reliability function after 136th failure from Equation 16.31:

$$R(t) = e^{\left[-0.00003497\,\left(141.903 - (137-1)\right)t\right]}$$

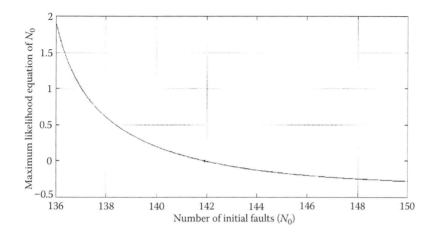

FIGURE 16.3
MLE of N_0.

Figure 16.4 shows the plot of the reliability after the 136th failure. Figure 16.4 suggests that the probability that the software operates without failure for 6000 s is approximately 0.3. The MTTF of the 137th fault is $\int\limits_0^\infty R(t)\,\mathrm{d}t = 4844.89$ or $\dfrac{4844.89}{60} = 80.7482$ minutes. While this number is not representative of a fielded system, it illustrates the role of SRGMs to demonstrate reliability growth as software faults are detected and corrected.

Figure 16.5 shows the plot of the MLE of the failure rate. Figure 16.5 shows that the failure rate decreases by a constant of size $\hat\phi$ at each failure time t_i. Early testing decreases the failure rate more quickly because there are more faults to detect. Later testing lowers the failure rate more slowly because fewer faults remain.

Figure 16.6 shows the plot of the MLE of the MTTF. Figure 16.6 indicates that the initial MTTF is low because there are frequent failures. The MTTF increases as faults are removed because fewer faults remain and the failure rate has decreased.

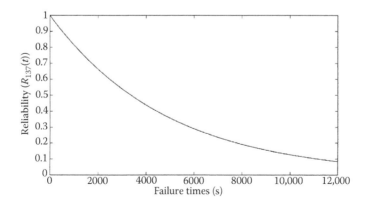

FIGURE 16.4
Reliability after 136th failure.

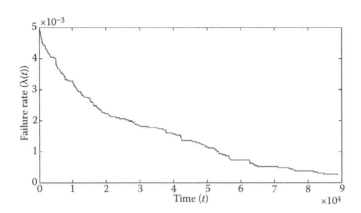

FIGURE 16.5
MLE of failure rate.

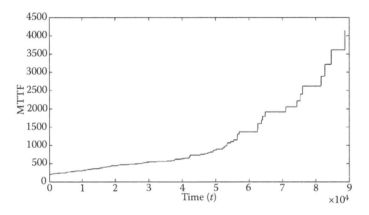

FIGURE 16.6
MLE of MTTF.

16.5 Nonhomogeneous Poisson Process Models

16.5.1 Counting and Poisson Process

A stochastic process $\{N(t), t \geq 0\}$ is a counting process [5] if $N(t)$ represents the cumulative number of events observed by time t. A counting process must satisfy the following properties:

- $N(t) \geq 0$.
- $N(t)$ is a non-negative integer.
- $t_1 < t_2$ implies $N(t_1) \leq N(t_2)$.
- For $t_1 < t_2$, $N(t_2) - N(t_1)$ is the number of events occurring in the interval $[t_1, t_2]$.

A counting process possesses *independent increments* if the number of events that occur in disjoint time intervals are independent. A counting process possesses *stationary increments* if the distribution of the number of events that occur in any time interval depends only on the length of the interval.

The *Poisson process* is a counting process with rate $\lambda > 0$, possessing the following properties:

1. $N(0) = 0$.
2. The process possesses independent increments.
3. The number of events in any interval of length t is Poisson with mean λt.

$$\Pr\left\{N(t_2) - N(t_1) = k\right\} = e^{-\lambda t} \frac{(\lambda t)^k}{k!} \tag{16.41}$$

Condition 3 implies that $E[N(t)] = \lambda t =: m(t)$, which is known as the MVF. It also follows that the time between events of a Poisson process are exponentially distributed, because the probability of no events in an interval $[t_1, t_2]$ of length t is

$$\begin{aligned}\Pr\left\{N(t_2) - N(t_1) = 0\right\} &= e^{-\lambda t} \frac{(\lambda t)^0}{0!} \\ &= e^{-\lambda t}.\end{aligned} \tag{16.42}$$

Hence,

$$F(t) = 1 - e^{-\lambda t}. \tag{16.43}$$

An alternative formulation of the Poisson process with rate λ is given by

1. $N(0) = 0$.
2. The process has independent increments:

$$\Pr\left\{N(\Delta t) = 1\right\} = \lambda \Delta t + o(\Delta t).$$

3. $\Pr\{N(\Delta t) \geq 2\} = o(\Delta t)$, where a function f is $o(\Delta t)$ if

$$\lim_{\Delta t \to 0} \frac{f(\Delta t)}{\Delta t} = 0. \tag{16.44}$$

Condition 1 states that the number of events at time $t = 0$ is 0 [10]. Condition 2 is as defined previously. Condition 3 states that the probability of an event in a small interval of length Δt is $\lambda \Delta t$, while condition 4 states that the probability of two or more events in a small interval is negligible.

16.5.2 Nonhomogeneous Poisson Process

A counting process is a *nonhomogeneous Poisson process* with intensity function $\lambda(t)$, $t \geq 0$ if

1. $N(0) = 0$.
2. $\{N(t), t \geq 0\}$ has independent increments:

$$\Pr\{N(t + \Delta t) - N(t) = 1\} = \lambda(t)\Delta t + o(\Delta t),$$

$$\Pr\{N(t + \Delta t) - N(t) \geq 2\} = o(\Delta t).$$

The *mean value function* of an NHPP is

$$m(t) = \int_0^t \lambda(x)\,\mathrm{d}x, \tag{16.45}$$

and it can be shown that

$$\Pr\{N(t_2) - N(t_1) = k\} = e^{-(m(t_2)-m(t_1))} \frac{\left(m(t_2) - m(t_1)\right)^k}{k!}. \tag{16.46}$$

Substituting the mean value function $m(t) = \lambda t$, Equation 16.46 simplifies to the Poisson process, also referred to as the *homogeneous Poisson process*.

16.5.3 NHPP Likelihood Function

The following theorem can be used to derive the failure times maximum likelihood estimates of NHPP SRGMs:

Theorem: Given n observed event times $T = \langle t_1, t_2, \ldots, t_n \rangle$, the likelihood function an NHPP with mean value function $m(t)$ and intensity function $\lambda(t)$ possessing parameters Θ is

$$L(\Theta|\mathbf{T}) = e^{-m(t_n)} \prod_{i=1}^{n} \lambda(t_i).$$

The log-likelihood function is

$$\ln\left(L(\Theta|\mathbf{T})\right) = \ln\left(e^{-m(t_n)} \prod_{i=1}^{n} \lambda(t_i) \right)$$

$$= -m(t_n) + \sum_{i=1}^{n} \ln\left(\lambda(t_i)\right). \tag{16.47}$$

The maximum likelihood estimate of a model with p parameters is obtained by solving the system of equations

$$\frac{\partial}{\partial \theta_i} \ln L(\Theta|T) = 0 \quad (1 \le i \le p).$$

(16.48)

16.5.4 Goel–Okumoto NHPP Software Reliability Growth Model

This section introduces the GO [11] SRGM. The mean value function is developed from its formulation as a differential equation and the maximum likelihood estimates for failure time and failure count data are derived.

The GO is formulated as

$$m(t + \Delta t) = b(a - m(t))\Delta t.$$

(16.49)

This expression suggests that the rate of fault detection in the time interval from $(t, t + \Delta t)$ is equal to the product of a constant fault detection rate (b), the number of faults remaining at time t ($a - m(t)$), and the length of the interval (Δt):

$$\frac{m(t + \Delta t)}{\Delta t} = b(a - m(t)),$$

$$\lim_{\Delta t \to 0} \frac{m(t + \Delta t)}{\Delta t} = \lim_{\Delta t \to 0} b(a - m(t)),$$

$$m'(t) = b(a - m(t)),$$

$$-\frac{du}{u} = b\left((u = a - m(t)), \ du = -m'(t)\right),$$

$$-\ln u = bt + c_0,$$

$$u = e^{-bt - c_0},$$

$$a - m(t) = e^{-bt}e^{-c_0},$$

$$m(t) = a - c_1 e^{-bt}.$$

(16.50)

TABLE 16.3

Some Generalizations of the GO SRGM

Model Name	MVF ($m(t)$)	Parameter Constraints
Hyper exponential	$a\sum_{i=1}^{k}p_i(1-e^{b_it})$	$a,b_i>0,\ 0<p_i<1,\ \sum_{i=1}^{k}p_i=1$
Weibull	$a(1-e^{-bt^c})$	$a,b,c>0$
Delayed S-shaped	$a(1-(1+bt)e^{-bt})$	$a,b>0$
Inflexion S-shaped	$\dfrac{a(1-e^{-bt})}{1+\psi e^{-bt}}$	$a,b>0,\ \psi(r)=\dfrac{1-r}{r},\ r\in(0,1]$
Bathtub	$a(1-e^{-bct^c e^{\phi t}})$	$a,b,\phi>0$ $0<c<1$

Given the initial condition $m(0)=0$, it follows that

$$m(0)=a-c_1e^{-b\times0}$$
$$0=a-c_1e^{0}\tag{16.51}$$
$$c_1=a.$$

Hence, the mean value function of the GO model is

$$m(t)=a-ae^{-bt}$$
$$=a(1-e^{-bt}),\tag{16.52}$$

where

$$a>0,\ b>0.$$

Note that all faults are detected as t tends to infinity

$$\lim_{t\to\infty}m(t)=a.\tag{16.53}$$

Table 16.3 provides a list of some generalizations of the GO SRGM.

16.5.4.1 Failure Times MLE

Recall that the definition of the mean value function is

$$m(t)=\int_{0}^{\infty}\lambda(x)\mathrm{d}x,\tag{16.54}$$

where $\lambda(t)$ is the instantaneous failure intensity at time t. It follows that the instantaneous failure intensity is

$$\frac{dm(t)}{dt} = \lambda(t). \tag{16.55}$$

Thus, the instantaneous failure intensity of the GO SRGM is

$$\lambda(t) = \frac{d}{dt}\Big[a(1-e^{-bt})\Big] \tag{16.56}$$
$$= abe^{-bt}.$$

Note that Equation 16.56 is monotonically decreasing with

$$\lim_{t\to\infty} \lambda(t) = 0. \tag{16.57}$$

From Equation 16.47, the log-likelihood function of the GO SRGM given the failure times is

$$\ln L\big(a,b\big|\mathbf{T}\big) = -a(1-e^{-bt_n}) + \sum_{i=1}^{n} \ln(abe^{-bt_i})$$
$$= -a(1-e^{-bt_n}) + \sum_{i=1}^{n} \ln(a) + \sum_{i=1}^{n} \ln(b) + \sum_{i=1}^{n} \ln(e^{-bt_i}) \tag{16.58}$$
$$= -a(1-e^{-bt_n}) + n\ln(a) + n\ln(b) - b\sum_{i=1}^{n} t_i.$$

Differentiating Equation 16.58 with respect to a,

$$\frac{\partial}{\partial a}\Big[\ln L\big(a,b\big|\mathbf{T}\big)\Big] = \frac{\partial}{\partial a}\Bigg[-a(1-e^{-bt_n}) + n\ln(a) + n\ln(b) - b\sum_{i=1}^{n} t_i\Bigg]$$
$$= \frac{\partial}{\partial a}\Big[-a(1-e^{-bt_n})\Big] + \frac{\partial}{\partial a}\Big[n\ln(a)\Big] + \frac{\partial}{\partial a}\Big[n\ln(b)\Big] - \frac{\partial}{\partial a}\Bigg[b\sum_{i=1}^{n} t_i\Bigg] \tag{16.59}$$
$$= -(1-e^{-bt_n}) + \frac{n}{a}.$$

Equating Equation 16.59 to zero and solving for a,

$$1-e^{-bt_n} = \frac{n}{a},$$

$$\hat{a} = \frac{n}{1-e^{-bt_n}}. \tag{16.60}$$

Differentiating Equation 16.60 with respect to b,

$$\frac{\partial}{\partial b}\Big[\ln L\big(a,b\big|\mathbf{T}\big)\Big] = \frac{\partial}{\partial b}\Bigg[-a(1-e^{-bt_n})+n\ \ln(a)+n\ln(b)-b\sum_{i=1}^{n}t_i\Bigg]$$

$$= \frac{\partial}{\partial b}\Big[-a+ae^{-bt_n}\Big]+\frac{\partial}{\partial b}\Big[n\ln(a)\Big]+\frac{\partial}{\partial b}\Big[n\ln(b)\Big]-\frac{\partial}{\partial b}\Bigg[b\sum_{i=1}^{n}t_i\Bigg] \quad (16.61)$$

$$= -at_n e^{-bt_n}+\frac{n}{b}-\sum_{i=1}^{n}t_i.$$

Equating Equation 16.61 to zero and substituting Equation 16.60 for a,

$$\frac{nt_n e^{-\hat{b}t_n}}{1-e^{-\hat{b}t_n}}+\sum_{i=1}^{n}t_i=\frac{n}{\hat{b}}. \quad (16.62)$$

Equation 16.62 cannot be manipulated to obtain a closed form expression for $\hat{b}$. Therefore, a numerical algorithm such as Newton's method is needed to identify the maximum likelihood estimate in practice.

16.5.4.2 Failure Counts MLE

In some instances, failure data are grouped into n consecutive, nonoverlapping, and possibly unequal intervals $[t_{i-1}, t_i]$, $1 \le i \le n$. The maximum likelihood estimates can also be determined from this vector of failure counts data $\langle \mathbf{T}, \mathbf{K}\rangle = ((t_1, k_1), (t_2, k_2),...,(t_1, k_1))$. The probability of k failures in an interval i is

$$\Pr\big\{m(t_i)-m(t_{i-1})=k_i\big\}=\frac{\big(m(t_i)-m(t_{i-1})\big)^{k_i}e^{-(m(t_i)-m(t_{i-1}))}}{k_i!}, \quad (16.63)$$

and the likelihood function of a vector of parameters Θ given n failure time intervals $\mathbf{T}$ and failure counts $\mathbf{K}$ is

$$L\big(\Theta\big|\mathbf{T},\mathbf{K}\big)=f\big(\mathbf{T},\mathbf{K}\big|m(t)\big)$$

$$=\prod_{i=1}^{n}\frac{\big(m(t_i)-m(t_{i-1})\big)^{k_i}e^{-(m(t_i)-m(t_{i-1}))}}{k_i!}. \quad (16.64)$$

The log-likelihood function is

$$
\ln L\big(\mathbf{\Theta}\,|\,\mathbf{T},\mathbf{K}\big) = \ln\left(\sum_{i=1}^{n}\frac{\big(m(t_i)-m(t_{i-1})\big)^{k_i}\,e^{-(m(t_i)-m(t_{i-1}))}}{k_i!}\right)
$$

$$
= \sum_{i=1}^{n}\ln\left(\prod_{i=1}^{n}\frac{\big(m(t_i)-m(t_{i-1})\big)^{k_i}\,e^{-\left(m(t_i)-m(t_{i-1})\right)}}{k_i!}\right) \tag{16.65}
$$

$$
\ln L\big(\mathbf{\Theta}\,|\,\mathbf{T},\mathbf{K}\big) = \sum_{i=1}^{n}k_i\ln\big(m(t_i)-m(t_{i-1})\big) - \sum_{i=1}^{n}\big(m(t_i)-m(t_{i-1})\big) - \sum_{i=1}^{n}\ln k_i!.
$$

To determine the log-likelihood function of the GO SRGM from Equation 16.65, first substitute the MVF into

$$
\begin{aligned}
m(t_i)-m(t_{i-1}) &= a(1-e^{-bt_i})-a(1-e^{-bt_{i-1}})\\
&= a(e^{-bt_{i-1}}-e^{-bt_i}).
\end{aligned} \tag{16.66}
$$

Substituting Equation 16.66 into Equation 16.65 and simplifying produces

$$
\begin{aligned}
\ln L\big(\mathbf{\Theta}\,|\,\mathbf{T},\mathbf{K}\big) &= \sum_{i=1}^{n}k_i\ln\big(a(e^{-bt_{i-1}}-e^{-bt_i})\big) - \sum_{i=1}^{n}\big(a(e^{-bt_{i-1}}-e^{-bt_i})\big) - \sum_{i=1}^{n}\ln k_i!\\
&= \sum_{i=1}^{n}k_i\ln(a) + \sum_{i=1}^{n}k_i\ln(e^{-bt_{i-1}}-e^{-ht_i}) + u\sum_{i=1}^{n}(e^{-bt_i}-e^{-bt_{i-1}})\\
&= \sum_{i=1}^{n}k_i\ln(a) + \sum_{i=1}^{n}k_i\ln(e^{-bt_{i-1}}-e^{-bt_i}) - a(1-e^{-bt_n}).
\end{aligned} \tag{16.67}
$$

The term $\sum_{i=1}^{n}\ln k_i!$ can be dropped because it is constant with respect to all model parameters. The final step follows from the telescoping series in the denominator since $t_0 = 0$. Furthermore, differentiating Equation 16.67 with respect to a,

$$
\frac{\partial}{\partial a}\Big[\ln L\big(a,b\,|\,\mathbf{T},\mathbf{K}\big)\Big] = \frac{\sum_{i=1}^{n}k_i}{a} - (1-e^{-bt_n}). \tag{16.68}
$$

Equating Equation 16.68 to zero and solving for a,

$$\frac{\sum_{i=1}^{n} k_i}{a} - (1 - e^{-bt_n}) = 0,$$

$$\hat{a} = \frac{\sum_{i=1}^{n} k_i}{(1 - e^{-bt_n})}. \tag{16.69}$$

Differentiating Equation 16.67 with respect to b,

$$\frac{\partial}{\partial b}\left[\ln L\left(a, b\,|\,\mathbf{T}, \mathbf{K}\right)\right] = \sum_{i=1}^{n} \frac{k_i(t_i e^{-bt_i} - t_{i-1} e^{-bt_{i-1}})}{e^{-bt_{i-1}} - e^{-bt_i}} - at_n e^{-bt_n}. \tag{16.70}$$

Equating Equation 16.70 to zero and substituting Equation 16.69 for a,

$$\sum_{i=1}^{n} \frac{k_i(t_i e^{-\hat{b}t_i} - t_{i-1} e^{-\hat{b}t_{i-1}})}{e^{-\hat{b}t_{i-1}} - e^{-\hat{b}t_i}} = \frac{t_n e^{-\hat{b}t_n} \sum_{i=1}^{n} k_i}{1 - e^{-\hat{b}t_n}}. \tag{16.71}$$

16.5.5 Goodness-of-Fit Measures

This section explains some additional theoretical and practical methods to assess competing software reliability models, including the Akaike information criterion (AIC) [12] and predictive sum of squares error (PSSE) [13].

16.5.5.1 Akaike Information Criterion

The AIC is

$$AIC = 2p - 2\ln L(\hat{\mathbf{\Theta}}), \tag{16.72}$$

where p is the number of model parameters and $\ln L$ is the maximum of the log-likelihood function. Given the AIC of two alternative models i and j, model i is preferred over model j if $AIC_i < AIC_j$. The AIC enables a more equitable comparison of models with different numbers of parameters because each model is "penalized'" two points for each of its parameters. Thus, a model with more parameters must achieve a higher maximum likelihood to offset this penalty factor.

16.5.5.2 Predictive Sum of Squares Error

The PSSE compares the predictions of a model to previously unobserved data. The PSSE for failure time data is

$$\text{PSSE} = \sum_{i=(n-k)+1}^{n} \left(\hat{m}(t_i) - i \right)^2, \tag{16.73}$$

where $\hat{m}(t_i)$ is the number of faults predicted by the fitted model and i the number of faults. The maximum likelihood estimates of the model parameters are determined from the first $n - k$ failure times. A common choice of k is approximately 10% of the available data, so that $\dfrac{k}{n} \approx 0.10$. The remaining data are used to calculate the prediction error.

16.5.5.3 Comparison of Goodness-of-Fit Measures

Figure 16.7 shows the three-dimensional plot of the failure times log-likelihood function of the GO model for the SYS1 [9] dataset consisting of $n = 136$ failures.

The maximum appears to occur for a between 140 and 145 and b between 3.3×10^{-5} and 3.5×10^{-5}, which can serve as initial estimates for an algorithm to identify the MLEs.

The parameter estimates of the failure times model are as follows:

- Fault detection rate: $\hat{b} = 0.0000342038$
- Number of initial faults: $\hat{a} = 142.881$
- Number of faults remaining: 6.827

$$(\hat{a} - n = 142.881 - 136 = 6.827)$$

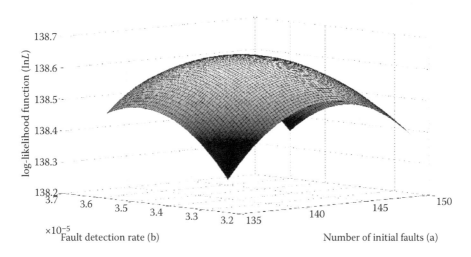

FIGURE 16.7
Log-likelihood function.

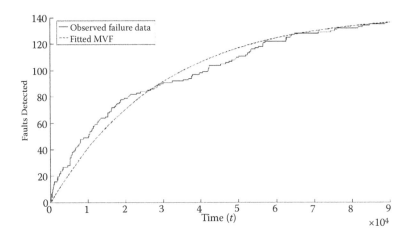

FIGURE 16.8
Fit of GO model to failure times data.

Figure 16.8 shows the fit of the GO model to failure times data along with the observed failure data.

Figure 16.8 indicates that the GO model underpredicts the data until approximately 25,000 time units and overpredicts afterward. However, in later stages, its predictions closely match the data well; hence, it is widely used on software failure datasets.

Figure 16.9 shows the fit of the GO model to failure counts data along with the observed failure count data for SYS1 dataset, which was converted to failure counts by grouping faults into $n = 25$ intervals of 1 hour.

Figure 16.9 indicates that the GO model underpredicts the data between 30,000 to 60,000 time units. Apart from this, it matches the observed data closely.

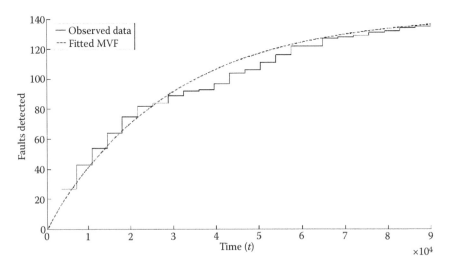

FIGURE 16.9
Fit of GO model to failure counts data.

The parameter estimates of the failure counts model are as follows:

- Fault detection rate: $\hat{b} = 0.0000342879$
- Number of initial faults: $\hat{a} = 142.827$
- Number of faults remaining: 117.827

$$\left(\hat{a} - \sum_{i=1}^{n} k_i = 142.827 - 136 = 6.827 \right)$$

Table 16.4 lists the AIC values for different SRGMs.

Table 16.4 also ranks the models based on their respective AIC values. The geometric model achieves the lowest AIC, indicating that it is the preferred model with respect to the AIC. The Weibull model comes close to the AIC of the Geometric, but the other three models do not.

Figure 16.10 illustrates the fit of the GO and Weibull models for 90% of the SYS1 dataset.

Figure 16.10 indicates that the GO model predicts the data better than the Weibull model, which is in contrast to the AIC assessment. Thus, no one measure of goodness of fit is sufficient to select a model and multiple measures should be considered.

TABLE 16.4

Ranking of SRGM according to AIC

Model Name	AIC	Ranking
Geometric	1937.03	1
Weibull	1938.16	2
JM	1950.53	3
GO	1953.61	4
Delayed S-shaped	2075.15	5

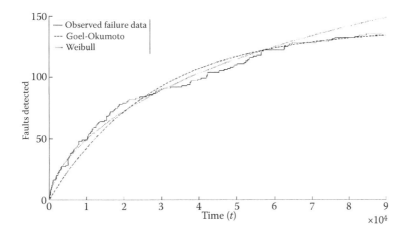

FIGURE 16.10
Comparison of GO and Weibull model predictions.

TABLE 16.5

Ranking of SRGM According to PSSE

Model Name	PSSE	Ranking
GO	23.17	1
JM	24.39	2
Weibull	74.94	3
Geometric	84.37	4
Delayed S-shaped	296.35	5

Table 16.5 reports the numerical values of the PSSE for each of the three NHPP models, which shows that the error of the GO model predictions is far less than the Weibull model.

16.6 Conclusions

This chapter presents an overview of software reliability and software reliability models. The concepts covered in this chapter will enable the audience to understand basic concepts of reliability and reliability mathematics. A summary of parameter estimation techniques, including LSE and MLE, were given. The two most common classes of software reliability models, namely, the failure rate and NHPP models, were then discussed. The illustrations demonstrated model application as well as some methods to assess their goodness of fit. The software failure and reliability assessment tool [14] implements many of the concepts discussed in this chapter. The tool and resources are available for free at http://sasdlc.org /lab/projects/srt.html.

Acknowledgments

This work was supported by the Naval Air Systems Command through the Systems Engineering Research Center, a Department of Defense University Affiliated Research Center under Research Task 139: Software Reliability Modeling and the National Science Foundation (Award 1526128).

References

1. H. Pham, *Software Reliability*. John Wiley & Sons Hoboken, NJ, 1993.
2. A. Amendola, L. Impagliazzo, P. Marmo, and F. Poli, Experimental evaluation of computer-based railway control systems. *Proceedings of International Symposium on Fault-Tolerant Computing*, pp. 380–384, 1997.

3. B. Littlewood and L. Strigini. Software reliability and dependability: A roadmap. *Proceedings of the Conference on the Future of Software Engineering.* Association for Computing Machinery, NY, USA, pp. 175–188, 2000.
4. M. R. Lyu, Software reliability engineering: A roadmap. *Proceedings of Future of Software Engineering,* Minneapolis, MN, 2007, pp. 153–170.
5. M. Xie, *Software Reliability Modelling.* Vol. 1. World Scientific, Singapore, 1991.
6. E. Elsayed, *Reliability Engineering,* 2nd Edition. John Wiley & Sons, Hoboken, NJ, 2012.
7. L. Leemis, *Reliability: Probabilistic Models and Statistical Methods,* 2nd Edition. 2009.
8. Z. Jelinski and P. Moranda, Software reliability research, *Statistical Computer Performance Evaluation.* Academic Press, New York, 1972, pp. 465–484.
9. M. Lyu (ed.), *Handbook of Software Reliability Engineering,* McGraw-Hill, New York, 1996.
10. K. Trivedi, *Probability and Statistics with Reliability, Queueing, and Computer Science Applications,* 2nd Edition. John Wiley & Sons, Hoboken, NJ, 2001.
11. A. Goel and K. Okumoto, Time-dependent error-detection rate model for software reliability and other performance measures, *IEEE Transactions on Reliability,* Vol. 28, no. 3, pp. 206–211, 1973.
12. H. Akaike, A new look at the statistical model identification. *IEEE Transactions on Automatic Control,* Vol. 19, No. 6, pp. 716–723, 1974.
13. L. Fiondella and S. Gokhale, Software reliability model with bathtub shaped fault detection rate, *Proceedings of Annual Reliability and Maintainability Symposium,* pp. 1–6, 2011.
14. V. Nagaraju, K. Katipally, R. Muri, L. Fiondella, and T. Wandji, An open source software reliability tool: A guide for users. *Proceedings of International Conference on Reliability and Quality in Design,* CA, pp. 132–137, 2016.

17

Applications of Formal Methods, Modeling, and Testing Strategies for Safe Software Development

Alessandro Fantechi, Alessio Ferrari, and Stefania Gnesi

CONTENTS

The challenges posed by the new scenarios of railway transportation (liberalization, distinction between infrastructure and operation, high speed, European interoperability, etc.) have a dramatic impact on the safety issues. This impact is counterbalanced by the growing adoption of innovative signaling equipment (the most notable example is the European Rail Traffic Management System/European Train Control System) and monitoring systems (such as onboard and wayside diagnosis systems). Each one of these devices includes some software, which in the end makes up the major part of their design costs; the malleability of the software is paramount for the innovation of solutions. On the other hand, it is notorious how software is often plagued by bugs that may threaten its correct functioning: how can the high safety standards assumed as normal practice in railway operation be compatible with such threats?

This chapter briefly summarizes the current answers to such a question. Although the question regards the whole software life cycle, we concentrate on those aspects that are peculiar to the development of safe railway-related software: in a sense, we consider that generic software engineering best practices are already known to the reader. In particular, in Section 17.1, we introduce the safety guidelines in effect for software development in this domain; in Section 17.2, we introduce the foundations of software testing; in Section 17.3, we introduce formal methods, with their applications in the railway domain in Section 17.4; and in Section 17.5, we introduce model-based software development.

17.1 CENELEC EN 50128 Standard, Principles, and Criteria for Software Development

In safety-critical domains, software development is often subject to a certification process, which in general has the aim of verifying that the products have been developed in conformity with specific software safety guidelines and domain-specific documents issued by national or international institutions.

The definition of these guidelines is a very long and complex process that needs to mediate between different stakeholders, such as the following:

- *Manufacturers* not only interested in keeping development costs low, but also paying attention to the safety of their products in order not to run the risk of losing clients because of accidents due to failures of their products

- *Certification bodies* that are responsible for certifying that the product is safe with a reasonably high certainty, according to the running laws and norms, and so tend to give less priority to production costs

- *Clients* (often themselves service providers to final users, as is the case of railway operators), who are interested in balancing cost containment with safety assurance

Due to the contrasting interests of the stakeholders, the guideline production is a slow process that goes through necessary trade-offs. Typically, a new edition of guidelines takes about 10 years: once issued, they become a reference standard for the particular domain until the next edition; also for this reason, the railway signaling sector has been historically reluctant to adopt technological innovation compared to other domains, especially for those functions that have significant impact on the safety of the railway.

17.1.1 Process and Product Certification

A first distinction to be done when speaking of computerized systems (and in particular, of software) certification is between *process* and *product certification.* Process certification is aimed to guarantee that the production process has followed the given guidelines and norms that have been adopted in order to guarantee that the process can deliver products of an expected quality level. Instead, product certification wants to guarantee that a particular product has been designed and developed according to determined quality guidelines.

Safety certification not only focuses more on *product* certification, but also usually requires that the development process does comply with some quality and maturity standards.

The Comité Européen de Normalisation Électrotechnique (CENELEC) safety standards for the railway sector (EN 50126, EN 50128, EN 50129) have much in common with the IEC 61508 standard for embedded systems, but they reflect the peculiar railway safety culture that has a history that goes back more than one century.

EN 50126 defines reliability, availability, maintainability, and safety concepts in relation to railway signaling. EN 50129 gives guidelines for the design of safety in hardware, while EN 50128 gives guidelines for software production. Notice that EN 50128, as well the other standards of the family, have been issued for railway signaling systems. Although their use has spread beyond, to a full range of railway applications, new specific standards are emerging, such as prEN 50657, which is intended for software on board of rolling stock.

17.1.2 Software Development Cycle

EN 50128 (CENELEC 2011) does not mandate any specific software development method, but rather it describes the generic phases that the software development process should follow and the attached documentation. One of the examples that are given about a possible software life cycle is the so-called V-model (Figure 17.1), where every design phase (left branch of the V) corresponds to a verification phase (right branch of the V). The dotted arrows represent the relation between the tests carried on in the right branch and the artifacts produced in each phase of the left branch.

The focus on verification and testing activities, typical of the safety guidelines, is well represented by the V-model, which graphically makes evident that verification and testing costs are comparable with the costs of design and development and that the former activities need to be prepared during the latter ones.

Indeed, EN 50128 does very little guidance on technical choices, e.g., how to develop a suitable software architecture for the composition of separate software components (apart from a specific section dedicated to the possible usage of object-oriented design), but is more concerned on which impact the different usual software engineering techniques can have on the safety of produced software. For this reason, we will concentrate in the following section on those techniques that can more directly help avoid the presence of bugs in the produced software.

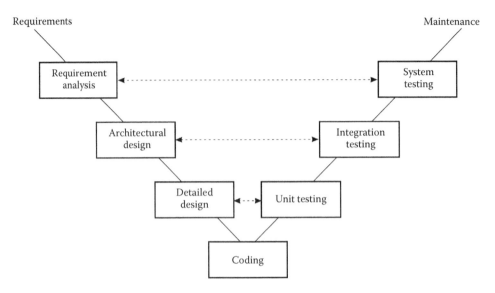

FIGURE 17.1
V-model for the software life cycle of safety-critical railway systems.

It is also worth noticing that EN 50128 requires that verification and testing activities are carried out independently and accurately defines the different roles involved in such activities, according to the safety integrity level (SIL) of the software component under development.

17.1.3 Safety Integrity Level

The *software* SIL is defined by EN 50128 as an attribute of a software component that indicates its required robustness degree in terms of protection with respect to software faults. SIL has a value between 0 and 4: the higher the SIL, the higher should be the assurance that the software is free from faults. The SIL is assigned to components by a system *safety assessment* and *SIL apportionment* process on the basis of the impact on safety that a failure of the component can have (see Chapter 5). Hence, the most serious consequences a failure of a component can have, the highest the SIL: SIL 0 is the level of a component with no effect on safety; SIL 4 is the level of a highly safety-critical component.*

The assignment of the SIL to a software component implies that it should be developed and verified using specific techniques that are considered suitable for that level. Although the techniques for software development in EN 50128 are relevant to cope with all types of software errors, special emphasis is on safety-relevant errors. EN 50128 lists the techniques in a series of tables related to the software development phases, classified on the strength of suggestion of their usage, graduated along the SIL, as follows:

- M: mandatory
- HR: highly recommended (which means that if not used, it should be justified in a proper document)

---

* Besides SIL 0, there is software without any relevance for safety, which is not in the scope of the EN 50128 standard.

- R: recommended
- —: no indication in favor or against
- NR: not recommended (which means that its usage should be justified in a proper document)

Every entry of the table, that is, every listed technique, gives a reference either to a description text or to a subtable which details the technique. Many tables also include notes that give indications on the recommended combination of techniques. We will give examples of recommendations in the following sections.

It is worth noticing that besides software SILs, the EN 50128 standard also provides classification of the tools that are used along the development of the software, including tools for testing and formal verification. According to the norm, tools of class T1 do not generate output that can directly or indirectly contribute to the executable code (including data) of the software; tools of class T2 support test or verification, and errors in the tools can fail to reveal defects in the software or in the design, but cannot directly generate errors in the executable software; and tools of class T3 generates outputs that can directly or indirectly contribute to the executable code (including data) of the safety-related system. For each class, the norm lists the evidence that shall be provided about the actual role of each tool in the process and about its validation.

17.2 Testing of Software

Testing activities consist of the systematic research of faults in the software. In the railway safety-critical domain, testing is fundamental to ensure system safety, and its cost is comparable to the cost of the actual software coding. In this chapter, we will first give some preliminary definitions that are useful in understanding the remainder of the sections, and then we will discuss the different types of testing activities, at different degrees of granularity (*unit testing, integration testing, system testing*) and in different development phases (*regression testing, mutation testing*) that are normally carried out in the railway domain. We give here a brief account of the foundations of testing, which is actually a separate discipline inside software engineering, with many reference textbooks (see, e.g., the book by Myers et al. [2011]).

17.2.1 Preliminary Definitions

A program P can be in principle represented as a function $P: D \rightarrow R$, in which D is the space of the input data and R is the space of the results. The correctness of a program could then be defined as a Boolean function $ok(P, d)$, applied on the program P, with the input data d. The function returns *true* if P produces the expected value for the input d, and *false* otherwise.

A *test* T (also called *test suite* or *test set*) is a subset of the input data D. Furthermore, any $t \in T$ is a *test case*. A program P is *correct* for a test T if for each $t \in T$, we have $ok(P, t)$ = true.

A test T is considered *passed* if it does not reveal any fault—i.e., if the program is correct for the test T—and it is considered *failed* otherwise.

A *selection criterion* C for a program P is a set of predicates on D. We say that a test T is *selected* if the following conditions hold:

- $\forall t \in T \exists c \in C$: $c(t)$ = true: This means that each test case satisfy at least one of the predicates of the selection criterion.
- $\forall c \in C \exists t \in T$: $c(t)$ = true: This means that each predicate selects at least one test case.

An *exhaustive test* is given by a criterion that selects all possible input values ($T = D$). A passed exhaustive test implies that the program is correct for all input values. Since it is usually impossible to exhaustively test a program, the selection of a test has the aim of approximating exhaustivity with the lowest possible cost (that is, number of test cases). Indeed, it can be demonstrated that the testing of a program can reveal faults, but cannot prove their absence.

Notice also that the preceding definition assume that $ok(P, d)$ is known, which means that for each test case, the requirements tell the expected outcome of the execution: this is the meaning of the horizontal links between the two sides of the V-model in Figure 17.1.

Given these minimal definitions, we can start discussing the different types of tests that are normally used in the railway domain.

17.2.2 Unit Testing

Unit testing aims to verify whether a code unit (i.e., a function, a module, or a class) is correct with respect to its expected behavior. Depending on the selection criteria adopted to define the test cases, we distinguish among *functional testing*, *structural testing*, and *statistical testing*.

17.2.2.1 Functional Testing

Functional testing, also known as *black box testing* or *requirement-based testing*, is performed on the basis of the functional requirements that determine which features are implemented by the unit under test. Functional testing does not look at the code of the unit, but only at its input/output behavior. The commonly used criteria for selecting the tests are as follows:

- *Equivalence class partitioning* (ECPT): The input domain of the unit is partitioned into equivalence classes, with the hypothesis that one test case for each class represents all the values for the same class.
- *Boundary value analysis* (BVAN): Test cases are selected based on the boundary values of the equivalence classes. This enables to check for typical programming errors, in which, e.g., a *less or equal* condition is erroneously replaced with a *less* condition.
- *Test case by error guessing*: Test cases are selected by domain experts, based on their intuition, experience on the code, and the application domain.

17.2.2.2 Structural Testing and Coverage Criteria

With structural testing, the selection criterion is directly derived from the structure of the code unit and, in particular, from its *flow graph*. The selected tests are those that exercise all the *structures* of the program. Hence, one must choose the reference structure (i.e., statement, branch, basic condition, compound condition, path), and a measure shall be used to indicate if all structures are exercised: this measure is called *coverage*:

- *Statement coverage* is pursued by choosing the tests according to their ability to cover the code statements, that is, the nodes of the flow graph. The statement coverage value is calculated as the ratio between the number of statements executed and the total number of statements. A test is selected by this criterion if its coverage value is equal to 1 (100%). This ensures that all statements are executed at least once.

- *Branch coverage* (or *decision coverage*) requires that all branches of the flow graph are executed; i.e., each branch belongs to at least one of the paths exercised by the test. The branch coverage value is calculated as the ratio between the number of branches executed and the total number of branches, and a test is selected if its coverage value is equal to 1. It is also referred as *decision coverage* since it exercises all the true/false outcomes of conditional statements, which are the sources of the branches in the flow graph.

- *Basic condition coverage* requires that all the basic Boolean conditions included in conditional statements are exercised, which means that for all the basic conditions, both true and false outcomes are exercised. This does not guarantee that branch coverage equals 1, since some *combinations* of basic condition values might not be exercised, and some branches might not be executed.

- *Compound condition coverage* requires that all possible combinations of Boolean values obtained from basic conditions are exercised at least once. This implies 2^n test cases, where n is the number of basic conditions.

- *Modified condition decision coverage* (MCDC) considers the Boolean values obtained from basic conditions in relation to their context in Boolean expressions. Boolean expressions are all those structures that fall in the syntactic category of Boolean expressions. This coverage is calculated as the ratio between the covered Boolean expressions and the total number of Boolean expressions. Here, it is useful to consider an example:

```
If ((x > 0)                    cond1
        && (y < -2             cond2
        ||   y == 0            cond3
        ||   y > 2)            cond4
   )
```

Here, we have five Boolean expressions, i.e., four basic conditions, and the overall Boolean expression. The MCDC criterion selects the test cases shown in Table 17.1.

Note that MCDC implies branch coverage, but with linear cost, instead of the exponential cost required by *compound condition coverage*.

- *Path coverage* requires that all paths of the program are exercised by the test. It is measured as the ratio between the number of paths exercised and the total number of paths. The total number of paths exponentially grows with the number of decisions that are independent and not nested, but such number becomes unbounded in case of cycles. Hence, when cycles are involved, the path coverage criterion considers a *finite* number of cycle executions (i.e., iterations). In this case, we speak

TABLE 17.1

Selected Test Cases according to the MCDC Criterion

Test Case	cond1	cond2	cond3	cond4	Result
1	F				F
2	T	T			T
3	T	F	T		T
4	T	F	F	T	T
5	T	F	F	F	F

about *k-coverage*, in which *k* is the number of iterations considered. Normally, *k* = 1, which means that two test cases are defined, one in which no iteration is performed (the decision is false) and one in which one iteration is performed. Another way to reduce the exponential number of paths to be considered is to evaluate the so-called McCabe number, defined as the number of paths in the flow graph that are linearly independent, which is equal to the number of decisions in the code, plus 1. In this case, a test is selected if the number of paths that are linearly independent and that are exercised by the test is equal to the *McCabe number*. Sometimes, paths can be *unfeasible*, in the sense that they cannot be exercised by any input. Unfeasible paths are common in railway safety-critical systems, due to the presence of defensive programming structures, which allow coping with hardware or software failures. Due to the presence of unfeasible paths, path coverage might never be satisfied by a test. Hence, when evaluating path coverage, it is reasonable to limit it to the feasible paths and to evaluate the nature of the unfeasible paths through code inspection. Alternatively, one may consider error seeding in the code to force the software to execute these unfeasible paths.

Figure 17.2 shows the relations between the different coverage criteria. The criteria are partially ordered. The criteria at the bottom of the figure are those that are weaker, but also less expensive. At the top of the figure appear the criteria that are stronger, but also more

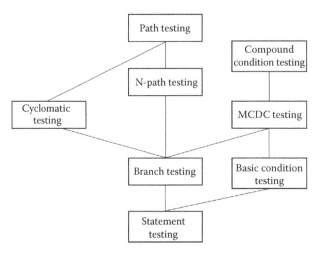

FIGURE 17.2
Relations between the different coverage criteria.

expensive. Other criteria have been proposed in the literature on testing; we have limited ourselves to the most common ones.

17.2.2.3 Statistical Testing

While in functional and structural tests the test data are provided by a deterministic criterion, in statistical tests, instead, they are random and can be based on the generation of pseudorandom test data according to an expected distribution of the input data to the program. Note that in both cases of nonstatistical test (functional and structural), the selection criteria are deterministic and define *ranges* of values for the test cases. The test data may be randomly chosen in these ranges: in practice, this corresponds to combining the statistical test with the earlier ones. A common way to conduct the testing is to first apply a functional test, a statistical test, or a combination of the two and then to measure the coverage according to the desired coverage criterion. If the resulting coverage is considered sufficient for testing purposes, the unit test ends; otherwise, new tests are defined to increase the coverage.

17.2.2.4 Performing Unit Tests

To practically perform unit tests, one should define the following:

- *Driver:* module that includes the function to invoke the unit under test, passing previously defined the test cases
- *Stub:* dummy module that presents the same interface of a module invoked by the unit under test
- *Oracle:* an entity (program or human user) that decides whether the test passed or failed
- *Code instrumentation:* inclusion in the code of instructions that allow seeing if the execution of a test actually exercises the structures of the code and that derives its coverage measure.

17.2.3 Integration Testing

After the coding phase, and after an adequate unit testing, it is appropriate to perform an *integration test* to verify the correctness of the overall program in order to be sure that there are no anomalies due to incorrect interactions between the various modules. Two methods are normally used: the *nonincremental* and the *incremental* approaches.

The *nonincremental* approach or *big bang test* assembles all the previously tested modules and performs the overall analysis of the system. The *incremental* approach instead consists of testing individual modules and then connecting them to the caller or called modules, testing their composition, and so on, until the completion of the system. This approach does not require all modules to be already tested, as required by the nonincremental approach. In addition, it allows locating interface anomalies more easily and exercising the module multiple times. With this approach, one may adopt a *top–down* strategy in incrementally assembling the components, which starts from the main module to gradually integrate the called modules, or a *bottom–up* strategy, which begins to integrate the units that provide basic functionalities and terminates with the main as last integrated unit.

Coverage criteria have also been defined for integration testing. In particular, a commonly used criterion is the *procedure call coverage*, which indicates the amount of procedure calls exercised by an integration test.

17.2.4 System Testing

When the integration test has already exercised all the functionality of the entire system, it may still be necessary to test certain global properties that are not strictly related to individual system features but to the system as a whole. This is the role of *system testing*, which typically includes the following:

- *Stress/overload test*: This checks that the system complies with the specifications and behaves correctly in overload conditions (e.g., high number of users, high number of connections with other systems).
- *Stability test*: This checks the correct behavior of the system even when it is used for long periods. For example, if dynamic memory allocation is used, this test checks that stack overflow errors occur only after an established period.
- *Robustness test*: The system is provided with unexpected or incorrect input data, and one must check its behavior. A typical example is typing random input on the user interface to check whether the interface shows some anomalous behavior.
- *Compatibility test*: This verifies the correct behavior of the software when connected to hardware devices to which it is expected to be compatible.
- *Interoperability test*: This verifies the correct behavior of the system when connected with other system of similar nature (e.g., other products provided by different vendors), once the communication protocol is established.
- *Safety test*: This verifies the correct behavior of the system in the presence of violations of its safety conditions; e.g., when a violation occurs, the system switches to fail-safe mode.

Specialized instruments are often used to perform system testing of railway control systems. In particular, ad hoc hardware and software simulation tools are normally employed to recreate the environment in which the software will be actually executed.

17.2.5 Regression Testing

When a novel version of the product is released, it is necessary to repeat the testing to check whether the changes to the product have introduced faults that were not present before. One speaks in this case of *regression testing*, which aims to minimize the cost of testing by using the tests performed on previous versions. This can be done by reusing the same drivers, stubs, and oracles and by repeating the test cases of the previous versions.

Before performing regression testing, an impact analysis is explicitly required by EN 50128. When the novel product is tested, one shall consider the changes introduced and assess their impact in terms of testing. A small change should introduce minimal effort in terms of additional tests. However, one should consider that structural tests are often not incremental. Hence, even a small change could greatly impact the coverage, and then one should define test cases that bring the coverage to the desired values. In case of even small changes to safety-critical railway systems, at least the safety tests have to be repeated completely.

17.2.6 Mutation Testing

Mutation testing helps evaluate the ability of the performed tests to reveal potential errors. To this end, faults are intentionally introduced in the code. An ideal test should reveal those faults. If the previously defined tests do not reveal those faults, the test is not adequate, and more sophisticated tests have to be defined. The intentional introduction of faults is also used to evaluate the robustness and safety of the software, and in this case, it is referred as *fault injection* or *fault-based testing*.

17.2.7 Testing according to EN 50128

We have already seen the importance of testing in the EN 50128 standard, exemplified by the V-model. Recommendation for the mentioned testing techniques varies according to the SIL:

- Functional/black box testing is considered mandatory by EN 50128 for software components at SIL 3/SIL 4 and highly recommended for software at SIL 0/SIL 1/ SIL 2. In particular, from SIL 1 to SIL 4, ECPT and BVAN are highly recommended.
- Statement coverage is highly recommended from SIL 1 to SIL 4, while the mentioned finer coverage measures are moreover highly recommended for SIL 3/SIL 4.

17.3 Formal Methods for the Development and Verification of Software

Testing cannot be used for definitely assuring the absence of bugs. In general, proving the absence of bugs is an undecidable problem; however, formal arguments can, in many cases, be used to demonstrate their absence. Nowadays, the necessity of formal methods as an essential step in the design process of industrial safety-critical systems is indeed widely recognized.

In its more general definition, the term *formal methods* encompasses all notations with a precise mathematical semantics, together with their associated analysis and development methods, that allow one to describe and reason about the behavior and functionality of a system in a formal manner, with the aim to produce an implementation of the system that is provably free from defects. The application of mathematical methods in the development and verification of software is very labor intensive and thus expensive. Therefore, it is often not feasible to check all the wanted properties of a complete computer program in detail. It is more cost effective to first determine what the crucial components of the software are. These parts can then be isolated and studied in detail by creating mathematical models of these sections and verifying them.

In order to reason about formal description, several different notations and techniques have been developed. We refer to Garavel and Graf (2013), Gnesi and Margaria (2013), and Woodcock et al. (2009) for more complete discussions over the different methods and their applications; in the following sections, we give a brief introduction to the main different aspects and concepts that can be grouped under this term.

17.3.1 Formal Specification

The languages used for formal specifications are characterized by their ability to describe the notion of internal state of the target system and by their focus on the description of how the operations of the system modify this state. The underlying foundations are in discrete mathematics, set theory, category theory, and logic.

17.3.1.1 B Method

The B method (Abrial 1996) targets software development from specification through refinement, down to implementation and automatic code generation, with formal verification at each refinement step: writing and refining a specification produces a series of *proof obligations* that need to be discharged by formal proofs. The B method is accompanied by support tools, such as tools for the derivation of proof obligations, theorem provers, and code generation tools.

17.3.1.2 Z Notation

The Z notation (Spivey 1989) is a formal specification language used for describing and modeling computing systems. Z is based on the standard mathematical notation used in axiomatic set theory, lambda calculus, and first-order predicate logic. All expressions in Z notation are typed, thereby avoiding some of the paradoxes of naive set theory. Z contains a standardized catalog (called the *mathematical toolkit*) of commonly used mathematical functions and predicates. The Z notation has been at the origin of many other systems such as, for example, Alloy* (Jackson 2012) and its related tool Alloy Analyzer, which adapts and extends Z to bring in fully automatic (but partial) analysis.

17.3.1.3 Automata-Based Modeling

In this case, it is the concurrent behavior of the system being specified that stands at the heart of the model. The main idea is to define how the system reacts to a set of stimuli or events. A state of the resulting transition system represents a particular configuration of the modeled system. This formalism, and the derived ones such as statecharts (Harel 1987) and their dialects, is particularly adequate for the specification of reactive, concurrent, or communicating systems and protocols. They are, however, less appropriate to model systems where the sets of states and transitions are difficult to express.

17.3.1.4 Modeling Languages for Real-Time Systems

Extending the simple automata framework gives rise to several interesting formalisms for the specification of real-time systems. When dealing with such systems, the modeling language must be able to cope with the physical concept of time (or duration), since examples of real-time systems include control systems that react in dynamic environments. In this regard, we can mention Lustre (Halbwachs et al. 1991), which is a (textual) synchronous data flow language, and SCADE[†] (Berry 2007), a complete modeling environment that provides a graphical notation based on Lustre. Both provide a notation for expressing synchronous concurrency based on data flow.

---

* http://alloy.mit.edu/alloy/.
† http://www.esterel-technologies.com.

Other graphical formalisms have proven to be suitable for the modeling of real-time systems. One of the most popular is based on networks of timed automata (Alur and Dill 1994). Basically, timed automata extend classic automata with clock variables (that continuously evolve but can only be compared with discrete values), communication channels, and guarded transitions.

17.3.2 Formal Verification

To ensure a certain behavior for a specification, it is essential to obtain a rigorous demonstration. Rather than simply constructing specifications and models, one is interested in proving properties about them.

17.3.2.1 Model Checking

A formal verification technique that has also recently acquired popularity in industrial applications is model checking (Clarke et al. 1999), an automated technique that, given a finite-state model of a system and a property stated in some appropriate logical formalism (such as temporal logic), checks the validity of this property on the model. Several temporal logics have been defined for expressing interesting properties. A temporal logic is an extension of the classical propositional logic in which the interpretation structure is made of a succession of states at different time instants. An example is the popular computation tree logic (CTL), a *branching time* temporal logic, whose interpretation structure (also called *Kripke structure*) is the *computation tree* that encodes all the computations departing from the initial states.

Formal verification by means of model checking consists of verifying that a Kripke structure M, modeling the behavior of a system, satisfies a temporal logic formula φ, expressing a desired property for M. A first simple algorithm for implementing model checking works by labeling each state of M with the subformulas of φ that hold in that state, starting with the ones having length 0, that is, with atomic propositions, then to subformulas of length 1, where a logic operator is used to connect atomic propositions, then to subformulas of length 2, and so on. This algorithm requires navigation of the state space and can be designed to show a linear complexity with respect to the number of states of M. One of the interesting features of model checking is that when a formula is found not to be satisfied, the subformula labeling collected on the states can be used to provide a *counterexample*, that is, an execution path that leads to the violation of the property, thus helping the debugging of the model.

The simple model-checking algorithm sketched earlier needs to explore the entire state space, incurring in the so-called exponential state space explosion, since the state space often has a size exponential in the number of independent variables of the system.

Many techniques have been developed to attack this problem: among them, two approaches are the most prominent and most widely adopted. The first one is based on a symbolic encoding of the state space by means of Boolean functions, compactly represented by binary decision diagrams (BDDs) (Bryant 1986). The second approach considers only a part of the state space that is sufficient to verify the formula, and within this approach, we can distinguish local model checking and bounded model checking: the latter is particularly convenient since the problem of checking a formula over a finite-depth computation tree can be encoded as a satisfiability problem and, hence, efficiently solved by current very powerful *SAT solvers* (Biere et al. 1999).

The availability of efficient model-checking tools able to work on large state space sizes has favored, from the second half of the 1990s, their diffusion in industrial applications. The most popular model checkers of an academic origin are as follows:

1. Symbolic Model Verifier (SMV)* (McMillan 1993), developed by the Carnegie Mellon University, for the CTL logic, based on a BDD representation of the state space
2. NuSMV[†] (Cimatti et al. 2002), developed by Fondazione Bruno Kessler, a reengineered version of SMV which includes a bounded model-checking engine
3. Simple Promela Interpreter (SPIN)[‡] (Holzmann 2003), a model checker for the LTL, developed at Bell Labs, for which the Promela language for the model definition has been designed

17.3.2.2 Software Model Checking

The first decade of model checking has seen its major applications in the hardware verification domain; meanwhile, applications to software have been made at the system level or at early software design. Later, applications within the model-based development (see Section 17.5) have instead considered models at a lower level of design, closer to implementation. But such an approach requires an established process, hence excluding software written by hand directly from requirements: indeed, software is often received from third parties, who do not disclose their development process. In such cases, direct verification of code correctness should be conducted, especially when the code is safety related: testing is the usual choice, but testing cannot guarantee exhaustiveness.

The direct application of model checking to code is, however, still a challenge, because the correspondence between a piece of code and a finite-state model on which temporal logic formulas can be proved is not immediate: in many cases, software has, at least theoretically, an infinite number of states, or at best, the state space is just huge.

Pioneering work on the direct application of model checking to code (also known as software model checking) has been made at National Aeronautics and Space Administration since the late 1990s by adopting in the time two strategies: first, by translating code into the input language of an existing model checker—in particular, translating into Promela, the input language for SPIN. Second, by developing ad hoc model checkers that directly deal with programs as input, such as Java PathFinder[§] (Havelund and Pressburger 2000). In both cases, there is the need to extract a finite-state abstract model from the code, with the aim of cutting the state space size to a manageable size: the development of advanced abstraction techniques has only recently allowed a large-scale application of software model checking. Java PathFinder has been used to verify software deployed on space probes. Some software model checkers, such as CBMC, hide the formality to the user by providing built-in default properties to be proven: absence of division by zero, safe usage of pointers, safe array bounds, etc. On this ground, such tools are in competition with tools based on abstract interpretation, discussed in the next section. It is likely that in the next years, software model checking will also gain growing industrial acceptance in the railway domain, due to its ability to prove the absence of typical software bugs, not only for

---

* http://www.cs.cmu.edu/~modelcheck/smv.html.
[†] http://nusmv.fbk.eu.
[‡] http://spinroot.com/spin/whatispin.html.
[§] http://javapathfinder.sourceforge.net.

proving safety properties, but also for guaranteeing correct behavior of non-safety-related software. Applying model checkers directly to production safety-related software calls for their qualification at level T2.

17.3.2.3 Abstract Interpretation

Contrary to testing, which is a form of dynamic analysis focusing on checking functional and control flow properties of the code, static analysis aims at automatically verifying properties of the code without actually executing it. In the recent years, we have assisted to a raising spread of abstract interpretation, a particular static analysis technique. Abstract interpretation is based on the theoretical framework developed by Patrick and Radhia Cousot in the 1970s (Cousot and Cousot 1977). However, due to the absence of effective analysis techniques and the lack of sufficient computer power, only after 20 years have software tools been developed for supporting the industrial application of the technology (e.g., Astrée* [Cousot et al. 2005], Polyspace†). In this case, the focus is mainly on the analysis of source code for runtime error detection, which means detecting variable overflow/underflow, division by zero, dereferencing of noninitialized pointers, out-of-bound array access, and all those errors that might then occur, bringing to undefined behavior of the program.

Since the correctness of the source cannot be decidable at the program level, the tools implementing abstract interpretation work on a conservative and sound approximation of the variable values in terms of intervals and consider the state space of the program at this level of abstraction. The problem boils down to solving a system of equations that represent an overapproximate version of the program state space. Finding errors at this higher level of abstraction does not imply that the bug also holds in the real program. The presence of false positives after the analysis is actually the drawback of abstract interpretation that hampers the possibility of fully automating the process (Ferrari et al. 2013a). Uncertain failure states (i.e., statements for which the tool cannot decide whether there will be an error or not) have normally to be checked manually, and several approaches have been put into practice to automatically reduce these false alarms.

17.3.2.4 Automated Theorem Proving

Another possibility for the automatic verification of systems properties is by theorem proving. Theorem provers favor the automation of deduction rather than expressiveness. The construction of proofs is automatic once the proof engine has been adequately parameterized. Obviously, these systems assume the decidability of at least a large fragment of the underlying theory. Theorem provers are powerful tools, capable of solving difficult problems, and are often used by domain experts in an interactive way. The interaction may be at a very detailed level, where the user guides the inferences made by the system, or at a much higher level, where the user determines intermediate lemmas to be proven on the way to the proof of a conjecture.

The language in which the conjecture, hypotheses, and axioms (generically known as *formulas*) are written as a logic, often not only classical first-order logic, but also nonclassical logic or higher-order logic. These languages allow a precise formal statement of the necessary information, which can then be manipulated by a theorem prover.

---

* https://www.absint.com/products.htm.
† https://mathworks.com/products/polyspace/.

The proofs produced by theorem provers describe how and why the conjecture follows from the axioms and hypotheses, in a manner that can be understood and agreed upon by domain experts and developers. There are many theorem prover systems readily available for use; the most popular are Coq* (Barras et al. 1997), HOL[†] (Nipkow et al. 2002), and PVS[‡] (Owre 1992).

Software verification is an obvious and attractive goal for theorem proving technology. It has been used in various applications, including diagnosis and scheduling algorithms for fault tolerant architectures and requirement specification for portions of safety-critical systems.

17.4 Railway Applications of Formal Methods and Formal Verification

For the past 30 years, formal methods, have promised to be the solution for the safety certification headaches of railway software designers. The employment of very stable technology and the quest for the highest possible guarantees have been key aspects in the adoption of computer-controlled equipment in railway applications. EN 50128 indicates formal methods as highly recommended for SIL 3/4 and recommended for SIL 1/2 in the software requirement specification phase, as well as in the software design phase. Formal proof is also highly recommended for SIL 3/4, recommended for SIL 1/2 in the verification phase, and highly recommended for SIL 3/4 for configuration data preparation.

Formal proof, or verification, of safety is therefore seen as a necessity. Moreover, consolidated tools and techniques have been more frequently selected in actual applications. We mention just a few of them, with no claim of completeness. Books by Flammini (2012) and Boulanger (2014) give more examples of railway applications.

17.4.1 Formal Specification

17.4.1.1 B Method

The B method has been successfully applied to several railway signaling systems. The SACEM (Système d'aide à la conduite, à l'exploitation et à la maintenance) system for the control of a line of Paris Réseau Express Régional (Da Silva et al. 1993) is the first acclaimed industrial application of B. B has been adopted for many later designs of similar systems by Matra (now absorbed by Siemens). One of the most striking applications has been the Paris automatic metro line 14. The paper by Behm et al. (1999) on this application of B reports that several errors were found and corrected during proof activities conducted at the specification and refinement stages. By contrast, no further bugs were detected by the various testing activities that followed the B development. The success of B has had a major impact in the sector of railway signaling by influencing the definition of the EN 50128 guidelines.

---

* https://coq.inria.fr.
† https://hol-theorem-prover.org.
‡ http://pvs.csl.sri.com.

17.4.2 Formal Verification

17.4.2.1 Model Checking

The revised EN 50128 in 2011 marked the first appearance of model checking as one of the recommended formal verification techniques in a norm regarding safety-critical software.

Indeed, model checking is now a common mean to gain confidence in a design, especially concerning the most critical parts. Safety verification of the logics of interlocking systems has been since many years an area of application of model checking by many research groups (see, e.g., Fantechi [2013], Hansen et al. [2010], James et al. [2014], and Vu et al. [2017]) and is still an open-research area due to the difficulty of verification of large designs due to raising state space explosion problems that overcome the capacity of model-checking tools.

17.4.2.2 Abstract Interpretation

Due to the recent diffusion of tools supporting static analysis by abstract interpretation, published stories about the application of this technology to the railway domain are rather limited, although static analysis is highly recommended for SILs 1–4 by EN 50128. Ferrari et al. (2013a) provide one of the few contributions in this sense, showing how abstract interpretation was proficiently used in conjunction with model-based testing, to reduce the cost of the testing activities of a railway signaling manufacturer by 70%.

17.5 Model-Based Development

Although several successful experiences on the use of formal methods in railway systems have been documented in the literature, formal methods are still perceived as experimental techniques by railway practitioners. The reasons are manifold. On the one hand, handling model checkers and theorem provers requires specialized knowledge that is often beyond the competencies of railway engineers. On the other hand, the introduction of formal methods requires radical restructuring of the development process of companies and does not allow to easily reuse code and other process artifacts that were produced with the traditional process. In addition, available formal tools are designed to be used by experts and rarely have engineering-friendly interfaces that could make their use more intuitive for practitioners. Hence, while it has taken more than 20 years to consolidate the usage of formal methods in the development process of railway manufacturers, the *model-based* design paradigm has gained ground much faster. Modeling has indeed the same degree of recommendation of formal methods by EN 50128. The defining principle of this approach is that the whole development shall be based on graphical model abstractions, from which implementation can be manually or automatically derived. Tools supporting this technology allows performance of *simulations* and *tests* of the system models before the actual deployment: the objective is not different from the one of formal methods, that is, detecting design defects before the actual implementation, but while formal methods are perceived as rigid and difficult, model-based design is regarded as closer to the needs of developers, which consider graphical simulation as more intuitive than formal verification. This trend has given increasing importance to tools such as MagicDraw,* IBM Rational Rhapsody,[†] SCADE, and the tool suite MATLAB®/Simulink®/Stateflow®.[‡] MagicDraw and Rhapsody provide capabilities for designing high-level models according to the well-known unified modeling language (UML) and provide support for systems modeling language (SysML), which is an extension of the UML for *system engineering*.

Instead, SCADE and the tool suite MATLAB/Simulink/Stateflow are focused on block diagrams and on the *statechart formalism* (Harel 1987), which is particularly suitable to represent the state-based behavior of embedded systems used in railways. While the former tools are oriented to represent the high-level artifacts of the development process—i.e., requirements, architecture, and design—the latter are more oriented toward modeling of the lower-level behavior of systems.

Without going into the details of each tool, in this section, we discuss the main capabilities offered by the model-based development paradigm and their impact in the development of railway safety-critical systems. The capability we discuss are *modeling*, *simulation*, *testing*, and *code generation*.

17.5.1 Modeling

Regardless of the tool adopted for modeling, it is first crucial to identify the right level of abstraction at which models are represented. The level of abstraction depends on what is the purpose of the model: if the model is used to define the *architecture* of the railway system, with different high-level subsystems interacting one with the other, the use of high-level languages such as SysML or UML is more suitable; instead, if the model is used to define the *behavior* of the system, and one wants to generate executable code from it, statecharts or UML state machines are more suitable for the goal. One key step in both cases is a clear definition of the interfaces between systems and between software components. For models that are going to be used for code generation, it is preferable to have interfaces that are composed solely of input and output *variables* and not function calls. This eases decoupling among model components and facilitates their testing, since one does not have to create complex stubs for external function calls. Another aspect to consider when modeling for code generation is the need to constrain the modeling language. Indeed, in railway systems, the generated code has to abide to the strict requirements of EN 50128, which are normally internally refined by the companies and that forbid the use of some typical code features, such as global variables, pointers, and dynamic memory allocation. To ensure that the generated code abides to the guidelines, the modeling language has to be restricted to a *safe* subset. For example, for Stateflow, a safe subset to be employed in railway applications was defined by Ferrari et al. (2013b).

17.5.2 Simulation

One of the main benefits of having graphical models is the possibility of animating the models, i.e., observing their dynamic behavior both in terms of input/output variables and in terms of internal states dynamics. In this way, models become *executable specifications*, and the modeler can observe the system behavior before the system is actually deployed, by providing appropriate stubs for input and output that simulate the *environment* of the specification. The simulation feature is available in all the mentioned platforms and is

---

* http://www.nomagic.com/products/magicdraw.html.
† http://www-03.ibm.com/software/products/en/ratirhapfami.
‡ http://www.mathworks.com/products/simulink/.

particularly useful in the debugging phase. Indeed, the modeler is able to *visualize* where, in the model, an error actually occurred and which is the context of execution of the model when the error occurred (i.e., the other states currently active in the model). This contextualized visualization is something that is not possible when debugging code and becomes useful in detecting faults in the model when a test on the model fails.

17.5.3 Model-Based Testing

As mentioned before in this chapter, testing is one of the key activities in the development of railway systems. When using a model-based paradigm, tests are normally executed on models, and existing tools give the possibility to automatically generate tests based on coverage criteria. For statecharts, coverage criteria are normally state coverage and transition coverage (equivalent to statement coverage and branch coverage for code testing). While automatic test case generation can allow easily reaching fixed coverage criteria, it does not ensure that the system actually complies with its requirements. For this reason, functional tests are defined and performed by means of *ad hoc* stubs that generate the appropriate input signals for the models.

If one wants to generate code from the models, one additional step is required: tests executed on the models have to be repeated on the generated code, and (a) the input/output behavior of the code has to match with the behavior of the models and (b) the resulting coverage for a test has to match between models and code. In this way, one ensures that the generated code behavior matches the behavior observed on models and that the code generator introduces no additional behavior. This step is often called *translation validation* (Conrad 2009) and was applied in the railway domain by Ferrari et al. (2013a).

17.5.4 Code Generation

With code generation, the code that will be deployed in the microprocessor(s) of the system is automatically generated from the models. Some modeling tools—normally those that allow to model statecharts or UML state machines—enable generation of the complete source code, while others—normally those that allow to define the architecture and high-level design of the software—generate *skeletons* for the classes or functions of the code, which have to be manually completed by the developer. Even in the case of complete generation, the developer normally has to manually define *glue* code, i.e., an adapter module, to attach the generated code to the drivers. It is worth remarking that, for example, the code generator from SCADE is qualified* for railway systems, which, in principle, implies that the models exhibit the same behavior of the code. In case that nonqualified code generators are used, one has to ensure model-to-code compliance by means of translation validation, as mentioned in the previous section.

17.6 Conclusions

This chapter has given just a brief account of techniques that can be used to produce safe software for railway systems. Looking again at the EN 50128 standard, one can see that for

---

* According to the *tool qualification* process of EN 50128:2011, the T3 most severe class of qualification is required for code generators.

space limits, we have not cited many techniques that are also commonly used (e.g., defensive programming and diversity) and we have not addressed the important distinction between *generic* and *specific* applications nor *data preparation* techniques, both used when a railway signaling application has to be deployed for a specific track or line layout. But we think that reading this chapter is a good introduction to the issue of developing safe software.

References

Abrial, J. R. 1996. *The B-Book*. Cambridge, MA: Cambridge University Press.

Alur, R., and Dill, D. L. 1994. A theory of timed automata. *Theoretical Computer Science*, 126(2):183–235.

Barras, B., Boutin, S., Cornes, C., Courant, J., Filliatre, J. C., Gimenez, E. et al. 1997. *The Coq Proof Assistant Reference Manual: Version 6.1, RT-203*. Le Chesnay, France: INRIA.

Behm, P., Benoit, P., Faivre, A., and Meynadier, J. M. 1999. Meteor: A successful application of B in a large project. In *Lecture Notes in Computer Science 1708*, eds: J. M. Wing, J. Woodcock, and J. Davies, 369–387. Berlin: Springer Verlag.

Berry, G. 2007. SCADE: Synchronous design and validation of embedded control software. In *Next Generation Design and Verification Methodologies for Distributed Embedded Control Systems*, eds: S. Ramesh and P. Sampath, 19–33. Dordrecht: Springer Netherlands.

Biere, A., Cimatti, A., Clarke, E., and Zhu, Y. 1999. Symbolic model checking without BDDs. In *Lecture Notes in Computer Science 1579*, ed.: W. R. Cleaveland, 193–207. Berlin: Springer Verlag.

Boulanger, J. L. 2014. *Formal Methods Applied to Industrial Complex Systems*. Hoboken, NJ: John Wiley & Sons.

Bryant, R. 1986. Graph-based algorithms for Boolean function manipulation. *IEEE Transactions on Computers*, 35(8):677–691.

CENELEC (Comité Européen de Normalisation Électrotechnique). 2011. EN 50128:2011: *Railway Applications—Communications, Signaling and Processing Systems—Software for Railway Control and Protection Systems*. CENELEC, Brussels.

Cimatti, A., Clarke, E., Giunchiglia, E., Giunchiglia, F., Pistore, M., Roveri, M., Sebastiani, R., and Tacchella, A. 2002. NuSMV 2: An OpenSource tool for symbolic model checking. In *Lecture Notes in Computer Science 2404*, eds: E. Brinksma and K. G. Larsen, 359–364. Berlin: Springer Verlag.

Clarke, E. M., Grumberg, O., and Peled, D. 1999. *Model Checking*. Cambridge, MA: MIT Press.

Conrad, M. 2009. Testing-based translation validation of generated code in the context of IEC 61508. *Formal Methods in System Design*, 35(3):389–401.

Cousot, P., and Cousot, R. 1977. Abstract interpretation: A unified lattice model for static analysis of programs by construction or approximation of fixpoints. In *Proceedings of the 4th ACM SIGACT-SIGPLAN Symposium on Principles of Programming Languages*, eds: R. M. Graham, M. A. Harrison, and R. Sethi, 238–353. New York: ACM Press.

Cousot, P., Cousot, R., Feret, J., Mauborgne, L., Miné, A., Monniaux, D., and Rival, X. 2005. The ASTRÉE analyzer. In *Lecture Notes in Computer Science 3444*, ed.: S. Sagiv, 21–30. Berlin: Springer Verlag.

Da Silva, C., Dehbonei, Y., and Mejia, F. 1993. Formal specification in the development of industrial applications: Subway speed control system. In *5th IFIP Conference on Formal Description Techniques for Distributed Systems and Communication Protocols*, eds: M. Diaz and R. Groz, 199–213. Amsterdam: North-Holland.

Fantechi, A. 2013. Twenty-five years of formal methods and railways: What next? In *Lecture Notes in Computer Science 8368*, eds: S. Counsell and M. Núñez, 167–183. Berlin: Springer Verlag.

Ferrari, A., Magnani, G., Grasso, D., Fantechi, A., and Tempestini, M. 2013a. Adoption of model-based testing and abstract interpretation by a railway signalling manufacturer. In *Adoption and Optimization of Embedded and Real-Time Communication Systems*, ed: S. Virtanen, 126–144. Hershey, PA: IGI Global.

Ferrari, A., Fantechi, A., Magnani, G., Grasso, D., and Tempestini, M. 2013b. The Metrô Rio case study. *Science of Computer Programming*, 78(7):828–842.

Flammini, F. 2012. *Railway Safety, Reliability, and Security: Technologies and Systems Engineering.* Hershey, PA: IGI Global.

Garavel, H., and Graf, S. 2013. *Formal Methods for Safe and Secure Computer Systems.* BSI Study 875. https://goo.gl/oIxlho (accessed October 28, 2016).

Gnesi, S., and Margaria, T. 2013. *Formal Methods for Industrial Critical Systems: A Survey of Applications.* Chichester, UK: Wiley-IEEE Press.

Halbwachs, N., Caspi, P., Raymond, P., and Pilaud, D. 1991. The synchronous data flow programming language LUSTRE. *Proceedings of the IEEE*, 79(9):1305–1320.

Hansen, H. H., Ketema, J., Luttik, B., Mousavi, M. R., van de Pol, J., and dos Santos, O. M. 2010. Automated verification of executable UML models. In *Lecture Notes in Computer Science 6957*, eds: B. K. Aichernig, F. S. de Boer, and M. M. Bonsangue, 225–250. Berlin: Springer Verlag.

Harel, D. 1987. Statecharts: A visual formalism for complex systems. *Science of Computer Programming*, 8(3):231–274.

Havelund, K., and Pressburger, T. 2000. Model checking Java programs using Java PathFinder. *International Journal on Software Tools for Technology Transfer*, 2(4):366–381.

Holzmann, G. 2003. *The Spin Model Checker: Primer and Reference Manual.* Boston, MA: Addison-Wesley Professional.

Jackson, D. 2012. *Software Abstractions: Logic, Language and Analysis.* Cambridge, MA: MIT Press.

James, P., Möller, F., Nguyen, H. N., Roggenbach, M., Schneider, S., Treharne, H., Trumble, M., and Williams, D. 2014. Verification of scheme plans using CSP||B. In *Lecture Notes in Computer Science 8368*, eds: S. Counsell and M. Nùñez, 189–204. Berlin: Springer Verlag.

McMillan, K. L. 1993. The SMV system. In *Symbolic Model Checking*, ed: K. L. McMillan, 61–85. New York: Springer.

Myers, G. J., Sandler, C., and Badgett, T. 2011. *The Art of Software Testing.* Hoboken, NJ: John Wiley & Sons.

Nipkow, T., Paulson, L. C., and Wenzel, M. 2002. *Isabelle/HOL: A Proof Assistant for Higher-Order Logic.* New York: Springer.

Owre, S., Rushby, J. M., and Shankar, N. 1992. PVS: A prototype verification system. In *Lecture Notes in Computer Science 607*, ed: D. Kapur, 748–752. Berlin: Springer Verlag.

Spivey, J. M., 1989. *The Z Notation: A Reference Manual.* Upper Saddle River, NJ: Prentice-Hall.

Vu, L. H., Haxthausen, A. E., and Peleska, J. 2017. Formal modelling and verification of interlocking systems featuring sequential release. *Science of Computer Programming*, 133:91–115.

Woodcock, J., Larsen, P. G., Bicarregui, J., and Fitzgerald, J. 2009. Formal methods: Practice and experience. *ACM Computing Surveys*, 41(4):19.

18

Practical Statistics and Demonstrations of RAMS in Projects

Joerg Schuette

CONTENTS

18.1 Preamble

Public guided transportation systems operate at very high levels of technical reliability, resulting correspondingly into very high operational availability for the customer.

Higher levels of automation, increasingly complex technical and contractual interfaces between subsystems, and higher appreciation of life cycle characteristics require, however, more and more precise definitions with respect to reliability and availability (as well as maintainability and safety). In practice, it turns out that the precise definitions of what technical reliability, availability, maintainability, and safety (RAMS) requirements shall actually "mean," how they may be demonstrated in a limited period, and how requirements may be broken down to subsystems are often disputed between consumer and supplier.

While common perception of the issues is often intuitively "clear," the precise (e.g., contractual) definitions require a basic appreciation of the relatively low statistical probabilities for the occurrence of potential failures or faults of equipment. On the other hand, the definitions and requirements must remain consistent and remain comprehensible for the practical engineer, so inadequately complex theoretical distributions and concepts are neither necessary nor appropriate in practice.

The chapter therefore intends to limit the overwhelming diversity of possible theoretical considerations to the few practically relevant measures that still have "obvious meaning" to the engineer, show some relations to other often used distributions, and highlight typical pitfalls in common "understanding" and encountered specifications. For the sake of comprehensibility, this chapter has been formulated in a typical engineering language rather than in overly formalized mathematical language; the associated risk of being occasionally less precise is willfully accepted. The subsets are not intended to substitute basic lecture textbooks, which shall be consulted if the reader feels unacquainted with the fundamentals of statistics/stochastics. Also, the focus clearly lies on examples of reliability and availability. While the generic methods and concepts are equally useful for maintainability and safety definitions as well, the specificities of these particular parameters are not much the subject of this chapter.

18.2 Basic Parameters and Models

In order to clarify the basic concepts and variables, it is convenient to start the analysis with one individual elementary component only. Consider as an example a limit switch component installed above a train door that shall switch an output power level to 60 V if activated by the closing door.

In transportation, these kinds of elementary components are traditionally conceived by high-quality materials and perform their intended function for long years. From time to time, the component may, however, "fail" nonetheless to perform the intended function, where "failure" shall mean, for simplicity here, all modes of unintended behavior of the component or ceasing to function entirely ("fault"). In the example case, not signaling the "Door Closed" state anymore while the door is actually closed may be considered as "the" failure of the limit switch.

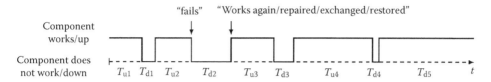

FIGURE 18.1
Typical up/down behavior of a repairable or exchangeable component.

Assume now that the component may correctly work for a certain "up time" T_{ui} until a failure is observed. Assume further that the component may be repaired or replaced after a certain "down time" T_{di} to work again as before. If an experiment is made with one component for a comparably long time (with "many" failures), then the up/down time distributions may look like as that sketched in Figure 18.1.

18.2.1 Basic Reliability and Availability Parameter Definitions

18.2.1.1 Reliability

Figure 18.1 representation of a long-time measurement example raises questions such as "Are the up times always similar or somehow distributed?" or "May at least meaningful parameters be derived from it that characterize the dependability of the component?"

To start with the second issue, the straightforward mean value calculation of the up times and down times already yields an often-used dependability indicator of the component in engineering practice:

$$\frac{1}{n}\sum_{k=1}^{n} T_{uk} = \text{MUT} = \text{MTTF}_{\text{meas}},$$

$$\frac{1}{n}\sum_{k=1}^{n} T_{dk} = \text{MDT} = \text{MTTR}_{\text{meas}}, \qquad (18.1)$$

$$\text{MTBF}_{\text{meas}} = \text{MDT} + \text{MUT},$$

with

MUT	Mean up time
MDT	Mean down time
MTTF	Mean time to failure
MTBF	Mean time between failures
MTTR	Mean time to repair (or restore or replace)

The abbreviations are basically self-explanatory; caution is, however, advised for the use of the notation "MTBF," which is often mixed up with or equaled to "MTTF" in practical language. While the MTTF denotes the mean time that a component works properly until it fails, the MTBF adds the mean time it needs to get the component up and working again after the failure (MTTR) and thus represents a time between two consecutive failures of the component. Since MTTF values (e.g., 100,000 hours) exceed the MTTR values (e.g., 1 hour) in practice by many orders of magnitude, the difference is often irrelevant.

Instead of indicating a "mean time" until the component fails, sometimes, the number n of failures of the component over a long period of operating time T is recorded, yielding a "rate" ("failures per hour"). This "failure rate" λ is a dependability parameter and fully equivalent to the MTTF (as it is formally just the reciprocal). Formally, the same applies to the MTTR, where instead the corresponding "repair rate" $\mu = 1/\text{MTTR}$ is indicating the maintainability or restoration capacity.

λ	Failure rate ($\lambda = 1/\text{MTTF}$)
μ	Repair rate ($\mu = 1/\text{MTTR}$)

Irrespective of whether we prefer to use the determined MTTF, MTBF, failure rate λ, or a "failure probability" $F(t)$ at a time, they all have in common that they refer to the capacity of one component to reliably serving its function, the "reliability" $R(\lambda, t)$ of the component.

$R(t)$	Probability of unfailed component at time t (reliability)
$F(t)$	Probability of failed component at time t

So far, the notion of failure has summarized for simplicity all kinds of unintended behavior, while in practice, only those failures that really impact the intended service of the component are of interest to the engineer. As these failures represent a subset of all failures, the mean time between them is longer than the MTTF and is often indicated as

MTBSF	Mean time between service failures

Finally, we had so far imagined an elementary component and attributed the preceding dependability parameters to it, while in the engineering practice, values such as the MTTF are similarly associated with higher aggregations of elementary components (such as assemblies, modules, and subsystems). The number of potential failure modes is then increasing with the complexity of the considered entity. For more detailed definitions and normative references, see other dedicated chapters of this handbook.

18.2.1.2 Availability

In transportation, the components or more aggregated entities form part of complex transport systems with many thousands of other different entities. Interacting properly together, they serve undisturbed (passenger traffic) operation services. Any particular failure of a component has a particular effect on the complete transport system to serve its purpose. For the system operator or passenger, the "availability" of the service function is therefore more relevant than the "technical" reliability of the component alone. As a practical convention in the industry, the average effective up time of a component after failure is therefore combined with the average time it is repaired or replaced (down time) and expressed, in general, as the ratio

$$A = \frac{\text{MUT}}{\text{MUT} + \text{MDT}} = \frac{\text{MTTF}}{\text{MTTF} + \text{MTTR}} = \frac{\text{MTBF} - \text{MTTR}}{\text{MTBF}} = \frac{\mu}{\lambda + \mu}, \tag{18.2}$$

$$A + U = 1 \Rightarrow U = \frac{\text{MTTR}}{\text{MTBF}} \overset{\text{MTTR} \ll \text{MTTF}}{\approx} \frac{\text{MTTR}}{\text{MTTF}} = \frac{\lambda}{\mu}, \tag{18.3}$$

with

A	Availability (dimensionless)
U	Unavailability (dimensionless)

The practical difficulties linked to the definition of *A* are, however, beyond and different from the pure definition of the preceding averaged indicators:

Adequate selection of the MTTFs	For a single component, it may be straightforward to employ the MTBSF instead; for aggregated subsystems, the adequate set of service affecting MTTFs must be defined.
The adequate selection of the MTTRs	The mean time to repair becomes, in general, more a "mean time to restore" to keep passenger traffic as much as possible flowing and "repairing" at a later time; "service restorations" are in this respect mainly operational actions.
Meaning of *A* and *U*	The parameters *A* and *U* give ratios of MTTFs and MTTRs, so even multiplying both latter by a factor of 10, for example, would not alter *A* or *U*, but may have an impact on the perceived passenger service once the event occurs.
Responsibility/implementation for *A* and *U*	Since an optimized service availability mixes both technically controlled MTTF parameters and operationally controlled restoration actions in the MTTR, responsibility for achieving availability optimization may require multiple parties.
Measurement period of "mean value" *A* and *U*	Since especially the unavailability *U* is intended to be a small number, based on rare events, adequately long periods of observation may be required to demonstrate the parameter.

Some practice examples and reflections on the matter are discussed further in this chapter. For normative references or more formal definitions, see other dedicated chapters of this handbook.

18.2.2 Concept of and Debate on Constant Failure Rates

In order to further consider the up and down time behaviors of components, it is useful to observe them over a very long period, much longer than an MTTF. Assume that it would be possible to implement one (or many identical) component(s) and starting from the very first produced batch to sample—say, every 10 failures—and estimate the measured failure rate λ. Then, distributions such as those in Figure 18.2 are obtained.

In general, three phases are distinguishable during such a lifetime (or "bathtub") curve:

- An initial phase where residual (design) problems ("early failure" or "infant mortality") lead to relatively high—but decreasing—failure rate
- A plateau over a longer time with a more or less constant failure rate
- An increase of the failure rate at the end of life of the components ("tear and wear")

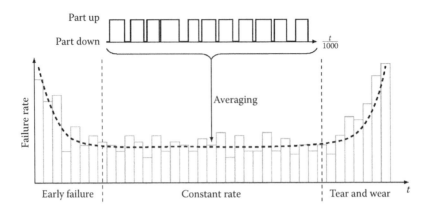

FIGURE 18.2
Failure rate behavior over a long (life) time.

While the infant mortality and the tear and wear phases depend on numerous systematic parameters such as materials, use, and construction, the phase of the "constant" failure rate (or MTTF) is of particular interest, as its statistics is relatively easily accessible to the engineer.

But what does "constant" failure rate mean? It means that the probability of a (still working or repaired) item to fail in a certain time period is the same in every such time period (thus, no aging effects). Should the item fail in any such of the time periods, it may be replaced or repaired. Afterward, the probability to fail shall again be the same in any such time period. If a component is of this kind, it can be shown that its "survival" probability follows an exponential distribution, as does the associated probability density function $f(t)$.

$R(t)$: Probability that a constant failure rate component is still working unfailed until time t	$R(t) = e^{-\lambda t}$
$F(t)$: Probability that a constant failure rate component failed until time t	$F(t) = 1 - R(t) = 1 - e^{-\lambda t}$
$f(t)$: Probability to observe a constant failure rate component fail at a time t (probability is $\Delta t \cdot f(t)$)	$f(t) = \lambda e^{-\lambda t}$
λ: Rate; relative decrease of probability to still work correctly per time	$\lambda = \dfrac{\dfrac{dR(t)}{dt}}{R(t)}$

18.3 Statistics of Multiple Components or Failures

Even assuming that failure rates are "constant," how could a failure rate of, say, $\lambda = 1.25 \times 10^{-6}$/h be measured during a typical demonstration period of 3 years of operating time (e.g., $\Delta t = 20,000$ hours)? If we have only one (or a few) components available, we would need over 100 years of operation to observe only one (expected) failure.

On the other hand, reliability relevant components are often implemented in batches of hundreds or more in transportation systems. If we consider, for example, a track-mounted positioning transponder (balise) with $\lambda = 1.25 \times 10^{-6}$/h, high-performance train localization may require five (or more) of these balises between two station tracks. So for a line of 40 station tracks, we may have around 200 balises installed. Therefore, at least some failures may be expected (e.g., 200 · 20,000 hours 1.25 × 10⁻⁶/h = 5 events in 3 operating years).

Provided the failure probabilities are identical and do not change if we measure a certain time Δt, we expect $\mu = n \cdot \lambda \cdot \Delta t$ failures from the sample, where n is the initial number of components. For this kind of samples, the number of actually observed failures k is also distributed around the expectation value μ; the respective distribution is called "Poisson distribution":

$$P_{PD}(k,\mu) = \frac{\mu^k}{k!} e^{-\mu}. \tag{18.4}$$

It shall first be confirmed that μ is in fact the expectation value of the distribution:

$$\sum_{k=0}^{\infty} k \cdot P_{PD}(k) = \sum_{k=0}^{\infty} k \cdot \frac{\mu^k e^{-\mu}}{k!} = \left(0 \cdot \frac{\mu^0 e^{-\mu}}{0!} \right)$$

$$+ \sum_{k=1}^{\infty} \mu \cdot \frac{\mu^{k-1} e^{-\mu}}{(k-1)!} = \mu \sum_{k=0}^{\infty} \frac{\mu^k e^{-\mu}}{k!} = \mu. \tag{18.5}$$

The fact that the distribution contains only the numbers k and μ facilitates the evaluation of MTTF or λ values and its probable (statistical) fluctuations quite considerably for the engineer by converting the effective "times" of N_c components into "event numbers" and vice versa ($\mu = N_c \cdot T \cdot \lambda$). Since every specific value k of the distribution is relatively small, adequate intervals around the expectation value must be used instead. If such an "acceptance interval" is chosen to small, then correct components may be dismissed just due to statistical fluctuations ("producer's risk α"). If selected to large, "wrong" components may ultimately find acceptance ("consumer's risk β").

18.3.1 Variance and Statistical Certainty of MTTF Measurements

For the practical engineer, it is often enough to know a "natural" interval or indicator of the inherent spread of a measured number of failures instead of a complete distribution. It is a common exercise in statistics to first determine a "mean" value from a number of measurements and then sum up the "deviations" of the individual measurements from this mean value. The normalized (averaged) sum of the deviation squares yields the variance of the sample. Its square root is called "standard deviation σ." Especially, the Poisson distribution yields a remarkable result in this respect:

$$\mathrm{Var}(k) = \sum_{r=0}^{\infty} (\mu - r)^2 \cdot P_{PD}(\mu, r) = \mu^2 \sum_{r=0}^{\infty} \frac{\mu^r e^{-\mu}}{r!} - 2\mu \sum_{r=0}^{\infty} r \frac{\mu^r e^{-\mu}}{r!} + \sum_{r=0}^{\infty} r^2 \frac{\mu^r e^{-\mu}}{r!}$$

$$= \mu^2 - 2\mu^2 + \left(0^2 \frac{\mu^0 e^{-\mu}}{0!} \right) + \mu \sum_{r=1}^{\infty} r \cdot \frac{\mu^{r-1} e^{-\mu}}{(r-1)!} = -\mu^2 + \mu \sum_{k=(r-1)=0}^{\infty} (k+1) \frac{\mu^k e^{-\mu}}{k!} \tag{18.6}$$

$$= -\mu^2 + \mu \left(\sum_{k=0}^{\infty} k \frac{\mu^k e^{-\mu}}{k!} + \sum_{k=0}^{\infty} \frac{\mu^k e^{-\mu}}{k!} \right) = -\mu^2 + \mu^2 + \mu = \mu \Rightarrow \sigma = \sqrt{\mathrm{Var}(k)} = \sqrt{\mu}.$$

The variance of the Poisson distribution is its expectation value itself (!), and the natural "standard" spread (σ) is the square root of the expectation value.

This facilitates even more the engineer's task with respect to failure number analysis. If we have reason to expect μ failures in a period, then values k may be expected within a natural fluctuation range $\sigma = \sqrt{\mu}$ around μ, i.e., $k \in \left[\mu - \sqrt{\mu}, \; \mu + \sqrt{\mu} \right]$ in a measurement.

18.3.2 Statistically Meaningful Specification and Interpretation of Failure Measurements

The previous discussions showed that one natural indication of the "uncertainty" of a failure number from a component is $\sigma = \sqrt{k}$, meaning that although we measure a value k, the true value μ cannot be assumed better than to be within the interval $k \pm \sqrt{k}$. Since the relative standard uncertainty is $\sqrt{k}/k = 1/\sqrt{k}$, the "precision" or statistical certainty with which we can determine the true μ depends on the number itself.

Assume that we want to demonstrate a specified MTTF = 100,000 hours of a control circuit. Contractually, the time period available to do this may be limited to 3 years of operation, e.g., $T = 25,000$ hours. If we have installed the controller in 16 trains, we would expect

$$\mu = N_{\text{C}} \cdot T \cdot \lambda = 16 \cdot 2.5 \times 10^4 \text{h} \cdot 10^{-5}/\text{h} = 4 \Rightarrow \sqrt{\mu} = 2 \Rightarrow k = 4 \pm 2,$$

$$\Delta\mu_{\text{rel}} = \pm \frac{\sqrt{\mu}}{\mu} = \pm \frac{1}{\sqrt{\mu}} = \pm 50\%.$$

In the example demonstration setup, if the sample size is not increased, we can do what we want; the expected stochastic "precision" in determining the "true" MTTF will not exceed 50%, so observed numbers between $k = 2$ and $k = 6$ should be expected. If we would have had only 4 components instead of 16, the relative uncertainty would have been even 100%.

The verification of supposed or specified reliability characteristics therefore becomes rather a question of precision of the measurement than of the result itself. An occasional pitfall is the idea to request high-precision/statistical power from a measurement (e.g., ±10%), where the expected failure statistics does not support it at all (e.g., 10 failures instead of the required 100). In transportation, failure numbers of one component type are rarely high enough to permit determinations with relative errors lower than 20% (see Section 18.3.3).

To complete the picture, it shall be noted that the statistical standard error of, e.g., 10% does not mean that the number of events within the error bar interval is 10% or anything near to it. The number of events in this $1 - \sigma$ environment is approximately 2/3 of all values and converges to 68.3% for high statistics (or expectation values), where the Poisson distribution converges towards a Gaussian distribution.

18.3.3 χ^2 Distribution and Continuous Confidence Intervals

The good congruency between Gaussian and Poisson distributions for higher numbers suggests consideration of the convolution of multiple sets of measurement values. Convoluting two Gaussian distributions again results in a Gaussian distribution where the mean value would be the sum of the two individual mean values. This is, however, not the case when convoluting the statistical variances σ^2 of multiple Gaussian distributions, which would instead yield the so-called χ^2-density ("chi-squared density"):

$$f_n(z = \chi^2) = \frac{e^{-\frac{1}{2}z} z^{\frac{n}{2}-1}}{2^{\frac{n}{2}} \Gamma\left(\frac{n}{2}\right)}, \quad \text{with } \Gamma(\alpha) = \int_0^\infty t^{\alpha-1} e^{-t} \, dt \quad (\alpha > 0). \tag{18.7}$$

The Gamma function $\Gamma(\alpha)$ is a kind of a generalized *factorial function* $(\alpha!)$ and reminds of the combinatorial effects of the convolution of squares.

Why is the χ^2 distribution so popular in reliability statistics? It shall first be noted that despite its complicated form, it is identical with the "conjugated" Poisson distribution (*Erlang density*) for twice the "degree of freedom (DoF)," meaning that the distribution of, e.g., the χ^2 density for $n = 14$ in Figure 18.3 also represents the probability distribution shape of the numbers of failures from a Poisson statistics with $\mu = 7$. The equality of numbers and squares of deviations is linked to the previous observation in Poisson statistics that "$\sigma^2 = \mu$" (cf. Equation 18.6). Different from the Poisson distribution, the χ^2 distribution is, however, a *continuous* distribution, which makes it convenient for the engineer to now specify *any* confidence interval (resolving one of the open issues of the preceding paragraphs). The right part of Figure 18.3 shows an examples of having initially observed seven failures of a component from an experiment and requesting now from the χ^2 distribution integral (of twice the DoF: DoF = 14) the limit values such that the probability to find the "true" value of the component within amounts to 80% (symmetrically):

$$r_m = 7, \ CL = 80\%, \ \alpha = \beta = 10\%, \ DoF = 14$$

Lower limit:

$$r = 14, \ r_{l,80\%} - \chi^2_{14,10\%} \approx 21, \ \text{Factor } f_l = \frac{14}{21} = 0.665 \Rightarrow r_{m,l,80\%} = \frac{1}{f_l} \cdot 7 = 10.5$$

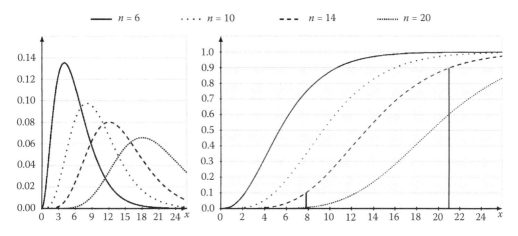

FIGURE 18.3
χ^2 Density function examples for $n = 6$, 10, 14, and 20 and their cumulative functions (*right*).

TABLE 18.1

MTTF Confidence Limit Multiplier (χ^2) Table for Selected Number of Failures

| Total Number of Failures | Confidence Intervals for MTTFs | | | | | |
| | 40% | | 60% | | 80% | |
	70% Lower Limit	70% Upper Limit	80% Lower Limit	80% Upper Limit	90% Lower Limit	90% Upper Limit
1	0.801	2.804	0.621	4.481	0.434	9.491
2	0.820	1.823	0.668	2.426	0.514	3.761
3	0.830	1.568	0.701	1.954	0.564	2.722
4	0.840	1.447	0.725	1.742	0.599	2.293
5	0.849	1.376	0.744	1.618	0.626	2.055
6	0.856	1.328	0.759	1.537	0.647	1.904
7	0.863	1.294	0.771	1.479	0.665	1.797
8	0.869	1.267	0.782	1.435	0.680	1.718
9	0.874	1.247	0.796	1.400	0.693	1.657
10	0.878	1.230	0.799	1.372	0.704	1.607
11	0.882	1.215	0.806	1.349	0.714	1.567
12	0.886	1.203	0.812	1.329	0.723	1.533
13	0.889	1.193	0.818	1.312	0.731	1.504
14	0.892	1.184	0.823	1.297	0.738	1.478
15	0.895	1.176	0.828	1.284	0.745	1.456
16	0.897	1.169	0.832	1.272	0.751	1.437
17	0.900	1.163	0.836	1.262	0.757	1.419
18	0.902	1.157	0.840	1.253	0.763	1.404
19	0.904	1.152	0.843	1.244	0.767	1.390
20	0.906	1.147	0.846	1.237	0.772	1.377
30	0.920	1.115	0.870	1.185	0.806	1.291

Upper limit

$$r = 14,\ r_{0,80\%} = \chi^2_{14,90\%} \approx 7.8,\ \text{Factor } f_0 = \frac{14}{7.8} = 1.795 \Rightarrow r_{m,0,80\%} = \frac{1}{f_0} \cdot 7 = 3.9.$$

Using the (fixed) measurement time T, these limit failure numbers may be converted into the *limit* MTTFs.

A widely used form of χ^2 integrals are tables (e.g., Table 18.1) in which the doubled DoF has already been accounted for and values are selected according to commonly used confidence levels. It shall be noted, however, that the tabled values represent the MTTF (or MTBF) multipliers, so applying it to failure number means that the engineer has to divide by the multiplier instead of multiplying; see Equation 18.17 for sample application.

18.4 Measurement and Acceptance of Reliability Parameters

Based on the previous (more theoretical) reflections, the currently used acceptance schemes for reliability parameters in transportation may now be (critically) comprehended and applied.

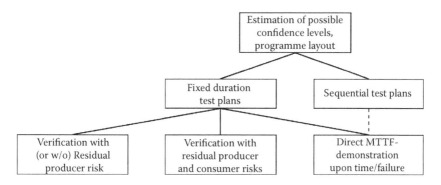

FIGURE 18.4
Overview of reliability demonstration programs and acceptance plans.

All substantiated schemes deployed today respond to the repeatedly emerging questions in transportation in adequate ways:

- From our typical demonstration period times and component numbers, what may we expect at all in terms of typical "failure numbers" and MTTFs?
- With what statistical certainty do we want to accept or reject the measurements with respect to previous specifications or target values?
- Confidence or acceptance interval definitions are based on "hypothesized" values, but how good are our "hypotheses"?
- What kind of risks shall ultimately remain acceptable, "only" producer risks or consumer risks as well? What may be adequate risk levels and discrimination rations?

Figure 18.4 shows an overview of adequate reliability programs/acceptance plans in transportation. It may not be surprising that the systematic organization of the adequate plans much resembles the processes of the (unsupported) *Military Handbook MIL-HDBK-781* (*Reliability Test Methods, Plans and Environments for Engineering Development, Qualification and Production*).

18.4.1 Demonstration Program Layout

Most reliability specifications in transportation are today still based on MTTF (or MTBF) Specifications for reliability critical components from multiple sources (e.g., from other specifications, experience, supplier information). An additional, more indirect, source of reliability requirements is typically defined in the framework of the availability definitions (see further).

The obligation of coherence demonstration of the built system with the specified reliability values has to be performed in a more or less limited period after system implementation, e.g., one or several years of operation (but not, for example, 10 years). In many cases, the demonstration time periods are not exceeding the warranty periods of the components (e.g., 3 years). In order to obtain a better feeling on what may be expected in terms of failure numbers during such a period, Table 18.2 assembles a selection of typical transportation component reliability requirements.

TABLE 18.2

Typical Transportation MTBFs and Expected Failure Numbers

Component	Specified Minimum MTBSF (h)	Number of Components	Number of Failures in 25,000 Hours of Demo Period
Wayside control unit (WCU)	200,000	16	2
Train detection device	100,000	74	19
Positioning balise	1,000,000	160	4
ATS console	100,000	8	2
Onboard control unit	40,000	42	26
Train obstacle detector	50,000	42	21
Guideway intrusion (platform)	50,000	32	16
Automatic train control communication network	200,000	16	2

To facilitate the further analysis, it may be concluded from the table that we have to expect typically between 1 and 25 service relevant failures of the implemented components per smaller system (e.g., subway line), rarely zero failure nor hundreds of failures. So, selecting example "failure statistics" of 2, 4, 9, 16, or 25 failures per 3-year period certainly covers a representative spectrum of what is actually observed. These failure numbers will scatter around the true expectation values as shown by the associated Poisson distributions in Figure 18.5 (drawn as a continuous line diagram), indicating that, especially for the lower numbers, the relative spread is quite considerable.

In order to obtain adequate acceptance boundary numbers (or confidence intervals or residual risk levels, respectively), several possibilities may be compared and are compiled in Table 18.3.

Most straightforward would be the standard deviation $\sqrt{\mu}$ of the Poisson distribution, leading to limit values around the expectation value of $\mu \pm \sqrt{\mu}$. It shall be noted, however, that this interval slightly exceeds the corresponding $1 - \sigma$ environment of a Gaussian distribution by approximately 5% for these low numbers.

Alternatively, we may directly search from the Poisson distributions limit values below and above the expected value such that between them, approximately 60% (or 80%) of all events may be found. The limit values of the "60% (or 80%) confidence interval" may

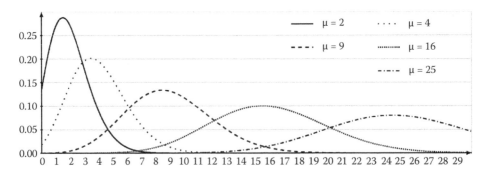

FIGURE 18.5

Statistical (Poisson) distribution of typical failure numbers in transportation.

TABLE 18.3

Comparison of Typical Statistical Intervals around the Expected Value μ

PD μ	$\sigma = \sqrt{\mu}$	$\mu \pm \sigma$	%	60% CL ~ PD Counting	60% χ^2 CL	80% CL ~ PD Counting	80% χ^2 CL
2	1.4	~1–3	>72	0.3–2.6	0.8–3.0	0.2–3.4	0.5–3.9
4	2	2–6	79	1.8–5.0	2.3–5.5	1.0–6.0	1.7–6.7
9	3	6–12	76	6.0–10.9	6.4–11.3	4.7–12.4	5.4–13.0
16	4	12–20	74	12.1–0.8	12.6–19.2	10.3–20.6	11.1–21.3
25	5	20–30	74	20.2–28.6	20.6–29.1	18.0–31.0	18.7–31.7

be found by directly counting the Poisson distribution or by reading the values from the graphical "integral" of the Poisson distributions (see Figure 18.6).

Instead of directly counting from the Poisson distributions, we may take the limit values equivalently from χ^2 integral tables as previously shown. The indicated values in Table 18.3 show good agreement with the "counted" coarse values from the discrete Poisson distributions. The corresponding χ^2 table is given as Table 18.1.

An alternative characteristic of the statistical certainty of our failure numbers in transportation is represented by the discrimination ratio $d = \mathrm{MTTF_0/MTTF_1}$, where the boundary values of the MTTFs are fixed before measurement and relate to limit failure numbers $r_0 = T/\mathrm{MTTF_0}$ and $r_1 = T/\mathrm{MTTF_1}$. Since the numbers from both limit hypotheses are Poisson distributed around r_0 and r_1, their respective "error bars" $r_0 + \sqrt{r_0}$ and $r_1 - \sqrt{r_1}$ may "touch" each other, yielding a condition for a meaningful discrimination ratio:

$$d = \frac{r_1}{r_0}, \; r_1 - \sqrt{r_1} \approx r_0 + \sqrt{r_0} \Rightarrow (d-1)r_0 \approx (\sqrt{d}+1)\sqrt{r_0}, \; \frac{\sqrt{d}+1}{d-1} \approx \sqrt{r_0},$$

$$r_0 = 4 \Rightarrow d \approx 2.2, \; r_0 = 9 \Rightarrow d \approx 1.8, \; r_0 = 25 \Rightarrow d \approx 1.45. \tag{18.8}$$

For our typical failure rates in transportation, adequate discrimination ratios shall therefore lie around $d = 2 \, (\pm 0.5)$.

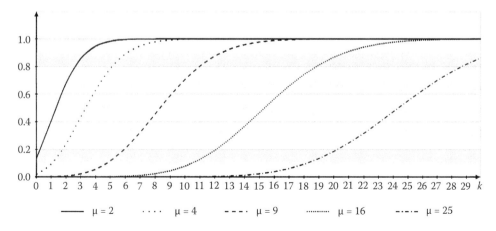

FIGURE 18.6
Integrals of the Poisson distributions with the 60% and 80% confidence intervals.

For transportation system projects (and especially control equipment), it may therefore be concluded as guideline:

Failure rates are typically between 1 and 50 per 3-year measurement period in a medium complexity transportation system (such as subway or regional transit line).

$T = 25,000$ hours operating time represents an adequate total demonstration period.

Confidence limit values for acceptance/rejection may then be determined from Poisson distributions or χ^2 integral tables, where 60% or 80% (or 90%) are adequate confidence intervals.

Adequate consumer/producer risk levels are consequently between 5% and 20%.

Discrimination ratios for hypothesis fixing shall adequately be selected close to $d = \text{MTTF}_0/\text{MTTF}_1 = 2$.

18.4.2 Fixed-Duration Acceptance Plans

The question of how much we may trust our hypothesis is often linked to the "substantiation" of the MTTF parameters by the suppliers. If the supplier has credible data from previous measurements or has conservatively predicted the MTTF from calculations, then the measured data may be interpreted as statistical fluctuation from the predicted expectation value, and a "fixed-time acceptance plan" can credibly be applied. The acceptance process becomes more of a "verification" that the observed MTTF is in fact the specified MTTF rather than a determination of an unknown MTTF. If the measured value is too remote from the limit values, confidence in the hypothesis becomes insufficient.

18.4.2.1 Fixed-Duration Acceptance Plan without Specific Risk Agreements

The most simple and straightforward utilized test plan is based on the required or "expected" failure number itself.

- Example:

  Have $N_C = 18$ components, specified (for each) $\lambda = 2 \times 10^{-5}/\text{h}$, measure $T = 25,000$ hours.

  Expect: $\mu = N_C \cdot \lambda \cdot T = 9$ failures.

  Accept: $k \le 9$; reject: $k > 9$ (see Figure 18.7).
  Appreciation of this test plan:

- Popular plan
- But in ~50% of all measurements (wrong) rejection of correct component
- Acceptance if $k \le \mu$ (e.g., $k = 8$ or 9), but also risk (~50%) that k comes from higher failure rate distribution, (wrong) acceptance of incorrect component, and
- Straightforward but statistically less adequate

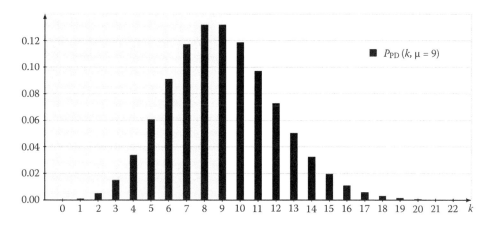

FIGURE 18.7
Expect $\mu = 9$ failures; accept if $k \leq 9$, and reject if $k > 9$.

18.4.2.2 Fixed-Duration Acceptance Plan with Specific Producer Risk

Rejecting all measured failure numbers $k > \mu$ as in the previous test plan leads to a high number of (wrongly) rejected good components (~50%) and, thus, a comparably high producer risk α. It is therefore common practice to allow for statistical fluctuations toward high failure numbers, at least, unless they do not exceed certain limits. The limits are identified as the boundary numbers of confidence intervals which include a certain percentage of all possible numbers. The confidence intervals ($\pm\sqrt{\mu}$, CL = 60%, CL = 80%) are predetermined, and the boundary values are found from χ^2 integral tables or directly by counting from the Poisson distribution.

Examples:

Have $N_C = 18$ components, specified (for each) $\lambda = 2 \times 10^{-5}/h$, measure $T = 25{,}000$ hours.

Expect: $\mu = N_C \cdot \lambda \cdot T = 9$ failures.

Consider the three following scenarios (Figure 18.8):

- CL = 80%; thus, $\alpha \approx 20\%$ and boundary $r_1 = 11$. Therefore, accept for $k \leq 11$ and reject if $k > 11$.
- Boundary $r_1 = \mu + \sqrt{\mu} = 12$ (i.e., CL ~ 87%, $\alpha \approx 12{,}5\%$). Therefore, accept for $k \leq 12$, and reject if $k > 12$.
- $\alpha \approx 10\%$, yielding CL = 90% and boundary $r_1 = 13$. Thus, accept if $k \leq 13$, and reject if $k > 13$.

Once the confidence level (or equivalently risk level) is defined, the limit numbers may be explicitly calculated from the Poisson distribution by iteratively solving

$$\sum_{k=0}^{r} \frac{\mu^k e^{-\mu}}{k!} = \frac{CL}{100\%}$$

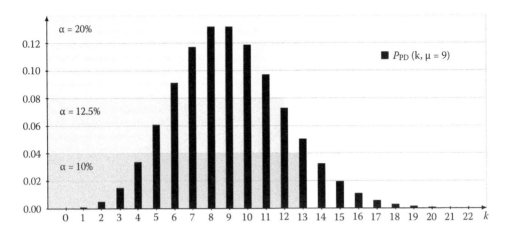

FIGURE 18.8
Different acceptance boundaries associated with specified producer risks.

with respect to r. For example, for $\alpha = 20\%$ (therefore, CL $= 80\%$) and $\mu = 9$, it is (starting with $r = 9$)

$$\sum_{k=0}^{9} \frac{9^k e^{-9}}{k!} = e^{-9} \cdot \left(\frac{9^0}{0!} + \frac{9^1}{1!} + \frac{9^2}{2!} + \frac{9^3}{3!} + \frac{9^4}{4!} + \frac{9^5}{5!} + \frac{9^6}{6!} + \frac{9^7}{7!} + \frac{9^8}{8!} + \frac{9^9}{9!} \right),$$

$$\approx 1.23 \cdot 10^{-4} \cdot (1 + 9 + 41 + 122 + 273 + 492$$

$$+ 738 + 949 + 1068 + 1068) \approx 0.588 < 0.8,$$

$$\sum_{k=0}^{10} \frac{9^k e^{-9}}{k!} \approx 0.706 < 0.8,$$

$$\sum_{k=0}^{11} \frac{9^k e^{-9}}{k!} \approx 0.803,$$

yielding $r = 11$ to apply the desired confidence level.

Since the discrete nature of the Poisson distribution permits only approximate values, the previously introduced χ^2 integral table (Table 18.1) may be used (for which, again, approximations are necessary to derive from it discrete boundary numbers, e.g., $r_1 = 9/0.796 = 11, 3 \approx 11$).

Appreciation of this acceptance plan:

- Also commonly used scheme
- Accounts adequately for statistical fluctuations
- Limits producer risk α (rejection of "good" component) to adequate values (e.g., $\alpha = 20\%$)

18.4.2.3 Fixed-Duration Acceptance Plan with Symmetric Producer and Consumer Risk

By limiting the producer risk α, failure numbers may get quite high, and the operators often wish to also limit their consumer risk β of accepting a number coming from an unacceptable Poisson distribution around the upper limit $r_0 = T/\mathrm{MTTF}_0$, where the respective risk boundaries (k_a for acceptance and k_r for rejection) may be calculated directly from the Poisson distributions:

$$\beta = \sum_{k=0}^{k_a} \frac{r_l^k e^{-r_1}}{k!}, \quad \alpha = \sum_{k=k_r}^{\infty} \frac{r_0^k e^{-r_0}}{k!} \quad \left(r_1 = \frac{T}{\mathrm{MTTF}_1}, r_0 = \frac{T}{\mathrm{MTTF}_0} \right). \tag{18.9}$$

The graphical representation of the issue is shown in Figure 18.9, where the limit values MTTF_1 and MTTF_0 are represented by the Poisson distributions of the failure numbers associated with them in a time interval T.

If the time T and the discrimination ratio $d = \mathrm{MTTF}_0/\mathrm{MTTF}_1$ are fixed, the resulting acceptance boundary values k_a and k_r are determined from the agreed risks α and β. Depending on the selected parameters d, α, and β, there may remain a "gap" between the two areas (rejection of alternative hypothesis and acceptance of null hypothesis), leaving us without decision. It may also happen that both hypotheses could be rejected by a measurement result depending on the parameter selections (see Figure 18.10).

In a fixed-duration test plan, we must, however, decide after the expiration of the test duration whether we accept or we reject a measurement, so the "acceptance" number k_a and the "rejection" number k_r shall only differ by exactly one event: $k_r = k_a + 1$. The numbers k_a and k_r are strictly related to the residual risks α and β and to the expected values $r_0 = \mu_0$ and $r_1 = \mu_1$, which are, in turn, related through the discrimination ratio and depend both in the same manner on the measurement time T. So the challenge is to find two Poisson distributions with a certain d such that their residual risk tails α and β "touch" at one discrete number, which then becomes the decision boundary between k_a and k_r. This exercise has not for all combinations of d, α, and β practical and precise solutions with discrete numbers, so often one parameter may have to be modified to make it properly work, either by graphical or table methods or from calculated approximations.

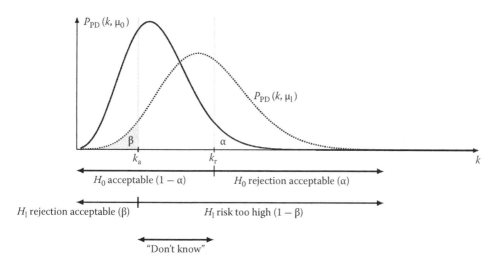

FIGURE 18.9
Fixing measurement time d and α/β may result in undecided situations.

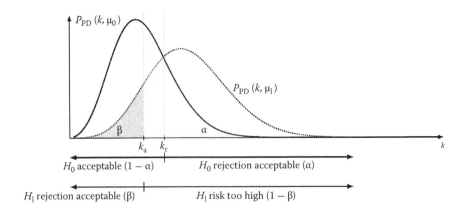

FIGURE 18.10
In fixed-duration plans, rejection of both hypotheses may also be a result.

For illustration of both methods, the preceding example is retained:

Have $N_C = 18$ components, specified (for each) $\lambda = 2 \times 10^{-5}$/h, measure $T = 25{,}000$ hours.
Expect: $\mu = N_C \cdot \lambda \cdot T = 9$ failures.

In a first example, we fix $d = 2$ (i.e., $r_0 = \mu = 9$, $r_1 = d\mu = 18$); see Figure 18.11. Since both the measurement time T and the discrimination ratio d are fixed, we are now looking for $\alpha = \beta$ such that the integral of the "upper" Poisson distribution $PD(r_0)$ reaches the value α exactly at the number r where the integral of the "lower" Poisson distribution $PD(r_1)$ (from right to left) reaches β (see "box" indication in right part of Figure 18.11). In the example, this is the case at $r \approx 12.5$ and the risks $\alpha = \beta \approx 10\%$, now leading to a clear criterion for acceptance or rejection: At CL $\approx 80\%$: Accept if $k \le 12$, and reject if $k > 12$.

Alternatively, we may have fixed α and β first (e.g., $\alpha = \beta = 10\%$) and adjusted d. The proper d (or r_1 value, respectively) may be found (approximately) by the following reflection: Whatever r_1 is, its lower part of the "90% error bar" is a multiple f of the natural error

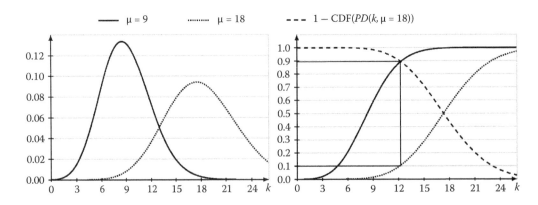

FIGURE 18.11
Poisson distributions for $\mu = 9, 18$ (*left*) and their integral functions (*right*).

bar $\sqrt{r_1}$, which should coincide with the upper part of the "90% error bar" around r_0 where the associated factor f is approximately the same:

$$r_0 + f \cdot \sqrt{r_0} \approx r_1 - f \cdot \sqrt{r_1}. \tag{18.10}$$

Since the factor f may be easily estimated from the χ^2 table, the corresponding value r_1 may be determined by

$$r_0 + f \cdot \sqrt{r_0} = \frac{r_0}{\chi^2_{\text{mult}}(r_{0\text{ll}}, 90\%)} = \frac{9}{0.693} \Rightarrow f \approx 1.3,$$

$$r_1 - f \cdot \sqrt{r_1} \approx 13 \Rightarrow r_1 \approx 18.6 \Rightarrow d \approx \frac{18.6}{9} = 2.07 \approx 2. \tag{18.11}$$

For the example case, this approach yields compatible values for the discrimination ratio ($d \approx 2$), and thus, the numeric acceptance/rejection results.

Repeating the process for $\alpha = \beta = 20\%$ (instead of 10%), the respective acceptance/rejection number would result into $k_a \leq 11$ by both methods.

Appreciation of this acceptance plan:

- Commonly used method
- Accounts adequately for statistical fluctuations
- Requires "approximate" determination of acceptance/rejection numbers

18.4.3 Sequential Acceptance Plan with Consumer/Producer Risk

Having fixed the discrimination ratio d and the consumer and producer risks, the preceding fixed-duration test plan requires quite an elaborate estimation work to define the suitable acceptance/rejection numbers $k_r = k_a + 1$, sometimes even hardly possible. Therefore, the question often emerges if the test could be performed by letting the measurement time slip, determine for a certain measurement time t the respective boundary values, and if a decision is not possible (e.g., as indicated in Figure 18.9), simply continue testing. In this case, testing would have to continue until the result is falling in the range of either clear acceptability or clear rejectability.

Instead of integrating each time the upper/lower limit Poisson distribution, this type of "sequential" test plan (sometimes also called "Wald test") directly calculates for a measured value r (as a general notation form of r_{meas}) the probability that r comes from the upper/lower limit Poisson distribution and sets these in relation:

$$PD(\text{MTTF}_1, r) = \frac{\left(\frac{t}{\text{MTTF}_1}\right)^r e^{-\left(\frac{t}{\text{MTTF}_1}\right)}}{r!}, \quad PD(\text{MTTF}_0, r) = \frac{\left(\frac{t}{\text{MTTF}_0}\right)^r e^{-\left(\frac{t}{\text{MTTF}_0}\right)}}{r!},$$

$$P(r, t) = \frac{PD_1(r)}{PD_0(r)} = \left(\frac{\text{MTTF}_0}{\text{MTTF}_1}\right)^r e^{-\left(\frac{1}{\text{MTTF}_1} - \frac{1}{\text{MTTF}_0}\right)t} = d^r e^{-\left(\frac{1}{\text{MTTF}_1} - \frac{1}{\text{MTTF}_0}\right)t}. \tag{18.12}$$

This probability ratio now calculates for every measurement time the "expectation values" t/MTTF of the lower/upper limit distribution and then expresses the probability that an actually measured value r comes from the lower limit Poisson distribution compared to the probability that it comes from the upper limit distribution. Since both distributions represent "hypotheses" (null and alternative), producer and consumer risks α and β had been defined to symmetrically limit the risks that the measured value r comes from one or the other hypothesis distribution. If the probability ratio is used instead of absolute probabilities, the residual risks must also be transformed into the respective ratios $A = (1 - \beta)/\alpha$ and $B = \beta/(1 - \alpha)$ to serve as acceptance/rejection boundaries:

$$B < P(r,t) < A$$

$$B < d^r e^{-\left(\frac{1}{MTTF_1} - \frac{1}{MTTF_0}\right) \cdot t} < A \qquad (18.13)$$

$$\text{Accept} \quad \text{Continue} \quad \text{Reject}$$

Equation 18.13 shows the three distinct regions the probability ratio may be found depending on the measured value r.

Testing may, however, not continue forever in practice, so the question arises: how long will the engineer have to test until either acceptance or rejection may be expected? The appropriate indicator is found by a similar reflection as in Equation 18.11. If r would be a matching result, then upper and lower risk tails (or the respective χ^2 boundary values $\chi^2(2r, 1 - \alpha)$ and $\chi^2(2r, \beta)$) would have to coincide with the expectation values of $\mu_0 = t/MTTF_0$ and $\mu_1 = t/MTTF_1$ or, in other words, $\chi^2(2r, 1 - \alpha)/\chi^2(2r, \beta) = MTTF_1/MTTF_{18}$. See Figure 18.12 for graphical representation.

In practice, the decision in Equation 18.13 is transformed into an equation, where the measurement value r is directly set in relation with the selected parameters and the measurement time (yielding the acceptance/rejection boundaries for the measurement):

$$\ln B < r \cdot \ln d - \left(\frac{1}{MTTF_1} - \frac{1}{MTTF_0}\right) \cdot t < \ln A$$

$$\Leftrightarrow \frac{\ln B}{\ln d} + \frac{1}{\ln d}\left(\frac{1}{MTTF_1} - \frac{1}{MTTF_0}\right) \cdot t < r < \frac{\ln A}{\ln d} \qquad (18.14)$$

$$+ \frac{1}{\ln d}\left(\frac{1}{MTTF_1} - \frac{1}{MTTF_0}\right) \cdot t$$

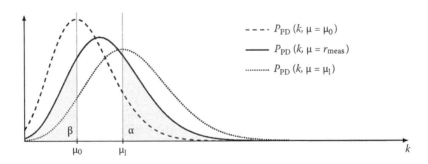

FIGURE 18.12
Coherence of risk areas and boundary values.

Example:

Have $N_C = 18$ components, specified (for each) $\lambda = 2 \times 10^{-5}/\text{h}$, measure $T = 25{,}000$ hours.
Expect: $\mu = N_C \cdot \lambda \cdot T = 9$ failures.
Fix $\alpha = \beta = 20\%$.

In this example, the test should be completed at six failures ($\chi^2(2r, 80\%)/\chi^2(2r, 20\%) = 1.57/0.76$) and yields the acceptance/rejection limits equation

$$5.2 \times 10^{-4}/\text{h} \cdot t - 2 < r < 2 + 5.2 \times 10^{-4}\ \text{h} \cdot t.$$

Figure 18.13 shows a graphical representation of examples (in the left is an example for acceptance, and in the right part is an example of rejection) and the development of the limits over time as straight curves. As in the example, it only very rarely takes times significantly over 10,000 hours to demonstrate acceptance or rejection.

Appreciation of the sequential test plan:

- Is not a fixed-duration test plan anymore, but has the potential to complete the result in either way (accept/reject) earlier than in preplanned fixed-duration tests
- Requires, however, a predetermined tabular or graphical test plan for each individual MTTF
- Takes adequately statistical effects into account
- Is well suited for support of multiple fixed-duration periods
- Requires "approximate" determination of acceptance/rejection numbers

18.4.3.1 Sequential Test Plans and Multiple Test Intervals

Despite their longer preparation work, sequential test plans have another specific advantage: As the truncation time (or time until we should have a clear accept/reject decision) is, in general, shorter than comparable fixed-duration test times, it may be used to subdivide

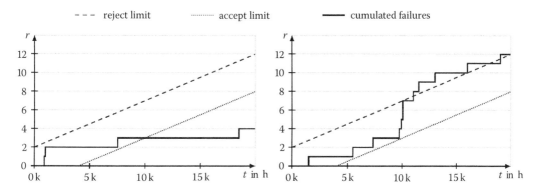

FIGURE 18.13
Sequential test example (random failures): *left*, for acceptance; *right*, for rejection.

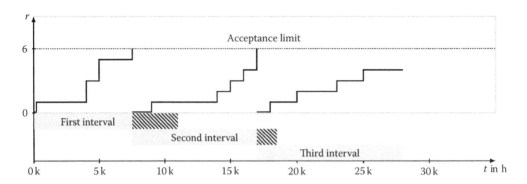

FIGURE 18.14
Example of multiple sequential test intervals within one demonstration program.

the total demonstration period into subintervals, e.g., three or four, each dimensioned to approximately 30–50% of the total time and coherent with the truncation times of an associated sequential test program.

The first interval of a sequential test program is at the commencement of the overall demonstration. Should an acceptance be reached before the expiration of this first interval, then the test may be concluded earlier. Should, however, the component be rejected in the first interval, then immediately after this first rejection, a second interval may be started. This process may be repeated several times until the maximum number of intervals is reached. Figure 18.14 shows an example of how such a test may develop, where in the following example, the three sequential test failure numbers are indicated as well as the maximum failure number that may be reached at all in each of the test intervals.

This process has the advantage of terminating early with success and, at the same time, giving the producer a second and/or third chance of recovery from initial infant mortality effects of the components. Restarting the test interval after rejection in a preceding interval is therefore often coupled to remedial action to be demonstrated by the supplier.

18.4.4 MTTF Determination upon Time or upon Failure

From the preceding test or acceptance plans, it may be noted that the MTTF (or MTBF, respectively) measurement itself is actually not the main objective of the procedure but rather the acceptance or rejection of hypotheses linked to specified values. Alternatively, a measurement of the MTTF value itself may be the objective and compared to a previously specified value. Acceptance is then linked to the question, whether the measured value lies within the confidence interval around a specified value or not.

Two methods are used for measurement of the MTTF:

- *Upon time*: This method is based (like in all previous considerations) on a set of n components during a fixed-duration interval T with repair of any failed component. The measured MTTF is then just the accumulated measurement time $n \cdot T$ divided by the number r of failures:

$$\mathrm{MTTF}_{\mathrm{Meas}} = \frac{n \cdot T}{r}. \tag{18.15}$$

- *Upon failure*: This method does not use a pre-set fixed time but consists of the following steps:
  - Start campaign with a specified number of components
  - On the failure of a component, note the time (t_1, t_2, ..., t_r) of failure occurrence but do not repair the component
  - Stop the measurement campaign upon observation of r failures

  Since the number of "active" unfailed components decreases in this campaign from failure to failure, the ratio [22a] is not adequate anymore. Instead, the likelihood of observing a specific pattern of failures at specific times (which is represented by the product of the individual reliabilities) is considered, and the parameter MTTF is determined such that it maximizes this likelihood L:

$$L(\text{MTTF}) = \prod_{i=1}^{r} \frac{1}{\text{MTTF}} e^{-\frac{t_i}{\text{MTTF}}} \prod_{j=r+1}^{n} e^{-\frac{t_r}{\text{MTTF}}} = \frac{1}{\text{MTTF}^r} e^{-\frac{1}{\text{MTTF}}\left(\sum_i^r t_i + (n-r)t_r\right)},$$

(18.16)

$$\frac{\mathrm{d}L}{\mathrm{dMTTF}} = 0 \Rightarrow \text{MTTF} = \frac{(n-r)t_r + \sum_{i=1}^{r} t_i}{r}.$$

To illustrate the second method, consider six components. We observe a first failure at $t_1 = 12{,}350$ hours, a second failure at $t_2 = 22{,}875$ hours, and stop the campaign upon a third failure at $t_3 = 30{,}250$ hours. Since the components had not been repaired, three components were active (unfailed) all 30,250 hours, while for the other three components, their unfailed times will have to be summed up; the resulting accumulated time of $T = 3t_3 + t_1 + t_2 + t_3 = 156{,}225$ hours divided by the three failures then yields an MTTF = 52,075 hours.

In guided transportation systems, MTTF measurements and demonstrations are, in general, entertained during operations of the equipment, so components are, in general, quickly repaired or replaced.

Once the MTTF measurement value is obtained, its possible statistical fluctuations within the selected confidence interval are accounted for by the χ^2 MTTF multiplier as per Table 18.1. The table values are conveniently converted into a Degree of Freedom and for the inverse of the failure number r, so it can directly be applied to the related MTTF value:

$$\text{ML} = \frac{2r}{\chi^2_{2r,(1-CL)/2}} \quad \text{(Lower limit MTTF – Multiplier)}$$

$$\text{MU} = \frac{2r}{\chi^2_{2r,(1+CL)/2}} \quad \text{(Upper limit MTTF – Multiplier)}$$

(18.17)

$$\text{MTTF}_{\text{Meas}} = \frac{T}{r} \Rightarrow \text{ML} \cdot \text{MTTF}_{\text{Meas}} \leq \text{MTTF}_{\text{True}} \leq \text{MU} \cdot \text{MTTF}_{\text{Meas}}.$$

In addition, the probability for $\text{MTTF}_{\text{True}}$ being at least/most the lower/upper MTTF limit is (CL + 100)/2.

TABLE 18.4

Upper/Lower Limit MTTFs for Several Measurements by Use of Table 18.1

r_{meas}	$\mathbf{MTTF_{meas}}$ (h)	$\mathbf{MU/MTTF_{UL=90\%}}$ (h)	$\mathbf{ML/MTTF_{LL=90\%}}$ (h)
8	56,300	1.718/96,700	0.680/38,300
9	50,000	1.657/82,900	0.693/34,700
10	45,000	1.607/72,300	0.704/31,700
11	40,900	1.567/64,100	0.714/29,200
12	37,500	1.533/57,500	0.723/27,100
13	34,600	1.504/52,000	0.731/25,300
14	32,100	1.478/47,400	0.738/23,700

Example:

Have $N_C = 18$ components, specified (for each) $\lambda = 2 \times 10^{-5}$/h, measure $T = 25,000$ hours.

Expect: $\mu = N_C \cdot \lambda \cdot T = 9$ failures.

Fix $\alpha = \beta = 10\%$.

Table 18.4 shows the confidence limit values of the MTTFs for measurements of 8–14 failures in the accumulated measurement time. Every individual failure number r is first used to determine the respective measured $MTTF_{Meas}$, then the specific upper and lower limit MTTF multiplier MU and ML are searched from Table 18.1 and multiplied to the $MTTF_{Meas}$, resulting in the 80% confidence interval limit values for the true MTTF. The first "measurement," which is not consistent anymore with the original specification of MTTF = 50,000 hours, is for $r = 14$; all values for $r \leq 13$ are consistent (and therefore acceptable). If the table is repeated for the 60% confidence interval, all values for $r \leq 11$ become acceptable. The process is therefore formally coherent with the previous acceptance test plans.

Appreciation of the measurement plan:

- The preceding MTTF measurement plan determines one measurement of an (a priori not well known) MTTF value upon fixed duration or upon failure.
- By application of the χ^2 distribution, it determines confidence interval limit values in which the true MTTF may be found (with the respective confidence).
- The process determines rather the consistency of the specified MTTF with an $MTTF_{Meas}$ (instead vice versa); the results are however equivalent for the purposes in transportation.
- The process takes adequately statistical fluctuations into account and is straightforward in application, thus a widespread method.

18.4.5 Acceptance Test Plan Comparison

Table 18.5 compiles the considered test or demonstration plans for the deployed examples (mostly nine expected failure events in a certain accumulated testing duration).

While it is easily noted that the numeric accept/reject limit numbers for most of the considered methods are the same, their meaning and interpretation is not, nor is the effort

TABLE 18.5

Comparison of Different Acceptance Test Plans

Acceptance Plan	Risk α	Risk β	Confid. CL	Discr. d	Accept (μ = 9)	Reject (μ = 9)	Appreciation
Fixed time, no specific risk	n.a.	n.a.	n.a.	n.a.	$r \leq 9$	$r > 9$	Simple and straight, high implicit risks α and β, not recommended
Fixed time with producer risk specification	E.g., 20%	n.a.	80%	n.a.	$r \leq 11$	$r > 11$	Practical, accounts for statistics (α), leaves unspecified β
Fixed time with producer and consumer risk	E.g., 20%	E.g., 20%	E.g., 60%	E.g., 1.6	$r \leq 11$	$r > 11$	Accounts for statistics with symmetric risks, less convenient
Sequential test plan	E.g., 20%	E.g., 20%	E.g., 60%	E.g., 2	Depend $r \leq 11$	Depend $r > 14$	Accounts for statistics with risks, less convenient, potential of early termination
MTTF measurement with CL χ^2	E.g., 20%	E.g., 20%	E.g., 60%	n.a.	$r \leq 11$	$r > 11$	Less hypothesis verification but consistency with measurement, accounts for statistics, practical

to prepare the test plans. In projects with tens or hundreds of individual MTTFs to be demonstrated, setting up and following sequential test plans requires the availability of a RAMS engineer in the organization for long times.

Since MTTF demonstrations are not the only instrument of evaluating RAMS parameters in the demonstration system, but availability and safety (and maintainability) are considered as well, the fixed-duration test plan with consumer risks or an MTTF determination within an 80% confidence limit is considered fully sufficient, while using the other test plans for specific purposes as described.

18.5 Specification and Demonstration of Availability Parameters

The (many) reliability values of transport system equipment may be aggregated into less technical but for the passenger (and operator) more relevant "availability" characteristics A of the system. For a single component, the availability A of this component had been defined by Equations 18.2 and 18.3 as the average time ratio the component may serve its purpose, expressed by the MTTF (or MTBF, respectively) and the MTTR that is required to make it operational again after a failure. Instead of the mean times themselves, the availability may be expressed by the rates associated with the mean times, where, formally, the concept behind are constant failure rates for both failure and repair.

$$A = \frac{MTTF}{MTTF + MTTR} = \frac{MTBF - MTTR}{MTBF} = \frac{\mu}{\lambda + \mu}; \; U = 1 - A = \frac{\lambda}{\lambda + \mu} \overset{\lambda \ll \mu}{\approx} \frac{\lambda}{\mu}. \quad (18.18)$$

While the concept of this single component parameter is relatively easy, the practical difficulties in transport come from the heterogeneity and complexity of a large electromechanical/civil system and its operation:

- We do not have one (or several similar) component(s) but hundreds to thousands of different components failing in all kinds of failure modes.

- In the event of any component failure, it is, in general, not directly "repaired" to restore the system, but multiple more general remedial actions are deployed to bring the system quickly back to regular operation (e.g., ignoring minor failures such as off compartment lights for some time and taking train out to yard at next occasion such as for some door failures, reset actions, degraded mode of operation such as for switch failures). So the "average" MTTR value becomes rather a more general "restoration" time (which may not be an exponentially distributed time anymore).

- Restoration times of failed components may have each a different impact on passenger traffic and punctuality, the ultimate metric for the operator. A failure of the relative train position determination system may have a completely different effect onto fleet capacity than a platform screen door failure, while both may have a similar "restoration" time.

- The availability of the transport system mixes responsibilities: the equipment supplier may well "guarantee" failure occurrence rates (MTTFs), but remedial action upon failure (MTTRs) resides, in general, within the operator's responsibility. From the pure "unavailability" of equipment, it cannot be seen any more if the operating engineer took twice as long to restore a failed component or if it failed twice as often as predicted.

Despite these complications, the concepts of Equations 18.2 and 18.3 or similar are still utilized to date in most contracts in one form or the other. The challenge consists especially in today's highly automated guided transport systems in trying to match the following:

- Traditional concepts (reliability block diagrams [RBDs], failure modes and effects analyses [FMEAs]), often required as a "reliability and availability prediction analysis" from the supplier

- More adequate models that reflect the operational perspective with a versatile metric accounting for all failure events, imposed as aggregated values onto the supplier by contracts and/or for "system availability monitoring/reporting" system

Both aspects are discussed in the following sections.

18.5.1 Reliability and Availability in Multiple Component Configurations

The formal question of how multiple different components with each λ_i and μ_i may be aggregated into system parameters λ_{Sys} and μ_{Sys} is extensively discussed in the literature and this handbook and is therefore just summarized in its main elements.

If two different availability-relevant components are combined in a system such that the complete configuration is considered "available" only if both components are available at a time, then the combination is often represented by a serial configuration of the two components in so-called RBDs.

Since the system becomes unavailable if either component fails, it is apparent that the failure rate of the system becomes the sum of both individual rates, and the system availability A_{Sys} is the product of both individual availabilities A_1 and A_2:

$$\lambda_{\text{Sys}} = \sum_i \lambda_i = \lambda_1 + \lambda_2, \ A_{\text{Sys}} = \prod_i A_i = A_1 \cdot A_2, \ U_{\text{Sys}} = 1 - A_{\text{Sys}}$$

$$= 1 - \prod_i (1 - U_i) \overset{U \ll 1}{\approx} \sum_i U_i = U_1 + U_2; \ U_{\text{Sys}} = \frac{\lambda_{\text{Sys}}}{\mu_{\text{Sys}}} \Rightarrow \mu_{\text{Sys}} = \frac{\lambda_1 + \lambda_2}{U_{\text{Sys}}}. \tag{18.19}$$

If two availability-relevant components are however combined such that the "system" remains available if either of both components (or both) is (are) available, then the configuration is considered a "redundant" (or "parallel") configuration. From the sample up/down time distributions, it is obvious that the system now becomes unavailable only if any one of the components has previously failed and is under repair while by chance, the second component also fails. Since it may fail at any time during the restoration of the first failed component, its repair/restoration is on average already completed to 50%, so the system repair time becomes only half the repair time of the component, and the "repair rate" doubles. The same may also happen, of course, for the second component failing first and the other during its repair. Weighing both situations leads to a system "repair rate" as the sum of the individual repair rates. Since the system becomes unavailable only if both components are unavailable, the system availability and the system "failure rate" calculation is straightforward:

$$\mu_{\text{Sys}} = \sum_i \mu_i = \mu_1 + \mu_2, \ U_{\text{Sys}} = \prod_i U_i = U_1 U_2,$$

$$U_{\text{Sys}} \approx \frac{\lambda_{\text{Sys}}}{\mu_{\text{Sys}}} \rightarrow \lambda_{\text{Sys}} = \mu_{\text{Sys}} \cdot U_{\text{Sys}}, \ A_{\text{Sys}} = 1 - \prod_i U_i,$$

$$\mu_1 = \mu_2, \ \lambda_1 = \lambda_2 \Rightarrow \mu_{\text{Sys}} = 2\mu_1, \ U_{\text{Sys}} = \frac{\lambda_1^2}{\mu_1^2}, \ \lambda_{\text{Sys}} = \frac{2\lambda_1^2}{\mu_1}. \tag{18.20}$$

Besides the serial or redundant configurations (or one-out-of-two, 1oo2), many other configurations may be found in (especially electronic) systems, for example, two-out-of-four (2oo4), one-out-of-three (1oo3), and four-out-of-six. Their calculations may be derived either from Markov model calculations, looked up in the literature, or respective chapters of this handbook. For typical guided transportation systems, virtually all configurations may be reduced to either serial or redundant configurations and thus be calculated step by step. The only occasional exception is the two-out-of-three (2oo3) configuration (e.g., in electronics), where the respective calculations are

$$A_{2\text{oo}3} = \sum_{m=2}^{3} \binom{3}{m} A^m (1 - A)^{(3-m)} = 3A^2 - 2A^3,$$

$$\mu_{2\text{oo}3} = 2\mu, \ \lambda_{2\text{oo}3} = 2\mu \frac{1 - A_{2\text{oo}3}}{A_{2\text{oo}3}} \approx \frac{6\lambda^2}{\mu}. \tag{18.21}$$

18.5.2 RBD Calculation and Availability Prediction Analysis

In order to evaluate and optimize the availability of the system, it is still common practice in railway projects to require from the supplier an RBD-based theoretical prediction and analysis of how the various system elements contribute to the unavailability of the transport system. Since modeling and calculation of RBDs is conceptually straightforward and described in detail elsewhere, only an example is considered here. As example, an onboard train control configuration may be considered at basic printed circuit board or equivalent level. A list of MTTFs and MTTRs may have been aggregated from an FMEA or previous databases. A list of such a configuration could look like Table 18.6.

Note that the indicated MTTFs may themselves represent aggregated values of multiple different failure modes and of numerous subelements, so the listed MTTFs are already aggregated values. Each listed component may occur several times in the configuration, and the RBD representation of the configuration may look like Figure 18.15.

The RBD is calculated step by step by using Equations 18.19 through 18.21 such that the final diagram is calculated as a serial block diagram, where all redundant parts are calculated and aggregated before (values indicated below blocks), yielding an MTTF = 10,000 hours for the whole configuration example.

While for operational purposes this aggregated value is of limited use for operations due to the variety of different operational effects of different failures, it may serve well as a representative availability indicator for the configuration. Also, for strength/weakness analysis, which quickly and straightforward identifies the major "unavailability" drivers, this analysis is useful.

In order to estimate the overall system availability, the RBD calculations/prediction analyses may also serve to put different partial RBDs into relation such as the following:

- Rolling stock (electromechanical configuration; doors; propulsion; heating, ventilation, and air conditioning; etc.)
- Automation and signaling control equipment (e.g., onboard, trackside, central)

TABLE 18.6

Typical MTTF/MTTR Example List Input to Transport RBDs

Element	MTTF (h)	MTTR (h)	U	A
Filter cards	600,000	0.75	1.25×10^{-6}	0.99999875
UPS-Battery/	250,000	0.75	3.00×10^{-6}	0.99999700
Power modules	350,000	0.75	2.14×10^{-6}	0.99999786
Input/Output cards	350,000	0.75	2.34×10^{-6}	0.99999766
Bus coupling	1,000,000	0.75	7.50×10^{-7}	0.99999925
Tacho C	800,000	1.50	1.88×10^{-6}	0.99999813
Transponder antenna	600,000	2.00	3.33×10^{-6}	0.99999667
Transmission cards	300,000	0.75	2.50×10^{-6}	0.99999750
Proc/comm cards	150,000	0.75	5.00×10^{-6}	0.99999500
Filter 1/2	50,000	0.75	1.50×10^{-5}	0.99998500
CPU channel	300,000	0.75	2.50×10^{-6}	0.99999750
Sync cards/memory	350,000	0.75	2.14×10^{-6}	0.99999786
Comparator	250,000	0.75	3.00×10^{-6}	0.99999700
Rack/slots	800,000	3.00	3.75×10^{-6}	0.99999625
Vehicle bus	300,000	0.75	2.50×10^{-6}	0.99999750
Front cabling	25,000	0.50	2.00×10^{-5}	0.99998000

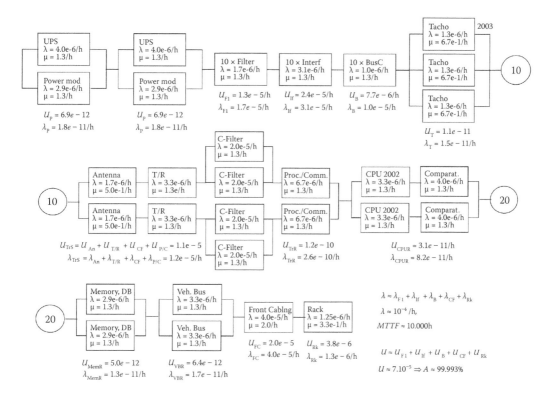

FIGURE 18.15
Typical example of partial multiple component configuration transport RBD.

- Traction power supply system (e.g., third rail, substations)
- Field elements (e.g., shelter gates, intrusion detectors, track vacancy detections, switches, signals, and hotbox detectors)
- Station equipments (e.g., passenger information, destination displays, platform protection doors, edge monitoring, and emergency stop handles)

The specific example of Figure 18.15 yielding an MTTF = 10,000 hours for one onboard controller equipment configuration (OCEC) means the following: Already the operation of 20 trains in a system leads to an OCEC-caused "unavailability event" every 500 hours somewhere in the system. Should all other subsystems of the five above lots have each a similar unavailability characteristic, it may accumulate to a disturbing event approximately every 100 hours (about 5 operating days). While theoretical values are often far above, this is in practice (e.g., for typical subway line) not an uncommon value.

Formally, also a complete system MTTR of 0.5 hours may translate a complete system MTTF of 100 hours into a complete "system availability" of A = 99.5%, whereas in specifications of new urban transport systems, it is rather common to define global requirements clearly above (up to A = 99.9%).

Today, the preceding process is often still considered as a traditional and viable option to also define procurement specification requirements for railway equipment and monitor operational conformance with the contractual requirements. Breaking a complete transport system down into the preceding sketched lots such as in Equation 18.22 offers not only distinct advantages but also disadvantages:

$$A_{\text{System}} = A_{\text{RS}} \cdot A_{\text{ATCSIG}} \cdot A_{\text{Station}} \cdot A_{\text{Track}} \cdot A_{\text{Power}} \geq 99.9\%. \qquad (18.22)$$

Since the calculation logic is relatively straightforward, and (small) partial system unavailabilities are simply summed up to the complete system availability, the inverse is equally easy: allocation of availability (or better "unavailability") budgets of partial systems. Since lots are often subcontracted to different suppliers, the homogenous logic of the calculation is in fact convenient and consistent.

Much less straightforward is however the system availability demonstration, since the preceding "metric" assumes that every unavailability event of any of the subsystems affects the same number of passengers, and inconvenience or "unavailability experience" by the passenger (e.g., delay) scales only linearly with the restoration time of the specific element. Obviously, both are rarely the case. Even if it would be so, the measurement of the duration of every individual restoration time and how many trains and stations may be affected in a specific situation is quite a challenging recording task and often contentious.

In summary, contractually demanding availabilities as suggested by the RBD logic in terms of "percentages" is a straightforward process and close to technical "availability prediction analyses" but leaves operationally oriented issues such as the following open for ambiguous interpretation:

- How often is (what kind of) staff intervention required to fix the system and keep traffic flowing?
- How often would passengers have to wait for a train much longer than scheduled or expected?
- How many other system elements (trains) are impacted by what kind of primary failure with what kind of secondary effects (e.g., bunching back in the system)?
- How often are passengers kept delayed for longer times (e.g., 1 headway time, 10 headway times, and one roundtrip time)?
- How often do failures lead to severe consequences (e.g., evacuations)?
- How are all these quality effects combined into one indicator/number (e.g., A = 99.9%)?

It has therefore become common practice to search for other operationally more adequate metrics as well in specification as in demonstration in procurement processes for railway/transit equipment. One critical issue in this context is the analysis of the passenger- and operator-relevant effects of failures and definition of an adequate metric to compare different unavailability events.

18.5.3 Failure Management, Service Restorations, and Delay Categories

Due to the high diversity of system failures and the equally diverse related repairs or other remedial actions to restore uncompromised traffic, the resulting spectrum of MTTRs must be organized into a finite set of operationally and maintainability sensible "failure categories" in order to be able to set up one common metric for all kinds of failures. This requires, in general, a specific operational analysis to derive what actions may typically be planned by the failure management of the operator. While an analysis of possible failures in the system

TABLE 18.7

Failure Log Protocol Example

Date	Duration (minutes)	Failure
12/02	9	Fire/smoke false alarm
12/07	42	OBCU shutdown
12/09	12	Platform screen door lock failure
12/11	9	Platform screen door lock failure
12/11	24	Platform screen door close failure
12/13	135	OBCU shutdown—train recovery
12/15	134	OBCU shutdown—train recovery
12/15	35	Platform control communication loss
12/16	12	Schedule invoke failure
12/16	30	Brake valve failure
12/17	80	WCU shutdown
12/17	27	WCU shutdown
12/17	13	Automated circuit breaker undervoltage
12/18	29	Platform control unit failure
12/18	39	Platform control unit failure
12/21	15	Platform control unit failure
12/21	23	Platform screen door failure

may yield stimulating input into the analysis, it is rather the organization and distribution of failure management staff and geographical distribution of the system elements themselves that determine the global operational availability. In other words, a high-speed intercity train positioning transponder failure with staff "only" located every 40 or 50 km requires a completely different failure management action than, for example, the same equipment failure in an unmanned (UTO = GOA4) subway operation with no staff onboard but staff in every second station. For illustration, Table 18.7 shows an excerpt of a failure protocol of a just implemented driverless urban guided transit system, while the failures and recoveries as such improved, of course, over time, diversity of the restoration times, persisted.

Furthermore, a common denominator for the effects of an event in a "failure category" must be elaborated such that it includes consequential (secondary) effects on other parts of the system. One possibility could be to compare the typical "still possible" remaining transport capacity of a railway corridor (e.g., in terms of PPHPD = passengers per hour per direction) after a failure to the "nominal" planned and scheduled capacity. In practice, mostly, the trip "delays" of all trains in the corridor are rather selected as initial measure before being transferred into other measures.

On the consequences, a simple specific example of a high-density/short headway subway line may illustrate the typical steps. In this simplified example, assume that we have only three general failure management actions available:

In the first category, C1, action shall consist in remote reset (or autoreset) of failures, so after a short time of less than 5 minutes, the system service is restored (with an average duration of $D1 = 3$ minutes).

The second category of failures, C2, may require staff intervention at site, which shall take between 5 and 20 minutes (with an average duration of $D2 = 13$ minutes).

The third category of failures, C3, shall contain failures which last longer than 20 minutes, so in a subway operation, it is often possible to set up a reduced traffic schedule where trains are routed "around" the incident location. There is no "average" or "typical" duration for this kind of failure, but the actual times must be recorded.

Since, in general, a failure of any of the categories does not only impact the affected (first) train but quickly propagates through the fleet, it must be determined how many of the trips are affected by the failure. This however depends on the "density" or headways of trains. Assume for simplicity only two "services," an off-peak-hour service with a 10-minute headway between two trains, and a peak-hour service with a 2.5-minute headway. A simplistic view of the situation is shown in Figure 18.16.

In order to find a common denominator of average delays induced by events of a failure category (here C1), easily countable parameters, which still adequately reflect delay and capacity loss, have to be found. The direct use of the delay times of affected primary and secondary trains requires further interpretations, since the direct sums of times do not necessarily reflect individual passenger uneasiness of a specific train (e.g., four trains delayed by 3 minutes versus one train delayed by 12 minutes). Instead, the primary and secondary delays must be translated into more discrete equivalent measures, e.g., lost departures, lost missions, or lost round trips, where above a certain delay of a train, the respective equivalent measure is counted "lost." This process has the clear advantage of quantified, discrete units that could be set in relation to the nominal ("planned") values, yielding again a ratio that can be interpreted as "unavailability" (or availability). One downside of this discretization is the fact that "delayed" trips/departures are actually not really "lost," which is however tolerated since a first delayed train coming substantially late is usually packed with passengers while the subsequent ("bunched") trains remain virtually empty, showing an equivalent characteristic to passengers as "lost" trips.

The sketch in Figure 18.16 shows a typical network (or network part) with trains flowing up and down the tracks "separated" by their planned headway. An incident at any location (e.g., a "stalled" train on guideway, train hold at platform, track circuit failure, and switch failure) holds a train for a representative time $D1$ on the track. If the time is small compared to the actual headway, it leads to small effects whatever the "effect" measure is (here it is "lost departures," e.g., 0.3 lost departures per incident). Once, the headway gets shorter, secondary trains pile up behind the primarily delayed train, leading to successive "lost departures."

Failure category **C1**: any failure between 0 min < D < 5 min (< **D1** >~3 min)
Off peak traffic consequence: 1 train delayed by < D > ~0.3 departures,
Total lost departures: 0.3

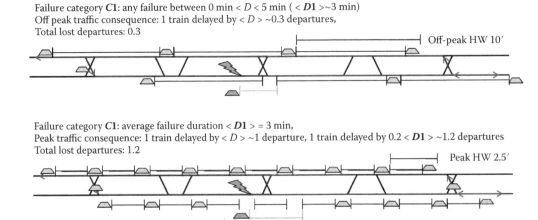

Failure category **C1**: average failure duration < **D1** > = 3 min,
Peak traffic consequence: 1 train delayed by < D > ~1 departure, 1 train delayed by 0.2 < **D1** > ~1.2 departures
Total lost departures: 1.2

FIGURE 18.16
Graphical representation of the impact on subway traffic of failure category C1.

Failure category **C2**: any failure between 5 min < D < 20 min, < **D2** > ~13 min
Off peak traffic consequence: 1 train delayed by < **D2** > ~1.3 Dep., 1 train delayed by < **D1** > ~0.3 Dep.
Total lost departures: 1.6

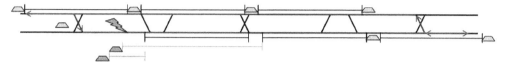

Failure category **C2**: any failure between 5 min < D < 20 min, < **D2** > ~13 min
Peak traffic consequence: 1 train by < **D2** > ~5 Dep., 1 train by 4 < **D1** > ~4Dep., 1 train by 3 < **D1** > ~3Dep., 1 train by 2 < **D1** > ~2
Dep., 1 train by < **D1** > ~1Dep., total lost departures: 15

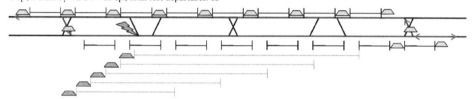

FIGURE 18.17
Graphical representation of the impact on subway traffic of the failure category C2.

Equivalently, if the incident is part of the "longer delay" category (e.g., C2, D2 example), the train bunching rapidly builds up in dense (compared to D2) headway operation as indicated in Figure 18.17.

Compiling the "effects" (here, lost departures) of failure of the categories, e.g., for 1 year, and "inducing" a spectrum of how often the failures of any category may occur per year and system (e.g., C1/C2/C3 = 600/60/6) yields accumulated results as indicated in Table 18.8.

The preceding sketches serve rather for illustration of the basic process, while it is obvious that the indicated train bunching may not be acceptable in many cases, and every railway or mass transit operator has specific track layouts, schedules, and failure/restoration management strategies (e.g., "safe haven" and "graceful degradation"). Therefore, the preceding simple diagrams must be replaced by more detailed complete system simulations, where "incidents" of the categories are randomly induced (location, causing subsystems, category) along the lines (mostly at home/exit signals), and system degradation/recovery is measured and optimized; Figure 18.18 shows an example for a European metro system by Technische Universität Dresden (TU Dresden).

Another sometimes required correction comes from "partial" failures. While a delay as such always meant a delay of one or more complete trains, sometimes only parts of the system (e.g., "1 out of 12 train/platform doors failed and is isolated") may become inoperable for some time. In general, the effects are accounted for by a correction factor (e.g., 1/12 of all train departures during a "delay" duration) determined by the ratio of full delay and partial delay failure sources (yielding a small percentage number correction to the overall "unavailability").

Further possibilities in consequence category evaluations consist in the definition of the acceptable ratios of occurrences as well as "penalizing" increasing delays by increasing factors of counted unavailability.

Irrespective of the definition details of the category and its metric is the fact that based on it, a spectrum of maximum acceptable numbers of failures (the delay spectrum) can be defined and verified by the transport operator.

TABLE 18.8

Compilation of a Simplified Delay Events Spectrum Impacts (per Year)

Mode	Consequence Metric	Nomin. Depart. Per year/30 Platforms	Loss per C1	Loss per C2	Loss per C3	Loss 600 C1	Loss 60 C2	Loss 6 C3 (1)	Σ	U/A
Off peak (e.g., 12 hours/day)	Lost departures	788,400	0.3	1.6	6/h	75/250	40/25	18/3 h	133	1.7×10^{-4}/99.98%
Peak (e.g., 8 hours/day)	Lost departures	2,102,400	1.2	15	360/h	420/350	525/35	1080/3 h	2025	9.6×10^{-4}/99.90%
Sum		2,890,800							2158	7.5×10^{-4}/99.92%

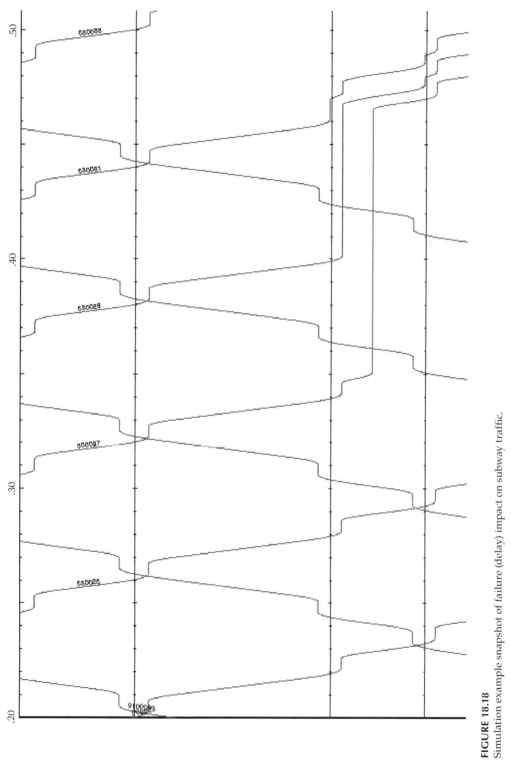

FIGURE 18.18
Simulation example snapshot of failure (delay) impact on subway traffic.

18.5.4 Delay Spectrum Definition and Demonstration

One of the major advantages of a delay spectrum definition as availability measure consists in disentangling failure/incident rates (i.e., "reliability") and restoration or repair times and thus disentangling responsibilities between supplier and operator. Thus, the "availability" requirement can be translated into a set of still acceptable failure or incident numbers per time interval (rates) and per category.

The process sketch in Figure 18.19 shows the relevant steps of how to reach consistent definitions, where the "gray-colored" steps predominantly fall into the competence domain of the operator and the "white-colored" steps predominantly fall into the competence domain of the supplier.

In practice, the relation between the delay spectrum and the technical system is established by the combination of a system FMEA at high level and a system RBD at high level. Figure 18.20 shows part of a system RBD example by TU Dresden for a refurbished subway line to give an idea about adequate level of details.

The quantified evaluation of the system RBD requires a system FMEA that is based on the delay categories, previously defined for the delay spectrum (e.g., C1, C2, C3) and breaks down every identified "failure" into the probabilities to occur in any of the categories. Based on the failure rates and the previously defined secondary consequences of every category (e.g., shift dependent), the system FMEA directly yields the consequences of every failure in the system in correct proportions. Figure 18.21 shows part of a system FMEA example by TU Dresden for this specific purpose.

The analyses not only yield (on paper) a possible verification of the targets in terms of number of acceptable failures but may also be used for possible target allocations (since the delay spectrum must also be established sooner or later). For this, first, an overall availability target is defined (e.g., 99.9%) and then translated into the decided metric indicator units (e.g., "lost departures" equivalents per "nominal departures" equivalents), and the unavailability units (e.g., "lost departures") spread over the consequence or delay

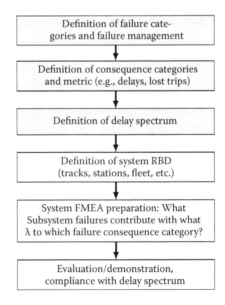

FIGURE 18.19
Process steps in availability demonstration by a delay spectrum.

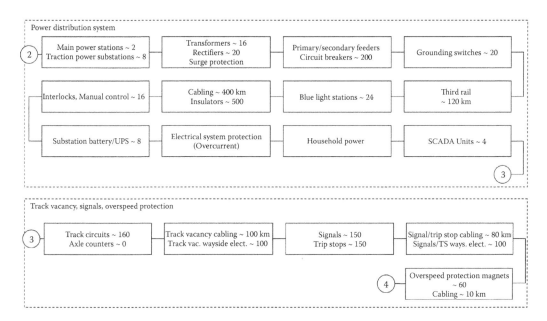

FIGURE 18.20
Part of a system RBD for availability requirement definitions at system level.

categories according certain distribution keys. Figure 18.22 shows typical steps of such an allocation process.

It shall be noted that the preceding considerations had been simplified for explanatory reasons. In practice, more or different types of modes may be selected. Also, there may be many more delay categories leading to delay spectra such as the one shown in Figure 18.23.

It shall be further noted that in practice, not only sophistications of the preceding process are encountered, but also clear complexity reductions.

On the demonstration of these "delay spectra" used in realization projects, two levels of demonstration are typically requested.

Independently of where the delay spectrum may have come from (top-down allocation from the operator or consultant, determined from existing operations data, etc.), the supplier is requested to demonstrate on paper that the equipment is likely to fulfill the required spectrum. Since the delay categories are defined by the spectrum itself, the task for the supplier is now reduced to an RBD and FMEA of the supplies, where the theoretical failure rates are determined and split/distributed into the delay categories (instead of calculating any restoration rates). Ultimately, the overall number of failure events per year is directly compared to the required numbers of the spectrum, yielding a clear decision base.

Secondly, the values have to be demonstrated in the field after commissioning the system. Demonstration itself thus becomes "only" a matter of event number recording instead of determining an individual "restoration time" and other consequences for each individual event. In order to account for initial (immaturity) effects, the demonstration is performed over multiple years (e.g., 3 years) broken down into several periods (e.g., 1-year periods, similar to Figure 18.14). Should now in the first period the numbers of the spectrum be exceeded by the built system failures before the first year has elapsed, this period is marked as "failed," and a second attempt period is directly started after. Final acceptance is reached if for a certain number of consecutive periods (e.g., two consecutive 1-year periods), the numbers of observed failure events are lower than the delay spectrum numbers.

| Failure source | | Failure | Comment/ effects | Failure category | Number/km equipment | MTBF (/unit, /km) | Events /yr | MTBU | Percentage peak/ off peak | Number roundtrips* delay D_1 | Number roundtrips* delay D_2 |
Subsystem	Equipment										
	Signals trip stops	Trip strop magnet failure trafo	50% Local remedy 50% 2 h run around	50% C2 50% C3	150	$1.4*10^6$ h	0.8	1	0.3*0.5 0.7*0.5	0 3.360	5.760 0
Vital peripherals		Signal wire failure	Redundant wire	100% C1	60	$1.3*10^4$ h	34.6	1	0.3 0.7	20.76 24.22	0 0
	Overspeed protection magnets/GUs	Wayside pulse gener./electr. cards failure	30% Remote reset 50% local remedy/HW Imp. 20% rund around Op.	30% C1 50% C2 20% C3	60	$9.7*10^4$ h	4.5	1	0.3*0.3 0.7*0.3	0.810 0.945	0 0
			30% Remote reset 50% local remedy/HW Imp. 20% rund around Op.	30% C1 50% C2 20% C3	60	$9.7*10^4$ h	4.5	1	0.3*0.5 0.7*0.5	2.025 1.575	2.025 1.575
			30% Remote reset 50% local remedy/HW Imp. 20% rund around Op. 2h	30% C1 50% C2 20% C3	60	$9.7*10^4$ h	4.5	1	0.3*0.2 0.7*0.2	0 7.560	12.960 0
		Int. electr. cards failure	Wrong speed feed back	100% C1	60	$4.4*10^4$ h	1.0	1	0.3 0.7	0.600 0.700	0 0

FIGURE 18.21
Snapshot of part of a system FMEA for entries into failure/delay categories.

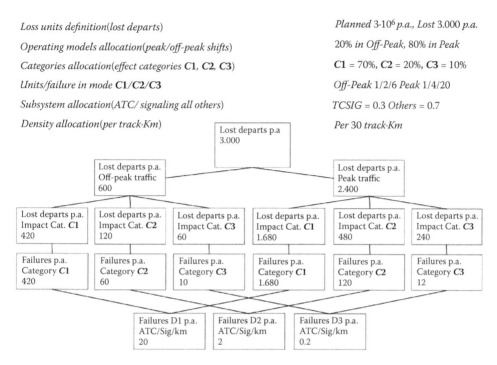

Loss units definition(*lost departs*) — Planned $3 \cdot 10^6$ *p.a.*, Lost 3.000 *p.a.*

Operating models allocation(*peak/off-peak shifts*) — 20% in *Off-Peak*, 80% in *Peak*

Categories allocation(*effect categories* **C1**, **C2**, **C3**) — **C1** = 70%, **C2** = 20%, **C3** = 10%

Units/failure in mode **C1/C2/C3** — *Off-Peak* 1/2/6 *Peak* 1/4/20

Subsystem allocation(*ATC/ signaling all others*) — TCSIG = 0.3 Others = 0.7

Density allocation(*per track·Km*) — Per 30 *track·Km*

FIGURE 18.22
Example top-down allocation of system "unavailability" allocation.

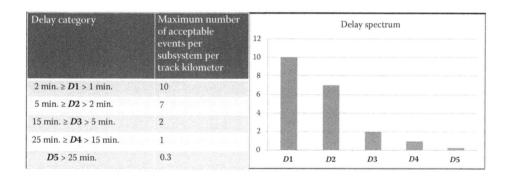

Delay category	Maximum number of acceptable events per subsystem per track kilometer
2 min. ≥ **D1** > 1 min.	10
5 min. ≥ **D2** > 2 min.	7
15 min. ≥ **D3** > 5 min.	2
25 min. ≥ **D4** > 15 min.	1
D5 > 25 min.	0.3

FIGURE 18.23
Typical delay spectrum presentation/notation in projects.

On statistical effects, the total numbers of the delay spectrum are, in general, high enough so that statistical uncertainties become negligible. Alternatively (especially for lower number categories), the variance is included into the (not to exceed) numbers in the delay spectrum itself.

In summary, the delay spectrum approach requires substantial previous failure (management) analyses but may account well for the specificities of a transit system. Also, this approach leaves less ambiguities open for interpretation, may be calculated backwards into an equivalent standard "availability" percentage, supports easy split of responsibilities in guaranteeing restoration within certain "delays" on one side, and guarantees a maximum number of failure events on the other side. As a consequence, it supports a clear accept/reject process in availability demonstration programs.

Finally, it shall be noted that also straightforward allocations of regular availability (percentage numbers) into several subsystems is still performed by some operators, e.g., by requesting a global "availability" (e.g., $A_{System} = 99.9\%$), measured in a breakdown into partial availabilities such as, for example, fleet availability A_{Fleet}, wayside equipment/station availability $A_{Station}$, and travel delay availability A_{Travel} so that $A_{System} = A_{Fleet} \cdot A_{Station} \cdot A_{Travel}$ or similar. In comparison to the preceding discussions, this approach leaves more ambiguities to the interpretation of the global value, but on the other hand more "freedom" to arbitrate effects of the partial values. Another relative disadvantage is the substantial effort in recording for every "event" the exact restoration/repair time and possible consequential effects and then aggregating the (possibly many) events. On the other hand, the precise recording of effects contributes to the precision of measurement and previous analyses (as for setting up a good "delay spectrum") may be kept comparatively easy. From the author's perspective, the "choice" between the two approaches depends to large extent on the capacities of the "owner/operator" to track in detail unavailability effects of failure events and the sophistication of the measurement tools.

19

Proven in Use for Software: Assigning an SIL Based on Statistics

Jens Braband, Heinz Gall, and Hendrik Schäbe

CONTENTS

19.1 Introduction

19.1.1 Motivation

Demonstration that a certain piece of software satisfies the requirement for a safety integrity level (SIL) is usually done by applying EN 50128 (CEN 2011). However, there is a wish to use experience in the field to qualify software. This is especially important for existing or commercial off-the-shelf (COTS) software. The railway standards on reliability, availability, maintainability, and safety EN 50126 (CEN 1999), EN 50128 (CEN 2011), and EN 50129 (CEN 2003) do not provide an approach for the use of proven-in-use arguments. Solely IEC 61508-7 (2010) provides some guidance.

The nature of software failures and the possibility to predict software failure behavior has attracted the interest of researchers and engineers. The first have tried to study the phenomenon, the second have been searching for a way to predict software failure behavior based on characteristics describing the software, preferably characteristics of the source code as metrics. In order to approach this problem, first of all one needs to understand the nature of software failure behavior. Obviously, this is different from hardware failure behavior. One main question is, can software failure behavior be described by probabilistic models.

If such a model is known, the next question is then how it can be used and whether it would be possible to use this model for prediction or to prove that a certain software falls, e.g., into a safety integrity level (SIL) regarding its failure rate (if this exists) and can therefore be judged to be qualified for this SIL.

Two of the authors have done some mathematical research on the question of statistical failure behavior of software and its existence under certain conditions (Schäbe and Braband 2016). For a simple software model, see also Schäbe (2013).

19.1.2 Problem

Proven-in-use arguments are needed when predeveloped products with an in-service history are to be used in different environments than those they were originally developed for. Particular cases may include the following:

- COTS components that are to be used in safety contexts
- Predeveloped products that have been certified by a different standard or against different requirements
- Products that have originally been certified to a lower level of integrity but are to be used for higher levels of integrity

A product may include software modules or may be stand-alone integrated hardware and software modules.

Also, there has been considerable confusion about the term *software reliability*. Therefore, the authors want to state the following very clearly: software does not fail; only physical systems can fail. So, unless software is executed on hardware, we cannot refer to failure.

19.1.3 Solution

The aim of this chapter is to base the argumentation on a general mathematical model, so that the results can be applied to a very general class of products without unnecessary limitations. This includes software for railway applications. The advantage of such an approach is also that the same requirements would hold for a broad class of products.

In this chapter, we will first discuss existing standards and approaches that have been used throughout the last 20 years.

Then, we will describe the mathematical model and show how it can be used. We will illustrate it with examples. Further, the authors will provide some of their personal experience and draw conclusions.

19.2 Existing Standards and Known Results

In this chapter, an overview of some existing results will be given, which cannot be exhaustive but provide only the main points.

Unfortunately, EN 50126 (CEN 1999), EN 50128 (CEN 2011), and EN 50129 (CEN 2003) do not give guidance for proven-in-use arguments for software.

In an earlier study (Gall, Kemp, and Schäbe 2000), two of the authors showed this based on the existing standards at that time: IEC 61508 (version from 1997/1998) as well as DIN V VDE 0801 (1990), DIN V VDE 0801 A1 (1994), the book by Hölscher (1984), KTA 3507 (1986), and Mü8004 (1995). The main results are reproduced in the following.

Two cases of application had to be distinguished:

1. Continuously used systems
2. On-demand systems

Statistical methods have been applied under the following regularity and application conditions (Gall, Kemp, and Schäbe 2000):

- Precise description of the unit and its tasks and functions
- Fault models to be considered
- Clear definition of application conditions and environment of use
- The sample used for the statistical evaluation must be representative
- An adequate statistical method must be chosen that reflects the type of use and the specific conditions of the units
- Strictly applying the computational rules of the chosen mathematical model

The statistical approach consists of three phases:

1. Data collection
2. Statistical inference of the data
3. Discussion of the results

Depending on the type of application, two different statistical models could be used according to Gall, Kemp, and Schäbe (2000). For case 1, this was the exponential distribution, and for case 2, the binomial distribution. The usual statistical methods were applicable. However, the estimation of the probability of failure on demand or the failure rate had to be done with statistical coverage of at least 95%.

The use of the exponential distribution is explained later on in detail in this chapter.

Currently, the actual version of IEC 61508-7 (2010) includes as annex D a proposal for the evaluation of software reliability based on a simple statistical model. However, this annex is discussed quite controversially and is currently under revision. EN 50129 (CEN 2003) also contains recommendations on how to demonstrate high confidence by use for systems, but without any further explanation. For example, for SILs 3 and 4, "1 million hours operation time, at least 2 years of experience with different equipment including safety analysis, detailed documentation also of minor changes during operation time" is recommended.

The topic itself is not new; see the book by Gall, Kemp, and Schäbe (2000) and the study by Littlewood and Strigini (1993), but most recent approaches have been based on elementary probability such as urn models which lead to very restrictive requirements for the system or software to which it has been applied by Ladkin and Littlewood (2015). Much earlier work has been performed with a different approach by Littlewood (1979).

19.3 Mathematical Model

19.3.1 General Considerations

In this section, we provide a simple model of a predeveloped product. It consists of software and hardware. Hardware is relatively easy to consider. It exhibits random failures and systematic failures. As for random failures, it is necessary to deal with aging, wear and tear, and comparable effects that lead to a possibly dangerous failure of the hardware. Since there are a large number of influences such as material, load, and use, the failure time is random. This is modeled by distributions such as the exponential distribution, Weibull distribution, and many other models, e.g., see the book by Barlow and Proschan (1975).

Furthermore, the hardware may contain systematic faults. We will treat those together with software errors. Normally, if a component has been developed according to a standard for functional safety, it is assumed that such methods have been used that systematic failures can be neglected, compared with random failures. However, for predeveloped components, this assumption might not hold. Therefore, we need to consider systematic failures of the hardware.

19.3.2 Software Model

For software, we will now use the following model. In order to simplify the situation, we will assume that there are no randomization procedures in the software, as reading system time, using uninitialized variables, and using data in memory cells and other data that might result in a random influence on the software. We also do not discuss temporal ordering in multicore processors or multithreat systems. These processes, also chaotic processes used, for example, for random number generation, could be discussed separately.

With these assumptions, the software could be imagined as a deterministic system. This system adopts a new state for each processor step. Obviously, the state of the software would depend on the previous states of the software. Here, *state* means the values of all registers, counters, pointers, etc. This means that we do not assume that the software fulfils the Markov property. In that case, i.e., if the software would fulfill the Markov property, the next state of the software would only depend on the previous step and not on the history. Now, we need to take into account the fact that the state of the software would also depend on the input data.

Hence, we have a space of input values E and a space of output values A. We can express the input values as a sequence E_i, $i = 1$, ... Now, the software provides a deterministic mapping $A_n = F(E_n, E_{n1}, ..., E_i, ..., E_1)$. This mapping is subjective. Now, there exist tuples of $n + 1$ elements $(A_n, E_n, E_{n-1}, ..., E_i, ..., E_1)$. Any tuple is either correct (C) or incorrect (N). So, the entire space S $(A_n, E_n, E_{n-1}, ..., E_i, ..., E_1)$ for all n is subdivided into two parts: C and N. Usually, C would have a much larger size than N. Size here means the number of points in the subspace. It needs to be understood that there would be a very large number of points in a space. That means, if we chose at random a point from the set of all possible $(A_n, E_n, E_{n-1}, ..., E_i, ... E_1)$, we would only rarely be in N.

Inputs E_1, E_2, etc., can be considered as random. However, their distribution influences how often the software status would belong to subspace N. It is not necessary that the distribution of E_1, E_2, etc., is a uniform distribution. Only the realizations of sequences E_1, E_2, ... must be such that the distribution of the occurrences of a state in N and C are the same under different circumstances. Of course, if sequence E_1 and E_2 always has the same distribution function, the distribution of the occurrence times of a state in N would be the same. Then, the interarrival times of states in N would be exponentially distributed since that is a rare event.

If the software states form a random walk in the space N $\cup$ C, then a state in N is achieved as a random peak.

In order to come to a result based on the complicated structure of the software, we assume the following:

- The software failures are decoupled so that we can assume that they can be considered as being independent of each other.
- The fraction of failures in the software is small so that they should occur only rarely.

When the software is then subject to exterior influence, it could be possible to model the occurrence of each of the N software errors with one counting process, say $X_i(t)$, $i = 1, ..., N$. This process counts the failures of the software and is a renewal process, see, e.g., the book by Barlow and Proschan (1975) for the mathematical explanation of a renewal process. In fact, it means that the interarrival times of a software failure would have an arbitrary distribution, and they would be independent. This choice can be motivated by the preceding two assumptions.

The overall counting process of all software errors (they show up as failures) is then

$$X(t) = \sum X_i(t),$$

where the total runs over *i*. Now, we can apply Grigelionis's theorem (1963):

If $E(X_i) = t/a_i$

$\sum 1/a_i \to \lambda$ for any fixed t

$\sup(X_i(B) \geq 1) \to 0$ as $n \to \infty$ for every bounded set B of the time interval $[0, \infty)$,

then $X(t)$ converges to a homogeneous Poisson process with intensity λ.

Practically, this means that a superposition of many processes where each of them gives a small contribution will come close to a Poisson process. This motivates many of the old software models using Poisson processes; see Section 10 in the book by Musa, Iannino, and Okumoto (1987). They are homogeneous Poisson processes if there are no changes, and then all interarrival times of the states in N are exponentially distributed. If the software changes, e.g., improves, the occurrence rate changes and a nonhomogeneous Poisson process is used.

19.3.3 Practical Meaning of Grigelionis's Theorem

What does Grigelionis's theorem practically mean?

- The software needs to subjected to an exterior influence that can be considered as random so that the software errors are randomly triggered.
- The occurrence probability of a single software error must be sufficiently small and the occurrence of different software errors must be considered as independent.
- The process of triggering software errors must be stationary.
- If a software failure occurs, the system is restarted and operates further on.

Note: If the software is changed, the stationarity assumption does not hold any more. It is then possible to find other models for reliability growth based, for example, on a non-homogeneous Poisson process, e.g., see the book by Musa, Iannino and Okumoto (1987).

Several variants of such limit theorems exist (Birolini 2013), e.g., the classic Palm–Khintchine theorem. Besides these simple models, other more complicated models exist that take into account a certain structure of the software, e.g., see the study by Littlewood (1979). He models that the software structure is modeled as a semi-Markov process, with software modules taking the role of states and control being transferred between the software modules. It is assumed that there may be failures within the modules with failure processes being a Poisson process. Also, the transfer of control may be subject to failure. The resulting process is quite complex, but again, the limiting process is a Poisson process. The major requirement on which the result is based is that the modules and the transfer of control are very reliable, meaning that failures are rare and the failure rates of both processes are much lower than the rate of switching between the software modules.

We will now treat systematic hardware failures in the same manner as software failures. A systematic hardware failure is always present in the hardware and waits until its activation by an exterior trigger. This is comparable with the activation of a software error. However, it needs to be taken into account that after the occurrence of a hardware fault, the system might fail and cannot be restored very easily to a working state. If it is repaired, comparable arguments as for the software might hold.

For many proven-in-use arguments, the exponential distribution is used, see, e.g., IEC 61508-7 (2010). This is then motivated by the Poisson process describing the failure counts of the systems under study.

It is necessary to distinguish between two aspects here.

A Poisson process is a failure counting process of a repairable system. That means a system is "repaired," e.g., by restarting it after a software failure, and since influence on the system is random and the detected software error is activated seldom, we can just neglect the effect of it being activated a second time. After the occurrence of a hardware failure, the system is minimally repaired, i.e., just the failed item is replaced.

The assumption of stationarity, however, does not allow aging to be used in the model.

The Poisson process has the property that the interarrival times are exponentially distributed, which makes the use of statistical models very easy.

Note: It is necessary to distinguish the Poisson counting process from an exponential distribution that is used for the lifetime statistics of a nonrepairable system; e.g., see the book by Ascher and Feingold (1984).

Now, a short note is provided on how to take into account possible internal sources of random influence as intentionally or unintentionally introduced random number generators. They can be treated in the setup as described earlier in the same manner as the exterior random influences.

19.4 How to Apply the Model

We have shown that for a continuously used system, the Poisson process is the right statistical model. Statistical inference for a Poisson process uses the exponential distribution. In this subsection, we show how this is applied. We refer to the existing annex D of IEC 61508, part 7.

The statistical approach is formally based on chapter D2.3 of IEC 61508-7 (2010).

The following prerequisites need to be fulfilled according to IEC 61508-7 (2010) in order to apply the procedure;

1. Test data distribution is equal to distribution during online operation.
2. The probability of no failure is proportional to the length of the considered time interval.

This is equivalent to the assumption that the failure rate of the software is constant over time. This can be ensured by equivalent and stable use conditions and the stability of the software (no changes in the software) and the stability of the underlying hardware and operating system:

3. An adequate mechanism exists to detect any failures which may occur.
4. The test extends over a test time t.
5. No failure occurs during t.

The reader might observe that these requirements mainly resemble the regularity conditions that we provided in the previous section.

Based on IEC 61508-7 (2010), D2.3.2, the failure rate can be computed as

$$\lambda = -\ln \alpha / t, \tag{19.1}$$

where α is the first kind error, which is frequently assumed to be 5%, and t is the test time accumulated during failure free operation.

With Equation 19.1, a failure rate is computed from observed lifetimes that are considered as random in mathematical statistics. Now, it could be possible that very large lifetimes have been observed by good chance so that the failure rate that would be computed would get small and give a wrong impression. Therefore, in this case, a larger value for the failure rate is taken that is larger by a factor of $-\ln(\alpha)$ than just a normal statistical estimate. This factor $-\ln(\alpha)$ is large, and only with a small probability would this value yield a too optimistic failure rate estimate. This small probability is an error of the first kind. A convenient choice for a is 0.05; see, e.g., the book by Mann, Schafer, and Singpurwalla (1974) or Misra (1992). Also, IEC 61508-7 (2010) uses this value.

IEC 61508-1 (2010) defines SIL and target rates for rates of dangerous failures in Table 19.1. For SIL 2, the rate of dangerous failures must be smaller than 10^{-6}/h. This is identical with the requirement for SIL 2 from Table A.1 of EN 50129 (CEN 2003), where the tolerable hazard rate per hour and per function (the same quantity as the rate of dangerous failures)

TABLE 19.1

SIL Table according to Table A.1 of EN 50129

SIL	Tolerable Hazard Rate per Hour and per Function
4	10^{-9} to $<10^{-8}$
3	10^{-8} to $<10^{-7}$
2	10^{-7} to $<10^{-6}$
1	10^{-6} to $<10^{-5}$

must be smaller than 10^{-6}/h. So, it has to be demonstrated that the failure rate of the software is smaller than the target value mentioned earlier.

Assume now that it has to be shown that the software is fit for use in SIL 1 systems. By setting $\alpha = 5\%$ and $\lambda = 10^{-5}$/h, we arrive at the requirement that

$$t > -\ln\alpha/\lambda = -\ln(0.1)/10^{-5} \text{ hours} = (2{,}996/10^{-5}) \text{ hours} = 299{,}600 \text{ hours}. \qquad (19.2)$$

Note that the choice of 5% is usual in statistics as can be seen from IEC 61508-7 (2010) and in the book by Misra (1992).

Therefore, it needs to be shown that at least about 300,000 hours of failure-free operation have been accumulated by the software in continuously running system to accept it as ready for an SIL 1 system.

It is worthwhile explaining how Equation 19.1 is derived.

Equation 19.1 is easily derived based on the following principles: Under certain regularity conditions (see IEC 61508-7 [2010] and the additional explanations in the preceding section), the software produces a process of failures which follows a Poisson process. It is important that the environment in which the software works can be seen as random, i.e., that quite different data are sent to the software for reaction on it so that many parts of the software are activated. Furthermore, the situation must be stable, i.e., no changes in the software (i.e., use of the same version of the software) or in the environmental conditions may occur.

If now the failures caused by the software form a Poisson process, the time to failure and the times between failures are distributed according to the exponential distribution, i.e.,

$$\Pr(T < s) = F(s) = 1 - \exp(-\lambda s). \qquad (19.3)$$

This means that the probability that until time s no failure has occurred is given by

$$\Pr(T \geq s) = F(s) = \exp(-\lambda s). \qquad (19.4)$$

Here T is the time of occurrence of the failure, where T is a random quantity. For application of the exponential distribution, see also IEC 60605-4 (2001), which also defines the data to be collected for a statistical analysis. The requirements for data collection in IEC 60605-4 (2001) coincide with the requirements for the collection of data for a proven-in use analysis as per IEC 61508-7 (2010), which are used in this report.

If now n systems have been observed with failure free times t_i, $i = 1, \ldots, n$, the likelihood function Lik of this observation is

$$\text{Lik} = \exp(-\lambda t_1)\exp(-\lambda t_2)\ldots\exp(-\lambda t_n) = \exp(-\lambda \Sigma t_i), \qquad (19.5)$$

where Σ is the sum over i from 1 to n.

Note that we have observed times $t_1, t_2, \ldots, t_n$, where no failures have occurred until these times. Therefore, the failure must occur later, i.e., in the future. Then, Equation 19.4 is the probability of this event. Since n of these events have occurred, we can multiply the probabilities, which yields the expression for Lik.

Now, one can observe that Lik is decreasing in λ. If now an upper bound of λ_U is chosen so that Lik equals a small value, say, α, then the argumentation is that the real value of λ

would lie with a probability of $1 - \alpha$ below λ_U. That means that λ_U is a worst-case estimate. This approach is called the fiducial approach, which is used in classical statistics to derive confidence intervals.

Therefore, we can set

$$\text{Lik} = \alpha = \exp(-\lambda \sum t_i) = \exp(-\lambda t), \tag{19.6}$$

with $t = \sum t_i$ as the accumulated total time on test. Resolving for λ_U, we arrive at

$$\lambda_U = -\ln(\alpha)/t. \tag{19.7}$$

Comparing Equation 19.6 with Equation 19.1, we see that the standard requires using a value for the failure rate that is the upper confidence level, i.e., a worst-case estimate that accounts for spread in the failure statistics.

Note that Table D1 from IEC 61508-7 (2010) requires a larger number of operational hours; however, this table is not derived from the formula in D2.3.2, which is the basic one in the standard The formula from D2.3.2 (Equation 19.1) can be derived using statistical methods—contrary to the table in D.1 of IEC 61508-7 (2010). When using a failure rate for software based on the statistical approach, this failure rate cannot be neglected compared with the failure rate of hardware. That means, if for the software a failure rate of 5×10^{-6}/h has been demonstrated, whereas for hardware a failure rate of $6 \; 10^{-6}$/h has been derived, the sum is 1.1×10^{-5}/h and such a system would not fulfill the SIL 1 requirement for the rate of dangerous failures. If the figures are smaller, e.g., 4×10^{-6}/h and 3×10^{-6}/h, the result—7×10^{-6}/h—would fall into the range for SIL 1.

The use of Table D1 in IEC 61507-1 would ensure that the software does not use more than 10% of the admissible rate of dangerous failures for a certain SIL.

19.5 Illustrating the Approach with Examples

Unfortunately, there are not many examples publicly available, and the ones we will describe in the following are not coming from the railway area. However, the problem of proven in arguments use for software is a general problem and not railway specific.

The Therac-25 was a radiation therapy machine which was involved in at least six incidents between 1985 and 1987, in which patients were given massive overdoses of radiation. The operational experience as reported by Thomas (1994) started in 1982 with a total of 11 units. The details have not been reported, but, for the example, we assume that each unit was in operation on average for less than half the period between 1983 and 1987, say, in total 20 unit-years. Assuming 250 operating days per year and 20 patients per day, we might assume 100,000 individual cases of treatment. Assuming 30 minutes of treatment per case, this would mean 50,000 hours of treatment. For the sake of simplicity of the argument, let us assume that all qualitative requirements (see Chapter 19.4.1 through 19.4.5) for proven-in-use arguments are fulfilled (which was probably not true in this case).

Let us assume that we would pursue a proven-in-use argument on June 2, 1985, the day before the first incident, and that half the reported operational experience would have been gained. This would amount to about 50,000 flawless operating hours, which is an order of magnitude far removed from the minimum number of operating hours demanded for an SIL 1 claim (IEC 61508-7 [2010]) (300,000 operating hours without dangerous failures would be required). For a comparison, let us regard the data collected until January 17, 1987. Here, we would have doubled the number of operating hours, but we had six failures and probably software modifications, so the results would be even worse.

Two particular software failures were reported as root causes of the incidents (see the study by Thomas [1994]):

1. One failure only occurred when a particular nonstandard sequence of keystrokes was entered on the terminal within 8 s. This sequence of keystrokes was improbable. It took some practice before operators were able to work quickly enough to trigger this failure mode.

2. In a second failure scenario, the software set a flag variable by incrementing it, rather than by setting it to a fixed nonzero value. Occasionally, an arithmetic overflow occurred, causing the flag to return to zero and the software to bypass safety checks.

These scenarios are similar in that there was a fault in the software from the very start, which was only triggered under a very particular environmental condition that is more or less of a random nature. So, it is reasonable to assume that the overall system behavior randomly appeared to an outsider, as it was triggered by an operational condition (e.g., particular treatment or particular set of parameters) and an error condition in the software which was triggered only with a particular probability (e.g., probability that the flag returned to 0). So, the overall failure behavior could be described by a possibly time-dependent failure rate $\lambda(t)$. Ariane 4 was an expendable launch system which had 113 successful launches between 1988 and 2003 (and three failures) (Lions 1996). The test flight of its successor Ariane 5 flight 501 failed on June 4, 1996, because of a malfunction in the control software. A data conversion caused a processor trap (operand error). The software was originally written for Ariane 4 where efficiency considerations (the computer running the software had an 80% maximum workload requirement) led to four variables being protected with a handler while three others, including the horizontal bias variable, were left unprotected because it was thought that they were "physically limited or that there was a large margin of error," (Lions 1996, Chapter 2.2). The software, written in Ada, was included in Ariane 5 through the reuse of an entire Ariane 4 subsystem. As a result of a qualitative proven-in-use argument, the extent of testing and validation effort was reduced. Ariane 5 reused the inertial reference platform, but the flight path of Ariane 5 considerably differed from that of Ariane 4. Specifically, greater horizontal acceleration of Ariane 5 caused the computers on both the backup and primary platforms to crash and emit diagnostic data misinterpreted by the autopilot as spurious position and velocity data.

Let us assume that the component worked error free in all 116 missions (the three losses had other causes) and that a mission would last on average 1 hour. It is clear that this operational experience is far removed from any reasonable SIL claim, but it is interesting to note that in this example, the change in operational environment would have invalidated any argument based on operational experience. In addition, the component was integrated into a different system, which would also have to be argued as insufficiently similar.

19.6 Experience

Besides these well-known examples, the authors have been part of at least three projects, where attempts have been made to qualify a piece of software according to a certain SIL.

The first example has been a piece of software in railway technology, where it could be shown that the software fulfilled the SIL 1 criteria. This was easy since there was a good and reliable database of all registered software faults and errors, none of which was dangerous. Also, the in-service time of a large number of installations was known. These data were easily accessible, since the railway infrastructure provider had them all at hand.

In a second example, it took 3 months to divide the registered failures into dangerous and safe ones. This required a lot of work, but has led to success at the end.

In a third example for a plant control system, the failures statistics was quite easy; there was also a sufficient number of installations with a sufficient number of in-service hours, but it was not possible to collect the evidence for the service hours. This was especially a pity, since at a first glance, the information was there, but the formal proof was not possible within reasonable time.

Unfortunately, the authors are not allowed to disclose details of these examples.

19.7 Summary and Conclusions

We have seen in the previous section that different and independent approaches lead to the same result, a limiting Poisson distribution, under very general requirements. And we are sure that if we put in more work, we could even expand the results.

The situation is very similar to the central limit theorems in statistics. If there is no dominant influence or considerable dependence among the random variables, their normalized sum then tends to a normal limit distribution under very general assumptions. If some of the assumptions do not hold or we combine them with other functions, other limiting distributions would then occur, e.g., extreme value distributions.

So, we are back to the well-known Poisson assumptions, but we can relax them due to the limit theorem quoted earlier. It is well known that the Poisson process arises when the following four assumptions hold:

1. Proportionality: The probability of observing a single event over a small interval is approximately proportional to the size of that interval.

2. Singularity: The probability of two events occurring in the same narrow interval is negligible.

3. Homogeneity: The probability of an event within a certain interval does not change over different intervals.

4. Independence: The probability of an event in one interval is independent of the probability of an event in any other nonoverlapping interval.

For a particular application, it is sufficient that these assumptions approximately hold and are plausible. The singularity requirement needs no discussion; it will be fulfilled in any system with reasonable operational experience. The proportionality and homogeneity assumptions may be weakened, still resulting in a Poisson process, as the limit theorems hold for even the superposition of general renewal processes. However, the components must be well tested and be reasonably error free as the limit theorems demand rare failure events. The independence and the stationarity requirements are more restrictive. They mean in particular that the environment must not significantly change and that the components regarded should be very similar. In particular, the Ariane 4 example has illustrated the problem concerning changes in the environment. And Therac 25 has shown that when the system itself is changed, the operational experience may be very limited and may not even be sufficient to claim SIL 1. In both examples, we have assumed a working complaint management system which would be another issue to be discussed.

Since many standards (e.g., see IEC 61508) and approaches use the Poisson process and the assumptions stated in the preceding section, we can now specify the requirements that must be necessarily fulfilled to apply such a proven-in-use approach:

- The main requirement is the random influence of the environment on the component and that random influence must be representative. This means that a component, which is used for a proven-in-use argument, must be in a typical environment for this type of component, and the environment must be such that the influences are typical for this type of use, including all the changes in it. So, all components must be operated at least in similar environments. And all compared components must be similar. In particular, this means that if bugs are fixed, the component has usually to be regarded as a different component.
- All failures must be recorded, and it must be possible to observe them.
- All lifetimes and times of use must be recorded. It must also be recorded when a unit is out of service for a certain time interval and when it is not used any more.

All these three points are nothing else than the requirements for a good statistical sample as can be found in good statistical textbooks, e.g., the study by Storm (2007), and they do not differ from earlier results such as those from Gall, Kemp, and Schäbe (2000) or Littlewood (1979).

It is very interesting that the conclusion was already known by Littlewood (1975): "This amounts to giving *carte blanche* to the reliability engineer to assume a Poisson distribution in such situations." And another conclusion is still valid: " ... it is common practice to make this assumption willy-nilly—the results contained here simply confirm the engineer's intuition." We wholeheartedly agree with this conclusion and so should standardization committees, making life for engineers as easy as possible with minimum requirements, but having the assessors ensure that these requirements are really fulfilled. That this may still not be an easy task is finally illustrated by examples.

Regarding practical application, the expert must be cautious. He/she will have to deal with two main problems:

- The mass of data which might need to be sorted out manually
- Providing evidence of absence of failures and the service times

So, the approach is feasible when the conditions stated in Section 19.4 are obeyed—but not always.

We were also able to show that the recent results are in line with many years of practice and the result of a former study. This, however, is not a surprise. General statistical laws as the law of large numbers or—in our case, Grigelionis's theorem—do not change together with new technical development. They are as general as, e.g., gravitation and keep on acting, so we can rely on them.

References

Ascher, H., and Feingold, H. (1984): *Repairable Systems Reliability: Modelling, Inference, Misconceptions and Their Causes*, Marcel Dekker, New York.

Barlow, R. E., and Proschan, P. (1975): *Statistical Theory of Reliability*, Holt, Rinehart and Winston, New York.

Birolini, A. (2013): *Reliability Engineering*, Springer, Heidelberg, 7th edition.

CEN (European Committee for Standardization). EN 50126/IEC 62278: *Railway Applications—The Specification and Demonstration of Reliability, Availability, Maintainability and Safety (RAMS)*, 1999, correction 2010. CEN, Brussels.

CEN. EN 50128: *Railway Applications—Communication, Signalling and Processing Systems—Software for Railway Control and Protection Systems*, 2011. CEN, Brussels.

CEN. EN 50129: *Railway Applications—Communication, Signalling and Processing Systems—Safety-related Electronic Systems for Signalling*, 2003. CEN, Brussels.

DIN (Deutsches Institut für Normung e.V.). DIN V VDE 0801: *Grundsätze für Rechner in Systemen mit Sicherheitsaufgaben*, 1990. DIN, Berlin.

DIN. DIN V VDE 0801 A1: *Grundsätze für Rechner in Systemen mit Sicherheitsaufgaben, Anhang 1*, 1994. DIN, Berlin.

IEC (International Electrotechnical Commission). IEC 61508-1: *Functional Safety of Electrical/Electronic /Programmable Electronic Safety-related Systems—Part 1: General Requirements*, 2010. IEC, Geneva.

IEC. IEC 61508-7: Functional Safety of Electrical/Electronic/Programmable Electronic Safety-related Systems—Part 7: Overview of Techniques and Measures, 2010. IEC, Geneva.

IEC. IEC 60605-4:2001: *Equipment Reliability Testing—Part 4: Statistical Procedures for Exponential Distribution: Point Estimates, Confidence Intervals, Prediction Intervals and Tolerance Intervals*, 2001. IEC, Geneva.

Gall, H., Kemp, K., and Schäbe, H. (2000): Betriebsbewährung von Hard- und Software beim Einsatz von Rechnern und ähnlichen Systemen für Sicherheitsaufgaben, Bundesanstalt für Arbeitsschutz und Arbeitsmedizin.

Grigelionis, B. (1963): On the convergence of sums of random step processes to a Poisson process, *Theory of Probability and Its Applications* 8, 177–182.

Hölscher, H. (1984): *Mikrocomputer in der Sicherheitstechnik*, Holger Hölscher, TÜV Rheinland Verlag GmbH, Köln.

KTA (Kerntechnischer Ausschuss) 3507: Sicherheitstechnische Regeln des KTA, Werksprüfung, Prüfung nach Instandsetzung und Nachweis der Betriebsbewährung für leittechnische Einrichtungen des Sicherheitssystems, Fassung 11/1986, KTA.

Ladkin, P., and Littlewood, B. (2015): *Practical Statistical Evaluation of Critical Software*, Preprint, http://www.rvs.uni-bielefeld.de/publications/Papers/LadLitt20150301.pdf.

Lions, J. L. (1996): *Ariane 5 Flight 501 Failure—Report by the Enquiry Board*, Enquiry Board.

Littlewood, B. (1975): A reliability model for systems with Markov structure, *Journal of Applied Statistics* 24, No. 2, 172–177.

Littlewood, B. (1979): A software reliability model for modular program structure, Supplement, *IEEE Transactions on Reliability*, 28, No. 3, 103–110.

Littlewood, B., and Strigini, L. (1993): Validation of ultra-high-dependability for software-based systems, *Communications of the ACM* 36, No. 11, 69–80.

Mann, N. R., Schafer, R. E., and Singpurwalla, N. D. (1974): *Methods for Statistical Analysis for Reliability and Life Data*, John Wiley & Sons, Hoboken, NJ.

Misra, K. B. (1992): *Reliability Analysis and Prediction*, Elsevier, Amsterdam.

Musa, J. D., Iannino, A., and Okumoto, K. (1987): *Software Reliability*, McGraw-Hill, New York.

Mü8004: Eisenbahn-Bundesamt, Zulassung von Sicherungsanlagen, Teil 12000, Erprobung von Sicherungsanlagen, 1995, Eisenbahnbundesamt.

Schäbe, H. (2013): A simple model of the software failure rate, *Proceedings of the 20th AR2TS Symposium*, May 21–23, 2013, pp. 434–440.

Schäbe, H., and Braband, J. (2016): Basic requirements for proven-in-use arguments, *Risk, Reliability and Safety: Innovating Theory and Practice*, L. Walls, M. Revie, and T. Bedford (Eds), 2017 Taylor & Francis, London, pp. 2586–2590.

Storm, R. (2007): Wahrscheinlichkeitsrechnung, mathematische Statistik und statistische Qualitätskontrolle, Hanser Fachbuch.

Thomas, M. H. (1994): The story of the Therac-25 in LOTOS, *High Integrity Systems Journal*, 1, No. 1, 3–15.

20

Target Reliability for New and Existing Railway Civil Engineering Structures

Miroslav Sykora, Dimitris Diamantidis, Milan Holicky, and Karel Jung

CONTENTS

20.1 Introduction

Risk and reliability acceptance criteria are well established in many industrial sectors such as the offshore, chemical, or nuclear industries. Comparative risk thresholds have thereby been specified to allow a responsible organization or regulator to identify activities which impose an acceptable level of risk on the participating individuals or on the society as a whole. The following sections complement the previous description of estimation and evaluation of risk and, particularly, the risk acceptance criteria analyzed in Chapter 8.

Presented target reliability criteria for new and existing railway engineering structures are based on acceptable societal risks. The structures under consideration are initially classified. Civil engineering structures such as bridges, station buildings, and tunnels are taken into account, while electrical and mechanical components such as signaling, ventilation, and electrification systems are not considered herein. The general concepts for risk acceptance are then briefly reviewed, particularly in their relation to the target reliability level. The difference between new and existing structures is highlighted and the target reliability levels in current standards are discussed.

20.2 Classification of Structures

Railway civil engineering structures considered herein cover a wide range of constructions and can be divided into the following:

1. Track-carrying structures such as bridges, viaducts, trestles, or earth structures
2. Ancillary structures such as tunnels, stations, towers, platforms, or loading docks

The aforementioned structures can be classified according to the design working life—for example, 50 years for station buildings and 100 years for bridges and tunnels. They can also be classified according to the possible consequences in case of failure, i.e., minor, medium, and large consequences, as shown in the following sections. The second type of classification is for the determination of the target reliability of importance: structures with large potential consequences are associated with higher target reliability. Figure 20.1 provides the classification of civil engineering structures according to various normative documents.

Focusing on railway bridges, tunnels, and buildings in stations and the most detailed classification according to ISO 2394 (ISO 2015), the following examples are provided for the different consequence classes (CCs):

- CC1 (description of expected consequences according to the International Organization for Standardization [ISO] standard—predominantly insignificant material damages): secondary station buildings
- CC2 (material damages and functionality losses of significance for owners and operators but with little or no societal impact; damages to the qualities of the environment of an order which may be completely restored in a matter of weeks; expected number of fatalities $N < 5$): bridges on local railway lines, common buildings stations
- CC3 (material losses and functionality losses of societal significance, causing regional disruptions and delays in important societal services over several weeks; damages to the qualities of the environment limited to the surroundings of the failure event and which may restored in a matter of weeks; $N < 50$): all structures not classified in CC1, CC2, CC4, and CC5
- CC4 (disastrous events causing severe losses of societal services and disruptions and delays at national scale over periods of the order of months; significant damages to the qualities of the environment contained at national scale but spreading significantly beyond the surroundings of the failure event and which may only be partly restored in a matter of months; $N < 500$): major bridges and all tunnels on major railway lines, key buildings of major stations
- CC5 (catastrophic events causing losses of societal services and disruptions and delays beyond national scale over periods in the order of years; significant damages to the qualities of the environment spreading significantly beyond national scale and which may only be partly restored in a matter of years to decades; $N > 500$): uncommon for railway engineering structures

ISO 2394:2015	1	2	3	4	5
	Low rise buildings where only few people are present, minor wind turbines, stables etc.	Most buildings up to 4 stories, normal industrial facilities, minor bridges, major wind turbines, smaller or unmanned offshore facilities, etc.	Most buildings up to 15 stories normal bridges and tunnels, normal offshore facilities, larger and or hazardous industrial facilities	High rise buildings, grandstands, major bridges and tunnels, dikes, dams, major offshore facilities, pipelines, refineries, chemical plants, etc.	Buildings of national significance, major containments and storages of toxic materials, major dams and dikes, etc.

UIC 777-2	**B** - Superstructures not supporting elevated structures such as:
	  - roadways
	  - road bridges
	  - railway bridges
	  - footbridges and similar structures
	
	Single-storey structures not providing long-term occupancy:
	  - parking areas,
	  - warehouses.
				A. Superstructures supporting elevated structures that are permanently occupied:
				- offices,
				- lodgigs,
				- business premises.
				Structures for short-term gathering of people:
				- theatres
				- cinemas.
				Class A also includes multi-storey structure subject to short-term occupancy:
				- multi-story car parks
				- warehouses.	

EN 1991-1-7 4.5 Accidental actions caused by derailed rail traffic	Massive structures that span across or near the operational railway such as bridges carrying vehicular traffic or single story buildings that are not permanently occupied or do not serve as a temporary gathering place for people.		Structures that span across or near to the operational railway that are either permanently occupied or serve as a temporary gathering place for people or consist of more than one story.		
EN1991-1-7 Annex A	CC1 (deemed to correspond to CC1 and CC2 in ISO 2394)	CC2a	CC2b	CC3 (ISO 2394 CC4 and CC5)	
		(ISO 2394 CC3)			

FIGURE 20.1

Classification of Civil Engineering Structures

(Continued)

	Single occupancy houses not exceeding 4 stories. Agricultural buildings. Buildings into which people rarely go, provided no part of the building is closer to another building, or area where people do go, than a distance of 5.5-times the building height.	5 story single occupancy houses. Hotels not exceeding 4 storys. flats, apartments and other residential buildings not exceeding 4 storys. Offices not exceeding 4 storys. Industrial buildings not exceeding 3 storys. Retailing premises not exceeding 3 storys of less than 1000 m² floor area in each story. single story educational buildings all buildings not exceeding two storys to which the public are admitted and which contain floor areas not exceeding 2000 m² at each story.	Hotels, flats, apartments, and other residential buildings greater than 4 stories but not exceeding 15 stories. Educational buildings greater than single story but not exceeding 15 stories. Retailing premises greater than 3 stories but not exceeding 15 stories. Hospital not exceeding 3 stories. Offices greater than 4 stories but not exceeding 15 stories. All building to which the public are admitted and which contain floor areas exceeding 2000 m² but not exceeding 5000 m² at each story. Car parking not exceeding 6 stories.	All buildings defined above as class 2 lower and upper consequences class that exceed the limits on area and number of storeys. All buildings to which members of the public are admitted in significant numbers. Stadia accommodating more than 5000 spectators buildings containing hazardous substances and /or processes
EN 1990	CC1 (deemed to correspond to ISO 2394 CCI)	CC2 (ISO 2394 CC2 and CC3)		CC3 (ISO 2394 CC4 and CC5)
	Agricultural buildings where people do not normally enter (e.g., storage buildings, greenhouses).	Residential and office buildings, public buildings where consequences of failure are medium (e.g., an office building).		Grandstands, public buildings where consequences of failure are high (e.g., a concert hall).

FIGURE 20.1 (CONTINUED)
Classification of Civil Engineering Structures

20.3 Risk Acceptance Criteria

As mentioned in the introduction and in Chapter 8, explicit risk acceptance criteria are commonly applied in many industrial sectors in order to provide either a quantitative decision tool for the regulator or a comparable requirement for the industry when dealing with the certification/approval of a particular structure or system. The following criteria are applied and analytically described, for example, by the Health and Safety Executive (HSE 2001) and Centraal Invorderings Bureau (CIB 2001) and in the previous chapters (mainly Chapter 8):

1. *Individual risk criteria:* No individual (or group of individuals) involved in a particular activity can be exposed to an unacceptable risk; a typical value is 10^{-6} per year. If a worker or a member of the public is found to be exposed to excessive risk, safety measures are adopted regardless of the cost–benefit effectiveness.

2. *Group risk criteria:* A certain activity must not produce high-frequency occurrences of large-scale accidents (i.e., with particularly severe consequences). This means that the unacceptable level of risk varies for different accident magnitudes. This principle attempts to capture a supposed sociopolitical aversion to large accidents and provides a regulatory basis (i.e., enforced investments in safety) in situations where the other criteria do not call for intervention.

The requirements based on criterion 1 can be significantly affected by the relative time fraction for which a person occupies or uses a structure. For railway civil engineering structures, it is assumed that this fraction is commonly low (exceptions may include most exposed workers), individual risk criteria become less important and group risk criteria dominate the derivation of target reliability values (JCSS 2001a; Steenbergen et al. 2015; Caspeele et al. 2016). These criteria include human, economic, and environmental criteria and are briefly reviewed in the following sections.

20.3.1 Human Risk

The group risk is often represented in the form of a numerical *F–N* curve where *N* represents the number of fatalities and *F* is the frequency of accidents with more than *N* fatalities (HSE 2001; CIB 2001). This curve shows the probability of exceedance as a function of the number of fatalities *N*, commonly using a double logarithmic scale (Jonkman, van Gelder, and Vrijling 2003):

$$1 - F_N(x) = P(N > x) = \int_x^\infty f_N(\xi)\,d\xi, \tag{20.1}$$

where $f_N(x)$ is the probability density function of number of fatalities per year, and $F_N(x)$ is the cumulative distribution function, the value of which gives a probability of less than *x* fatalities per year. A simple measure for human risk is the annual expected value of the number of fatalities, which is frequently used to compare alternative projects in terms of their inherent risk with respect to human safety.

Typical *F–N* curves reported in the literature show different patterns for the same industrial activity in various countries or for different industrial activities in the same country. The following general formula has been proposed to represent the group human risk acceptance criterion:

$$F \le aN^{-k}, \tag{20.2}$$

where a and k are predefined constants that can be related to statistical observations from natural and human-made hazards (Vrouwenvelder, Leira, and Sykora 2012). Note that F is cumulative frequency. Some natural hazards show relationships with k slightly smaller than unity, while most human-made hazards are described by a relationship with $k > 1$. From statistical observations the constants a and k widely vary depending on the type of hazard and the type of technical activity. It was proposed to set the constants in such a way that the curve envelops the curves for most natural hazards and some of the more common manmade hazards (JCSS 2001a). For acceptable risks related to structural failures, the constant would be around $a = 10^{-6}$ and for marginally acceptable risks $a = 10^{-4}$; $k = 1$ represents risk-neutral curves, and $k > 1$ describes curves with risk aversion and $k < 1$ curves with risk proneness. The case of $k < 1$ leads to infinitely large expected losses (in terms of lives or cost) and is therefore not acceptable.

The constant a represents the frequency of occurrence of events with one and more fatalities; commonly, annual values are considered. Its value should be consistent with the reference system to which the criterion in Equation 20.2 is applied. The reference system can range from a group of structures to an individual structural member and can include other structure-specific and industry-specific parameters (Tanner and Hingorani 2015). For railway infrastructures, a national scale is assumed; for details, see CIB (2001); Tanner and Hingorani (2015); and Vrijling, van Gelder, and Ouwerkerk (2005).

Based on the F–N curves, the so-called ALARP—as low as reasonably possible—region can be defined by two limits (CIB 2001). The area above the upper limit represents the risk that is not tolerated in any circumstances, while the risk below the lower limit is of no practical interest. Such acceptability curves have been developed in various industrial fields, including the chemical and the transportation industries.

In the ALARP principle, the width between the upper and lower bound curves is of importance. This width is often two orders of magnitudes, allowing for excessive flexibility in practical cases. Examples of fatality criteria (F–N curves) are given in Figure 20.2, following the recommendations provided in the following:

- ISO 2394:1998 (superseded in 2015): $a = 10^{-2}$ per annum and $k = 2$—Examples provided as a risk acceptance criterion in structural design to avoid accidents where large numbers of people may be killed, deemed to be associated with a collapse of the whole structure

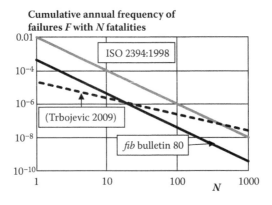

FIGURE 20.2
F–N curves relating expected fatalities (N) from an accidental event or failure and the cumulative annual frequency of occurrence (F) of events with more than N fatalities.

- The study by Trbojevic (2009): $a = 2 \times 10^{-5}$ per annum and $k = 1$—Criterion for a structure, the collapse of which endangers 100 persons
- *fib* COM3 TG3.1 (Caspeele et al. 2016): $a = 5 \times 10^{-4}$ per annum and $k = 2$—Criterion derived for a main load-bearing member of a road bridge in order to maintain safety levels associated with current best practice.

Note that the ISO and *fib* COM3 recommendations are based on the notion that N represents the expected number of fatalities in an event, and thus, a limiting probability does not have the cumulative character assumed in Equation 20.2. However, the numerical difference between these concepts is small, and thus, the ISO and *fib* COM3 curves are considered as F–N curves. The ISO recommendation is applied in some countries; for instance, Belgium and the Netherlands use it for the events with $N \geq 10$ fatalities.

The criteria in Figure 20.2 provides values ranging across more than two orders of magnitude. This is attributed to different reference systems for which these criteria are applicable. While the example in ISO 2394:1998 is considered to be applicable for large groups of structures (the level of the applicability of standards), the other two recommendations are focused on individual buildings (Trbojevic 2009) or key structural member of a bridge (*fib* Caspeele et al. 2016). Hence, the reference system is of key importance when establishing the criteria for human safety.

It is noted that human safety does involve not only fatalities but also injuries. In many studies, injuries are related to fatalities by using a multiplicative factor, for example, 0.1 for moderate injury and 0.5 for major injury. Based on this simple procedure, weighted fatalities can be obtained.

Diamantidis, Zuccarelli, and Westhäuser (2000) derived human risk acceptance criteria for the railway industry by assuming that the safety inherent in the traditional railways in the last two or three decades is acceptable. The safety target was therefore derived by analyzing recent risk history of the railway system in terms of frequency of occurrence of accidents and the extent of their consequences. The procedure generally used to estimate the risk associated with railway transport is based on the analyses of the frequency of occurrence of given consequences for a given accident. Diamantidis, Zuccarelli, and Westhäuser (2000) thereby classified consequences as follows:

- Medium consequences: 1–10 fatalities with an average value of 3 fatalities
- Severe consequences: 11–100 fatalities with an average value of 30 fatalities
- Catastrophic consequences: 101–1000 fatalities with an average value of 300 fatalities

The evaluation of the probability can be performed by assuming that accidents occur according to a Poisson process (Melchers 2001). In general, a Bayesian approach by which expert judgments can be combined with statistical data is recommended, due to the limited experience with accidents (Holicky, Markova, and Sykora 2013). This approach was applied to data of recorded accidents of the Italian, Austrian, and German railways (Diamantidis, Zuccarelli, and Westhäuser 2000). The obtained results are considered valid for a first definition of an acceptable safety level for the western European railway systems and are comparable to computed values for various tunnel projects. The results and the recommended acceptance criteria can be summarized as follows:

- Events of medium consequences are associated with an annual probability of 10^{-9} per train-kilometer
- Events of severe consequences are associated with an annual probability of 10^{-10} per train-kilometer
- Events of catastrophic consequences are associated with an annual probability of 10^{-11} per train-kilometer.

Example: Consider a railway tunnel with a length of 50 km and a daily traffic flow of 200 trains in both directions. The acceptable return periods associated with accidental events are derived from the aforementioned results as follows:

- Number of train-kilometers per year: $50 \times 200 \times 365 = 3.65 \times 10^6$
- Accidents associated with medium consequences: Acceptable annual probability of an accident $p \leq 3.65 \times 10^6 \times 10^{-9} = 0.00365$; acceptable return period $T \geq 1/p \approx 300$ years
- Severe consequences: 3000 years
- Catastrophic consequences: 30,000 years

The derived acceptable return period for medium consequences is approximately within the range of the working life of important infrastructures, such as a long tunnel. For catastrophic consequences, the return period is of the same order of magnitude as is considered when designing industrial plants.

The recommendations in Figure 20.2 can be represented in a so-called risk matrix. For that purpose, qualitative hazard probability levels suitable for use can be defined as shown in Table 20.1 (Diamantidis 2005), which provides an illustrative example for road tunnels. Hazard severity levels of accidental consequences are defined in Table 20.2.

TABLE 20.1

Hazard Probability Levels

Class	Frequency	Range (Events per year)
A	Frequent	>10
B	Occasional	1–10
C	Remote	0.1–1
D	Improbable	0.01–0.1
E	Incredible	0.001–0.01

TABLE 20.2

Hazard Severity Levels

Class	Severity Category	Human Losses
1	Insignificant	–
2	Marginal	Injuries only
3	Critical	1
4	Severe	10
5	Catastrophic	100

TABLE 20.3

Example of a Risk Acceptability Matrix

	1	2	3	4	5
A	ALARP	NAL	NAL	NAL	NAL
B	ALARP	ALARP	NAL	NAL	NAL
C	AL	ALARP	ALARP	NAL	NAL
D	AL	AL	ALARP	ALARP	NAL
E	AL	AL	AL	ALARP	ALARP

Note: AL: acceptable level; ALARP: as low as reasonably practicable; NAL: not acceptable level.

The hazard probability levels and the hazard severity levels can be combined to generate a risk classification matrix. The principle of the risk classification matrix is shown in Table 20.3. The railway authority is usually responsible for defining the tolerability of the risk combinations contained within the risk classification matrix.

The risk matrix procedure was often applied to assess risk in the transportation industry and is particularly useful in cases when limited accidental data are available (Diamantidis 2005). Recommendations on the use of risk matrices were provided (Duijm 2015).

20.3.2 Economic and Environmental Risk

Besides human safety, economic risk plays an important role in decision-making. Economic losses are direct consequences related, for example, to the repair of initial damage; replacement of structure and equipment; and indirect consequences such as loss of production, temporary relocation, rescue costs, or loss of reputation (Section 20.4). Analogous to the *F–N* curve and the expected number of fatalities, *F–D* curves, with *D* being the economic damage, can be established. Criteria for economic losses need to be specified considering case-specific conditions.

Environmental consequences can be presented in terms of permanent or long-term damage to terrestrial, freshwater, marine habitats, and groundwater reservoirs. Thereby, the parameter of damage can be the damaged area. A different parameter was selected (Norsok 1998) in which the recovery time from the accident defines the damage. The overall principle implies that recovery following environmental damage shall be of insignificant duration when compared to the expected period (return period) between such occurrences of damage; see Table 20.4 (Norsok 1998). The recommendations for environmental impact assessment for rail transport infrastructure are provided in the report of Mainline Consortium (2013).

TABLE 20.4

Environmental Risk Acceptance Criteria

Environmental Damage Category	Average Recovery	Acceptable Frequency Limit
Minor	6 months	<1 event per 10 years
Moderate	2 years	<1 event per 40 years
Significant	5 years	<1 event per 100 years
Serious	20 years	<1 event per 400 years

It appears difficult to combine human, economic, and environmental risk acceptance criteria due to the different metrics used. However, if these metrics are expressed in monetary values, decision criteria can be derived.

20.4 Optimization of Structural Reliability

20.4.1 Reliability Measure

The term *reliability* is often vaguely used and deserves some clarification. Commonly the concept of reliability is conceived in an absolute (black and white) way—the structure is or is not reliable. In accordance with this approach, the positive statement is frequently understood in the sense that a failure of the structure will never occur. Unfortunately, this interpretation is an oversimplification. The interpretation of the complementary (negative) statement is usually understood more correctly: failures are accepted as a part of the real world, and the probability or frequency of their occurrence is then discussed. In fact, in the design, it is necessary to admit a certain small probability that a failure may occur within the intended life of the structure. Fundamental requirements in current standards such as Eurocodes include the statement that a structure shall be designed and executed in such a way that it will, during its intended life, with appropriate degrees of reliability and in an economic way, sustain all actions and influences likely to occur during execution and use, and remain fit for the use for which it is required.

The basic reliability measures include probability of failure and reliability index; see EN 1990 (CEN 2002). The probability of structural failure p_f can be generally defined for a given limit state of structural member such as bending or shear resistance as

$$p_f = P\{Z(\mathbf{X}) < 0\}. \tag{20.3}$$

The limit state (performance) function $Z(\mathbf{X})$ is formulated in such a way that the reliable (safe) domain of a vector of basic influencing variables $\mathbf{X} = X_1, X_2, \ldots, X_n$ such as action effects, material properties, or geometry corresponds to the inequality $Z(\mathbf{X}) > 0$, while the failure domain, to the complementary inequality $Z(\mathbf{X}) < 0$. A simple example of $Z(\mathbf{X})$ describes the basic relationship between a load effect E and resistance R:

$$Z(\mathbf{X}) = R - E. \tag{20.4}$$

It is noted that the exceedance of the limit state is related to the failure of a component of a structure. The structure is represented by a system of components and might survive when a component (column, slab, beam, etc.) fails. Probabilistic models for the influencing variables can be found in the literature and can be implemented for the calculation of failure probability. General guidance for probabilistic modeling provides the probabilistic model code of the Joint Committee on Structural Safety (JCSS 2001b); recommendations for the reliability analysis of railway bridges are given by Wisniewski, Casas, and Ghosn (2009); Casas and Wisniewski (2013); and Moreira et al. (2016) and by the reports of the Sustainable Bridges project (http://www.sustainablebridges.net). System reliability methodologies can be applied to compute the reliability of a structural system.

Instead of failure probability p_f, the reliability index β is used as an equivalent measure to p_f:

$$p_f = \Phi(-\beta) \approx 10^{-\beta}, \tag{20.5}$$

where Φ is the distribution function of standardized normal distribution; see, for example, the books by Melchers (2001) and Holicky (2013a).

Reliability acceptance criteria can be defined in terms of acceptable failure probability p_t or acceptable reliability index β_t. These quantities are recommended as reasonable minimum requirements, and it is emphasized that p_t and β_t are formal conventional quantities only and may not correspond to the actual frequency of failures.

In reliability analysis of a structure, it is generally required that

$$p_f \leq p_t \tag{20.6}$$

or, equivalently, in terms of reliability index,

$$\beta \geq \beta_t, \tag{20.7}$$

where p_t is the specified design (target) failure probability corresponding to the target reliability index β_t. The target reliability is different for each structure depending on the potential consequences of failure and thus reflects the aforementioned risk acceptability criteria. Therefore, it is appropriate to classify structures bearing this in mind as emphasized in Section 20.2 and thus reflect the risk acceptance criteria summarized in Section 20.3. The optimum level of target reliability is discussed next.

20.4.2 Cost Optimization

Life cycle cost aspects in railway management are presented in Chapter 14. The optimization of the target failure probability is addressed in this section (ISO 2394 [ISO 2015]) and indicates that the target level of reliability should depend on a balance between the consequences of failure and the costs of safety measures. From an economic point of view, the objective is to minimize the total working-life cost. The expected total costs C_{tot} may be generally considered as the sum of the expected structural cost, costs of inspections and maintenance, and costs related to failure (malfunction) of a structure. The decision parameter(s) d to be optimized in structural design may influence resistance, serviceability, durability, maintenance, inspection strategies, etc. Examples of d include shear or flexural resistances and stiffness of a girder to control deflections. In this chapter, the decision parameter is assumed to represent structural resistance affecting ultimate limit states. Moreover, the benefits related to the use of the structure that, in general, should be considered in the optimization are assumed to be independent of a value of the decision parameter.

The structural cost consists of the following:

- Cost C_0 independent of the decision parameter (surveys and design, temporary and assembly works, administration and management, etc.)
- Cost $C_1(d)$ dependent on the decision parameter; normally, the linear relationship can be accepted $C_1 \times d$

In general, the former cost significantly exceeds the latter, i.e., $C_0 \gg C_1 \times d$; see the study by Rackwitz (2000) and ISO 15686-5 (ISO 2008).

The failure cost C_f—the cost related to consequences of structural failure—may include, depending on a subject concerned (ISO 2394 [ISO 2015]), the following:

- Cost of repair or replacement of the structure
- Economic losses due to nonavailability or malfunction of the structure, as discussed before
- Costs of injuries and fatalities that can be expressed, e.g., in terms of compensations or insurance cost
- Unfavorable environmental effects (CO_2 emissions, energy use, release of dangerous substances, as related to the aforementioned environmental risk)
- Other (loss of reputation, introducing undesirable nonoptimal changes of design practice)

The estimation of the failure cost is a very important, but likely the most difficult, step in cost optimization. According to ISO 2394 (ISO 2015), not only direct consequences of failure (those resulting from the failures of individual components), but also follow-up consequences (related to malfunction of a whole structure in a railway network) should be included. Detailed guidance for failure consequence estimation is provided by Mainline Consortium (2013). Particularly for a public authority as a decision maker, human compensation costs can be expressed by a Life Quality Index (LQI)-based societal value of a statistical life SVSL (Narasimhan et al. 2016). The SVSL corresponds to the amount which should be compensated for each fatality. The LQI is discussed in the following.

For consistency, the structural and failure costs need to be expressed on a common basis. This is achieved by converting the expected failure costs, related to a working life t, to the present value (Holicky 2013b):

$$E[C_f(t,d)] \approx C_f p_f(d) Q(t,d), \tag{20.8}$$

where C_f is the present value of the failure cost; $p_f(d)$ is the failure probability related to a basic reference period for which failure events can be assumed independent—typically 1 year as is assumed hereafter (Holicky, Diamantidis, and Sykora 2015; Diamantidis, Holicky, and Sykora 2016); and Q is the time factor.

The expected total costs are expressed as

$$E[C_{tot}(t;d)] = C_0 + C_1 \times d + C_f p_f(d) Q(t,d). \tag{20.9}$$

The optimum value of the decision parameter d_{opt} (optimum design strategy) is then obtained by minimizing the total cost. Apparently, d_{opt} is independent of C_0. Following the JCSS probabilistic model code JCSS 2001b, annex G of ISO 2394 (ISO 2015), indicates the target reliabilities based on economic (monetary) optimization. More details are provided in the recent papers by the authors (Holicky 2013b; Holicky 2014; Sykora et al. 2016).

The time factor Q is obtained for a sum of the geometric sequence (Holicky 2013b; Holicky 2014):

$$Q(t_{ref}, d) = \frac{1 - \left[\left(1 - p_f(d)\right)/(1+q)\right]^{t_{ref}}}{1 - \left[\left(1 - p_f(d)\right)/(1+q)\right]},$$ (20.10)

where q is the annual discount rate for which ISO 15686-5 (ISO 2008) assumes values between 0 and 0.04. ISO 2394 (ISO 2015) suggests 0.01–0.03 for most Western economies and 0.05–0.08 for most Asian economies. Lentz (2007) discussed in detail discounting for private and public sectors, indicating values around 0.05 for a long-term average of q, while Lee and Ellingwood (2015) suggested lower values for investments over multiple generations. As an example of public administration recommendations, the UK Treasury in its Green Book assumes a discount rate of 3.5% for public sector projects, particularly for existing infrastructures. For time periods longer than 30 years, the use of declining long-term discount rates is advised –3.5% for up to 30 years and linear decrease to 1.0% for 300 years. In general, a high discount rate leads to favoring options with low capital cost, short life, and high maintenance costs.

The LQI (Nathwani, Pandey, and Lind 2009) and several other metrics were derived to support decisions related to allocations of available public resources between and within various societal sectors and industries. The LQI is an indicator of the societal preference and capacity for investments in life safety expressed as a function of gross domestic product, life expectancy at birth, and ratio of leisure to working time; see ISO 2394 (ISO 2015).

The ISO standard provides detailed guidance on how the preferences of society with regard to investments into health and life safety improvements can be described by the LQI concept. The target level is derived by considering the costs of safety measures, the monetary equivalent of societal willingness to save one life, and the expected number of fatalities in the event of structural failure. Essentially, this approach combines economic and human safety aspects. Compared with economic optimization, it should lead to lower target reliability indices, as only the human consequences of structural failure are taken into account, while other losses such as economic and environmental costs are disregarded. In principle, the LQI approach weighs the value of the expected casualties of a certain activity against the economic value of the activity. In such an analysis, the danger to which the people are subjected might vary on an individual basis within the group of people affected, which may be deemed unethical. Examples of application of the LQI approach are provided by Sykora et al. (2016); Smit and Barnardo-Vijloen (2013); and Fischer, Barnardo-Viljoen, and Faber (2012).

20.4.3 Example: Optimum Strategy of Railway Bridge Design

The application of the procedure is illustrated by the reliability optimization of a key structural member of a railway bridge with a given design working life *T*. The structural member under consideration is exposed to permanent and railway traffic loads. Deterioration is not taken into account. The decision parameter is the ratio of resistance of the member over the resistance required by Eurocodes for design using the partial factor method and considering the required target reliability level (Caspeele et al. 2013; Sykora, Holicky, and Markova 2013). The failure probability is obtained as follows:

$$p_f(d) = P[K_R R(d) - K_E(G + Q) < 0].$$ (20.11)

TABLE 20.5

Models for Basic Variables

Variable	X	Dist.	X_k	μ_X/X_k	V_X	Reference
Resistance	R	lognormal	0.81	1.23	0.13	Casas and Wisniewski (2013)
Permanent load (self-weight and ballast)	G	normal	0.19	1.00	0.10	Wisniewski, Casas, and Ghosn (2009)
Traffic load (1 y.)	Q	Gumbel (max.)	0.57	0.80	0.10	Wisniewski, Casas, and Ghosn (2009)
Resistance model uncertainty	K_R	lognormal	1.00	1.00	0.10	Casas and Wisniewski (2013); Nowak (1999)
Load effect model uncertainty	K_E	lognormal	1.00	1.00	0.10	JCSS (2001b); Nowak (1999)

Note: X_k: characteristic value; μ_X: mean; V_X: coefficient of variation.

Characteristic values and probabilistic models of the basic variables are given in Table 20.5. When the requirements of Eurocodes are satisfied ($d = 1$), the reliability index is 3.5 for a reference period of 100 years.

The total cost optimization is based on the following assumptions:

1. The failure costs can be estimated using the available data (Kanda and Shah 1997):
   - $C_f/C_{str} \leq 3$ for small economic consequences of failure—Consequence class CC1 according to EN 1990 (CEN 2002)
   - $C_f/C_{str} \leq 20$ for considerable consequences (CC2)
   - $C_f/C_{str} \leq 50$ for very great consequences (CC3), where C_{str} is the initial structural cost comprising non-construction (site costs, property management, administration, etc.) and construction (temporary works, construction of a bridge, etc.) costs; for details, see ISO 15686-5 (ISO 2008).

2. Based on the economic considerations provided by Ang and De Leon (2005); McClelland and Reifel (1986); and Lee, Cho, and Cha (2006) and focusing on flexural resistance of steel girders, the initial cost $[C_0 + C_1(d)]/C_{str}$ can be estimated as follows:

$$[C_0 + C_1(d)]/C_{str} \approx 0.03d + 0.97, \text{ for } 0.66 < d < 1.75, \tag{20.12}$$

   where the term $0.03d$ expresses the contribution of cost of the optimized girder to the initial structural cost.

3. The discount rate ranges from $q = 0\%$ to $q = 3\%$.

As Equation 20.12 is derived for steel members, it provides a lower estimate of $C_0 + C_1(d)$ for most structures, which then yields conservative estimates of target reliability levels.

The total cost given in Equation 20.9 is optimized with respect to the decision parameter d, considering the failure probability obtained from Equation 20.11. Figure 20.3 shows the variation of the total cost and of reliability index with the decision parameter d for CC2 and CC3.

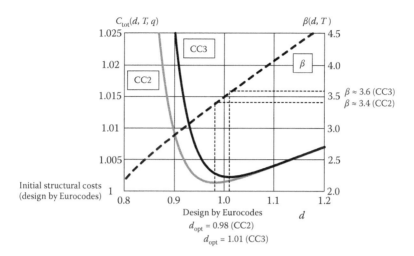

FIGURE 20.3
Variation of the total cost and reliability index with the decision parameter d for design working life $T = 100$ years and annual discount rate $q = 0.03$.

Considering purely economic criteria, it can be concluded that

- CC2: The optimum upgrade strategy corresponds well to the design based on Eurocodes ($d_{opt} \approx 1$); the corresponding optimum reliability level, related to a period of 100 years, is $\beta \approx 3.4$.
- CC3: The optimum upgrade strategy slightly exceeds the requirements of Eurocodes and the optimum reliability level increases to $\beta \approx 3.6$.

Figure 20.4 shows the variation of target reliability indices with a design working life T for CC2 and CC3. Two values of the annual discount rate—0 and 0.03—are taken into account. It is observed that for large infrastructure projects with long working lifetimes, the effect of the discount rate is moderate (0.35 in terms of reliability index) and should not be neglected.

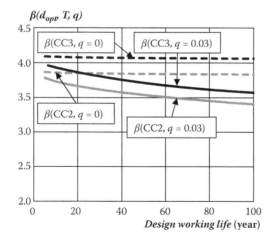

FIGURE 20.4
Variation of target reliability indices with design working life T for CC2 and CC3.

20.5 Implementation in Codes

Target reliability criteria are implemented in various international and national standards. In EN 1990 (CEN 2002) and ISO 2394 (ISO 2015), the index β is generally used as a measure of the reliability. The target reliability levels in codes of practice provide criteria for limit states that do not account for human errors, i.e., the target levels should be compared with the so-called notional reliability indicators; see ISO 2394 (ISO 2015). The target levels are differentiated in regard to various parameters. It is shown here that the target reliability can be specified by taking the following into account:

1. *Costs of safety measures*: Costs of safety measures should reflect efforts needed to improve structural reliability considering properties of construction materials and characteristics of investigated failure modes. The relative cost of safety measures significantly depends on the variability of load effects and resistances (JCSS 2001b; Vrouwenvelder 2002).

2. *Failure consequences*: Failure consequences are understood to cover all direct and indirect (follow-up) consequences related to failure including human, economic, and environmental impacts as discussed in Section 20.3. When specifying these costs, the distinction between ductile or brittle failure (warning factor), redundancy, and possibility of progressive failure (collapse) should be taken into account. In this way, it would be possible to consider the system failure in component design. However, such an implementation is not always feasible in practice, and therefore, consequence classes are specified with respect to the use of the structure in EN 1990 (CEN 2002) and to the number of persons at risk in ASCE 7-10 (American Society of Civil Engineers 2013). Detailed classification with respect to the importance of a structure and expected failure consequences provides the Australian and New Zealand standard AS/NZS 1170.0 (Standards Australia 2002).

3. *Time parameters*: Target levels are commonly related to a reference period or a design working life. The reference period is understood as a chosen period used as a basis for statistically assessing time-variant basic variables and the corresponding probability of failure. The design working life is considered here as an assumed period for which a structure is to be used for its intended purpose without any major repair work being necessary. The concept of reference period is therefore fundamentally different from the concept of design working life. Obviously, target reliability should always be specified together with a reference period considered in reliability verification. ISO 2394 (ISO 2015) indicates that the remaining working life can be considered as a reference period for the serviceability and fatigue limit states, while a shorter reference period might be reasonable for the ultimate limit states. When related to failure consequences, it is proposed here to refer to lifetime probabilities if economic consequences are decisive. When human safety is endangered, other reference periods such as 1 year are commonly accepted, as discussed in the example in the previous section.

EN 1990 (CEN 2002) recommends the target reliability index β for two reference periods (1 and 50 years); see the example for medium consequences of failure in Table 20.6. These target reliabilities are intended to be used primarily for the design of members of new structures. The two β values given in Table 20.6 are provided for two reference periods

TABLE 20.6

Target Reliability Indices according to Selected Standards

Standard	Failure Consequences	Reference Period	β in standard
EN 1990 (CEN 2002)	Medium	50 years (1 year)	3.8 (4.7)
ISO 2394 (ISO 2015)—economic optimization	Class 3	1 year	3.3/4.2/4.4[a]
ISO 2394 (ISO 2015)—LQI	–	1 year	3.1/3.7/4.2[a]
ISO 13822 (ISO 2010)	Moderate	Minimum standard period for safety	3.8

[a] High/moderate/low relative costs of safety measures, respectively.

used for reliability verification and should approximately correspond to the same reliability level:

- $\beta = 3.8$ should thus be used provided that probabilistic models of basic variables are related to the reference period of 50 years.
- The same reliability level should be approximately reached when $\beta = 4.7$ is applied using the related models for 1 year and failure probabilities in individual yearly intervals (basic reference periods for variable loads) are independent.

Considering an arbitrary reference period t_{ref}, the reliability level is derived from the annual target in accordance with EN 1990 (CEN 2002) as follows:

$$\beta_{tref} = \Phi^{-1}\left\{\left[\Phi(\beta_1)\right]^{tref}\right\},$$ (20.13)

where Φ^{-1} is the inverse cumulative distribution function of the standardized normal distribution, and β_1 is the target reliability index related to the reference period $t_{ref} = 1$ year.

Note that Equation 20.13 should be used with caution as the full independency of failure events in subsequent years (reference periods) is frequently not realistic.

20.5.1 Example: Target Reliability Levels for Different Working Lifetimes and Reference Periods

In this example, a railway bridge classified in CC2 (considerable economic consequences of failure) is to be designed considering a working life $T = 100$ years. The span of the bridge is 30 m. An appropriate target reliability level should be based on economic optimization for an annual discount rate $q = 0.015$; minimum human safety levels should not be exceeded.

Interpolation between $q = 0$ and 0.03 in Figure 20.4 leads to economically optimal $\beta \approx 3.6$ for $q = 0.015$ and $T = 100$ years. The *fib* bulletin (Caspeele et al. 2016) suggests for road bridges with a collapsed length of 10 m an annual reliability index of 3.3 to meet group risk criteria. In the absence of widely accepted recommendations for railway bridges, this suggestion is adopted here for the considered railway bridge.

When a reference period for reliability analysis is taken equal to the design working life (i.e., 100-year maxima of the traffic load are considered), the economically optimal value 3.6 is directly applicable, while the value for human safety is adapted using Equation 20.13:

$$\beta_{100} = \Phi^{-1}\left\{\left[\Phi(3.3)\right]^{100}\right\} = 1.7.$$

Apparently, it is more efficient to design for higher reliability levels than those required by the human safety criteria. For $t_{ref} = 100$ years, the target reliability index of 3.6 would then be selected. Note that for existing bridges, human safety criteria often dominate target reliabilities (Sykora et al. 2016); see the following.

When a reference period for reliability analysis is taken different from the design working life, the procedure needs to be modified. For a reference period of 1 year (considering annual maxima of the traffic load in reliability analysis), the annual reliability index 3.3 for human safety is compared with the economic optimum adjusted as follows:

$$\beta_{100} = \Phi^{-1}\left\{\left[\Phi(\beta_1)\right]^{100}\right\} = 3.6; \ \beta_1 = 4.7.$$

When compared to EN 1990 (CEN 2002), more detailed and substantially different recommendations are provided in the probabilistic model code by JCSS (2001b) and in ISO 2394 (ISO 2015). Recommended target reliability indices are also related to both the consequences and the relative costs of safety measures, however, for the reference period of one year. ISO 2394 (ISO 2015) gives target levels based on economic optimization and acceptance criteria using the LQI as exemplified in Table 20.6. The consideration of costs of safety measures is particularly important for existing structures. According to the ISO standard, the target level for existing structures apparently decreases as it takes relatively more effort to increase the reliability level compared to a new structure. Consequently for an existing structure, one may use the values of one category higher, i.e., instead of moderate, consider high relative costs of safety measures.

In ASCE 7-10 (American Society of Civil Engineers 2013) buildings and other structures are classified into four risk categories according to the number of persons at risk. Category I is associated with few persons at risk and category IV with tens of thousands. For all loads addressed by the standard excluding earthquake, ASCE 7-10 (American Society of Civil Engineers 2013) aims to reach target annual reliability of 3.7 for category I up to 4.4 for category IV. The Canadian Standards Association uses a different and slightly more detailed approach for bridges than the aforementioned documents by including additional factors such as inspectability.

ISO 13822 (ISO 2010) related to the assessment of existing structures indicates four target reliability levels for different consequences of failure (the ultimate limit states): small—2.3; some—3.1; moderate—3.8; and high—4.3. The related reference period is a minimum standard period for safety (e.g., 50 years).

The aforementioned values were derived for structural components. However, structural members designed by such a procedure may have sufficient reliability beyond these local failure conditions due to section and internal redundancies, possible alternative load paths, and/or formation of plastic hinges. Therefore, failures of single members often have significantly lower consequences compared to the failure of the main part of the structure (system failure).

It thus appears important to distinguish between member and system failure (Bhattacharya, Basu, and Ma 2001) when deriving target reliability levels. For instance, ASCE 7-10 (American Society of Civil Engineers 2013) requires target reliability index by about 0.5 greater for progressive collapse than for a member or connection failure (Hamburger 2013). This difference was also proposed for bridges to distinguish between single- or multiple-path failures (Kaszynska and Nowak 2005). It consequently seems to be appropriate to increase the target reliability index by 0.5 when the reliability of a system is verified. This recommendation relates to ductile systems with alternative load paths (parallel systems), as is relevant to many civil engineering structures. However, as the system behavior of structures can hardly be classified into idealized types of systems such as series, parallel ductile, or parallel brittle, detailed discussion on system targets is beyond the scope of this chapter.

20.5.2 Example: Optimization of Distance to Support

In addition to specification of target reliability levels, the principles of economic optimization and human safety criteria can support other reliability- and risk-related design decisions, as is demonstrated in the following example. It is focused on a bridge pier located in the vicinity of a railway line (Figure 20.5) and the related danger of accidental impact of a train on the pier. The lateral distance D the between centerline of the nearest track and the face of the pier is optimized considering the provisions of UIC 777-2 (International Union of Railways 2002). A brief description demonstrates the key principles; more details are provided by Jung and Marková (2004).

The train intensity on the double-track line is 200 trains per day. Preliminary risk analysis is based on the event tree method; Figure 20.6 provides the results of the analysis of probabilities and consequences for a train speed of 120 km/h and a minimum lateral $D = 3$ m. The conditional probabilities are specified following the guidance provide based on the recommendations of UIC 777-2 (International Union of Railways 2002):

- Derailment covers the situations when a railway vehicle runs off or lifts off the track or runs on two different pairs of tracks.
- Impact occurs when a railway vehicle hits an object other than a railway vehicle within the track clearance, e.g., running into a buffer stop or hitting an obstacle in the track clearance.
- Secondary collision refers to the impact of a railway vehicle on another railway vehicle.

The optimized distance D affects the probability of the impact of a derailed train on the bridge.

Cost optimization attempts to counterbalance the following:

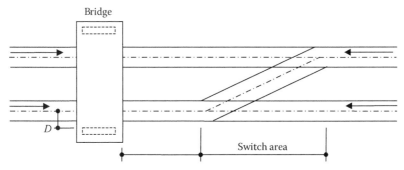

FIGURE 20.5
Plan view of the bridge.

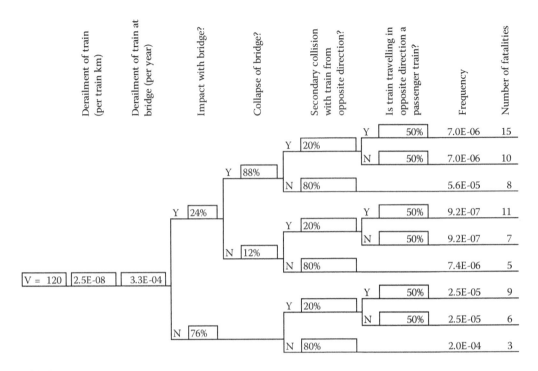

FIGURE 20.6
Analysis of probabilities and consequences for train speed of 120 km/h and $D = 3$ m.

- Cost of the bridge over the railway line, increasing with a span, i.e., with the lateral distance D
- Failure consequences covering economic losses and compensation costs related to harm to human health due to the impact of a train on the bridge pier; these consequences substantially decreasing with increasing D are specified for a design working life of the bridge of 100 years.

Cost of the bridge and failure consequences are based on the guidance provided by UIC 777-2 (International Union of Railways 2002).

The optimum bridge design strategies are as follows (rounded to meters):

- $D_{opt} = 3$ m for train speed of 120 km/h
- $D_{opt} = 3$ m for 160 km/h
- $D_{opt} = 9$ m for 230 km/h
- $D_{opt} > 15$ m for 300 km/h

Figure 20.7 shows the variation of total risk with the lateral distance D for working life of the bridge of 100 years and train speed of 230 km/h.

It remains to be verified whether the optimum design strategy complies with the criteria on human safety. Following the guidance given in Section 20.3.1, Figure 20.8 displays the F–N curves for different train speeds and $D_{opt} = 9$ m. For each train speed, probabilities of events for which N fatalities are expected are estimated; see Figure 20.6.

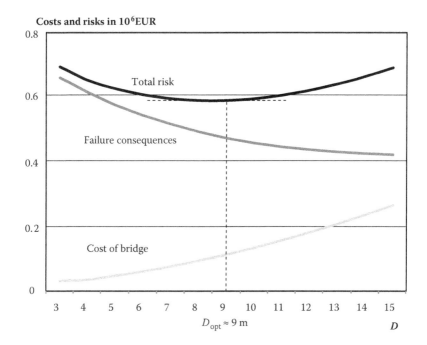

FIGURE 20.7
Variation of total risk with the lateral distance D for a working life of the bridge of 100 years and a train speed of 230 km/h.

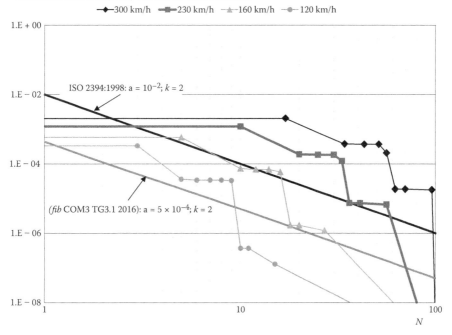

FIGURE 20.8
F–N curves for different train speeds, D_{opt} = 9 m, and acceptance criteria according to ISO 2394:1998 and COM3 TG3.1 (Caspeele et al. 2016).

Also plotted in Figure 20.8 are the acceptance criteria according to (a) the previous ISO recommendation ISO 2394:1998, which included *F–N* curves and (b) *fib* COM3 TG3.1 (Caspeele et al. 2016). For a train speed of 230 km/h, the human safety criteria suggest that human safety risks exceed a commonly acceptable level and further safety measures to mitigate human safety risks are required. Options include increasing the lateral distance beyond the optimum value or increasing the distance between the bridge and switch area.

20.6 Difference between New and Existing Structures

According to ISO 13822 (ISO 2010) lower target reliability levels can be used if justified on the basis of societal, cultural, economic, and sustainable considerations. Risk acceptance criteria and related target reliabilities for structural design and assessment of existing structures are discussed in scientific literature. Recent contributions revealed that the differences between the assessment of existing structures and structural design—considering higher costs of safety measures for existing structures as a key one—are often inadequately reflected (Diamantidis and Bazzurro 2007; Steenbergen and Vrouwenvelder 2010; Steenbergen et al. 2015). The following remarks may be useful when specifying target reliability levels for the assessment of existing structures:

- The approaches discussed in the previous sections apply for both new and existing structures.
- It is uneconomical to require the same target reliabilities for existing structures as for new structures. This requirement is consistent with regulations accepted in nuclear and offshore industry, for buildings in seismic regions, for bridges in the United States and Canada, etc.
- Minimum levels for human safety are commonly decisive for existing structures, while economic optimization dominates the criteria for design of new structures.
- Two target levels are needed for the assessment of existing structures—the minimum level below which a structure should be repaired and the optimum level for repair.
- Sykora et al. (2016) and Steenbergen et al. (2015) indicated that the target reliability indices considered for structural design can be tentatively reduced for the assessment of existing structures by
  - $\Delta\beta \approx 1.5$–2 for the minimum level below which a structure should be repaired
  - $\Delta\beta \approx 0.5$–1 for the optimum upgrade strategy; however, a common requirement to upgrade to meet design criteria leads to an economically similar solution and can often be justified by the need for balanced reliability levels in a railway line or part of a network.

20.7 Conclusions

The target reliability levels recommended in various normative documents for civil engineering structures, also applicable to railway structures such as station buildings, bridges, or tunnels, are inconsistent in terms of the values and the criteria according to which the appropriate values are to be specified. In general, optimum reliability levels should be obtained by considering both relative costs of safety measures and failure consequences over a required working life; the minimum reliability for human safety should also be considered. The following conclusions are drawn:

- The design strategy and respective codified target reliability levels are mostly driven by economic arguments, as it is affordable and economically optimal to design for higher reliability than that required by human safety criteria.
- It is uneconomical to require the same reliability levels for existing and new structures.
- Decisions made in assessment can result in the acceptance of an actual state or in the upgrade of a structure; two reliability levels are thus needed:
  1. The minimum level below which the structure is considered unreliable and should be upgraded—This level is often dominated by human safety criteria.
  2. The target level indicating an optimum upgrade strategy that is close to the levels required in design.
- Economically optimal target reliabilities primarily depend on failure consequences and costs of safety measures; the influence of a working life and discount rate is less significant. However, for large infrastructure projects with long design lifetimes, the effect of discounting should be considered.
- Economic optimization should also take into account societal and environmental consequences; for detailed guidance, see Diamantidis, Holicky, and Sykora (2017).

References

American Society of Civil Engineers. 2013. ASCE 7-10: *Minimum Design Loads for Buildings and Other Structures*. Reston, VA: American Society of Civil Engineers.

Ang, A. H.-S., and D. De Leon. 2005. Modeling and analysis of uncertainties for risk-informed decisions in infrastructures engineering. *Structure and Infrastructure Engineering* 1 (1): 19–31.

Bhattacharya, B., R. Basu, and K. Ma. 2001. Developing target reliability for novel structures: The case of the mobile offshore base. *Marine Structures* 14 (1–2): 37–58.

Casas, J. R., and D. Wisniewski. 2013. Safety requirements and probabilistic models of resistance in the assessment of existing railway bridges. *Structure and Infrastructure Engineering* 9 (6): 529–545.

Caspeele, R., R. Steenbergen, M. Sýkora, D. L. Allaix, W. Botte, G. Mancini, M. Prieto, P. Tanner, and M. Holický. *Partial Factor Methods for Existing Concrete Structures (fib bulletin 80)*. Fédération internationale du béton (fib), fib COM3 TG3.1, 2017. 129 p. ISSN 1562-3610. ISBN 978-2-88394-120-5.

Caspeele, R., M. Sykora, D. L. Allaix, and R. Steenbergen. 2013. The design value method and adjusted partial factor approach for existing structures. *Structural Engineering International* 23 (4): 386–393.

CEN (European Committee for Standardization). 2002. EN 1990: *Eurocode—Basis of Structural Design*. Brussels: CEN.

CIB (Central Invorderings Bureau). 2001. CIB TG 32: *Risk Assessment and Risk Communication in Civil Engineering* (Report 259). Rotterdam: CIB.

Diamantidis, D. 2005. Risk analysis versus risk acceptability in major European tunnel projects. *Asia Pacific Conference on Risk Management and Safety, Hong Kong.*

Diamantidis, D., and P. Bazzurro. 2007. Safety acceptance criteria for existing structures. *Special Workshop on Risk Acceptance and Risk Communication, Stanford University, Stanford, CA, March 26–27, 2007.*

Diamantidis, D., F. Zuccarelli, and A. Westhäuser. 2000. Safety of long railway tunnels. *Reliability Engineering and System Safety* 67 (2): 135–145.

Diamantidis, D., M. Holicky, and M. Sykora. 2016. Risk and reliability acceptance criteria for civil engineering structures. *Transactions of the VSB—Technical University of Ostrava, Civil Engineering Series* 16 (2): 1–10.

Diamantidis, D., M. Holicky, and M. Sykora. 2017. Target reliability levels based on societal, economic and environmental consequences of structural failure. *Proceedings of the International Conference on Structural Safety and Reliability 2017, Vienna, CRC Press/Balkema, August 6–10, 2017.*

Duijm, N. J. 2015. Recommendations on the use and design of risk matrices. *Safety Science* 76: 21–31.

Fischer, K., C. Barnardo-Viljoen, and M. H. Faber. 2012. Deriving target reliabilities from the LQI. *Proceedings of the Life Quality Index Symposium, Technical University of Denmark, Lyngby, August 21–23, 2012.*

Hamburger, R. O. 2013. *Provisions in Present US Building Codes and Standards for Resilience* (NIST Technical Note 1795). Gaithersburg, MD: National Institute of Standards and Technology.

Holicky, M. 2013a. *Introduction to Probability and Statistics for Engineers.* Berlin: Springer-Verlag.

Holicky, M. 2013b. Optimisation of the target reliability for temporary structures. *Civil Engineering and Environmental Systems* 30 (2): 87–96.

Holicky, M. 2014. Optimum reliability levels for structures. In *Vulnerability, Uncertainty, and Risk,* 184–193. Reston, VA: American Society of Civil Engineers.

Holicky, M., J. Markova, and M. Sykora. 2013. Forensic assessment of a bridge downfall using Bayesian networks. *Engineering Failure Analysis* 30 (0): 1–9.

Holicky, M., D. Diamantidis, and M. Sykora. 2015. Determination of target safety for structures. *Proceedings of the 12th International Conference on Applications of Statistics and Probability in Civil Engineering, Vancouver, BC, July 12–15, 2015.*

HSE (Health and Safety Executive). 2001. *Reducing Risks, Protecting People.* Norwich: HSE.

International Union of Railways. 2002. *UIC 777-2: Structures Built over Railway Lines, Construction Requirements in the Track Zone.* 2nd ed. Paris: International Union of Railways.

ISO (International Organization for Standardization). 2010. ISO 13822: *Bases for Design of Structures—Assessment of Existing Structures* (ISO TC98/SC2). Geneva: ISO.

ISO. 2008. ISO 15686-5: *Buildings and Constructed Assets—Service-Life Planning—Part 5: Life-Cycle Costing.* 1st ed. Geneva: ISO.

ISO. 2015. ISO 2394: *General Principles on Reliability for Structures.* 4th ed. Geneva: ISO.

JCSS (Joint Committee on Structural Safety). 2001a. *Probabilistic Assessment of Existing Structures,* Dimitris Diamantidis (ed.). Paris: Joint Committee on Structural Safety, RILEM Publications SARL.

JCSS. 2001b. *JCSS Probabilistic Model Code* (periodically updated, online publication). Lyngby: JCSS.

Jonkman, S. N., P. H. A. J. M. van Gelder, and J. K. Vrijling. 2003. An overview of quantitative risk measures for loss of life and economic damage. *Journal of Hazardous Materials* 99 (1): 1–30.

Jung, K., and J. Marková. 2004. Risk assessment of structures exposed to impact by trains. *5th International PhD Symposium in Civil Engineering—Proceedings of the 5th International PhD Symposium in Civil Engineering, Delft.* pp. 1057–1063.

Kanda, J., and H. Shah. 1997. Engineering role in failure cost evaluation for buildings. *Structural Safety* 19 (1): 79–90.

Kaszynska, M., and A. S. Nowak. 2005. Target reliability for design and evaluation of bridges. *Proceedings of 5th International Conference on Bridge Management.*

Lee, K.-M., H.-N. Cho, and C.-J. Cha. 2006. Life-cycle cost-effective optimum design of steel bridges considering environmental stressors. *Engineering Structures* 28 (9): 1252–1265.

Lee, J. Y., and B. R. Ellingwood. 2015. Ethical discounting for civil infrastructure decisions extending over multiple generations. *Structural Safety* 57: 43–52.

Lentz, A. 2007. *Acceptability of Civil Engineering Decisions Involving Human Consequences* (PhD Thesis). Munich: Technical University of Munich.

Mainline Consortium. 2013. *Deliverable 5.4: Proposed Methodology for a Life Cycle Assessment Tool (LCAT)*. Lyngby: COWI.

McClelland, B., and M. D. Reifel. 1986. *Planning and Design of Fixed Offshore Platforms* (Chapter 5 Reliability Engineering and Risk Analysis by B. Stahl), pp. 59–98. Berlin: Springer.

Melchers, R. E. 2001. *Structural Reliability Analysis and Prediction*. 2nd ed. Chichester: John Wiley & Sons.

Moreira, V. N., J. Fernandes, J. C. Matos, and D. V. Oliveira. 2016. Reliability-based assessment of existing masonry arch railway bridges. *Construction and Building Materials* 115: 544–554.

Narasimhan, H., S. Ferlisi, L. Cascini, G. De Chiara, and M. H. Faber. 2016. A cost–benefit analysis of mitigation options for optimal management of risks posed by flow-like phenomena. *Natural Hazards* 81: 117–144.

Nathwani, J. S., M. D. Pandey, and N. C. Lind. 2009. *Engineering Decisions for Life Quality: How Safe is Safe enough?* London: Springer-Verlag.

Norsok. 1998. *Risk and Emergency Preparedness Analysis. Annex C. Methodology for Establishment and use of Environmental Risk Acceptance Criteria* (Norsok Standard, Z-013., Rev. 1, informative), Oslo: Norsok.

Nowak, A. S. 1999. *Calibration of LRFD Bridge Design Code* (NCHRP Report 368). Washington, DC: Transportation Research Board.

Rackwitz, R. 2000. Optimization—The basis of code-making and reliability verification. *Structural Safety* 22 (1): 27–60.

Smit, C. F., and C. Barnardo-Vijloen. 2013. Reliability based optimization of concrete structural components. *Proceedings of the 11th International Probabilistic Workshop, Brno, Litera, November 6–8, 2013*.

Standards Australia. 2002. AS/NZS 1170.0: *Structural Design Actions—General Principles*. Sydney: Standards Australia.

Steenbergen, R. D. J. M., M. Sykora, D. Diamantidis, M. Holicky, and A. C. W. M. Vrouwenvelder. 2015. Economic and human safety reliability levels for existing structures. *Structural Concrete* 16 (September 2015): 323–332.

Steenbergen, R. D. J. M., and A. C. W. M. Vrouwenvelder. 2010. Safety philosophy for existing structures and partial factors for traffic loads on bridges. *Heron* 55 (2): 123–139.

Sykora, M., M. Holicky, and J. Markova. 2013. Verification of existing reinforced concrete bridges using the semi-probabilistic approach. *Engineering Structures* 56 (November 2013): 1419–1426.

Sykora, M., M. Holicky, K. Jung, and D. Diamantidis. 2016. Target reliability for existing structures considering economic and societal aspects. *Structure and Infrastructure Engineering* 13 (1): 181–194.

Tanner, P., and R. Hingorani. 2015. Acceptable risks to persons associated with building structures. *Structural Concrete* 16 (3): 314–322.

Trbojevic, V. M. 2009. Another look at risk and structural reliability criteria. *Structural Safety* 31 (3): 245–250.

Vrijling, J., P. van Gelder, and S. Ouwerkerk. 2005. Criteria for acceptable risk in the Netherlands. In *Infrastructure Risk Management Processes*, 143–157. Reston, VA: American Society of Civil Engineers.

Vrouwenvelder, A. C. W. M. 2002. Developments towards full probabilistic design codes. *Structural Safety* 24 (2–4): 417–432.

Vrouwenvelder, T., B. J. Leira, and M. Sykora. 2012. Modelling of hazards. *Structural Engineering International* 22 (1): 73–78.

Wisniewski, D. F., J. R. Casas, and M. Ghosn. 2009. Simplified probabilistic non-linear assessment of existing railway bridges. *Structure and Infrastructure Engineering* 5 (6): 439–453.

Section 2

21

Methodology and Application of RAM Management along the Railway Rolling Stock Life Cycle

Olga Fink and Simone Finkeldei

CONTENTS

21.1 Introduction and Framework Overview

Today's railway systems are operated under dynamically changing load and stress factors. They are increasingly equipped with complex systems and are usually highly interconnected and interdependent. Particularly, the influencing factors on the degradation and deterioration of systems are dynamic and are often difficult to predict, and their distinct influence is very challenging to quantify. Concurrently, the requirements on the system behavior and the reliability of the system functions have also considerably increased. The pressure to reduce costs and increase the productivity of the railway systems has also been continuously increasing in the last years. However, the customer expectations on the quality of the transport service, such as punctuality, availability of comfort systems, and information in the case of disruptions and delays, also have been increasing.

An essential characteristic of railway vehicles is the lifetime of several decades (when taking modernization and refit into consideration, it can be up to 50 years). The long lifetime is inherently linked to dynamic changes in requirements and operating environment. Therefore, holistic considerations of reliability, availability, and maintainability (RAM) and life cycle costs (LCCs) have been gaining importance. This requires not only their optimization during the concept and design phase but also a proactive monitoring of the achieved reliability and availability and the evolution of the maintenance and operation costs during the life cycle of the system.

The general goal of a railway system is to produce a defined level of railway operation at a given time, with a defined quality of service under tolerable residual risk and within defined cost limits. Thereby, RAM parameters characterize the long-term operation of a system. They are achieved by an integrated RAM/LCC management along the entire life cycle that comprises among others the application of established engineering concepts, RAM/LCC analyses, modeling and simulation, sensitivity and uncertainty analyses, continuous monitoring, and the demonstration of the fulfillment of the defined requirements during the operation phase. An essential part of RAM/LCC management is a close collaboration between the stakeholders along the entire life cycle of the railway vehicles. This enables not only knowledge building on system behavior during operation but also continuous improvements of the operated fleet and of the new developments.

The framework proposed in this chapter for managing RAM/LCC along the railway rolling stock life cycle comprises several steps and, most importantly, feedback processes and is displayed in Figure 21.1. While RAM parameters of a system comprise only a part of the LCCs, the focus of this chapter is primarily on the LCCs influenced by the RAM parameters. The holistic framework integrates the points of view of the different stakeholders involved in the life cycle of the system (Figure 21.1). The proposed framework is a guideline of the chapter and covers the relevant tasks and supporting tools for RAM/LCC management.

The life cycle phases in the framework (Figure 21.1) are defined in a general way and are derived from the EN 60300-3-3 standard (EN 60300-3-3:2014-09 [CEN 2014]). They can be matched to the life cycle phases of the EN 50126 standard (EN 50126:1999-09 [CEN 1999]). A successful RAM/LCC management comprises a proactive planning and continuous assessment of the RAM/LCC characteristics not only during the concept and development phase but also during the subsequent phases and particularly during the main operation and maintenance phase. The integration of RAM/LCC requirements in contracts has gained importance and has become an integral part of RAM/LCC management. The contractual liability has also made it necessary to demonstrate the fulfillment of the contractual requirements on RAM parameters in operation.

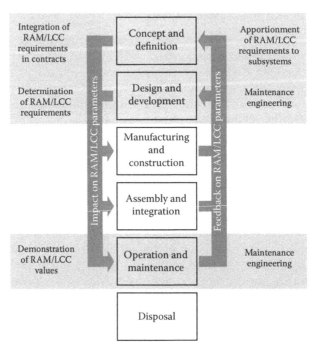

FIGURE 21.1
Framework for RAM/LCC management.

A key task of RAM engineering during the design and development phase is the definition of RAM requirements and their apportionment to the subsystems. Moreover, continuous improvement is only possible if the feedback from field operation is integrated into the RAM engineering process. The feedback from operation requires a systematic analysis of failure and maintenance data. The feedback can be used to optimize the systems and the maintenance plans. Only a cooperative RAM/LCC management between the operator, the main supplier, and the subsuppliers enables a consistent optimization toward more reliable and available systems with cost-efficient maintenance processes. The single steps within the process comprise the structure of the chapter (Figure 21.1).

21.2 Integration of RAM/LCC Requirements in Contracts and the Demonstration of Their Contractual Fulfillment

RAM parameters determine to a large extent the costs incurring during the operation phase (including operating and maintenance costs, failure and unavailability costs, and costs caused by operational disturbances). However, these costs and the pertinent RAM parameters are determined at a relatively early stage of the concept and design process (see Figure 21.2). Generally speaking, at the end of the design and development phase, approximately 70–85% of the total LCCs are fixed, even though they are only realized at a later point in time during the operation and maintenance phase (Figure 21.2). This is one of the essential reasons why parameters and features determining LCCs are increasingly becoming part of contractual requirements, including requirements on RAM parameters, energy consumption, and cleaning. This implies that the determining parameters need to be analyzed at very early life cycle phases and need to be subsequently integrated, monitored, and analyzed in the following life cycle phases. Additionally, the interdependency of the RAM system characteristics and the maintenance requirements need to be integrated into the engineering processes.

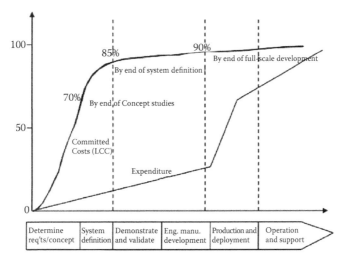

FIGURE 21.2
Degree of cost commitment and expenditure.

To ensure that the manufacturers integrate the relevant system characteristics influencing the LCCs of the vehicles early in the design phase, a close cooperation between the operator and the manufacturer is required.

In recent years, the integration of specific requirements in contracts on defined RAM parameters has significantly increased. There are different ways of defining the RAM/LCC requirements on fleet level. They are always related to operating conditions, mission profile, environment, cost structures, priorities, and strategy of the operator and are influenced by the operational experience with the existing fleets. A suboptimal allocation of RAM/LCC requirements may result in local optimization of the defined values. This is particularly true because the RAM characteristics of the system are interrelated and, for example, only optimizing reliability may result in high-maintenance expenditures. The assessment of the predicted RAM/LCC parameters during the tendering phase comprises a significant part of the decision-making process. RAM requirements are often subject to contractual penalties after the contracts are fixed.

Since operators are defining the requirements on the RAM characteristics, their main focus is the impact of failures on the operation. Therefore, the requirements often comprise different categories for operational disturbances, such as, for example, immediate stop of the train, stop at the next station/strategic point, or taking the train out of operation at the end of the day. Operational disturbances can also be classified in delay intervals to capture the impact of the failures on the timetable and the punctuality, such as, for example, delays exceeding 5 minutes. Figure 21.3 gives an example of hierarchical system reliability and availability requirements (see Section 21.3). While the operator typically specifies requirements on the operational level, the manufacturer primarily focuses on the inherent system reliability and availability characterized by the features of the technical system.

In addition to failures affecting the operation, there are also failures only affecting the comfort (e.g., a failure of the air conditioning) or primarily affecting the maintenance costs (e.g., failure of a redundant component).

Besides requirements on (operational/inherent) reliability and availability, requirements on system and component maintainability are also important. The requirements on maintainability typically include dedicated values for mean time to restore (MTTR) for preventive and corrective maintenance tasks, respectively. In some cases, also the time in which a defined percentage of components need to be restored is specified, such as, for example, that 90% of the components need to be restored within 3 hours.

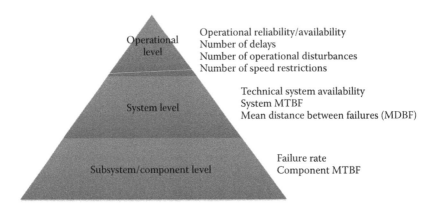

FIGURE 21.3
Example of hierarchical reliability and availability requirements.

If RAM requirements are defined in contracts, they also need to be demonstrated on the fleet during the operation. Because the requirements are not deterministic, and reliability tests, such as accelerated lifetime testing, are insufficient to demonstrate the reliability under real operating conditions; the fulfillment of the requirement can only be demonstrated during operation. However, due to contractual requirements, the demonstration period cannot last the entire lifetime of the system but comprises a defined limited period of time that is typically not congruent to the warranty period.

To achieve a smooth demonstration of the compliance with contractual requirements, the acceptance criteria and process of demonstration need to be defined and agreed at a very early stage of the project. RAM demonstration is often confused with verification or validation of system requirements. Certainly, the verification of RAM analyses performed during the design and development phases and the fulfillment of the RAM requirements through the analyses are an integral part of the verification process. However, since the nature of the RAM characteristics is stochastic, they cannot be validated during the system validation. For example, test dismantling is an integral part of system validation. It is primarily used to validate the maintainability and the suitability of the defined maintenance processes and the maintenance support resources.

The demonstration of RAM parameters under operation is a problem of statistical hypothesis tests. If reliability needs to be demonstrated, a one- or two-sided test for the demonstration of a constant failure rate (or a mean time between failures) can be performed (Birolini 2007). The general process is similar to that of statistical process control. The producer's and operator's risks need to be defined. With this input, the hypothesis can be tested.

Additionally, sequential reliability tests can be performed (Meeker and Escobar 1998). For performing a sequential reliability test, the alternative hypothesis needs to be defined, representing the maximum acceptable limit for reliability. The ratio between the null hypothesis and the alternative hypothesis is also referred to as the design ratio. Similarly, as for the non-sequential hypothesis tests, the producer's and the operator's risks need to be defined.

Figure 21.4 presents an example of a sequential reliability test with the cumulated test time and the cumulated number of occurring failures. The lower boundary line represents the acceptance boundary for the null hypothesis, and the upper boundary line represents

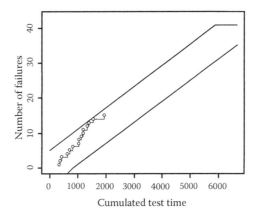

FIGURE 21.4
Sequential reliability test.

the acceptance boundary for the alternative hypothesis. The figure demonstrates exemplarily a possible realization of the test (dotted line).

The main advantage of sequential reliability tests is that the test duration is on average shorter than with simple two-sided tests. The required cumulated test time depends on the alternative hypothesis, respectively the ratio between the two hypotheses, the producer's and the operator's risk.

Not only is the demonstration of contractual requirements important but also the monitoring of the achieved RAM characteristics. During the operation phase, the monitoring is particularly important for optimizing maintenance strategies and identifying weak points within the system.

21.3 Determination of RAM/LCC Requirements and Their Apportionment to Subsystems

The invitation to tender of rolling stock operators requests either the commitment to defined RAM/LCC target values or the designation of RAM/LCC values by the rolling stock manufacturer comprising the commitment to these. The committed values are part of the contractual agreement between manufacturer and operator (see Section 21.2) and thereby become requirements for the development. In the last years, the RAM/LCC values have become increasingly important for the assessment of the different bidders and, hence, for winning the contract. As a consequence, target values have to be competitive, to gain a market share. Rolling stock manufacturers, on the one side, study the market needs and, on the other side, analyze the competitors' and their own products to define the RAM/LCC target values that are stated in the offer.

The design responsibility for the subsystems and components does typically not reside with one division of the rolling stock manufacturer. Several divisions of the manufacturer and external suppliers for subsystems or components are contributing the RAM/LCC values of their products to calculate the RAM/LCC values on system and operational level. The roles along the supply chain as per Figure 21.5 are responsible for the fulfillment of RAM/LCC requirements of the systems in their scope of supply and the apportionment of RAM/LCC target values to their supplier.

RAM/LCC requirements have to be apportioned to the different subsystems and further to each component for the following reasons:

1. To ensure their consideration during product development.
2. RAM requirements are input values for the design and development phase (Figure 21.1) that have to be passed through by every subsystem or component (EN 50126: 1999-09 [CEN 1999]).
3. To reduce the risk of the apportioning legal entity/division, RAM requirements have to be considered in contracts between the rolling stock manufacturer and subsystem/component supplier.

Passing on of RAM/LCC requirements along the supply chain can be implemented by requirement documents, which have to be committed to by the receiver of the document.

The requirement documents ensure that the commercial risk of each stakeholder along the supply chain is transferred to its supplier in case it was not decided to keep the

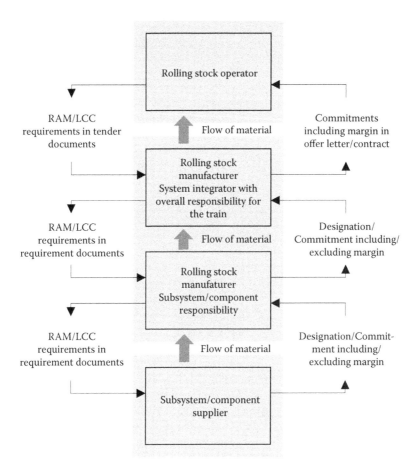

FIGURE 21.5
Roles along the supply chain.

responsibility in-house to prevent a higher product price. In order to reduce the inherent risk by committing to the requested RAM/LCC requirements buffers are added to the estimated RAM/LCC values. This leads to the situation that the committed values are not transparent. As a consequence, the committed values from the internal and external suppliers are treated like net values, and additional buffers are added. Especially within one company (as indicated by the gray areas in Figure 21.1), it is worth considering not to request commitments and provide indicative RAM/LCC values for the subsystem to be able to have competitive overall RAM/LCC values for the offer. One further approach is to ask the external suppliers to provide two values for each RAM/LCC parameter, namely, the committed and the expected values. This increases the transparency and therewith the competitiveness.

The apportionment of the RAM/LCC requirements to the subsystems and components as described before is a top–down procedure. In fact, it is essential to allocate the RAM/LCC requirements observing the capabilities of the subsystem or component to prevent high costs for redesign; consequences on other requirements/cost areas and for not fulfilling the RAM/LCC requirements are effectuated in all phases of the product life cycle.

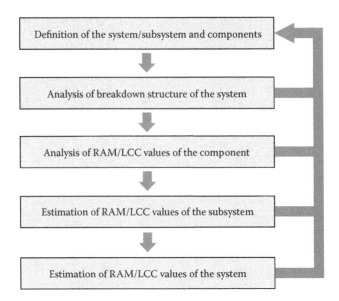

FIGURE 21.6
Bottom–up analysis to determine RAM/LCC values of a system.

Hence, a bottom–up analysis is recommendable to determine the achievable and realistic RAM/LCC values for the system (Figure 21.6). Based on the breakdown structure of the system the RAM/LCC values of each component have to be estimated. The successive consolidation determines the RAM/LCC values of the subsystem and the system. In case, the RAM/LCC requirements for the component, subsystem, or system are not fulfilled, the definition of the system needs to be changed and a feedback regarding the apportionment of the RAM/LCC requirements should be provided.

The apportionment of RAM requirements has to be accompanied by environmental conditions, the rolling stock shall be operated in; and by technical requirements, the respective system has to fulfill as these factors have a huge impact on the RAM values of each system. A systematic interface and requirement management for both internal interfaces and interfaces to external customers ensure that the environmental conditions and technical requirements are correctly addressed, and hence, the results of the RAM determination are as close to praxis as possible. However, the environmental conditions and technical requirements and the dependency of the RAM values on these factors are not always determinable, and the effect of variations of these factors is often not calculable.

The determination of RAM/LCC values of one component can be acquired through theoretical calculations or via practical investigations. The practical investigations can be either performed directly in the field, e.g., analysis of components that have been used in the field, or in a laboratory, e.g., endurance tests on a test bench.

Field data are the most realistic way to reflect the reality and consequently are of high value for the railway industry. However, typically, these data are not available for the rolling stock manufacturer and component supplier beyond the warranty phase. Even data from the warranty phase are not systematically available. Only in the case of maintenance contracts or agreed partnerships, access to maintenance data is available for the manufacturer to a high extent.

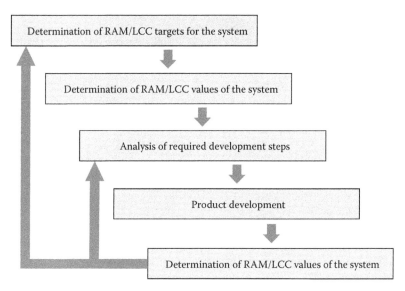

FIGURE 21.7
Iterative strategy to control RAM requirements.

To draw conclusions from the number of spare parts that are supplied by the manufacturer and component supplier is hardly possible in praxis when gaining experience as stock size, operator specific maintenance concepts, reason for the component exchange and corresponding mileage, and operating hours are not known by the manufacturer and supplier.

To ensure a holistic approach, both described methods, namely, top–down and bottom–up, have to be applied. This demands for an iterative strategy (see Figure 21.7) comprising the feedback regarding achievable RAM/LCC values of a system and development steps including additional RAM analysis which must be performed to meet the RAM requirements.

21.4 Maintenance Engineering

Maintenance engineering is the discipline of applying engineering concepts to ensure the availability, reliability, safety, and cost effectiveness of technical systems. Maintenance engineering is closely related to reliability engineering. Particularly, the systematic analysis of failure behavior enables knowledge transfer and optimization of maintenance strategies and concepts (Figure 21.8).

While the maintainability of a system is one of its inherent characteristics, maintenance ensures that the system is able to fulfill the defined function under the defined operating conditions, with defined maintenance support resources, safely and reliably.

Even though maintenance starts after the systems are commissioned, the planning of the maintenance already needs to be taken into consideration in the concept phase. This is particularly important for condition-based and predictive maintenance. Conventionally, mechanical wear has been one of the main parameters determining the maintenance needs for many railway rolling stock components. Additionally, preventive maintenance activities

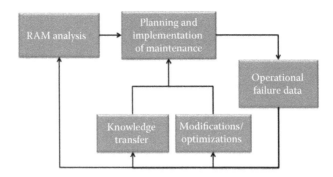

FIGURE 21.8
RAM and maintenance analyses.

tended to be scheduled based on a limited number of usage parameters, such as mileage and operating hours, or even the calendar time, leading to costly and excessive maintenance downtime. Today, purely mechanical components have been to a large extent replaced by complex mechatronic devices, systems that combine mechanical, electronic, and information technology, with very different degradation and failure characteristics. This makes the planning of maintenance tasks more challenging. To overcome some of the complexity issues and to reduce the time for finding the location of the fault and isolating its cause, diagnostics systems have been implemented to support the maintenance personnel in identifying the root cause of the failure and performing the optimal remedying actions.

Falling costs and increased reliability of sensors and data storage and transmission have allowed condition monitoring to gain importance. As an example, assessing the condition of a train previously required an on-site inspection, while now, a large volume of data generated over time can be used for remote decision making in maintenance operations. Data generated by remote monitoring and diagnostic systems provide information on the processes and state changes occurring in the individual components or subsystems. This is in contrast to more commonly applied approaches that use assumptions that are valid for average systems under average operating conditions.

However, the sensing devices also have their challenges: If sensors with short lifetimes are applied to monitor systems with long life spans, such as, for example, acceleration sensors applied in bogies to monitor their condition. A particular challenge arises if part of the condition monitoring function becomes safety relevant. In this case, there are specific requirements on their design and control, and it needs to be included in the safety case.

The condition of the systems can be assessed by integrated monitoring and diagnostic devices and the condition of some of the components, which can also be partly done by wayside monitoring devices. Generally, the wayside monitoring devices have been primarily designed for safety reasons. However, nowadays, increasingly, the data on the condition of the system is also used for maintenance planning.

Condition-based and predictive maintenance enables railway operators to mitigate failures; avoid excessive maintenance; and enable optimally scheduled maintenance interruptions, reducing downtime, increasing system resilience, and adding business value. While condition-based maintenance implies that the maintenance task is scheduled based on condition monitoring activity, which can range from inspections and online condition monitoring systems, predictive maintenance implies predicting the remaining useful life, anticipating the failure, reducing the impact of the failure, and determining the optimal point in time for maintenance intervention.

Recently, there has also been the term of prescriptive maintenance. It has been used in many cases with the same definition as predictive maintenance: that is, considering the preemptive actions to prevent the occurrence of a predicted fault. However, we define prescriptive maintenance as going one step beyond the prediction of the remaining useful lifetime of the systems (IBM 2013). Prescriptive maintenance comprises a proactive control of the operating parameters to prolong the lifetime of the system, depending on operational requirements, cost structures, the specific user priorities, time tables, etc. This requires the expert knowledge on the possible control parameters and their influence on the effective useful lifetime of the system.

Also, the term of prognostics and health management (PHM) has been used within a similar context. PHM aims to provide users with an integrated view of the health state of a component or an overall system (Vachtsevanos 2006). PHM comprises fault detection, diagnostics, and prognostics. The main task of diagnostics is fault isolation and identification of failure root causes. Prognostics can be defined as the process of assessing the health state of the system and the prediction of its evolution in time. This process includes detecting an incipient fault and predicting the remaining useful lifetime of a component. However, PHM does not end with the prediction of the remaining useful life. Health management is the process of taking timely and optimal maintenance actions based on outputs from diagnostics and prognostics, available resources, and operational demand. PHM focuses on assessing the impact of failures and minimizing impact and loss with maintenance management (Lee et al. 2014).

For determining maintenance plans, two approaches have been commonly applied:

- Reliability-centered maintenance (RCM)
- Risk-based maintenance (RBM)

RCM is a systematic approach for developing maintenance strategies. RCM aims to maintain a function and not primarily maintaining a specific system. RCM supports the search for the best maintenance strategy to reduce the probability of a failure and its severity. The maintenance tasks are selected based on the wear mechanisms, the fault evolution, and progression.

The goal of RCM is to reduce the maintenance costs, by focusing on the most important functions of the system and avoiding or removing maintenance actions that are not strictly necessary.

Traditionally, there are seven questions in the core of RCM (Moubray 1997):

1. What are the functions and associated performance standards of the equipment in its present operating context?
2. In what ways can it fail to fulfill its functions?
3. What is the cause of each functional failure?
4. What happens when each failure occurs?
5. In what way does each failure matter?
6. What can be done to prevent each failure?
7. What should be done if a suitable preventive task cannot be found?

Since the last question addresses the question of design changes, it is particularly important to already start with the RCM analysis during the design and development phase.

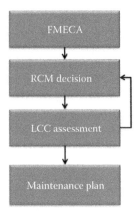

FIGURE 21.9
RCM process.

Generally, the first four questions of the RCM process are also part the failure mode, effects, and criticality analysis (FMECA). Therefore, RCM can be integrated into the RAM analyses that are performed during the design process. RCM does not consider the costs of a maintenance task. Therefore, an additional step needs to be integrated into the process: the LCC assessment of the defined maintenance tasks (Figure 21.9). If the maintenance costs justify the benefits on the gains in operational reliability, safety, and the availability of the comfort function, the maintenance task is integrated into the overall maintenance plan; otherwise, an alternative needs to be defined.

The core part of the RCM process is a decision logic that integrates the different criticality considerations (which can comprise consequences in terms of operational reliability, safety, comfort, and economic damages). Additionally, the different fault evolution and progression patterns need to be integrated into the decision logistic. One additional consideration is if the failure is hidden and how it can the made visible.

The second commonly applied approach for determining the maintenance plan is the RBM. RBM particularly focuses on maintenance that reduces risks (Figure 21.10).

FIGURE 21.10
RBM process.

Risk assessment and risk reduction measures are in the center of the RBM. However, not only safety risks are considered, but also economical and operational risks.

Both methods, RCM and RBM, are partly overlapping and are systematic methods for deriving maintenance programs. RCM does not only consider the reliability but also safety, environmental, and economic aspects, and it assesses the criticality of the failing functions. RBM, on the other hand, does consider not only safety risks but also economic, environmental, and operational risks. One of the distinguishing characteristics is that RCM also takes the different types of fault evolution and progression into consideration.

In practice, components of both analyses have been connected to integrate the most important characteristics and to enable optimal decisions.

21.5 Discussion of the Determination of RAM Requirements and Their Conflicting Goals

Due to deviating requirements from customers and authorities in different countries, no offer is equal to another. As a consequence, the overall RAM/LCC values for a train have to be determined for each offer. Additionally, environmental impacts influence the achievable RAM/LCC values. Nevertheless, usually similar products from the rolling stock manufacturer are in operation; thus, the experience is available that serves as the basis for upcoming offers. For newly developed systems, the increased uncertainty leads to a higher commercial risk for the rolling stock manufacturer in the case that the committed RAM/LCC targets are subject to penalties (see Section 21.2). RAM/LCC target values are defined in the concept and definition phase (see Figure 21.1), whereas the system is—especially for new products—not or only partly defined.

Additionally, the complexity for the stakeholders along the supply chain (see Figure 21.5) has increased as they are confronted with input parameters of RAM/LCC requirements that can be divided into two groups: On the one hand, the group of parameters that can be influenced by the railway industry and, on the other hand, those which cannot be influenced by them.

Passing on RAM/LCC requirements that are not applicable for all stakeholders and requesting the complete set of RAM/LCC documentation increases the overall project costs, as it increases the complexity for the affected stakeholder and the coordination efforts.

For example, suppliers without knowledge of the complete system cannot judge the exchange times of their components including the accessibility especially when they do not define the interfaces. Another example is to expect from the component suppliers to adapt their RAM/LCC documentation for each and every project by requesting to provide the failure rates classified according to the project specific failure categories that are predefined by the rolling stock operator.

Furthermore, a balanced apportionment of RAM/LCC requirements to subsystems and components that takes the capabilities of the subsystems and components into account supports the reduction of project costs, as efforts for redesign and the risk of the insight that the requirements cannot be met on the component, subsystem, or system level are lower. Furthermore, a redesign in order to achieve RAM/LCC target values can be contradictory to specified product properties.

An example is the enlargement of the wheel diameter in order to reduce the frequency of wheel exchange and therewith the availability and LCC of the product. This can cause a discrepancy with the defined installation space and the limits for unsprang masses.

Additionally, RAM/LCC requirements themselves can be contradicting, e.g., the fulfillment of operational reliability target values by a system architecture containing a lot of redundancies increases the product reliability and therewith maintenance costs. A product for a special application case offers a good balance between the sale price, RAM/LCC requirements, and potentially related penalties.

References

Birolini, A. 2007. *Reliability Engineering: Theory and Practice*. 5th ed. Berlin: Springer.

CEN (European Committee for Standardization). 1999. EN 50126:1999-09: *Railway Applications—The Specification and Demonstration of Reliability, Availability, Maintainability and Safety (RAMS)* (refer to IEC 62278 for the international standard). CEN: Geneva.

CEN. 2014. EN 60300-3-3:2014-09: *Dependability Management—Part 3-3: Application Guide—Life Cycle Costing*. CEN: Geneva.

IBM. 2013. Descriptive, predictive, prescriptive: Transforming asset and facilities management with analytics (White paper (external)-USEN). Available from http://www-01.ibm.com/common/ssi/cgi-bin/ssialias?infotype=SA%26subtype=WH%26htmlfid=TIW14162USEN.

Lee, J., Wu, F., Zhao, W., Ghaffari, M., Liao, L., and Siegel, D. 2014. Prognostics and health management design for rotary machinery systems—Reviews, methodology, and applications. *Mech Syst Signal Process*. 42:314–334.

Meeker, W. Q., and Escobar, L. A. 1998. *Statistical Methods for Reliability Data*. Hoboken, NJ: Wiley.

Moubray, J. 1997. *Reliability-Centred Maintenance*. Second ed. Oxford: Butterworth-Heinemann.

Vachtsevanos, G. 2006. *Intelligent Fault Diagnosis and Prognosis for Engineering Systems*. Hoboken, NJ: John Wiley & Sons.

22

IT Security Framework for Safe Railway Automation

Jens Braband

CONTENTS

22.1 Introduction

Recently, reports on information technology (IT) security incidents related to railways have increased as well as public awareness. For example, it was reported that on December 1, 2011, "hackers, possibly from abroad, executed an attack on a Northwest rail company's computers that disrupted railway signals for two days" [1]. Although the details of the attack and its consequences remain unclear, this episode clearly shows the threats to which railways are exposed when they rely on modern commercial off-the-shelf communication and computing technology. However, in most cases, the attacks are denial-of-service attacks leading to service interruptions, but, so far, not safety-critical incidents. But also other services, such as satellite positioning systems, have been shown to be susceptible to IT security attacks, leading to a recommendation that global navigation satellite system services should not be used as stand-alone positioning services for safety-related applications [2].

What distinguishes railway systems from many other systems is their inherent distributed and networked nature with tens of thousands of kilometers of track length for large operators. Thus, it is not economical to provide complete protection against physical access to this infrastructure, and as a consequence, railways are very vulnerable to physical denial-of-service attacks leading to service interruptions.

Another feature of railways distinguishing them from other systems is the long lifespan of their systems and components. Current contracts usually demand support for over 25 years, and history has shown that many systems, e.g., mechanical or relay interlockings, last much longer. IT security analyses have to take into account such a long lifespan. Nevertheless, it also should be noted that at least some of the technical problems are not railway specific, but are shared by other sectors such as air traffic management [3].

The number of publications and presentations related to IT security in the railway domain are increasing. Some are particularly targeted at the use of public networks such as Ethernet or the Global System for Mobile communication for railway purposes [4], while others, at least rhetorically, pose the question, "can trains be hacked?" [5] As mentioned earlier, some publications give detailed security-related recommendations [2]. In railway automation, harmonized safety standards were elaborated more than a decade ago; up to now, no harmonized IT security requirements for railway automation exist.

This chapter starts with a discussion of the normative background and then defines a reference architecture which aims to separate IT security and safety requirements as well as certification processes as far as possible. A short overview of the basic concepts of ISA99/IEC 62443 is given, and it is finally discussed how these concepts could be adapted to the railway automation domain.

22.2 Normative Background

In railway automation, there exists an established standard for safety-related communication, European Norm (EN) 50159 [6]. The first version of the standard was elaborated in 2001. It has proven quite successful and is used in other application areas, e.g., industry automation. This standard defines threats and countermeasures to ensure safe communication in railway systems. So, at an early stage, the standard established methods to build a safe channel (in security, called "tunnel" or "conduit") through an unsafe environment. However, the threats considered in EN 50159 arise from technical sources or the environment rather than from humans. The methods described in the standard are partially able to protect the railway system also from intentional attacks, but not completely. Until now, additional organizational and technical measures have been implemented in railway systems, such as separated networks, to achieve a sufficient level of protection.

The safety aspects of electronic hardware and systems are covered by EN 50129 [7]. However, security issues are taken into account by EN 50129 only as far as they affect safety issues, but, for example, denial-of-service attacks often do not fall into this category. Questions such as intrusion protection are only covered by one requirement in Table E.10 (unauthorized access). Nevertheless, EN 50129 provides a structure for a safety case which explicitly includes a subsection on protection against unauthorized access (both physical and informational).

On the other hand, industrial standards on information security exist. Here, we can identify the following standards:

- ISO/IEC 15408 [8] provides evaluation criteria for IT security, the so-called common criteria [9–11]. This standard is solely centered on information systems and has, of course, no direct relation to safety systems.
- The ISA99/IEC 62443 series [12] is a set of 11 standards currently elaborated by the Industrial Automation and Control System Security Committee of the International Society for Automation (ISA). The drafts are usually elaborated in ISA committees in the United States and then introduced in the normal commenting and voting process of the International Electrotechnical Commission (IEC). This standard is not railway specific and focuses on industrial control systems.

It is dedicated to different hierarchical levels, starting from concepts and going down to components of control systems.

A more comprehensive overview on existing information security standards is presented in BITKOM/DIN [13].

Railways are certainly critical national and international infrastructures, so recently, national governments, e.g., the United States and Germany, as well as the European Union have identified the problem. They have defined clear policies to support the implementation of industry-defined sector-specific IT security standards.

How can the gap between information security standards for general systems and railways be bridged? The bridge is provided by the European Commission Regulation on common safety methods no. 402/2013 [14]. This Commission Regulation mentions four different methods to demonstrate that a railway system is sufficiently safe:

1. By evaluation that the risk is broadly acceptable (based on expert judgment)
2. By following existing rules and standards (application of codes of practice)
3. By similarity analysis, i.e., showing that the given (railway) system is equivalent to an existing and used one
4. By explicit risk analysis, where risk is assessed explicitly and shown to be acceptable

We assume that from the process point of view, security can be treated just like safety, meaning that threats would be treated as particular hazards. Using the approach under item 2, common criteria [8] or ISA99/IEC 62443 [12] may be used in railway systems, but particular tailoring would have to be performed due to different safety requirements and application conditions. By this approach, a code of practice that is approved in other areas of technology and provides a sufficient level of security that can be adapted to railways. This ensures a sufficient level of safety.

However, the application of the general standards ISO/IEC 15408 [8] or ISA 99 [12] requires tailoring them to the specific needs of a railway system. This is necessary to cover the specific threats associated with railway systems and possible accidents and to take into account specific other risk-reducing measures already present in railway systems, such as the use of specifically trained personnel.

This finally leads to a kind of "IT-security-for-safety approach," where the IT security objectives and processes are embedded into the technical safety report from EN 50129; see Figure 22.1. Other security objectives can also be described in that structure; however, the puzzle is not complete today and needs further railway-specific supporting standards and guidelines.

Based on the common criteria approach, a first standard has been developed by the German standardization committee Deutsche Kommission für Elektrotechnik (DKE) [11], based on the common criteria and a specific protection profile tailored for railways, considering railway-specific threats and scenarios and yielding a set of IT security requirements. Assessment and certification of such a system can be carried out by independent expert organizations. Safety approval in Germany could then be achieved via the governmental organizations Federal German Railways Office (Eisenbahn-Bundesamt, EBA) for railway aspects and Federal German Office for Security in Information Technology (Bundesamt für Sicherheit in der Informationstechnik, BSI) for IT security aspects, which has accepted the protection profile.

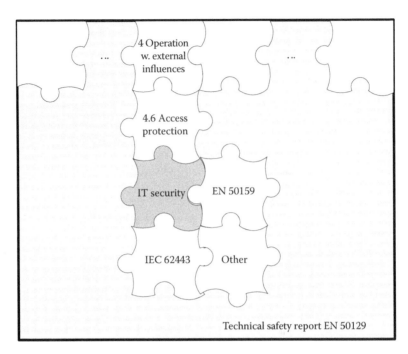

FIGURE 22.1
Embedding on IT security in the technical safety report from EN 50129.

22.3 Reference Architecture

The selected reference architecture starts from architecture B0 in the Comité Européen de Normalisation Électrotechnique (CENELEC) standard EN 50159, which aims at the separation of safety and security concerns. This concept can be illustrated by the onion skin model, where a security shell is placed between the railway signaling technology (RST) application and network layers. It is similar to a layer-of-protection approach. This security shell would be the security target of evaluation (TOE) according to the common criteria (see Figure 22.2).

Based on this onion skin model, a reference model (see Figure 22.3) has been chosen, in which the RST applications are contained in a zone (A or B in Figure 22.3). It is assumed that if communication between the zones were through a simple wire (as a model for a simple and proprietary communication means), then all safety requirements of EN 50159 would be fulfilled. Communication between the two zones will be through a tunnel or conduit in an open network. This matches well with the zone and conduit model in ISA 99/IEC 62443 [12].

In order to implement the conduit, additional security components have to be provided which are the physical implementations of the TOE. In Figure 22.3, the user is a generic representative of security management, which could have many different physical implementations, ranging from manual on-site to automated centralized management.

Based on these experiences in a first study in German DKE standardization committee 351.3.7, several IT security standards have been evaluated, and the best match to the requirements of the railway domain has been found with the ISA 99/IEC 62443 series.

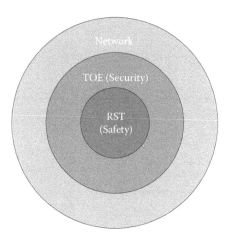

FIGURE 22.2
Onion skin model.

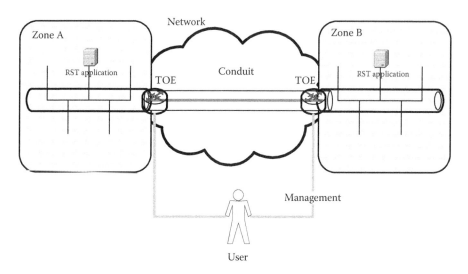

FIGURE 22.3
Zone and conduit reference architecture.

So the decision was taken to try to adapt these standards to the railway domain. For this purpose, a single additional guideline acting as an "adapter" for ISA 99/IEC 62443 to the railway domain has been issued [15].

Recently, similar activities have been performed in CENELEC [16] with almost the same result that sector-specific standards would be necessary to adapt the ISA99/IEC 62443 series to the railway domain; see Figure 22.4. Here it is proposed to have one overarching standard related to operator and global system integrator activities, which is then supplemented by a signaling-specific standard targeting the subsystem integration and product supplier roles.

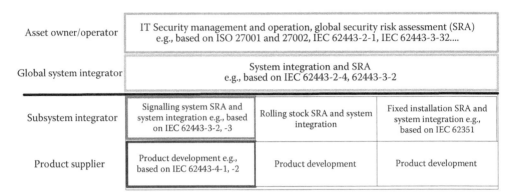

Asset owner/operator	IT Security management and operation, global security risk assessment (SRA) e.g., based on ISO 27001 and 27002, IEC 62443-2-1, IEC 62443-3-32....		
Global system integrator	System integration and SRA e.g., based on IEC 62443-2-4, 62443-3-2		
Subsystem integrator	Signalling system SRA and system integration e.g., based on IEC 62443-3-2, -3	Rolling stock SRA and system integration	Fixed installation SRA and system integration e.g., based on IEC 62351
Product supplier	Product development e.g., based on IEC 62443-4-1, -2	Product development	Product development

FIGURE 22.4
CENELEC proposal for the integration of ISA 99/IEC 62443.

22.4 Overview of ISA 99/IEC 62443 Standards

Currently, 11 parts are planned in this standard series covering different aspects for industrial automation and control systems (International Association of Classification Societies [IACS]; the main stakeholders are addressed in brackets):

- General (all):
  - 1-1: Terminology, concepts, and models
  - 1-2: Master glossary of terms and abbreviations
  - 1-3: System security compliance metrics
- Policies and procedures (railway operators)
  - 2-1: Requirements for an IACS security management system
  - 2-3: Patch management in the IACS environment
  - 2-4: Requirements for IACS solution suppliers
- System (system integrators)
  - 3-1: Security technologies for IACS
  - 3-2: Security levels (SLs) for zones and conduits
  - 3-3: System security requirements and SLs
- Components (suppliers of IT security products)
  - 4-1: Product development requirements
  - 4-2: Technical security requirements for IACS products

The documents are at different stages of development, some being already international standards, while others are at the first drafting stage. This leads, in particular, to problems when the documents build on each other, e.g., part 3-3 with detailed security requirements is in the final voting round, but it builds on part 3-2 which defines the SLs and has not been issued yet.

The fundamental concept of the standard is to define foundational requirements (FRs) and SLs which are a "measure of confidence that the IACS is free from vulnerabilities and functions in the intended manner." There are seven groups of FR:

1. Identification and authentication control
2. Use control
3. System integrity
4. Data confidentiality
5. Restricted data flow
6. Timely response to events
7. Resource availability

Each FR group has up to 13 subrequirement categories which are tailored according to the SL.

22.5 Security Levels

The default SL assignment for each zone and conduit is based on the attacker capability only:

- SL1: Casual or unintended
- SL 2: Simple means—low resources, generic skills, and low motivation
- SL 3: Sophisticated means—moderate resources, IACS-specific skills, and moderate motivation
- SL 4: Sophisticated means—extended resources, IACS-specific skills, and high motivation

The default assignment can be changed based on the results of a threat and risk analysis. However, it must be noticed that IT security threats cannot be quantified; as for intentional attacks exploiting vulnerabilities, there exists no meaningful way to define probabilities or rates. The IT security threats must be treated similarly to systematic faults in safety.

For each FR, a different SL may be assigned. There is a distinction between a target SL (as derived by threat and risk analysis), a design SL (capability of the solution architecture), and finally the achieved SL (as finally realized). If either the design SL or the achieved SL does not match the target SL, then additional measures have to be implemented (e.g., organizational) as compensation.

Taking into account the fact that there may also be no IT security requirement (SL 0), an SL assignment results in a seven-dimensional vector with $5^7 = 78.125$ possible different assignments. Based on the SL assignment, a standardized set of IT security requirements can be found in part 3-3, which would be a great advantage of the approach.

Note that there is no simple match between SL and the safety integrity levels (SILs) applied in safety standards. However, the definition of SL1 is very similar to requirements in the safety field, as also safety-related systems also have to address topics such as operator error, foreseeable misuse, or effects of random failure. So we can conclude that a safety-related system (SIL > 0) should also fulfill the technical requirements of IEC 62443-3-3 [17], as the threats which SL1 addresses are also safety hazards.

A concise comparison shows that there are some differences in detail. IEC 62443-3-3 contains 41 requirements for SL1, of which more than half are directly covered by safety standards such as EN 50129 or EN 50159, and about a quarter are usually fulfilled in railway safety systems representing good practice. However, another quarter of the requirements are usually not directly addressed in EN 50129 safety cases. The main reasons are that these requirements do not fall under the "IT security for safety" category but address availability requirements in order to prevent denial of service or traceability requirements; the following are examples:

- SR 6.1: The control system shall provide the capability for authorized persons and/or tools to access audit logs on a read-only basis.
- SR 7.1: The control system shall provide the capability to operate in a degraded mode during a denial-of-service event.
- SR 7.2: The control system shall provide the capability to limit the use of resources by security functions to prevent resource exhaustion.
- SR 7.5: The control system shall provide the capability to switch to and from an emergency power supply without affecting the existing security state or a documented degraded mode.

The current proposal is to include all SL1 requirements from IEC 62443-3-3 in the system requirement specification of any safety-related signaling system. In this way, no additional SL1 IT security certification would be necessary. Finally, these requirements should find their place in the EN 50126 standards series.

22.6 Basic IT Security Approach for Railway Automation Applications

For the sake of brevity, we are focusing on system aspects in this chapter. The first step after system definition would be to divide the system into zones and conduits (see Figure 22.3) according to the following basic rules:

- The system definition must include all hardware and software objects.
- Each object is allocated to a zone or a conduit.
- Inside each zone, the same IT security requirements are applicable.
- There exists at least one conduit for communication with the environment.

The next step is the threat and risk analysis resulting in SL assignment to each zone and conduit. Here, railway applications might need procedures different from industry automation as factories and plants are usually physically well protected and are not moving.

It should be noted that in this process for the threats identified, all four risk acceptance principles may be applied. In particular, the SL can be derived by different means, e.g., there could be a code of practice defining SL for particular architectures, or the SL could be taken over from a reference system.

The SL represents the effort which must be made so that the system effectively withstands attacks by these kinds of attackers. Only attackers who considerably exceed this effort might be able to overcome the IT security countermeasures. Different kinds of attackers on railway assets have already been researched [18], and the results were compatible with the classification proposed here.

It should be noted that according to IEC 62443, the assessment would have to be carried out for all seven FRs. However, from a railway safety point of view, some of the FRs have only little safety impact so that FRs such as "data confidentiality" or "resource availability" should always be assigned SL1. Also, it can be argued that there is no real reason to distinguish between the other FRs, as they are not independent, and it is proposed to allocate the same SL to all five remaining FRs. This would lead to only four classes for railway signaling applications:

- SL1 = (1,1,1,1,1,1,1), usually fulfilled by safety-related applications
- SL2 = (2,2,2,1,2,2,1)
- SL3 = (3,3,3,1,3,3,1)
- SL4 = (4,4,4,1,4,4,1)

As soon as the SL is assigned, standardized requirements from IEC 62443-3-3 can be derived. These requirements would be taken over to the railway automation domain without any technical changes. They would define the interface to use precertified IT security components for the railway automation domain.

Finally, the correct implementation of the IT security countermeasures according to IEC 62443-3-3 must be evaluated similar to the validation of safety functions.

22.7 Conclusion

This chapter has reviewed the normative situation in railway automation with respect to IT security. A proposed architectural concept has been presented aiming at the separation of safety and security aspects, as far as possible. This is achieved by integrating safety-related security requirements into the safety process and the safety case.

It has been explained that this approach matches well with the planned ISA 99/IEC 62443 series of standards, but that some aspects such as SL allocation need to be adapted. If this adaptation were successful, then railway automation could reuse the ISA 99/IEC 62443 series, in particular, their standardized IT security requirements, and would not have to create its own IT security standards. However, the work presented is still ongoing.

References

1. Sternstein, A. *Hackers Manipulated Railway Computers, TSA Memo Says.* Nextgov. http://www.nextgov.com/nextgov/ng_20120123_3491.php?oref=topstory, 2012.
2. Thomas, M. Accidental systems, hidden assumptions and safety assurance, in Dale, C., and Anderson, T., (eds) *Achieving System Safety, Proceedings of 20th Safety-Critical Systems Symposium,* Springer, Berlin, 2012.
3. Johnson, C. Cybersafety: Cybersecurity and safety-critical software engineering, in: Dale, C., and Anderson, T., (eds) *Achieving System Safety, Proceedings of 20th Safety-Critical Systems Symposium,* Springer, Berlin, 2012.
4. Stumpf, F. Datenübertragung über öffentliche Netze im Bahnverkehr—Fluch oder Segen? *Proceedings of Safetronic 2010,* Hanser, Munich.
5. Katzenbeisser, S. Can trains be hacked? *28th Chaos Communication Congress,* 2011.
6. CEN (European Committee for Standardization). EN 50159: *Railway Applications, Communication, Signaling and Processing Systems—Safety-Related Communication in Transmission Systems,* CEN, Brussels, September 2010.
7. CEN. EN 50129: *Railway Applications, Communication, Signaling and Processing Systems—Safety-Related Electronic Systems for Signaling,* CEN, Brussels, February 2003.
8. ISO (International Standards Organisation)/IEC (International Electrotechnical Commission) 15408: Information Technology—Security Techniques—Evaluation Criteria for IT Security, 2009.
9. *Common Criteria for Information Technology Security Evaluation: Part 2: Functional Security Components,* Common Criteria Recognition Agreement members Version 3.1, Revision 3, July 2009.
10. *Common Criteria for Information Technology Security Evaluation: Part 3: Assurance Security Components,* Common Criteria Recognition Agreement members Version 3.1, Revision 3, July 2009.
11. DIN (Deutsches Institut für Normung e.V.) DIN V VDE V 0831-102: *Electric Signaling Systems for Railways—Part 102: Protection Profile for Technical Functions in Railway Signaling* (in German), DIN, Berlin, 2013.
12. ISA (International Society of Automation) 99: *Standards of the Industrial Automation and Control System Security Committee of the International Society for Automation (ISA) on Information Security,* see http://isa99.isa.org/Documents/Forms/AllItems.aspx.
13. BITKOM (Bundesverband Informationswirtschaft, Telekommunikation und Neue Medien e. V.)/DIN Kompass der IT-Sicherheitsstandards Leitfaden und Nachschlagewerk 4. Auflage, 2009.
14. European Commission. Commission Implementing Regulation (EU) No. 402/2013 of April 30, 2013 on the common safety method for risk evaluation and assessment and repealing Regulation (EC) No 352/2009, Official Journal of the European Union, L121/8, 2013.
15. DIN. DIN V VDE V 0831-104: *Electric Signaling Systems for Railways—Part 104* (in German), DIN, Berlin, 2015.
16. CENELEC (European Comittee for Electrotechnical Standardisation) SC9XA: *Final Report of SGA16 to SC9XA,* 2016.
17. International Electrotechnical Commission. IEC 62443-3-3: Industrial Communication Networks—Network and System Security—Part 3-3: System Security Requirements and Security Levels, International Electrotechnical Commission, Brussels, 2013.
18. Schlehuber, C. *Analysis of Security Requirements in Critical Infrastructure and Control Systems* (in German), Master thesis, Technische Universität Darmstadt, Darmstadt, 2013.

23

Reliability Modeling and Analysis of European Train Control System

Yiliu Liu and Lei Jiang

CONTENTS

23.1 Introduction

The European Train Control System (ETCS) is a standard specification in signaling and train controlling proposed by the European Commission, aiming to improve safety, reliability, and interoperability of railway lines throughout European (ERA, 2014).

A brief historical review can indicate the background that the ETCS was initialized. Since the very early days, railway lines have always been divided into sections or blocks, and only one train is permitted in each block at a time to avoid collisions. Humans were firstly employed to stand along with lines and gave hand signals to a train driver whether he/she can go ahead or not. And then, with the development of technology, electrical telegraph, telephone, optical signaling, and mechanical signaling were also respectively introduced in railways, to transmit the information that a train had passed a specific block or the block was cleared. In 1980s, the concept of automatic train protection (ATP) systems appeared on the scene. Such kinds of system allows automatic braking, so as to effectively prevent trains from possible overspeeding or exceedance of stop signals (Palumbo et al., 2015).

Over the years, European countries independently developed at least 15 ATP systems, but these systems are incompatible and produced by different suppliers. In order to break the barriers to the cross-border railway traffic in Europe, the European Union initialized the European Rail Traffic Management System (ERTMS) standardization program in order to develop a common interoperable railway platform. The ERTMS/ETCS, or simply ETCS, is one component in the ERTMS program specified for control and signaling systems. Another important component of the ETRMS is called the global system for mobile communications—railways (GSM-R), which can provide voice and data communication between the track and the train.

Nowadays, the ETCS has been the technical specification of European legislation managed by the European Railway Agency (ERA). All new, upgraded, or renewed tracks in the European Union and the associated countries should adopt the ETCS. As highlighted in the report by Palumbo et al. (2015), the implementation of the ETCS can lead to advantages of safety, cost, maintenance, accessibility, and interoperability. Moreover, many railway networks outside Europe, especially those with high-speed railways, have also referenced the ETCS, e.g., China train control system (Ning et al., 2010) and Japanese train control system have many common points with the ETCS.

The ETCS is a typical safety-critical system, meaning that its failure can result in harm to people, economic loss, and/or environmental damage (Rausand, 2011). And therefore it is necessary to ensure that the system is highly reliable from the beginning. In this chapter, our focus will be reliability modeling and analysis methods for the ETCS.

The remainder of this chapter is organized as follows: Section 23.2 introduces the principles and structure of the ETCS, and then Section 23.3 presents several reliability analysis approaches. Case studies for ETCS level 2 are carried out in Section 23.4, and results based on different approaches are discussed. The summary and suggestions are presented in Section 23.5.

23.2 European Train Control System

23.2.1 Basic Elements of ETCS

Generally, the ETCS has two subsystems: onboard subsystem and trackside subsystem. The onboard system is installed on a train, controlling the movement of the train against a permitted speed profile with the information received from the trackside subsystem (Flammini et al., 2006). In the onboard subsystem, the main elements include (ERA, 2014) the following:

- European vital computer (EVC)—the core computing module in preventing overspeeding of a train
- Balise transmission module (BTM)—a module for intermittent transmission between track and train, processing signals received from the onboard antenna, and retrieving application data messages from a balise
- Train interface unit (TIU)—an interface to exchange information and issue commands to the rolling stock

- Radio transmission module (RTM)—a wireless module to receive and process GSM-R signals
- Driver machine interface (DMI)—a human–machine interface to provide the train driver with controlling information, including permission speed, actual speed, target speed, target distance, and alarms
- Juridical recording unit (JRU)—a unit to store the most important data for accident analysis and the information of operation of onboard equipment
- Odometer (ODO)—a device to measure the movement of the train (speed and distance)

In the trackside system, the main elements include the following:

- Euro-balise—A device installed between rail lines, to collect and store data related to the infrastructure, e.g., speed limits, position references, etc.
- Lineside electronic units (LEU)—An interface between the Euro-balise and interlocking, to receive information from the interlocking and send the appropriate telegrams to the Euro-balises in concordance with the lineside signaling.
- Radio block center (RBC)—A computer-based system that elaborates messages to be sent to the train, on the basis of the information received from external trackside systems and the onboard subsystems. RBC provides movement authorities to allow the movement of the train
- Key management center (KMC)—The role of the KMC is to manage the cryptographic keys and secure the communications between the ERTMS/ETCS entities.

At different levels of the ETCS, there are also some optional components, which will be introduced in the next subsections. Figure 23.1 provides the basic structure of the ETCS with the two subsystems.

In addition, it should be noted that although the interlocking module is not regarded as a part of the ETCS, it is always playing an important role in the signaling system.

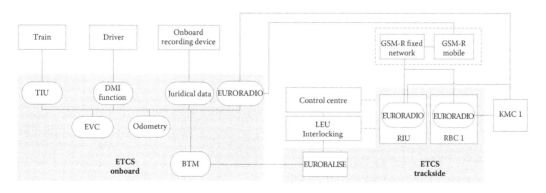

FIGURE 23.1
Basic structure of the ETCS. (Adapted from ERA, *ERTMS/ETCS—Class1 System Requirements Specification Issue 3.4.0, Subset-026*, ERA, Valenciennes and Lille, 2014.)

23.2.2 Application Levels of ETCS

ETCS is specified as four or five application levels dependent on equipment adopted and the approaches of information transmission. Individual railway administrations can consider their strategies, trackside infrastructure, and required performance, so as to select the appropriate ETCS applications (Leveque, 2008).

The five application levels of the ETCS can be summarized as follows:

- ETCS level 0—No ETCS device is equipped. There is no exchange of information between the train and the trackside, and the train driver has to observe the trackside signals. Movement authority (MA) is given by lineside optical signals or other means of signaling.

- ETCS level NTC—If the train is equipped with a specific transmission module (STM) but operates in a national system, the rail line can be regarded as the ETCS level STM line. Level NTC is used for trains equipped with national train control and speed supervision systems. Train detection and train integrity supervision are achieved by trackside signaling equipment.

- ETCS level 1—The trackside subsystem with Euro-balises is equipped to monitor the track occupancy. The MA is transmitted from the trackside via LEU to the train, and the onboard subsystem continuously monitors and controls the maximum speed. Due to the spot transmission of data, the train must travel over the beacon to obtain the next MA signal. Radio in-fill unit is an optional component in the trackside subsystem at ETCS level 1, and it can send a message to train by GSM-R transmission.

- ETCS level 2—The movement of the train is controlled based on radio. RBC is the main trackside equipment and a centralized safety unit at ETCS level 2: It receives the position information of a train from the trackside subsystem and then transmits the MA via GSM-R to the train. At ETCS level 2, it is still the trackside subsystem that monitors the track occupancy and the train integrity. However, signals are displayed to the driver through the train DMI at this level, rather than by fixed light lamps.

    The Euro-balises level 2 are passive positioning beacons or electronic milestones, and the train can identify its position between two balises with sensors. The onboard subsystem can continuously monitor the maximum speed of the train.

- ETCS level 3—No fixed signaling devices are required at ETCS level 3. The movement of the train is based on the full radio-based control. The MA is given to the driver based on the actual distance of a train from the previous.

The ETCS application levels are downward compatible, meaning that level 3 equipped trains are able to operate at level 1 and 2, while level 2 equipped trains can operate in level 1. Figure 23.2 illustrates the operations of trains at different levels, and here level NTC is ignored.

Currently, ETCS level 2 is the most widely used version in Europe, and level 3 is still under development. Therefore, in the following sections of this chapter, we put most attention to ETCS level 2.

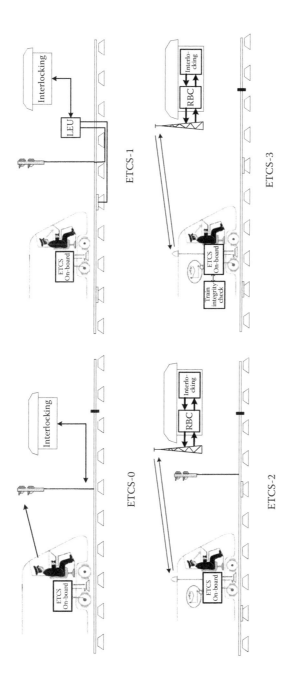

FIGURE 23.2
Illustrations of four levels of ETCS.

23.2.3 Failures of ETCS

In the ERTMS/ETCS reliability, availability, maintainability, and safety (RAMS) requirement specialization (EUG, 1998), three main types of ETCS failure modes are defined:

- Immobilizing failure—a failure which causes the ETCS to be unable to safely control two or more trains
- Service failures—a failure which causes the nominal performance of one or more trains to decrease and/or the system to be unable to safely control at most one train
- Minor failure—a failure that results in excessive unscheduled maintenance and cannot be classified in the preceding defined failure conditions

For each of these failure modes, the specification has defined the required indicators (Flammini, 2006), and at the same time, it is also necessary to identify the contributions of different elements in the trackside subsystem and in the on-board subsystem to these failure modes.

23.2.4 RAMS Requirements and Relevant Standards

Due to the safety criticality of railway signaling systems, when the ERTMS/ETCS was firstly initialized, the requirements for RAMS of ETCS were specified as the ERTMS/ETCS RAMS requirements specification (EUG, 1998). The specification includes four chapters, and in Chapter 2, the guidance on RAM target apportionment and the elements in a RAM improvement program are provided, including reliability modeling, and prediction, failure modes, effects, and criticality analysis, critical item lists, software reliability estimation, reliability preliminary test, and reliability demonstration test.

The ERTMS/ETCS RAMS requirement specification is developed on the basis of a series of European and international standards. Among these standards, the following are the latest versions of European standards in railway industries, and they are proposed by the European Committee for Electrotechnical Standardization:

- EN 50126: *Railway Applications—The Specification and Demonstration of Reliability, Availability, Maintainability and Safety (RAMS)* (CENELEC 1999)
- EN 50128: *Railway Applications—Software for Railway Control and Protection Systems* (CENELEC 2001)
- EN 50129: *Railway Applications—Communications, Signaling and Processing Systems— Safety-related Electronic Systems for Signaling* (CENELEC 2003)

International standard IEC 62278 (IEC 2002) was developed and released by the International Electrotechnical Standardization (IEC), which can be regarded as an international version of EN 50126 in terms of contents. The standard provides railway authorities and railway support industry with a common understanding and approach to the management of RAMS throughout all phases of the life cycle of a railway application. China's national standard GB/T 21562-2008 is developed on the basis of IEC 62278.

Considering the ETCS as a safety-critical system, the international standard IEC 61508 (IEC 2010) is a generic standard for electrical, electronic, and programmable safety-critical systems, which plays as the basis to develop relevant standards for all sectors. In the part 6 of IEC 61508, several reliability modeling methods have been mentioned, and they are the methods to be introduced in the following sections of this chapter.

In these RAMS specification and standards, several reliability parameters are commonly used, including success probability ($R(t)$), failure probability ($F(t)$), failure rate ($z(t)$ or λ if it is constant), mean time to failure (MTTF) and mean time between failures (MTBF). It should be noted that the time in MTTF and MTBF is not necessarily calendar time, and in fact, it can be operational time, cycle, or operation distance, e.g., kilometer.

23.3 Reliability Modeling and Analysis Methods

In this section, the commonly used reliability modeling methods in relevant standards and literature will be briefly reviewed. It should be noted that we only focus on the quantitative analysis approaches. More details about these methods can be found in Chapters 3 and 4 of this handbook.

Every involved method can be regarded including two elements: the modeling framework, which is used to represent a system, and the algorithm, which is behind the representation and is used to carry out the reliability analysis.

23.3.1 Fault Tree Analysis

Fault tree analysis (FTA) is one of the most important techniques in reliability engineering, by linking individual failures with the system failure. This method is based on a top-down logic diagram always with a tree shape as shown in Figure 23.3a.

FTA is a deductive method, and the analysis starts with a potential critical event of the system, e.g., a failure, and this event is called the TOP event of the tree. When checking the cause of this TOP event, it can be found that the system failure resulted from component failures or the combinations of component failures, and these component failures are basic events in the fault tree. Two logic gates are always employed in FTA to connect the TOP and the basic events: One is the OR gate (as G_0 in Figure 23.3a), and the other is the AND gate (as the G_1 in Figure 23.3). The OR gate means that the output event of the gate will occur if any of the input events occurs, and meanwhile, the AND gate means that the output event of the gate can occur only if all of the input events occur.

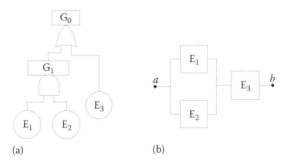

(a)   (b)

FIGURE 23.3
Simple examples of (a) fault tree and (b) RBD.

A fault tree can support qualitative and quantitative analyses. The analysis can start from defining the TOP event or the system failure of interest, and the next step is to identify the direct causes of the TOP event and connect them with logic gates. Then, the identified causes are regarded as effects, and their causes need to be identified. The preceding steps can be repeated until basic events are found, meaning that the reliability or event data are easily found.

In the qualitative analysis of fault trees, a commonly used approach for small size models is to identify minimal cut sets. A cut set is a set of basic events whose (simultaneous) occurrence ensures that the TOP event occurs, while a cut set is said to be minimal if the set cannot be reduced without losing its status as a cut set. For the example in Figure 23.3, the minimal cut sets of the fault tree include $\{E_1, E_2\}$ and $\{E_3\}$.

In the quantitative analysis, the occurrence of basic event i at time t is a random variable $Y_i(t)$ with two possible values of 1 (occur) and 0 (not occur), $q_i(t)$ can be used to denote the occurrence probability of the event (the unreliability of component i in many cases), and then we have

$$\Pr\big(Y_i(t)=1\big)=q_i(t) \quad \text{for } i=1,2,\ldots,n. \tag{23.1}$$

If the basic event i is one of n input events of an AND gate, the occurrence probability of the output event ($Q_s(t)$) of the gate can be written as

$$Q_s(t)=\prod_{i=1}^{n} Q_i(t). \tag{23.2}$$

While for an OR gate, the occurrence probability of the output event at time t is

$$Q_s(t)=1-\prod_{i=1}^{n}\big(1-Q_i(t)\big). \tag{23.3}$$

Both EN 50128 (CENELEC 2001) and EN 50129 (CENELEC 2003) have recommended FTA for the RAMS analysis in railway applications. Flammini et al. (2006) have evaluated system reliability of ETCS by fault trees, in consideration of the modeling capacity of FTA for large-scale systems. More symbols and information about FTA can be found in the standard IEC 61025 (IEC 2006).

23.3.2 Reliability Block Diagram

Another commonly used method in reliability analysis is reliability block diagram (RBD), which is a success-oriented network with two basic elements including functional blocks (rectangles) and connections (lines) between blocks (as shown in Figure 23.3b). Different from a fault tree, an RBD illustrates how a specified system function is fulfilled with the connections of components. It should be noted the RBD is the logical connections that are required to function, rather than the physical layout of a system.

A basic RBD can also be regarded as a Boolean model, with each block having two states: functional and failed. A system is considered as functioning if it is possible to pass through its RBD model from one end point to the other end point (e.g., from a to b in Figure 23.3b).

The RBD has two basic structures: series structure and parallel structure. In a series structure, the system is functioning if and only if all of its n components are functioning, while in the parallel structure, the system is functioning if at least one of its n components is functioning. It can be found that a parallel structure is equivalent to the AND gate in a fault tree, and a series structure is equivalent to the OR gate.

In the qualitative analysis with RBD, it is helpful to identify the ways in which the system can realize its function. A set of components that by functioning ensure that the system is functioning is called a path set, and it is said to be minimal if the path set cannot be reduced without losing its status. For example, in Figure 23.3, the sets $\{E_1, E_3\}$ and $\{E_2, E_3\}$ are two minimal path sets.

In a quantitative analysis, the state of component i at time t can be regarded as a random variable $X_i(t)$ with two possible values: 0 (failed) and 1 (functional). We are interested in the following probability:

$$\Pr\big(X_i(t)=1\big)=R_i(t) \quad \text{for} \quad i=1,2,\dots,n, \tag{23.4}$$

where $R_i(t)$ is called as the reliability or survivor probability of component i at time t.

For a series structure with n component, its survivor function is

$$R_s(t)=\prod_{i=1}^{n} R_i(t). \tag{23.5}$$

while for a parallel structure with n components, its survivor function is

$$R_s(t)=1-\prod_{i=1}^{n}\big(1-R_i(t)\big). \tag{23.6}$$

The RBD is also mentioned in EN 51026 (CENELEC 1999) and EN 50129 (CENELEC 2003). In fact, as a classic reliability modeling approach, the RBD has been used in the railway industry for several decades, e.g., analysis of electrical substation in high-speed railway system (Cosulich et al., 1996) and performance analysis of train services (Chen, 2003). More details about RBD models can be found in the latest version IEC 61078 (IEC 2016).

23.3.3 Bayesian Belief Network

A Bayesian network (BN), or a Bayesian belief network (BBN), is a graphical model illustrating the causal relationships between causes and outputs of a system. A BN consists of two parts, a directed acyclic graph (DAG) with a finite set of nodes and a set of directed edges and the conditional probability table (CPT) associated with each node. Therefore, a BBN can be written as $N = \langle(V, E), P\rangle$, where (V, E) represents the nodes and edges of DAG, and P is a probability distribution of each node. The directed edges between the nodes present the causal relationship among them. For example, in Figure 23.4, nodes E_1 and E_2 are used to denote the causes of the event represented by E_3, and so E_1 can be called as a parent node of E_3, while E_3 is the child node of E_1. In a BBN, nodes without parent nodes are called root nodes, nodes without children nodes are called leaf nodes, and all the others are intermediate nodes.

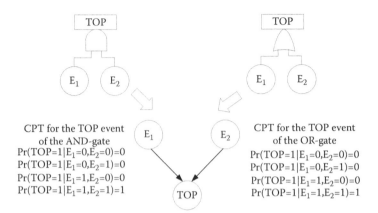

FIGURE 23.4
Translation from a fault tree to a BBN.

When we consider three events A, B, and C, and we know that event A has occurred if the following equation is satisfied:

$$\Pr\left(B \cap C \mid A\right) = \Pr\left(B \mid A\right) \cdot \Pr\left(C \mid A\right) \tag{23.7}$$

Events B and C are said to be conditionally independent, meaning that the two nodes are independent when we know the status of their parent node (Rausand, 2011). The quantitative analysis of a BBN can be carried out under the assumption that each node in the network is conditionally independent when we know the status of its all parents.

If we generally consider a BBN with n nodes, the probability that $X_i = x_i$ (for i=1, 2, …, n) is

$$\Pr(X_1 = x_1 \cap \cdots \cap X_n = x_n) = \prod_{i=1}^{n} \Pr\left(X_i = x_i \mid \text{Parent}(X_i)\right). \tag{23.8}$$

A fault tree model can be translated into a BBN model without losing any details. Figure 23.4 shows how AND and OR gates are translated into BBN models. It can be found that the TOP events of the two logic gates can be translated into a BBN with the same DAG but different CPTs.

A BBN is a powerful technique in analyzing the complex system with uncertainties, and it has been adopted by several researchers in railway signaling applications. In the paper by Flammini et al. (2006), the authors have modeled the ETCS from a BBN for reliability analysis. Mahboob et al. (2012) have carried out safety analysis of a DMI with a BBN, and more recently, Castillo et al. (2016) performed a BBN-based safety analysis for railway lines involving many elements. Although no international standard of BBNs exists, numerous guidelines and tutorials can be easily found on the internet.

23.3.4 Markov Method

Both the FTA and RBD methods are very effective tools in the structural reliability analysis of systems, but it is difficult for them to analyze some dynamic behaviors of systems,

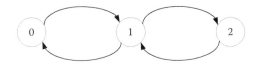

FIGURE 23.5
Simple example of the Markov diagram.

such as maintenances and restorations from faulty states. Some state transition models are therefore introduced in the system analysis, and one of the most widely used is the Markov method.

The states and the transitions between different states of a system are illustrated by a state transition diagram, called a Markov diagram, in the continuous time Markov model, which is commonly used in reliability analysis. For example, Figure 23.5 shows a Markov diagram of a small system with two components, where the circles represent system states: 0, two components functional; 1, one component functional and one failed; and 2, two component failed. Directed arcs are the transitions between states. The transitions of 0 to 1 and 1 to 2 mean that components fail, and the transitions of 2–1 and 1–0 mean that components are restored.

The calculation behind a Markov diagram is always based on the assumption that the state transitions are stochastic processes following the Markov property, or Markov processes: The future states of the system are only dependent on the present state and are not influenced by how the process arrived in the present state. All transitions in a Markov process are exponentially distributed in terms of transition time.

The introduction of the quantitative analysis based on the Markov diagram is ignored in this chapter due to the limitation of pages. Readers can find the specific knowledge in Chapter 8 of the book by Rausand and Høyland (2004). In addition, cases of reliability analysis for safety-critical systems (Liu and Rausand, 2011, 2013) and railway applications can also be found in many literatures. In the railway applications, several researchers, e.g., Chandra and Kumar (1997), Krenzelok et al. (2010), and Zhang et al. (2010), have adopted Markov methods to analyze the reliability and safety of signaling systems.

23.3.5 Petri Net and Extensions

An alternative state transition model, a Petri net, is also a graphical tool, with better performance in terms of modeling flexibility. According to the terminology specified in the standard IEC 62551 (IEC 2013), places (circles in Figure 23.6) and transitions (bars in Figure 23.6) are two static elements of a Petri net, and they are connected by arcs. Tokens are dynamic elements which represent the movable resources in the system. Tokens can reside in the places, and the distribution of tokens in the places is regarded as a system state.

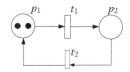

FIGURE 23.6
Simple example of Petri net.

Tokens can be delivered by arcs, and each arc is assigned a multiplicity to denote its token deliverability. When the number of tokens in all input places of a transition is at least equal to the multiplicities of the associate arcs to the transition, the transition is enabled. And then, the transition can be fired to change the distribution of tokens in places.

Figure 23.6 provides a simple example of Petri net, which can be used to represent the same system in Figure 23.5. When two tokens are in place p_1, both of the two components are functional. At this moment, transition t_1 is enabled. If t_1 is fired, one token is removed from p_1, and then a new token is released to p_2, since the two arcs p_1 to t_1 and t_1 to p_2 have a multiplicity of 1 (1 is default in drawing if no specific digit is added). So, the new state is that one token in p_1 and one in p_2, meaning that one component is failed, and one is still functional. In fact, the firing of t_1 models the failure of a component. If t_1 is fired again, no token is in p_1 and two tokens are in p_2, meaning that two components are failed. On the other side, if t_2 is fired, one token is removed from p_2, representing that one component is repaired.

A firing time (delay from enabled to fired) can be assigned to each transition. If all transitions in a Petri net model have zero firing time or exponentially distributed firing time, the model is actually equivalent to a Markov process, and the numerical analysis approaches suitable for Markov models can be adopted. While deterministic transitions and other nonexponential transitions are involved in a Petri net model, Monte Carlo simulation is needed to capture the useful information.

A Petri net has many extensions, such as colored Petri net and stochastic activity net. In IEC 61508, an RBD-driven Petri net is recommended for reliability analysis. In such kinds of modeling techniques, two formulas are added in an ordinary Petri net model:

- Predicate (identified by "?" or "??")—to control the enabling condition of a transition
- Assertion (identified by "!" or "!!")—to update one variable when the transition is fired

More details of basic Petri net models can be found in IEC 62551 (IEC 2013), and details of RBD Petri nets are in part 6 of IEC 61508 (IEC 2010). Relevant literatures include Petri net–based analysis of fixed-block railway signaling systems (Durmus et al., 2015), colored Petri net safety analysis of ETRMS (Barger et al., 2009), and RBD-driven Petri net analysis of safety critical systems (Liu and Rausand, 2016).

23.4 Modeling and Analysis for ETCS Level 2 System

In this chapter, the onboard and trackside subsystems under ETCS level 2 will be taken as examples to illustrate the implementation of the preceding methods. The reason why level 2 is selected lies in its wide applications in Europe now. Firstly, in the reliability analysis based on system structures, the Boolean methods, e.g., FTA, RBD, and BN, are presented while ignoring the repairs after failures. And then, the transition-based methods, e.g., Markov method and Petri net, will be adopted by assuming that all components are repairable.

23.4.1 Structural Modeling and Analysis of the Onboard Subsystem

On the side of onboard, the subsystem consists of EVC, RTM, TIU, EURORADIO, BTM, BTM antenna, DMI, and ODO. These devices communicate with each other via Process Field Bus (PROFIBUS), which is a standard for fieldbus communication. The subsystem has a distributed structure and a modular design, in order to avoid the interactive effects between different components/modules. Modules of EVC, RTM, TIU, EURORADIO, odometer, and PROFIBUS are based on the hot standby redundancy, where the primary and backup channels simultaneously run, while BTM, BTM antenna and, DMI have the cold standby redundancy, meaning that the backup system is only called upon when the primary system fails. The subsystem structure is illustrated in Figure 23.7. JRU is not considered in this section since it has no obvious influence on the system reliability.

In order to carry out the reliability analysis, some assumptions are needed:

- The failures of all components in the system follow the exponential distribution, meaning that component i has a constant failure rate λ_i.
- All the components are repairable, and the times for component i to be repaired are in the following exponential distribution, with the repair rate μ_i.
- Cold standby redundancies are regarded as no-redundancy since they cannot immediately shift in case of the master component fails.

We set the failure of an ETCS-2 onboard subsystem as the event of interest and build a fault tree (as shown in Figure 23.8) based on cause analysis. The construction of the tree model stops at the levels of component failures where reliability information is available. The data of the basic events (listed in Table 23.1) are collected in surveys and literature reviews.

For the fault tree, it is not difficult to identify all minimum cut sets: {X1}, {X2}, {X3}, {X4, X5}, {X6, X7}, {X8, X9}, {X10, X11}, {X12, X13}, and {X14, X15}. With the exponential distribution, the occurrence probability of the basic event X1 in the operation time of 10^4 hours is

$$F_{X1}(10^4) = 1 - e^{-0.5 \times 10^{-5} \times 10,000} = 4.8771 \times 10^{-2}.$$

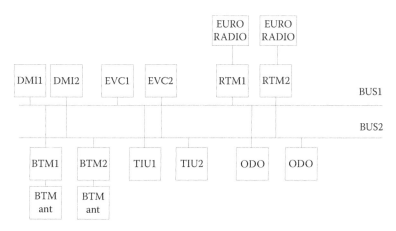

FIGURE 23.7
Onboard subsystem structure of ETCS level 2.

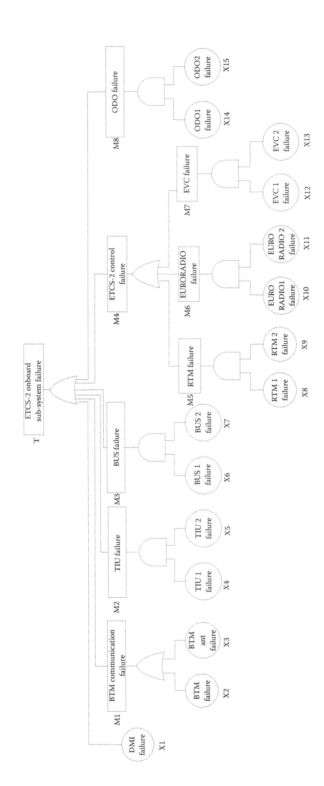

FIGURE 23.8
Fault tree of the ETCS level onboard subsystem.

TABLE 23.1

Reliability Parameters of Basic Events

Basic Event	Equipment	Failure Rate (10^{-5}/h)
X1	DMI	0.50
X2	BTM	0.20
X3	BTM antenna	0.07
X4, X5	TIU1, 2	2.10
X6, X7	BUS1, 2	0.60
X8, X9	RTM1, 2	1.80
X10, X11	EURORADIO 1, 2	1.20
X12, X13	EVC1, 2	1.49
X14, X15	ODO1, 2	0.003

Similarly, the failure probabilities of X2 and X3 are 1.9801×10^{-2} and 0.6976×10^{-2}, respectively, while the MTTF of X1 is equal to $1/(0.5 \times 10^{-5}) = 2 \times 10^5$ hours.

For the cut set {X4, X5}, its failure probability can be expressed as

$$F_{\{X4,X5\}}(t) = (1 - e^{-\lambda_4 t})(1 - e^{-\lambda_5 t}) = 1 - e^{-\lambda_4 t} - e^{-\lambda_5 t} + e^{-(\lambda_4 + \lambda_5)t}.$$

Within 10^4 hours, the failure probability of the cut set is 3.5878×10^{-2}. Similarly, we can calculate the failure probabilities of all the other cut sets as 0.3391×10^{-2}, 2.7136×10^{-2}, 1.2787×10^{-2}, 1.9163×10^{-2}, and 8.9973×10^{-8}. For the entire fault tree, the occurrence probability of the TOP event, or the hazard probability, is the sum of the probabilities of all minimum cut sets, so it is calculated as 0.1739.

However, it should be noted that the preceding calculation does not consider that all the components can be repairable. In fact, the maintenance strategy is influential on the failure probability. Assume that after a failure, the train still needs to be operated for 4 hours until it is delivered to a workshop. Under this assumption, given that X1 occurs, the train operates under hazard for 4 hours, namely, the mean time to restoration service is 4 hours, the sojourn probability in hazard of the system is $F_{X1}(10^4) \times 4/10^4 = 1.9508 \times 10^{-5}$.

In case X4 fails, the failure of X5 in another channel of the parallel TIU 1 in the next 4 hours can result in the exposure to hazard. Suppose the second failure is evenly distributed in the 4 hours; then the mean down time is therefore 2 hours, and the hazard probability is

$$H = \binom{2}{1} F_{X4}^2(10^4) \cdot 4 \cdot 4/2 = 7.0599 \times 10^{-9}.$$

It can be found that the contribution of such parallel structures to the system hazard is much lower than those single modules, so the system hazard probability can be approximated based on the X1, X2, and X3, as 3.0219×10^{-5}.

Another parameter of interest is the availability/unavailability of the system. We can assume that the train with failures will be repaired with the mean repair time of 8 hours. For X1, the component unavailability can be obtained as

$$\bar{A} = \frac{\text{MDT}}{\text{MTTF} + \text{MDT}} = \frac{\text{MTTRS} + \text{MRT}}{\text{MTTF} + \text{MTTRS} + \text{MRT}} = \frac{4 + 8}{20,000 + 4 + 8} = 5.9964 \times 10^{-4}.$$

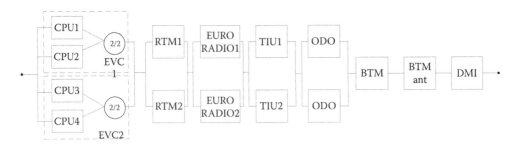

FIGURE 23.9
RBD model for the onboard subsystem of ETCS level 2.

The system unavailability comes from the unfunctionality of the onboard system when the train is operating and the repair time. Once one component in any module fails, the repair is assumed to result in unavailability of the train. Therefore, for the three single modules, the unavailable time of each failure is 12 hours, and for other components in parallel structures, the unavailable time is approximated as 8 hours. The system unavailability is the sum of all component availability as 0.1153%. In practice, standby trains can be put into operation when others are in repair, so the actual operation unavailability will be lower than this value.

With the fault tree, it is not difficult to develop an equivalent RBD model as shown in Figure 23.9. In this RBD model, we present more information in the EVC module, which has a double 2-out-of-2 (2oo2) structure, meaning that both EVC1 and EVC2 have an internal 2oo2 voting mechanism, and meanwhile, EVC1 and EVC2 play as hot standby for each other. The cycles with 2/2 denote the voting mechanism of 2oo2.

23.4.2 Bayesian Analysis of the Onboard Subsystem

Based on the fault tree in Figure 23.8, a BN model is also built as shown in Figure 23.10.

The nodes in the BN are actually minimum cut sets in the fault tree of Figure 23.8. For those root nodes, their failure rates have been listed in Table 23.1 or obtained in the last subsection. For the intermediate nodes, e.g., M1, their CPTs can be generated with the mechanisms for AND and OR gates shown in Figure 23.4. It is also possible to calculate the hazard probability of the subsystem, namely, the occurrence probability of the node T in Figure 23.10, as 0.1619 within 10^4 hours.

A backward analysis can be conducted based on the BN. We assume that the hazard has occurred, and then track back which nodes has more contributions to the hazard.

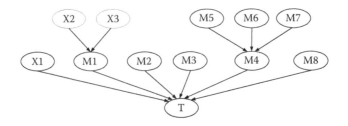

FIGURE 23.10
BN for the onboard subsystem.

TABLE 23.2

Posterior Probability of Root Nodes

Node	Posterior Probability
X1	0.301169
X2	0.122275
X3	0.043078
M2	0.221553
M3	0.02094
M5	0.167569
M6	0.078962
M7	0.118335
M8	0.000001

The posterior probabilities of all root nodes can be calculated as listed in Table 23.2, and the nodes with higher posterior probabilities can be regarded as the vulnerable parts of the system. According to Table 23.2, we can find that X1, M2, and M5 are three most vulnerable modules. Such an analysis provides some clues for reliability improvement of the system.

23.4.3 Availability Analysis with Transition-Based Methods

Transition-based methods, such as the Markov model, are powerful in analyzing small systems but with more dynamical behaviors. Here, we take the repairable double 2oo2 EVC module as an example. Figure 23.11 illustrates a Markov state transition diagram for the EVC module.

Table 23.3 describes all states in the Markov diagram: state 0 means that all Central Processing Units (CPUs) in the EVC module are functional. If one of the four CPUs fails, the system moves to state 1. We use λ to denote the failure rate of one CPU, and so the transition rate from state 0 to state 1 is 4λ. With one CPU failure, e.g., in the channel of EVC1, the module is still working with the functioning EVC2. And then, if another CPU in EVC1 fails, the system goes in state 3, but it keeps working because EVC1 and EVC2 are in parallel in terms of functionality. However, if a failure occurs in one of the two CPUs of EVC2, the module will stop working, meaning that the system transits from state 1 to state 2 with the rate of 2λ. Suppose the rate to service after the second failure is μ_1, the system can move from state 2 to 4 for repair. The rate to service after the failure from states 3 to 4 is μ_1; from states 4 to 0, the transition follows the repair rate μ_3.

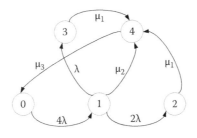

FIGURE 23.11
Markov model of the double 2oo2 EVC module.

TABLE 23.3

State Descriptions of the Markov Model

State	Description
0	No failure
1	One failure in EVC1 or EVC2
2	Both EVC1 and EVC 2 have a failure
3	Two CPUs in EVC1 or EVC2 are in failure
4	Repair

The steady sojourn probabilities $(P_0, P_1, P_2, P_3, P_4)$ of the system in the five states can be determined from the matrix equation (see more explanations in the book Rausand and Høyland, [2004, p. 315]):

$$(P_0, P_1, P_2, P_3, P_4) \cdot \begin{pmatrix} -4\lambda & 4\lambda & 0 & 0 & 0 \\ 0 & -(3\lambda + \mu_2) & 2\lambda & \lambda & \mu_2 \\ 0 & 0 & -\mu_1 & 0 & \mu_1 \\ 0 & 0 & 0 & \mu_1 & -\mu_1 \\ \mu_3 & 0 & 0 & 0 & -\mu_3 \end{pmatrix} = (0,0,0,0,0).$$

And also we know that

$$P_0 + P_1 + P_2 + P_3 + P_4 = 1.$$

With these equations, we can obtain the steady-state probabilities. For example, when $\lambda = 7.45 \times 10^{-6}$/h, $\mu_1 = 1/2$, $\mu_2 = 1/4$, and $\mu_3 = 1/8$, the hazard probability, namely, the sojourn probability of the model in state 2, is 3.5506×10^{-9}, while the unavailability of this module is the sojourn probability in states 2 and 4, as 2.3832×10^{-4}.

For such an EVC module, the unavailability can be calculated by considering the failure probability of one channel when the other one in failure. In this way, the approximation result is 3.0717×10^{-9}. The two results are rather close, but it should be noted that the Markov method considers more situations.

Figure 23.12 illustrates an RBD-driven Petri net model which is an alternative of the Markov model in Figure 23.11 in terms of the analysis of the double 2oo2 EVC module. Two tokens in p_1 are used to denote the two CPUs in EVC1, and once the transition t_1 is

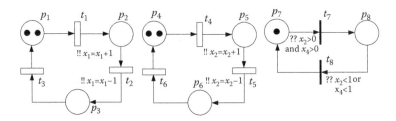

FIGURE 23.12
RBD-driven Petri net for the EVC module.

fired, one token in p_1 is removed and one token is released in p_2. We can set a variable x_2 to denote the number of tokens in p_2, and so the assertion of t_1 increased the value of x_1 with 1. Such a change of the token distribution means a failure of a CPU. The firing of the transition t_2 can absorb one token from p_2 and deposit one in p_3, meaning that the failed CPU is delivered to repair. The assertion of t_2 is $x_1 = x_1 - 1$. The transition rate of t_1 is dependent on the variable x_1.

The submodel in the middle of Figure 23.12 is for EVC2, with the same dynamics of that for EVC1; we use x_2 to represent the number of tokens in p_5, and the transition rate of t_4 relies on the variable x_1. Given that at least one token resides in p_2, and meanwhile, one token is in p_5, there are failures in both of the channels of the EVC module, and so the module stops working. In this case, the enabling condition/predicate of t_7 ($x_1 > 0$ and $x_2 > 0$) is satisfied, and then it will be fired in no time to move the token from p_7 to p_8, since it is an immediate transition. If any channel is repaired ($x_1 < 1$ or $x_2 < 1$), the token in p_8 will be removed.

It is possible to run Monte Carlo simulations based on this model in order to capture the average unavailability of the module, which is the sojourn probability of one token in p_6. With 10^7 iterations in the reliability analysis software GRaphical Interface of reliability Forecasting, developed by TOTAL (GRIP), the hazard probability is obtained as 3.1154×10^{-9}. It can be found that the simulation result is similar with that by the numerical analysis, and it can be regarded as the verification of the Markov model.

If we want to obtain the unavailability of the module, we can add a similar structure of p_7 and p_8, to capture the sojourn probabilities of tokens in p_2, p_3, p_5, and p_6.

The RBD-driven Petri net model in Figure 23.6 can be used for modeling parallel structures in the ETCS onboard subsystem.

23.4.4 Models of Trackside Subsystem and Operations

In this subsection, we present the fault tree model for the trackside subsystem in Figure 23.13, but due to page limitations, further analysis of this subsystem is not presented. In fact,

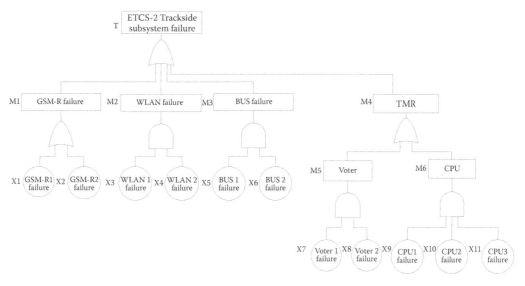

FIGURE 23.13
Fault tree of ETCS-2 trackside subsystem.

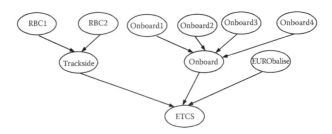

FIGURE 23.14
BN of an operating ETCS.

the analysis for the trackside subsystem is similar to that for the onboard subsystem. The only difference is that the CPU adopts triple modular redundancy.

It is also possible to analyze some characteristics in the system operation. Suppose four trains and two RBCs are operating in an ETCS. By combining the BNs of ETCS onboard and trackside subsystem, a whole picture can be obtained by one BN model in Figure 23.14.

23.5 Conclusion and Suggestions

This chapter has reviewed several commonly used modeling approaches and presented their applications in reliability analysis of the ETCS level 2 system. Based on the case studies, some preliminary suggestions can be provided to engineers and analysts in railway RAMS when they need to select suitable methods for specific systems:

- For large-scale systems with components having two states, fault trees and RBDs are effective, and a BN can be considered.
- A BN is more flexible in modeling and can be efficiently updated when new information is introduced.
- For small systems with dynamics and other complex behaviors, the Markov model is helpful. If the Markov property is satisfied, numerical analysis is possible to be carried out.
- A Petri net can be an alternative of the Markov model, especially for the cases where the system states are not identified. Some extensions of Petri net, e.g., RBD-driven Petri net, are manageable tools for large-scale systems.
- A Monte Carlo simulation is suitable for most reliability analyses, but the authors recommend that the simulation can be combined with some graphical models in order to facilitate the communication and verification.

In recent years, formal languages have been introduced in the reliability analysis of complex systems; for example, Xie et al. (2013) have proposed a unified modeling language-based model to estimate the reliability of the automatic train protection and block systems. Such kinds of models are more effective and flexible for the analysis of large complex systems, but the modeling complexity and maintenance difficulty also increase.

References

Barger, P., Schon, W., and M. Bouali. 2009. A study of railway ERTMS safety with colored Petri nets. In *Proceeding of European Safety and Reliability Conference (ESREL 2009)*, eds: S. Martorell, C. G. Soares, and J. Barnett, 1303–1309. Boca Raton, FL: CRC Press.

Castillo, E., Grande, Z., and A. Calvino. 2016. Bayesian networks-based probabilistic safety analysis for railway lines. *Computer-Aided Civil and Infrastructure Engineering* 31(9): 681–700.

CENELEC (European Committee for Electrotechnical Standardization). 1991. EN 50126: *Railway Applications—The Specification and Demonstration of Dependability, Reliability, Availability, Maintainability and Safety (RAMS)*. Brussels: CENELEC.

CENELEC. 2001. EN 50128: *Railway Applications—Software for Railway Control and Protection Systems*. Brussels: CENELEC.

CENELEC. 2003. EN 50129: *Railway Applications—Communications, Signaling and Processing Systems—Safety-Related Electronic Systems for Signaling*. Brussels: CENELEC.

Chandra, V., and K. V. Kumar. 1997. Reliability and safety analysis of fault tolerant and fail safe node for use in a railway signalling system. *Reliability Engineering and System Safety* 57(2): 177–183.

Chen, H. K. 2003. New models for measuring the reliability performance of train service, In *Proceeding of 14th European Safety and Reliability Conference (ESREL 2003)*, eds: T. Bedford and P. H. A. J. M. Van Gelder, 397–401. Lisse: A. A. Balkema.

Cosulich, G., Firpo, P., and S. Savio. 1996. Electrical susbstation dependability analysis for high speed railway systems: A simplified approach. In *Computers in Railway Vol 2: Railway Technology and Environment*, eds: A. J. Brebbia, R. J. Hill, S. Sciutto, and S. Sone, 139–148. Southampton: WIT Press.

Durmus, M. S., Takai, S., and M. T. Soylemez. 2015. Decision-making strategies in fixed-block railway signaling systems: A discrete event systems approach. *IEEJ Transactions on Electrical and Electronic Engineering* 10(2): 186–194.

EUG (ERTMS User Group). 1998. *ERTMS/ETCS RAMS Requirements Specification*. Brussels: EUG.

ERA (European Railway Agency). 2014. *ERTMS/ETCS—Class1 System Requirements Specification Issue 3.4.0, Subset-026*. Valenciennes and Lille: ERA.

Flammini, F., Marrone, S., Mazzocca, N., and V. Vittorini. 2006. Modeling system reliability aspects of ERTMS/ETCS by fault trees and Bayesian networks. In *Proceedings of European Reliability and Safety Conference (ESREL 2006)*, eds: C. G. Soares and E. Zio, 2675–2683. London: Taylor & Francis Group.

IEC (International Electrotechnical Commission). 2002. IEC 62278: *Railway Applications—Specification and Demonstration of Reliability, Availability, Maintainability and Safety (RAMS)*. Geneva: IEC.

IEC. 2006. IEC 61025: *Fault Tree Analysis*. Geneva: IEC.

IEC. 2010. IEC 61508: *Functional Safety of Electrical/Electronic/Programmable Electronic Safety-related Systems, parts 1–7*. Geneva: IEC.

IEC. 2013. IEC 62551: *Analysis Techniques for Dependability—Petri Net Techniques*. Geneva: IEC.

IEC. 2016. IEC 61078: *Reliability Block Diagram*. Geneva: IEC.

Krenzelok, T., Bris, R., Klatil, P., and V. Styskala. 2009. Reliability and safety of railway signalling and interlocking devices. In *Proceeding of European Safety and Reliability Conference (ESREL 2009)*, eds: S. Martorell, C. G. Soares, and J. Barnett, 2325–2330. Boca Raton, FL: CRC Press.

Leveque, O. 2008. *ETCS Implementation Handbook*. Paris: International Union of Railways.

Liu, Y. L., and M. Rausand. 2011. Reliability assessment of safety instrumented systems subject to different demand modes. *Journal of Loss Prevention in the Process Industries* 24: 49–56.

Liu, Y. L., and M. Rausand. 2013. Reliability effects of test strategies on safety-instrumented systems in different demand modes. *Reliability Engineering and System Safety* 119: 235–243.

Liu, Y. L., and M. Rausand. 2016. Proof-testing strategies induced by dangerous detected failures of safety-instrumented systems. *Reliability Engineering and System Safety* 145: 366–372.

Mahboob, Q., Kunze, M., Trinckauf, J., and U. Maschek. 2012. A flexible and concise framework for hazard quantification: An example from railway signalling system, In *Proceedings of 2012 International Conference of Quality, Reliability, Risk, Maintenance, Safety Engineering*, eds: H. Z. Huang, M. J. Zuo and Y. Liu, 49–54 Chengdu, IEEE.

Ning, B., Tang, T., Qiu, K., Gao, C., and Q. Wang. 2010. CTCS—Chinese train control system. In *Advanced Train Control Systems*, ed. B. Ning. Southampton: WIT Press.

Palumbo, M., Ruscigno, M., and J. Scalise. 2015. The ERTMS/ETCS signalling system: An overview on the standard European interoperable signaling and train control system. http://www.railway signalling.eu/wp-content/uploads/2016/09/ERTMS_ETCS_signalling_system_revF.pdf.

Rausand, M. 2011. *Risk Assessment: Theory, Methods and Applications*. Hoboken, NJ: John Wiley & Sons.

Rausand, M., and A. Høyland. 2004. *System Reliability Theory: Models, Statistical Methods and Applications*, 2nd edition. Hoboken, NJ: John Wiley & Sons.

Xie, G., Hei, X. H., Mochizuki, H., Takahashi, S., and H. Nakamura. 2013. Safety and reliability estimation of automatic train protection and block system. *Quality and Reliability Engineering International* 30(4): 463–472.

Zhang, Y., Guo, J., and L. Liu. 2010. State-based risk frequency estimation of a rail traffic signal system, In *Proceeding of 12th International Conference on Computer System Design and Operation in the Railway and Other Transit Systems*, eds: B. Ning, and C. A. Brebbia, 805–813. Southampton: WIT Press.

24

Designing for RAM in Railway Systems: An Application to the Railway Signaling Subsystem*

Pierre Dersin, Alban Péronne, and René Valenzuela

CONTENTS

24.1 Definitions

The general definitions can be found in the IEC 60050-192 [1], but the following definitions are added, which are relevant to the present case study:

- *Service availability*: This is the availability which corresponds to the probability of a train fulfilling its mission without loss of function.

- *Asset availability*: This is the availability which corresponds to the probability of the train being available to fulfill its missions.

---

* From Péronne, A., and Dersin, P., Annual Reliability & Maintainability Symposium, Comparison of high-availability automation networks, 2011. © 2011 IEEE. With permission.

24.2 Introduction

Modern technical systems, such as those found in manufacturing, power generation, aerospace, defense or high-speed trains, heavily rely on complex fault-tolerant communication networks which impact their sustainability. For historical reasons, different architectures are used in different applications, but a rationalization and standardization effort is ongoing, in particular, under the auspices of the International Electrotechnical Commission (IEC).

The present work compares different solutions, described in the new IEC 62439 [2] standard on high-availability automation networks (multifunction vehicle bus [MVB] and various types of Ethernet redundant architectures), from a reliability and availability viewpoint, and makes design recommendations.

REMARK: The comparison between different Ethernet redundant architectures has been addressed too but will not be treated here.

The authors have been motivated by onboard network applications in high-speed trains, but the results are of a wider applicability.

It is desired to compare MVB and Ethernet rings from a reliability, availability, and maintainability (RAM) point of view. More precisely, two complementary perspectives are relevant, corresponding to different phases of the day of a train:

1. Service reliability and service availability
2. Intrinsic reliability and asset availability

In the first perspective, the relevant phase is train operation (i.e., revenue service). The question asked is, "during train operation, what is the frequency of events that cause a loss of function?" The corresponding performance measure is the service failure rate λ_S or the mean time to service failure (MTTF_S). A corresponding availability measure, "service availability" (A_S), can also be defined, where the mean logistic delay (MLD) in fact corresponds to the logistic time to take the train to the depot:

$$\frac{\text{MTTF}_S}{\text{MTTF}_S + \text{MLD}} \tag{24.1}$$

In the second perspective, the entire day is the relevant period, which therefore encompasses both the operating time and the time spent in depot/workshop. The question asked is, "what is the probability of the train being available to fulfill its missions?" The corresponding measure is availability, which is the result of how often failures occur and how quickly they are repaired. By failure here we mean any event that causes the train to be taken to the workshop, including those without impact on service (for instance, a failure of one channel in a redundant item of equipment). Thus, the availability is measured by

$$\frac{\text{MTTF}_{CM}}{\text{MTTF}_{CM} + \text{MLD} + \text{MACMT}}, \tag{24.2}$$

where MTTF_{CM} is the mean time to failure (MTTF) (any failure requiring corrective maintenance) and MACMT is the mean active corrective maintenance time (diagnostics, removal and refit).

The first analysis (service reliability/availability) is best conducted with respect to a train function. The second analysis (intrinsic reliability/asset availability) is best conducted at the train or train subsystem (asset) level because when the train is in the depot, it is not available for commercial service (for any function).

24.3 Comparison of MVB and Ethernet from Service Reliability Point of View

24.3.1 Overview of MVB and Ethernet Technologies

The MVB is a serial communication bus which, in the railway context, is used at rail car level. It connects programmable equipment (end devices) with each other and directly attaches to simple sensors and actuators. It is not intended for communication between vehicles—this is the domain of the train bus. The MVB allows for redundant layout with both twisted wire pairs and optical fiber. This increases availability. The redundancy can be limited to critical segments of the bus for cost reasons.

An Ethernet is a family of frame-based computer networking technologies for local area networks (LANs). The combination of the twisted pair versions of Ethernet for connecting end devices to the network, along with the fiber optic versions for site backbones, is the most widespread wired LAN technology.

Up to now, Ethernet protocol in railway vehicles has been mainly used in a few passenger entertainment system applications, but its use is spreading to other subsystems as well. Compared with conventional MVB, Ethernet devices reduce costs and increased functionalities.

24.3.2 Network Architectures and Service RAM Assumptions

The present chapter is focused on the RAM analysis of two representative functions:

- Propulsion: Sending a traction order from the manipulator in the front cab to the power control equipment (PCE) in the rear cab (Figures 24.1 and 24.2).
- Display: Displaying information from input/output in the rear cab, on the driver display unit (DDU) in the front cab (Figures 24.1 and 24.2).

For the MVB solution, function 1 (propulsion) involves the following elements:

- The T215 input/output module redundant devices in the front cab
- The bus (which is present in each end device) and the two main processor unit (MPU) redundant devices
- The PCE in the rear cab

Function 2 (display) involves the following elements:

- The T215 (front cab)
- The bus (which is present in each End Device) and the two MPU redundant devices
- The DDU in the front cab

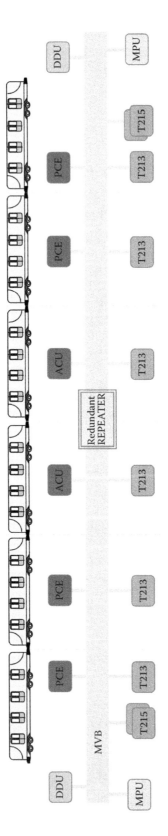

FIGURE 24.1
MVB network architecture.

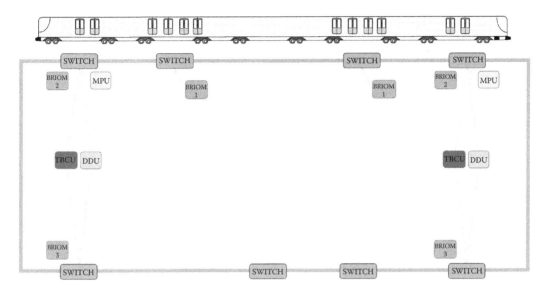

FIGURE 24.2
Ethernet network architecture.

The reliability and maintainability parameters are listed in Table 24.1.

REMARK: MDT is the service mean down time: the mean time to restore function once lost; in this case, the faulty train is removed from the track and replaced with a spare one MDT = MLD.

For the Ethernet solution, function 1 (propulsion) involves the following elements:

- The basic remote input/output module 1 (BRIOM1) or BRIOM2 redundant devices in the front cab
- The Ethernet ring made up of switches (consist ring switches [CRS]) and the two MPU redundant devices
- The PCE (rear cab)

TABLE 24.1

Input R&M Figures for MVB

Item	Failure Rate (h^{-1})	MTTF (Hours)	MLD (Hours)
ED-BUS	1.48×10^{-6}	674,157	3.0
REP	2.86×10^{-6}	350,000	3.0
MPU	1.25×10^{-5}	80,000	3.0
T213	8.33×10^{-6}	120,000	3.0
T215	1.25×10^{-6}	80,000	3.0
PCE	2.78×10^{-5}	36,000	3.0
DDU	1.67×10^{-5}	60,000	3.0
ED-BUS	1.48×10^{-5}	674,157	3.0

TABLE 24.2

Input Reliability and Maintainability (R&M) Figures for Ethernet

Item	Failure Rate (h^{-1})	MTTF (Hours)	MLD
CRS	1.52×10^{-6}	659,330	3.0
CRS-CRS	1.20×10^{-6}	830,454	3.0
CRS-ED	1.12×10^{-6}	893,857	3.0
MPU	1.25×10^{-5}	80,000	3.0
BRIOM 1	9.52×10^{-6}	105,000	3.0
BRIOM 2	1.54×10^{-5}	65,000	3.0
PCE	2.78×10^{-5}	36,000	3.0
DDU	1.67×10^{-5}	60,000	3.0

Function 2 (display) involves the following elements:

- The BRIOM1 in the rear cab
- The Ethernet ring made up of switches and the two MPU redundant devices
- The DDU in the front cab

The reliability and maintainability parameters are listed in Table 24.2.

REMARK: For the needs of the reliability model, the switches have been split; therefore, the following components are considered:

- CRS: This is the core of the switch
- CRS-CRS: This consists of two switch interface modules and a cable
- CRS-ED: This consists of a switch interface module, a cable, and an end device interface module

The MTTF of a switch is assumed equal to 200,000 hours; an apportionment has been performed to identify the failure rate of each typical element (i.e., core and interface module) of a switch.

24.3.3 Service RAM Modeling of the Two Solutions

In order to calculate service reliability and service availability at train level, the breakdown into train consists has been taken advantage of: the train-level service failure rate is equal to the sum of consist-level failure rates and the ring failure rate, and the train-level availability is the product of consist-level availabilities and ring availability.

In order to obtain each consist-level service reliability and availability, a Markov model has been constructed for each consist. The availability is obtained as the sum of probabilities of operating states; the service failure rate is obtained in the usual way for Markov models [3].

Paradigms for reliability modeling of repairable systems include Markov processes, stochastic Petri nets, and semi-Markov processes. Solving models expressed through these paradigms generally involve Monte Carlo simulation. In some cases, depending on the performance metrics being evaluated, analytical solutions exist to Markov models. The main constraint in Markov models is the fact that the only allowed time-to-failure and restoration time probability model is the exponential model. In the case of electronic components,

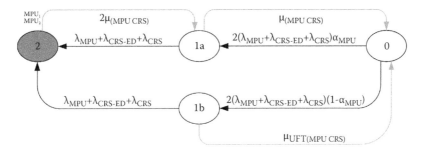

FIGURE 24.3
Example of Markov model for MPU redundant module (1-out-of-2 with imperfect detection rate).

exponential failure models represent reality fairly accurately. On the other hand, exponential restoration times are not considered to be in good agreement with reality. However, if one is interested in comparing performance of steady measures of two different systems, then this limitation of the modeling framework does not introduce any bias in the results.

Comparisons between Markov and semi-Markov models can be found in the studies by Malefaki et al. [4] and Volovoi [5].

In the Ethernet-related model, as far as the ring is concerned, a truth table is used, which refers to all the other submodels (one for each consist). This is in order to avoid double counting: the unavailability of the ring is obtained by adding all probabilities of the degraded states other than the states of complete function loss.

Figure 24.3 shows the Markov model for the main MPU redundant module. The 1-out-of-2 (1oo2) redundant configuration has been modeled by a four-state model, to take into account the possibility of nondetection of the first failure. State 0 is the nominal state. State 1a corresponds to one detected failed channel (system is in a 1-out-of-1 configuration); the detection probability is denoted as αMPU. The transition rate from state 0 to state 1a is therefore the product by αMPU of the sum of the failure rates of, respectively, the MPU, the CRS, and the CRS-ED (with a factor 2 to account for the two MPUs). State 1b corresponds to one undetected failed channel. The transition rate from state 0 to state 1b is therefore the product of the same failure rates by the complementary probability $(1 - P_{0 \to 1a})$.

When the channel failure has not been detected by the built-in test, it is assumed to be detected during the next preventive inspection. The average time to detection is thus the mean fault detection time), which is equivalent to half the interval between two consecutive inspections [6]. The corresponding transition rate in the Markov model is then μUFT = 1/MUFT.

Table 24.3 shows the service availability results of the Ethernet solution per function.

REMARK: Dual homing is a network topology in which a device is connected to the network by way of two independent access points (points of attachment, see Figure 24.2).

Table 24.4 shows the service reliability results of the Ethernet solution per function.

24.3.4 Service RAM Results for the Networks Alone

Table 24.5 shows the reliability results of both solutions when the contributions of the end devices are not included.

The difference between the results of the two solutions is mostly due to the common-cause failure estimation, based on an approximate approach, the beta-factor method

TABLE 24.3

Service Availability Results (End Devices Included)

Element/Function	Propulsion	Display
Car l: CRS & BRIOM (1oo2)	4.57×10^{-9}	1.25×10^{-10}
Car 1: DDU	NA	5.34×10^{-5}
Cars 1 and 6: CRS and MPU (1oo2)	4.12×10^{-9}	4.12×10^{-9}
Car 6: CRS dual homing (PCE)	4.14×10^{-11}	NA
Car 6: PCE	8.67×10^{-5}	NA
Car 6: BRIOM 1	NA	3.19×10^{-5}
Train: CRS ring	4.81×10^{-9}	4.81×10^{-9}
Total service unavailability	8.67×10^{-5}	8.53×10^{-5}
Total service availability	99.9913%	99.9915%

TABLE 24.4

Service Reliability Results (End Devices Included)

Element/Failure Rate	Propulsion	Display
Car 1: CRS & BRIOM (redundant)	1.52×10^{-9}/h	4.17×10^{-11}/h
Car 1: DDU	NA	1.78×10^{-5}/h
Car 1 and 6: CRS & MPU (redundant)	1.37×10^{-9}/h	1.37×10^{-9}/h
Car 6: CRS Dual Homing (PCE)	1.38×10^{-11}/h	NA
Car 6: PCE	2.89×10^{-5}/h	NA
Car 6: BRIOM 1	NA	1.06×10^{-5}/h
Train: CRS RING	1.60×10^{-9}/h	1.60×10^{-9}/h
Total λ_S	2.89×10^{-5}/h	2.84×10^{-5}/h
Total MTTF_S	34,601 h	35,173 h

TABLE 24.5

Service Reliability Results (End Devices Not Included)

Solution	MVB		Ethernet	
Function	Propulsion	Display	Propulsion	Display
MTTF_S	3.25×10^8 h	3.31×10^8 h	2.52×10^7 h	2.52×10^7 h

outlined in IEC 61508 [7]. The MVB is assumed to be more sensitive to the presence of common-cause failures than is of Ethernet. However, the difference is not significant.

24.3.5 Service RAM Results for the Networks and End Devices

Table 24.6 shows the reliability results of both solutions when the contributions of the end devices are included.

Table 24.5 shows the results not including the end device. It confirms that the contribution of the network to both unreliability and unavailability is not significant overall. The conclusion is that service RAM figures for both solutions are not significantly different for the two solutions and that the key lever to improve the train reliability is the reliability performance of the end devices.

TABLE 24.6

Service Reliability Results (End Devices Included)

Solution	MVB		Ethernet	
Function	Propulsion	Display	Propulsion	Display
MTTF$_S$	34,600 h	35,172 h	35,947 h	34,238 h

24.4 Comparison of MVB and Ethernet from Asset Availability Point of View

24.4.1 Asset RAM Modeling of the Two Solutions

In order to model the Ethernet and the MVB networks in similar ways, the concept of $\lambda_{Ethernet}$ and λ_{Medium} has been introduced. In the case of Ethernet, this "network failure rate" is obtained from the failure rates of switches, links, and dedicated parts of the end devices. In the case of MVB, it is the failure rate of one medium and dedicated parts of the end devices.

A distinction has been made in the analysis between the end devices and the network: a more thorough analysis has been deemed necessary for the network.

End devices and networks are considered independent, and therefore, the overall system (asset) availability is the product of all end device availabilities and the network availability, as illustrated in the reliability block diagrams of Figures 24.4 and 24.5.

The availability of each end device is straightforwardly obtained from MTTF and MDT using the standard formula. In order to calculate network availability, Markov models have been constructed, as they allow the modeling of detection and localization rate as well as maintenance policy.

When a first fault occurs, it is always assumed to be detected (state 1ab); the train is brought to the depot, not immediately but after some time (deferred repair), of average MAD$_{deferred}$ (state 3ab); then the cause is localized and the item is repaired rather quickly (average time MACMT1) (Figure 24.6).

A fault may be wrongly detected and a false alarm is then generated (state 1c); the train is brought to the depot, not immediately but after some time (deferred repair), state 3c; then time is spent to localize the cause (average time MACMT3), but it is a no-fault-found event. After the loss of one Ethernet path (leading to state 1ab), the loss of the second path leads to total failure (state 2). There are also common-cause failures, which immediately lead to full function loss.

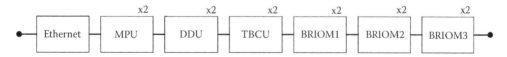

FIGURE 24.4
RBD for train equipped with Ethernet solution.

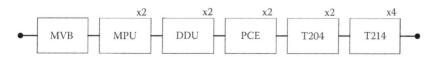

FIGURE 24.5
RBD for train equipped with MVB solution.

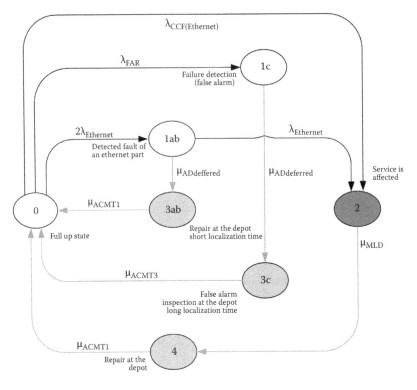

FIGURE 24.6
Ethernet network Markov model.

For MVB, a supplementary parameter $p_{\text{Localization}}$ has been introduced, which is the probability of a quick fault localization (Figure 24.7). The situation is similar to the case of two different failure modes, one of which is easily localizable and the other is not. Therefore, the system goes into one of two different states. In state 3a, a short time is spent to localize the fault (average time MACMT1); in state 3b, a long time is spent to localize the fault (average time MACMT2).

24.4.2 RAM Assumptions

From the device reliability and device contribution to the network, the Ethernet path and medium failure rates have been derived.

For the MVB this leads to

$$\text{MTTF}_{\text{CM}} = \text{MTTF}_{\text{Medium}} = 120192\,[\text{h}] = \frac{1}{\lambda_{\text{Medium}}}. \tag{24.3}$$

For the Ethernet:

$$\text{MTTF}_{\text{CM}} = \text{MTTF}_{\text{Ethernet}} = 28345\,[\text{h}] = \frac{1}{\lambda_{\text{Ethernet}}}. \tag{24.4}$$

The reason for the difference is the relatively poor reliability of the switches (MTTF = 200,000 hours assumed for one CRS) and the number of switches in the networks considered.

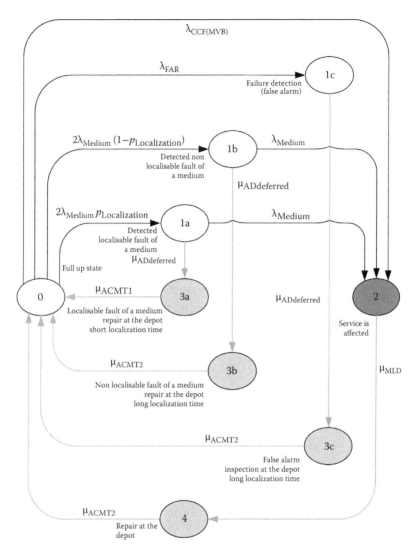

FIGURE 24.7
MVB network Markov model.

In view of what is understood of the MVB (the connector and bus problem), a higher common-cause failure rate has been taken than for the Ethernet (beta factor 2% versus 1%).

The beta factor method defined in IEC 61508-6 Ed.2 [8] has been used in order to estimate the common-cause failures of Ethernet and MVB redundant solutions.

Ethernet is assumed to be better regarding the following questions:

- Complexity/design/application/maturity/experience
  - Are inputs and outputs protected from potential levels of overvoltage and overcurrent?
  - Does a cross-connection between channels preclude the exchange of any information other than that used for diagnostic testing or voting purposes?
  - Is the system simple, for example, no more than 10 inputs or outputs per channel?

- Diversity/redundancy
  - Is maintenance on each channel carried out by different people at different times?
- Environmental control
  - Are all signal and power cables separate at all positions?
- Procedures/human interface
  - Do the system diagnostic tests report failures to the level of a field-replaceable module?
  - Does the system have high diagnostics coverage (>99%) and report failures to the level of a field-replaceable module?
- Separation/segregation
  - Are all signal cables for the channels routed separately at all positions?
  - Are the logic subsystem channels on separate printed circuit boards?
  - If the sensors/final elements have dedicated control electronics, is the electronics for each channel indoors and in separate cabinets?
  - If the sensors/final elements have dedicated control electronics, is the electronics for each channel on separate printed-circuit boards?

MVB is assumed to be better regarding the following questions:

- Complexity/design/application/maturity/experience
  - Is the design based on techniques used in equipment that has been used successfully in the field for >5 years?
  - Is there more than 5 years of experience with the same hardware used in similar environments?

The grand scores are 79 for Ethernet and 62.5 for MVB; therefore, it respectively yields $\beta =$ 1% for Ethernet and $\beta = 2\%$ for MVB.

In comparison to the Ethernet, the MVB is a well-known technical solution in railway application, but it is too much integrated with poor separation/segregation and diagnostic capabilities.

REMARK: The false alarm rate (FAR) is represented in the models, but it will not be taken into account in the computation because its value cannot be estimated at this stage. Therefore, MACMT3 is not used in the computation.

The nominal assumptions taken for maintenance parameters are the following: MACMT1 = 4 hours; MACMT2 = 80 hours; $p_{\text{Localization}} = 10\%$ (where $p_{\text{Localization}}$ is the probability that the maintainer will be successful on localizing the failure).

24.4.3 Asset RAM Results of the Two Solutions

In summary, one could say that with an Ethernet, failures occur more often but are usually diagnosed and, therefore, repaired more quickly. Results are summarized in Table 24.7.

It is seen that, unlike in the first part of the study dealing with service availability, the contribution of the network to asset availability is a key factor: about 30% for the Ethernet

TABLE 24.7

Asset RAM Results

Solution	MVB	Ethernet
MTTF_{CM}	59,511 h	11,505 h
MDT	72.5 h	4 h
A	99.878%	99.965%

and 62% for MVB. Therefore, the choice of network technology will have a more substantial impact than for service availability.

24.4.4 Sensitivity Analysis

The result of the comparison will therefore depend on the relative values of the reliability and maintainability parameters, on which there is a fair amount of uncertainty; therefore, a sensitivity analysis is useful with respect to those parameters:

- $\text{MTTF}_{\text{Medium}}$ and $\text{MTTF}_{\text{Ethernet}}$
- $\lambda_{\text{CCF(MVB)}}$ and $\lambda_{\text{CCF(Ethernet)}}$
- MACMT1, MACMT2 (and MACMT3)
- $p_{\text{Localization}}$
- (λ_{FAR})

Figure 24.8 shows that the difference in availability between the two networks shrinks at higher medium/Ethernet reliability levels.

REMARK: MVB and Ethernet horizontal lines represent the availability levels corresponding to nominal parameter values.

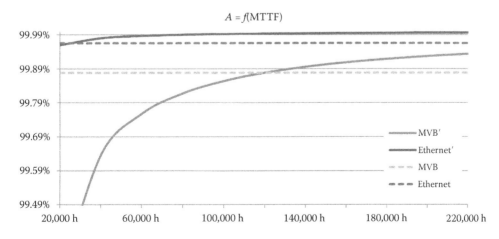

FIGURE 24.8
Asset availability as a function of network MTTF_{CM}.

Figure 24.9 shows that an Ethernet is more available than an MVB only as long as MACMT1 does not exceed 12 hours. Otherwise, the main advantage of Ethernet (better maintainability) is not sufficient to compensate for its lesser reliability.

However, Figure 24.10 shows that if the MVB localization rate is higher, then the Ethernet advantage still prevails as long as MACMT1 is less than 7 hours.

Finally, Figure 24.11 shows that if MACMT2 is made to vary, the advantage of Ethernet prevails as long as MACMT2 is higher than 28 hours (the nominal conditions corresponds to MACMT2 = 80 hours).

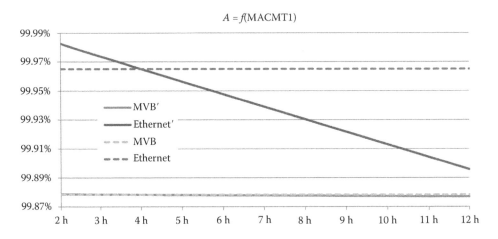

FIGURE 24.9
Asset availability as a function of MACMT1 (with $p_{\text{Localization}} = 10\%$, nominal value).

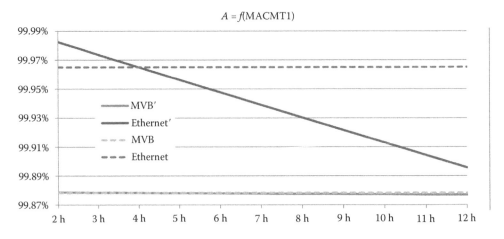

FIGURE 24.10
Asset availability as a function of MACMT1 (with $p_{\text{Localization}} = 60\%$, nominal value).

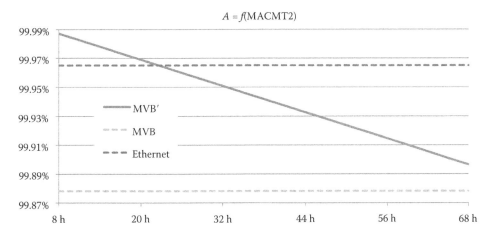

FIGURE 24.11
Asset availability as a function of MACMT2.

24.5 General Conclusions

This analysis has adopted two different perspectives in comparing Ethernet and MVB architectures. In the first, where service reliability is the prime concern, the two architectures are virtually indistinguishable because performance is overwhelmingly determined by the end device reliability: the key lever is then the end device reliability. (This conclusion would probably be modified in favor of an Ethernet if common-cause failures were thoroughly modeled.)

In the second perspective, that of asset availability, Ethernet is shown to clearly outperform MVB, primarily due to its much higher diagnostic capabilities.

Markov models have proven to be a useful tool in describing those networks, as they lend themselves to explicit description of maintenance policies.

References

1. IEC. IEC 60050-192: *International Electrotechnical Vocabulary—Dependability*, 2015, IEC, Geneva.
2. IEC (International Electrotechnical Commission). IEC 62439: *High Availability Automation Networks*, 2008, IEC, Geneva.
3. L. M. Maillart. Introduction to Markov chain modeling, analysis and optimization. *RAMS 2009, Annual Reliability and Maintainability Symposium, Fort Worth, TX*, 2009.
4. S. Malefaki, N. Limnios, and P. Dersin. *Reliability of Maintained Systems under a Semi-Markov Setting*, 2014.
5. V. Volovoi. Deferred maintenance of redundant systems, Invited Paper. *Mathematical Methods of Reliability, Grenoble France, July 3–6, 2017*, 2017.
6. P. Dersin. Achieving availability cost effectively in complex systems. *RAMS 2010, Annual Reliability and Maintainability Symposium, San José, CA*, 2010.
7. IEC. IEC 61508: *Functional Safety of Electrical/Electronic/Programmable Electronic Safety-related Systems*, 2010, IEC, Geneva.
8. IEC. IEC 61508-6 Ed.2: *Functional Safety of Electrical/Electronic/Programmable Electronic Safety-Related Systems*, 2010, IEC, Geneva.

25

Fuzzy Reasoning Approach and Fuzzy Analytical Hierarchy Process for Expert Judgment Capture and Process in Risk Analysis

Min An and Yao Chen

CONTENTS

25.1 Introduction

Many risk assessment techniques currently used in the railway industry are comparatively mature tools, which have been developed on the basis of probabilistic risk analysis (PRA), for example, fault tree analysis, event tree analysis, Monte Carlo simulation, consequence analysis, and equivalent fatality analysis (LUL, 2001; Railway Safety, 2002; Metronet, 2005; An et al., 2011). The results of using these tools heavily rely on the availability and accuracy of the risk data (LUL, 2001; Railway Safety, 2002; Metronet, 2005). However, in many circumstances, these methods often do not cope well with uncertainty of information. Furthermore, statistical data do not exist, and these must be estimated on the basis of expert knowledge and experience or engineering judgment. Therefore, railway risk analysts often face circumstances where the risk data are incomplete or there is a high level of uncertainty involved in the risk data (An et al., 2011). Additionally, railways are a traditional industry, whose history extends for at least two centuries. The existing databases contain a lot of data and information; however, the information may be both an excess of other information that cannot be used in risk analysis and a shortage of key information of major failure events. In many circumstances, it may be extremely difficult to conduct PRA to assess the failure frequency of hazards, the probability, and the magnitude of their possible consequences, because of the uncertainty in the risk data. Although some work has been conducted in this field, no formal risk analysis tools have been developed and applied to a stable environment in the railway industry (Chen et al., 2007). Therefore, it is essential to develop new risk analysis methods to identify major hazards and assess the associated risks in an acceptable way in various environments where those mature tools cannot be effectively or efficiently applied. The railway safety problem is appropriate for examination by fuzzy reasoning approach (FRA) and fuzzy analytical hierarchy process (fuzzy-AHP).

The FRA method provides a useful tool for modeling risks and other risk parameters for risk analysis involving risks with incomplete or redundant safety information. Because the contribution of each hazardous event to the safety of a railway system is different, the weight of the contribution of each hazardous event should be taken into consideration in order to represent its relative contribution to the risk level (RL) of the railway system. Therefore, the weight factor (WF) is introduced, which indicates the magnitude of the relevant importance of a hazardous event or hazard group to its belongings in a risk tree. Modified fuzzy-AHP has been developed and then employed to calculate the WFs (Chen et al., 2011). This has been proven to facilitate the use of fuzzy-AHP and provide relevant reliable results. This chapter presents a development of a railway risk assessment system using FRA and modified fuzzy-AHP. The outcomes of risk assessment are represented as risk degrees, defined risk categories of RLs with a belief of percentage, and risk contributions. They provide safety analysts, managers, engineers, and decision makers with useful information to improve safety management and set safety standards.

25.2 Fuzzy Expression

A fuzzy set is an extension of a classical set (Zadeh, 1965; Bojadziev and Bojacziev, 1997). Instead of using a binary membership function (MF) in a classical set, which only tells whether an object belongs to the set or not, fuzzy set applies a continuous graded MF, which also indicates how much the object is close to the set. A fuzzy number is a parametric representation of a convex and normalized fuzzy set.

A fuzzy number is defined in the universe R as a *convex* and *normalized* fuzzy set, and it is a parametric representation of the fuzzy set. There are various fuzzy numbers available such as triangular fuzzy number, trapezoidal fuzzy number, and bell-shaped fuzzy number. However, triangular and trapezoidal fuzzy numbers are the most widely used in the engineering risk analysis because of their intuitive appeal and perceived computational efficacy.

A trapezoidal fuzzy number can be defined as $A = (a, b, c, d)$, and the MF of its corresponding fuzzy set $\tilde{A}$ is defined as

$$\mu_A(x) = \begin{cases} (x - a/b - a), & x \in [a, b], \\ 1, & x \subset [b, c], \\ (x - d/c - d), & x \in [c, d], \\ 0, & \text{otherwise.} \end{cases}$$

where four real numbers (a, b, c, and d are also vertex values of a trapezoidal MF with satisfaction of the relationship $a \le b \le c \le d$) determine the x coordinates of the four corners of a trapezoidal MF.

25.3 Fundamentals of FRA

FRA has been developed based on the concept of fuzzy sets and inference mechanism, which generalizes an ordinary mapping of a function to a mapping between fuzzy sets (Lee, 2005; An et al., 2000a, 2000b). The inference mechanism is based on the compositional rule of inference, and the result is derived from a set of fuzzy rules and given inputs. It indicates that FRA possesses the ability to model a complex nonlinear function to a desired degree of accuracy.

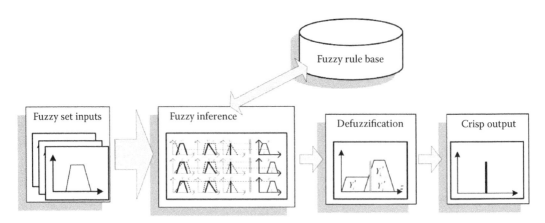

FIGURE 25.1
Fuzzy inference process.

Linguistic variables are the generalization of ordinary variables, whose values are words or sentences in natural or artificial languages rather than real numbers. They are defined by the following quintuple (Lee, 2005): Linguistic variable = $(x, T(x), U, M)$, where x is the name of variable; $T(x)$ is the set of linguistic terms which can be a value of the variable; U is the set of universe of discourse which defines the characteristics of the variable; M is the semantic rules which map terms in $T(x)$ to fuzzy sets in U.

A major component of an FRA model is fuzzy rulebase, which consists of a set of fuzzy rules, and these rules are expressed in the form of If–then statements. For example, the following is a fuzzy If–then rule: If x is A and y is B, then z is C, where A, B, and C are linguistic terms; "If x is A and y is B" is called the *antecedent* (*If part*) of the rule and "then z is C" is called the *consequent* (*then part*). If there are more than one argument in the If part, then AND or OR operation needs to be applied.

A fuzzy If–then rulebase is the core of the fuzzy inference process for formulating the mapping from given input fuzzy sets to an output fuzzy set. Once the rulebase is established, a fuzzy inference process can be carried out as shown in Figure 25.1. Fuzzy set inputs are directly input into the fuzzy inference system to determine which rules are relevant to the current situation, and the results from inference of individual rules are then aggregated to the output result from inference with current inputs (Lee, 2005; An et al., 2006, 2007).

25.4 Fundamentals of Fuzzy-AHP

AHP has been widely used in multiple criteria decision-making (Satty, 1980; Vaidya and Kumar, 2006). It decomposes a complex problem into a hierarchy, in which each level is composed of specific elements. AHP enables decision makers to analyze a complex problem in a hierarchy with quantitative and qualitative data and information in a systematic manner. However, it is difficult for experts to express their judgments if they are not confident in relative importance between two alternatives. The applications of Fuzzy-AHP may therefore solve such a problem.

A fuzzy-AHP approach can be simplified as a process of calculating weight factors of alternatives at a level in a hierarchy or of a certain criterion (Murtaza, 2003; Huang et al., 2006; An et al., 2007; Chen et al., 2007). The process is described as follows:

Step 1: Establishment of a fuzzy-AHP estimation scheme

Fuzzy-AHP determines WFs by conducting a pairwise comparison. The comparison is based on an estimation scheme of intensity of importance using linguistic terms. Each linguistic term has a corresponding fuzzy number.

Step 2: Construction of fuzzy comparison matrix

The pairwise comparison matrix can be developed on the scheme in step 1. Each element of matrix presents the preference intensity of one event over another.

Step 3: Calculation of fuzzy weights

The fuzzy WF of each element can be calculated based on fuzzy operations.

Step 4: Defuzzification and normalization

As the fuzzy WFs are presented in terms of fuzzy numbers, it is necessary to convert fuzzy numbers into crisp values. This can be done via various defuzzification methods.

25.5 Railway Safety Risk Model with FRA and Fuzzy-AHP

A risk assessment is a process that can be divided into five phases: the problem definition phase, data and information collection and analysis phase, hazard identification phase, risk estimation phase, and risk response phase (An et al., 2007). This process provides a systematic approach to the identification and control of high-risk areas. According to this effective process, a risk assessment model based on FRA and fuzzy-AHP approach for a railway system is proposed, as shown in Figure 25.2, where EI stands for expert index and UFN stands for uniform format number. The algorithm of the risk model consists of five phases: preliminary phase, FRA risk estimation phase, fuzzy-AHP risk estimation phase, and risk response phase.

25.5.1 Preliminary Phase

Risk assessment begins with problem definition, which involves identifying the need for safety, i.e., specific safety requirements. The requirements regarding railway safety at different levels, e.g., hazardous event level, hazard group level, and the railway system level, should be specified and made, which may include sets of rules and regulations made by national authorities and classification societies; deterministic requirements for safety, reliability, availability, and maintainability; and criteria referring to probability of occurrence of serious hazardous events and the possible consequences (An et al., 2007; Chen et al., 2007).

Once the need for safety is established, the risk assessment moves from problem identification to data and information collection and analysis. The aim of data and information collection and analysis is to develop a good understanding of what serious accidents and incidents have occurred in a particular railway system over the years and generate

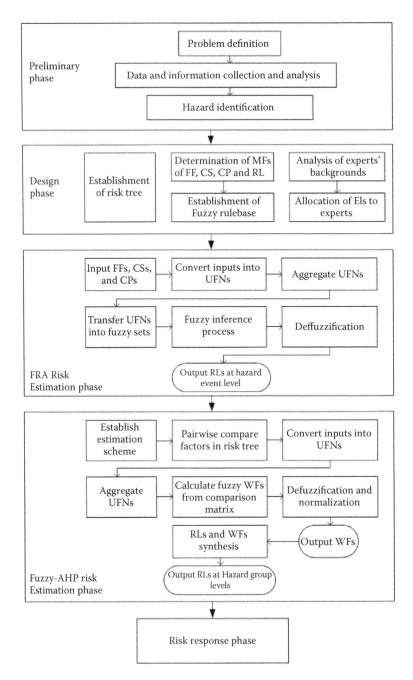

FIGURE 25.2
Risk assessment model.

a body of information. If the statistic data do not exist, expert and engineering judgments should be applied. The information gained from data and information collection will then be used to define the standards of qualitative descriptors and associated MFs of risk parameters, i.e., failure frequency (FF), consequence severity (CS), consequence probability (CP), and RL.

The purpose of hazard identification is to systematically identify all potential hazardous events associated with a railway system at each required level, e.g., hazardous event level and hazard group level, with a view of assessing their effects on railway system safety. Various hazard identification methods, such as a brainstorming approach, check-list, "what if?," hazard and operability, and failure mode and effect analysis, may be used individually or in combination to identify the potential hazardous events for a railway system (HSE, 2001; Chen et al., 2007). The hazard identification can be initially carried out to identify hazardous events and then progressed up to hazard group level and finally to the system level. The information from hazard identification will then be used to establish a risk tree.

25.5.2 Design Phase

Once the risk information of a railway system is obtained in the preliminary phase, the risk assessment moves from the preliminary phase to the design phase. On the basis of information collection, the tasks in the design phase are to develop a risk tree and MFs of FF, CS, CP, and RL and a fuzzy rulebase.

25.5.2.1 Development of Risk Tree

There are many possible causes of risks that have an impact on railway system safety. The purpose of the development of a risk tree is to decompose these risk contributors into adequate details in which risks associated with a railway system can be efficiently assessed (Chen et al., 2007; An et al., 2008). A bottom–up approach is employed for the development of a risk tree. Figure 25.3 shows a typical risk tree that can be broken down into hazardous event level, hazard group level, and system level. For example, hazardous events of E_1, E_2, …, E_n at the hazardous event level affect the RL of S-HG$_1$ at the subhazard group level; the RLs of S-HG$_1$, S-HG$_2$, …, S-HG$_n$ contribute to the RL of HG$_2$ at the hazard group level; and RLs of HG$_1$, HG$_2$, …, HG$_n$ contribute to the overall RL of a railway system at the system level.

25.5.2.2 Establishment of Fuzzy Rulebase

Fuzzy rulebases are basically built through the study of engineering knowledge, historical incident, and accident information. Human experts have a good intuitive knowledge

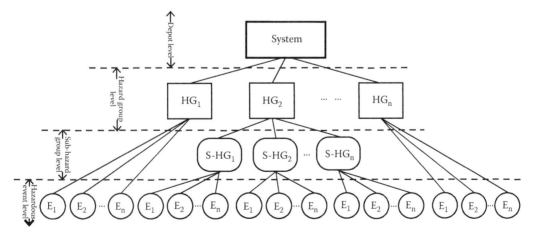

FIGURE 25.3
Example of a risk tree.

of the system behavior and the risks involved in various types of failures. A fuzzy rule-base is established in terms of qualitative descriptors of input parameters: FF, CS, CP, and the output RL. The qualitative descriptors are characterized by fuzzy sets which are derived from experimental data or past information or characteristics of input or output variables, such as FF, CS, CP, and RL. The fuzzy sets are defined in the universe of discourse and described by MFs. Currently, there are several geometric mapping functions widely adopted, such as triangular, trapezoidal, and S-shaped MFs. However, triangular and trapezoidal MFs are the most frequently used in railway risk analysis practice (An et al., 2006, 2007). For example, in the railway safety risk assessment, input parameters FF, CP, and CS and output RL are constructed by trapezoidal MFs, which are determined according to characteristics of inputs or output variables, where the input parameter FF is defined as "remote," "rare," "infrequent," "occasional," "frequent," "regular," and "common"; CS is defined as "negligible," "marginal," "moderate," "critical," and "catastrophic"; CP is defined as "highly unlikely," "unlikely," "reasonably unlikely," "likely," "reasonably likely," "highly likely," and "definite"; and the output RL is defined as "low," "possible," "substantial," and "high." Rules using defined variables in the fuzzy rulebase are in If–then format. For example, rule 1 is "If FF is 'remote' and CS is 'negligible' and CP is 'highly unlikely,' then RL is 'low.'"

25.5.2.3 Allocation of EIs to Experts

In practice, risk assessment usually involves a number of experts from different backgrounds or disciplines with essential experience regarding railway safety. Thus, these experts may have different impacts on the final decision. EI is therefore introduced into the risk model to distinguish experts' competence. The EI of ith experts in n experts can be obtained by (Huang et al., 2006):

$$EI_i = \frac{RI_i}{\sum\limits_{j=1}^{n} RI_j}, \tag{25.1}$$

where RI_i is of relevant importance to ith expert according to experience, knowledge, and expertise, which takes a value in the universe of [1, 9]. RI is defined in a manner that "1" means less importance, whereas "9" means most importance. Obviously, it is necessary to review EIs when the topic or the circumstance has changed.

25.5.3 FRA Analysis

In the FRA risk estimation phase, each risk is assessed at hazardous event level based on FF, CS, and CP to calculate its RL. However, railway risk analysts often face the circumstances where the risk data are incomplete or there is a high level of uncertainty involved in the risk data. A flexible method for expressing expert and engineering judgments is proposed (An et al., 2006, 2008; Chen et al., 2007). The uniform format number (UFN) is introduced to capture and convert expert and engineering subjective judgments. As described earlier in this chapter, this allows imprecision or approximate information to be used in the risk analysis process. There are six steps in calculating the RLs of hazardous events that are described in the following.

25.5.3.1 Step 1: Inputting of FFs, CSs, and CPs

The input data can usually be gathered from historical data; however, in many circumstances, the data may not exist or uncertainty may be involved in the risk data. Experts may provide their judgments on the basis of their knowledge and expertise for each hazardous event (Herrera et al., 2001, 2009). The input values can be a precise numerical value, a range of numerical values, or a linguistic term. For example, if adequate information is obtained and the risk factor is quantitatively measurable, an expert is likely to provide a precise numerical value. However, experts sometimes find that it is hard to give numerical values due to the uncertainties involved or because the hazardous event is quantitative immeasurable, and then a range of numerical values, a linguistic term, or a fuzzy number can be used in the proposed model (An et al., 2006; Chen et al., 2007; Herrera et al., 2009), e.g., "CP is 60–70%," "CS is around 3–7 and most likely to be 5 in the universe of [0, 10]," and "FF is average."

25.5.3.2 Step 2: Conversion of Inputs into UFNs

As described in step 1, because the input values of hazardous events derived from experts' judgments are crisp, e.g., a numerical value, a range of numerical values, a fuzzy number, or a linguistic term, the UFN is employed to convert these experts' judgments into a uniform format for the composition of final decisions. UFN is developed based on a fuzzy trapezoidal number as it can represent most expert judgments.

A UFN can be defined as $A = \{a, b, c, d\}$, and its corresponding MF indicates the degree of preference and is defined as

$$\mu_A(x) = \begin{cases} (x - a/b - a), & x \in [a, b], \\ 1, & x \in [b, c], \\ (x - d/c - d), & x \in [c, d], \\ 0, & \text{otherwise.} \end{cases} , \tag{25.2}$$

where four real numbers (a, b, c, and d) with satisfaction of the relationship $a \le b \le c \le d$ determine the x coordinates of the four corners of a trapezoidal MF. It should be noted that a numerical value, a range of numerical values, a fuzzy number, and a linguistic term can be converted as simplified UFN. Table 25.1 shows the possible expert judgments and their corresponding UFNs.

TABLE 25.1

Experts' Judgments and Corresponding UFNs

Expert Judgment	Input Values	Input Type	UFNs
"… is *a*"	{*a*}	Numerical value	{*a*,*a*,*a*,*a*}
"… is between *a* and *b*"	{*a*,*b*}	A range number	{*a*, (*a* + *b*)/2, (*a* + *b*)/2, *a*}
"… is between *a* and *c* and most likely to be *b*"	{*a*,*b*,*c*}	Triangular fuzzy number	{*a*,*b*,*b*,*c*}
"… is between *a* and *d* and most likely between *b* and *c*"	{*a*,*b*,*c*,*d*}	Trapezoidal fuzzy number	{*a*,*b*,*c*,*d*}
"… is *rare*"	Rare	Linguistic term	Rare MF {*a*,*b*,*c*,*d*}

Each UFN at this stage represents an opinion to one of the risk parameters of a hazardous event, which is given by an expert in a risk assessment on the basis of available information and personal subjective judgment.

25.5.3.3 Step 3: Aggregation of UFNs

The aim of this step is to apply an appropriate operator to aggregate individual judgments made by individual experts into a group judgment of each hazardous event. On the basis of experts' EIs calculated in the design phase, their judgments can be aggregated according to a weighted trapezoidal average formula (Bojadziev et al., 1997). Assume m experts involved in the assessment and n experts providing nonzero judgments for a hazardous event i, the aggregated UNF A^i can be determined by

$$A^i = \{a^i, b^i, c^i, d^i\} = \left\{ \frac{\sum_{k=1}^{m} a_k^i EI_k}{\sum_{k=1}^{n} EI_k}, \frac{\sum_{k=1}^{m} b_k^i EI_k}{\sum_{k=1}^{n} EI_k}, \frac{\sum_{k=1}^{m} c_k^i EI_k}{\sum_{k=1}^{n} EI_k}, \frac{\sum_{k=1}^{m} d_k^i EI_k}{\sum_{k=1}^{n} EI_k} \right\},$$

$$A_k^i = \left\{ a_k^i, b_k^i, c_k^i, d_k^i \right\}, \tag{25.3}$$

where a_k^i, b_k^i, c_k^i, and d_k^i are the numbers of UFN A_k^i that represent the judgment of the kth expert for hazardous event i. EI_k stands for kth expert's EI.

25.5.3.4 Step 4: Transfer of UFNs into Fuzzy Sets

The FRA allows imprecision or approximate information, for example, expert and engineering judgments in the risk assessment process, which provides a useful tool for modeling risks and other parameters for risk analysis where the risk data are incomplete or include redundant information (Chen et al., 2007). FRA here is employed to deal with UFNs of each hazardous event. In FRA, the aggregated UFNs of FF, CS, and CP are converted into matching fuzzy sets before being inputted into a fuzzy inference system. Assume A_{FF}^i, A_{CS}^i, and A_{CP}^i are three UFN of FF, CS, and CP of hazardous event i, respectively. Their corresponding fuzzy sets $\tilde{A}_{FF}^i$, $\tilde{A}_{CS}^i$, and $\tilde{A}_{CP}^i$ are defined as

$$\tilde{A}_{FF}^i = \left\{ \left(u, \mu_{A_{FF}^i}(u)\right) \middle| u \in U = [0, u_n], \mu_{A_{FF}^i}(u) \in [0,1] \right\}, \tag{25.4}$$

$$\tilde{A}_{CS}^i = \left\{ \left(u, \mu_{A_{CS}^i}(v)\right) \middle| v \in V = [0, v_n], \mu_{A_{CS}^i}(v) \in [0,1] \right\}, \tag{25.5}$$

$$\tilde{A}_{CP}^i = \left\{ \left(w, \mu_{A_{CP}^i}(w)\right) \middle| w \in W = [0, w_n], \mu_{A_{CP}^i}(w) \in [0,1] \right\}, \tag{25.6}$$

where $\mu_{A_{FF}^i}$, $\mu_{A_{CS}^i}$, and $\mu_{A_{CP}^i}$ are trapezoidal MFs of A_{FF}^i, A_{CS}^i, and A_{CP}^i, respectively, and u, v, and w are input variables in the universe of discourse U, V, and W of FF, CS, and CP, respectively.

25.5.3.5 Step 5: Fuzzy Inference Process

During the fuzzy inference process, these fuzzy sets of aggregated UFNs are then input to the fuzzy inference system to decide which rules are relevant to the current situation and then calculate the fuzzy output of RL. The overall process is developed on the basis of the Mamdani method (Lee, 2005). The rules are stored in the rulebase, which contains expert judgments and historical information. Relations between input parameters FF, CS, and CP and output RL are presented in a form of If–then rules. Supposing the ith rule in the rulebase is defined as

$$R_i: \text{If } u \text{ is } \tilde{B}_{FF}^i \text{ and } v \text{ is } \tilde{B}_{CS}^i \text{ and } w \text{ is } \tilde{B}_{CP}^i, \text{then } x \text{ is } \tilde{B}_{RL}^i, \; i = 1, 2, \ldots, n, \tag{25.7}$$

where u, v, w, and x are variables in the universe of discourse U, V, W, and X of FF, CS, CP, and RL, respectively, and $\tilde{B}_{FF}^i$, $\tilde{B}_{CS}^i$, $\tilde{B}_{CP}^i$, and $\tilde{B}_{RL}^i$ are qualitative descriptors of FF, CS, CP, and RL, respectively. The calculation of the fire strength α_i of rule R_i with inputting fuzzy sets $\tilde{A}_{FF}^i$, $\tilde{A}_{CS}^i$, and $\tilde{A}_{CP}^i$ using fuzzy intersection operation is given by

$$\alpha_i = \min \Big[\max \big(\mu_{A_{FF}'}(u) \wedge \mu_{B_{FF}^i}(u) \big), \max \big(\mu_{A_{CS}'}(v) \wedge 1_{B_{CS}^i}(v) \big),$$
$$\max \big(\mu_{A_{CP}'}(w) \wedge \mu_{B_{CP}^i}(w) \big) \Big], \tag{25.8}$$

where $\mu_{A_{FF}'}(u)$, $\mu_{A_{CS}'}(v)$, and $\mu_{A_{CP}'}(w)$ are the MFs of fuzzy sets $\tilde{A}_{FF}^i$, $\tilde{A}_{CS}^i$, and $\tilde{A}_{CP}^i$, respectively, and $\mu_{B_{FF}^i}(u)$, $\mu_{B_{CS}^i}(v)$, and $\mu_{B_{CP}^i}(w)$ are the MFs of fuzzy sets $\tilde{B}_{FF}^i$, $\tilde{B}_{CS}^i$, and $\tilde{B}_{CP}^i$ of qualitative descriptors in rule R_i. After the fuzzy implication, the truncated MF $\mu_{B_{RL}^i}'$ of the inferred conclusion fuzzy set of rule R_i is obtained by

$$\mu_{B_{RL}^i}'(x) = \alpha \wedge \mu_{B_{RL}^i}'(x), \tag{25.9}$$

where α_i is the fire strength of rule R_i, $\mu_{B_{RL}^i}'(x)$ is the MF of qualitative descriptor $\tilde{B}_{RL}^i$, and x is an input variable in the universe of discourse X.

The firing strength is implicated with the value of the conclusion MF and the output is a truncated MF. The corresponding fuzzy sets of the truncated MF, which represent the implication output fuzzy sets of rules, are aggregated into a single fuzzy set. The MF $\mu_{B_{RL}^i}'(x)$ of output fuzzy set after aggregation using fuzzy union (maximum) operation is denoted by

$$\mu_{B_{RL}^i}'(x) = \overset{n}{\underset{i=1}{V}} \mu_{B_{RL}^i}'(x), \tag{25.10}$$

where $\mu_{B_{RL}^i}'$ is the MF of conclusion fuzzy set of rule R_i and n is the total number of rules in the rulebase.

25.5.3.6 Step 6: Defuzzification

As the output from the fuzzy inference system is a fuzzy set, defuzzification is used to convert the fuzzy result into a matching numerical value that can adequately represent RL. The center of area method (Lee, 2005) is employed for defuzzification. Assume that the output fuzzy set obtained from the fuzzy inference system is $\tilde{B}'_{RL} = \left\{ \left(x, \mu'_{B_{RL}}(x)\right) \middle| x \subset X, \mu'_{B_{RL}}(x) \subset [0,1] \right\}$, the matching crisp value RL. The RL^i of hazardous event i can be calculated by

$$RL^i = \frac{\displaystyle\sum_{j=1}^{m} \mu'_{B^i_{RL}}(x_j) \cdot x_j}{\displaystyle\sum_{j=1}^{m} \mu'_{B^i_{RL}}(x_j)}, \tag{25.11}$$

where m is the number of quantization levels of the output fuzzy set.

25.5.4 Fuzzy-AHP Analysis

Because the contribution of each hazardous event to the overall RL is different, the weight of the contribution of each hazardous event should be taken into consideration in order to represent its relative contribution to the RL of a railway system. The application of fuzzy-AHP may also solve the problems of risk information loss in the hierarchical process in determining the relative importance of the hazardous events in the decision making process, so that risk assessment can be progressed from hazardous event level to hazard group level and finally to a railway system level. A fuzzy-AHP analysis leads to the generation of WFs for representing the primary hazardous events within each category. There are six steps to calculate WFs as described in the following.

25.5.4.1 Step 1: Establishment of Estimation Scheme

Fuzzy-AHP determines WFs by conducting pairwise comparison. The comparison is based on an estimation scheme, which lists the intensity of importance using qualitative descriptors. Each qualitative descriptor has a corresponding triangular MF that is employed to transfer expert judgments into a comparison matrix (An et al., 2007). Table 25.2 describes qualitative descriptors and their corresponding triangular fuzzy numbers for risk analysis at railway depots. Each grade is described by an important expression and a general intensity number. When two risk contributors are of equal importance, it is considered (1,1,2). Fuzzy number (8,9,9) describes that one risk contributor is absolutely more important than the other. Figure 25.4 shows triangular MFs (solid lines) with *equal importance*, (1,1,2); *weak importance*, (2,3,4); *strong importance*, (4,5,6); *very strong importance*, (6,7,8); and *absolute importance*, (8,9,9). The other triangular MFs (dashed lines) describe the corresponding intermediate descriptors between them.

25.5.4.2 Step 2: Pairwise Comparison of Factors in Risk Tree

"HG$_1$," "HG$_2$," …, and "HG$_n$" as shown in Figure 25.3 are the risk contributors that contribute to the overall RL of a railway system. Assume two risk contributors HG$_1$ and HG$_2$; if HG$_1$ is of very strong importance more than HG$_2$, a fuzzy number of (6,7,8) is then assigned

TABLE 25.2

Fuzzy-AHP Estimation Scheme

Qualitative Descriptors	Description	Parameters of MFs (Triangular)
Equal importance (EQ)	Two risk contributors contribute equally to the shunting event	(1,1,2)
Between equal and weak importance (BEW)	When compromise is needed	(1,2,3)
Weak importance (WI)	Experience and judgment slightly favor one risk contributor over another	(2,3,4)
Between weak and strong importance (BWS)	When compromise is needed	(3,4,5)
Strong importance (SI)	Experience and judgment strongly favor one risk contributor over another	(4,5,6)
Between strong and very strong importance (BSV)	When compromise is needed	(5,6,7)
Very strong importance (VI)	A risk contributor is favored very strongly over the other	(6,7,8)
Between very strong and absolute importance (BVA)	When compromise is needed	(7,8,9)
Absolute importance (AI)	The evidence favoring one risk contributor over another is of the highest possible order of affirmation	(8,9,9)

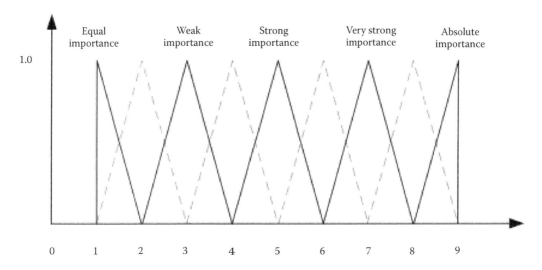

FIGURE 25.4
MFs of qualitative descriptors in fuzzy-AHP estimation scheme.

to HG_1 based on the estimation scheme as shown in Table 25.2. Obviously, risk contributor HG_2 has a fuzzy number of (1/8,1/7,1/6). Suppose n risk contributors; there are a total of $N = n - 1$ pairs which need to be compared due to the benefits of applying improved fuzzy-AHP.

25.5.4.3 Step 3: Conversion of Inputs into UFNs

As described in steps 1 and 2, because the values of risk contributors are crisps, e.g., a numerical value, a range of numerical values, a linguistic term, or a fuzzy number, the

FRA is employed again to convert these values into UFNs according to Table 25.1. A series of UFNs can be obtained to correspond to the scores and the scales of the defined risk contributors in the risk tree.

25.5.4.4 Step 4: Aggregation of UFNs

Usually, there are a number of experts in the risk assessment group and their judgments may be different. Therefore, UFNs produced at step 3 need to be aggregated into a group UFN for each risk contributor. The process is same as described in FRA analysis at step 3.

25.5.4.5 Step 5: Calculation of Fuzzy WFs from Comparison Matrix

The aggregated UFNs are then used to construct a comparison matrix. Suppose C_1, C_2, ..., C_n are risk factors in a hazard group p; $A_{i,j}$ is the aggregated UFN representing the quantified judgment on C_i compared to C_j, and C_i is more important than C_j. By applying the improved fuzzy-AHP, there will be $n - 1$ aggregated UFN, i.e., $A_{1,2}, A_{2,3},..., A_{i,j}, A_{i,j+1},...,$ $A_{n-1,n}$. Then, an entire comparison matrix for the hazard group p can be developed. The pairwise comparison between C_i and C_j in the hazard group p thus yields an $n \times n$ matrix defined as

$$M = [A_{i,j}] = \begin{bmatrix} A_{1,1} & A_{1,2} & \cdots & A_{1,n} \\ A_{2,1} & A_{2,2} & \cdots & A_{2,n} \\ \vdots & \vdots & \ddots & \vdots \\ A_{n,1} & A_{n,2} & \cdots & A_{n,n} \end{bmatrix}, \quad i,j = 1,2,...,n, \tag{25.12}$$

$$A_{i,j} = \left\{ a_{i,j}, b_{i,j}, c_{i,j}, d_{i,j} \right\}, \quad A_{j,i} \left\{ 1/a_{i,j}, 1/b_{i,j}, 1/c_{i,j}, 1/d_{i,j} \right\},$$

where $a_{i,j}$, $b_{i,j}$, $c_{i,j}$, and $d_{i,j}$ are the numbers of UFN $A_{i,j}$.

Then, the WFs can be calculated by the geometric mean technique (Bojadziev and Bojacziev, 1997). The UFN geometric mean $\bar{A}_i$ of the ith row in the comparison matrix is defined as

$$\bar{A}_i = \left\{ \bar{a}_i, \bar{b}_i, \bar{c}_i, \bar{d}_i \right\} = \left\{ \sqrt[n]{\prod_{j=1}^{n} a_{i,j}}, \sqrt[n]{\prod_{j=1}^{n} b_{i,j}}, \sqrt[n]{\prod_{j=1}^{n} c_{i,j}}, \sqrt[n]{\prod_{j=1}^{n} d_{i,j}} \right\}, \tag{25.13}$$

$$W_i = \left\{ a_i, b_i, c_i, d_i \right\} = \left\{ \frac{\bar{a}_i}{\sum_{j=1}^{n} \bar{d}_j}, \frac{\bar{b}_i}{\sum_{j=1}^{n} \bar{c}_j}, \frac{\bar{c}_i}{\sum_{j=1}^{n} \bar{b}_j}, \frac{\bar{d}_i}{\sum_{j=1}^{n} \bar{a}_j} \right\}, \tag{25.14}$$

where W_i is the fuzzy WF of C_i.

25.5.4.6 Step 6: Defuzzification and Normalization

Because the outputs of geometric mean methods are fuzzy WFs, a defuzzification is adopted to convert fuzzy WFs to matching crisp values in which the fuzzy AHP employs a defuzzification approach proposed by Bojadziev and Bojacziev (1997). The crisp value w_i' of fuzzy WF can be calculated by

$$w_i' = \frac{a_i + 2(b_i + c_i) + d_i}{6}. \tag{25.15}$$

The final WF of C_i is obtained by

$$WF_i = \frac{w_i'}{\sum\limits_{j=1}^{n} w_j'}. \tag{25.16}$$

25.5.4.7 Step 7: Calculation of RLs of Subhazard Groups

Once the WFs of risk contributors are obtained, the overall RL at the subhazard group can be calculated by the synthesizing of WF and RL for each hazardous event produced in the FRA risk estimation phase. The RL of a subhazard group S-HG$_i$ is defined by

$$RL_{\text{S-HG}_i} = \sum\limits_{i=1}^{n} RL_{C_i} WF_{C_i}, \ i = 1, 2, ..., n, \tag{25.17}$$

where RL_{C_i} and WF_{C_i} are the RL and WF of C_i.

Similarly, the $WF_{\text{S-HG}_i}$ of subhazard groups and the WF_{HG_i} of hazard groups can be obtained by repeating steps 1 to 7. The RLs of hazard groups and the overall RL of a railway system can be obtained by

$$RL_{\text{HG}_i} = \sum\limits_{i=1}^{n} RL_{\text{S-HG}_i} WF_{\text{S-HG}_i}, \ i = 1, 2, ..., n, \tag{25.18}$$

$$i = 1, 2, \cdots, n \ RL_{\text{System}} = \sum\limits_{i=1}^{n} RL_{\text{HG}_i} WF_{\text{HG}_i}, \ i = 1, 2, ..., n, \tag{25.19}$$

where RL_{HG_i} and WF_{HG_i} are the RL and WF of the ith hazard group HG$_i$, $RL_{\text{S-HG}_i}$ and $WF_{\text{S-HG}_i}$ are the RL and WF of the ith subhazard group, and RL_{System} is the overall RL of the railway system.

25.5.5 Risk Response Phase

The results produced from the risk estimation phases may be used through the risk response phase to assist risk analysts, engineers, and managers in developing maintenance and operation policies. If risks are high, risk reduction measures must be applied or the depot operation has to be reconsidered to reduce the occurrence probabilities or to control the possible consequences. However, the acceptable and unacceptable regions are usually divided by a transition region. Risks that fall in this transition region need to be reduced to as low as reasonably practicable (ALARP) (Railway Safety, 2002).

25.6 Case Study: Risk Assessment of Shunting at Hammersmith Depot

25.6.1 Introduction

In this section, an illustrated case example on the risk assessment of shunting at the Hammersmith depot is used to demonstrate the proposed risk assessment methodology. The case materials have been collected from industry (Metronet, 2005). The input parameters are FF, CP, and CS of hazardous events. The outputs of risk assessment are RLs of hazardous events, hazard groups, and the overall RL of shunting at the Hammersmith depot with risk scores located from 0 to 10 and risk categorized as "low," "possible," "substantial," and "high" with a percentage belief. In the FRA risk estimation phase, the RLs of hazard groups are calculated using the FRA based on the aggregation results of each hazardous event belonging to the particular hazard group. In the fuzzy-AHP estimation phase, the overall RL of shunting at Hammersmith depot is obtained on the basis of the aggregation of the RLs of each hazard group contribution weighted by using fuzzy-AHP method.

25.6.2 Hazard Risk Identification at Hammersmith Depot

Hammersmith depot is one of the largest depots in London Underground. Historical data of accidents and incidents have been recorded over the past 10 years. In this case, the historical accident and incident databases have been reviewed in Hammersmith depot. Seven hazard groups and 17 subhazard groups have been identified and defined, and each subhazard group consists of a number of hazardous events (Metronet, 2005), which are described as follows:

1. The derailment hazard group (DHG) includes two subhazard groups, i.e., typical outcome (minor injury) and worst-case scenario (major injury), which have been identified based on the previous accidents and incidents.

2. The collision hazard group (CHG) consists of four subhazard groups, i.e., collision between trains in a worst-case scenario (fatality), collision between trains with a typical outcome (multiple minor injuries), collision hazard of worst-case scenario (fatality), and collision hazard of typical outcome (minor injury).

3. The train fire hazard group (TfHG) only has one subhazard group, i.e., train fire typical outcome, which covers minor injury, as it is believed that a train would not catch fire fast enough to endanger a driver more than through smoke inhalation.

4. The electrocution hazard group (EHG) has two subhazard groups, typical outcome (fatality) and best case scenario (major injury), which cover a number of hazardous events, for example, contact with the conductor rail while entering/leaving the cab, contact with the conductor rail while walking to the train, and plugging in gap jumper leads if the train has stalled/gapped.

5. The slips/trips hazard group (SHG) includes three subhazard groups, i.e., minor injury, major injury, and fatality.

6. The falls from height hazard group (FHG) consists of three subhazard groups, i.e., minor injury, major injury, and fatality, which cover falls from height such as when a shunter leaves the train cab.

7. The train strikes person hazard group (TsHG) has been identified based on the record in the past 10 years into two subhazard groups—major injury and fatality. The hazardous events in these two subhazard groups include a train striking an authorized person, including other depot workers (e.g., ground shunter) or track-side staff, and a train striking an unauthorized person, e.g., trespassers.

A risk tree has been developed for risk analysis of shunting at Hammersmith depot, as shown in Figure 25.5. Risk assessment is initially carried out from hazardous events and then progressed up to subhazard group level, hazard group level, and finally to depot level. The qualitative descriptors of FF, CS, CP, and RL have been developed for the analysis of shunting at Hammersmith depot, and the FRA is employed to estimate the RL of each hazardous event in terms of FF, CS, and CP. The definition of FF defines the number of times an event occurs over a specified period, e.g., number of events/year. The qualitative descriptors of FF are defined as "remote," "rare," "infrequent," "occasional," "frequent," "regular," and "common," and their meanings are presented in Table 25.3.

CS describes the magnitude of the possible consequence in terms of the number of fatalities, major and minor injuries resulting from the occurrence of a particular hazardous event. The qualitative descriptors of CS are defined as "negligible," "marginal," "moderate," "critical," and "catastrophic," and their meanings are shown in Table 25.4, where major and minor injuries are calculated in terms of equivalent fatalities.

CP defines the FF that failure effects that will happen given the occurrence of the failure. One may often use such qualitative descriptors as "highly unlikely," "unlikely," "reasonably unlikely," "likely," "reasonably likely," "highly likely," and "definite." Table 25.5 shows the evaluation criteria of CP and the corresponding qualitative descriptors.

The qualitative descriptors of RL are defined as "low," "possible," "substantial," and "high." Their definitions, which are generally similar to those described in EN 50126, EN 50129, and GE/GN 8561 (Railway Safety, 2002), are listed in Table 25.6. The risk score is defined in a manner that the lowest score is 0, whereas the highest score is 10. Similar to the input qualitative descriptors of FF, CS, and CP, the trapezoidal MFs are used to describe the RL. The results of RLs can be expressed either as a risk score located in the range from 0 to 10 or as risk category with a belief of percentage.

Because three parameters, FF, CP, and CS, are used to determine the RLs of hazardous events, the rulebase consists of 245 If–then rules for this study. Figure 25.6 shows five rule matrices. It can be seen that each matrix consists of 49 rules with a particular qualitative descriptor of CS. For example, the rule at the top left of the matrix of CS = Negligible would be expressed as follows:

> If FF is *remote* and CP is *highly unlikely* and CS is *negligible*, then RL is *low*.

In this case, five experts with high qualifications regarding this subject are involved in the risk assessment group. EIs are allocated to experts on the basis of their background and experience as shown in Table 25.7. For example, expert E5 has less experience; therefore, he/she has the lowest EI = 0.16.

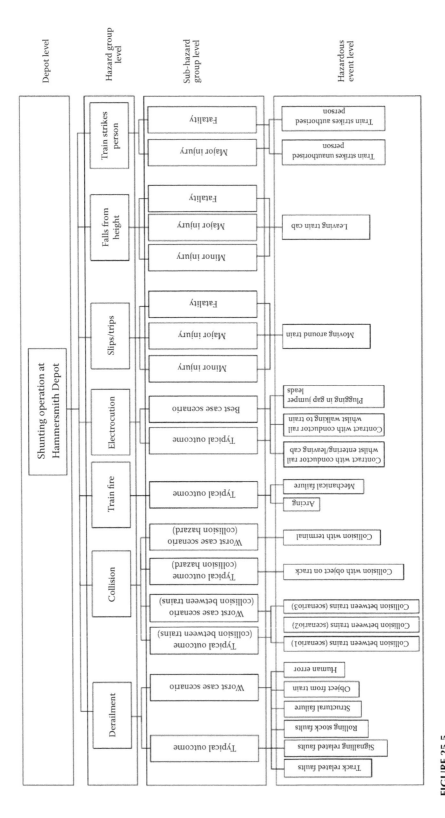

FIGURE 25.5
Hazard identification at different levels for Hammersmith depot.

TABLE 25.3

Definitions of Qualitative Descriptors of FF

Index	Qualitative Descriptors	Description	Approximate Numerical Value (Event/Year)	Parameters of MFs (Trapezoid)
1	Remote	<1 in 175 years	2.00×10^{-3}	$0, 0, 2 \times 10^{-3}, 6 \times 10^{-3}$
2	Rare	1 in 35 years to 1 in 175 years	1.00×10^{-2}	$2 \times 10^{-3}, 6 \times 10^{-3}, 1.5 \times 10^{-2}, 4 \times 10^{-2}$
3	Infrequent	1 in 7 years to 1 in 35 years	5.00×10^{-2}	$1.5 \times 10^{-2}, 4 \times 10^{-2}, 8 \times 10^{-2}, 2 \times 10^{-1}$
4	Occasional	1 in 1 ¼ years to 1 in 7 years	2.50×10^{-1}	$8 \times 10^{-2}, 2 \times 10^{-1}, 5 \times 10^{-1}, 1.25$
5	Frequent	1 in 3 months to 1 in 1 ¼ years	1.25	$5 \times 10^{-1}, 1.25, 2.25, 5.25$
6	Regular	1 in 20 days to 1 in 3 months	6.25	2.25, 5.25, 10.25, 31.25
7	Common	1 in 4 days to 1 in 20 days	31.25	10.25, 31.25, 100, 100

TABLE 25.4

Definitions of Qualitative Descriptors of CS

Index	Qualitative Descriptors	Description	Numerical Value (Event/Year)	Parameters of MFs (Trapezoid)
1	Negligible	No injury and/or negligible damage to the system	0–0.1	$0, 0, 9 \times 10^{-3}, 2 \times 10^{-2}$
2	Marginal	Minor system damage and/or minor injury	0.1–2	$9 \times 10^{-3}, 2 \times 10^{-2}, 1 \times 10^{-1}, 2 \times 10^{-2}$
3	Moderate	Failure causes some operational dissatisfaction and/or major injury	2–5	$1 \times 10^{-1}, 2 \times 10^{-1}, 4 \times 10^{-1}, 5 \times 10^{-1}$
4	Critical	Major system damage and/or severe injury	5–10	$4 \times 10^{-1}, 5 \times 10^{-1}, 9 \times 10^{-1}, 2$
5	Catastrophic	System loss and/or fatality	>10	$9 \times 10^{-1}, 2, 5, 5$

TABLE 25.5

Definitions of Qualitative Descriptors of CP

Index	Qualitative Descriptors	Description	Parameters of MFs (Trapezoid)
1	Highly unlikely	The occurrence likelihood of accident is highly unlikely.	$0, 0, 1.5 \times 10^{-1}, 2 \times 10^{-1}$
2	Unlikely	The occurrence likelihood of accident is unlikely but possible given the occurrence of the failure event.	$1.5 \times 10^{-1}, 2 \times 10^{-1}, 2.5 \times 10^{-1}, 3 \times 10^{-1}$
3	Reasonably unlikely	The occurrence likelihood of accident is between likely and unlikely.	$2.5 \times 10^{-1}, 3 \times 10^{-1}, 3.5 \times 10^{-1}, 4.25 \times 10^{-1}$
4	Likely	The occurrence likelihood of accident is likely.	$3.5 \times 10^{-1}, 4.25 \times 10^{-1}, 5.75 \times 10^{-1}, 6.5 \times 10^{-1}$
5	Reasonably likely	The occurrence likelihood of accident is between likely and highly likely.	$5.75 \times 10^{-1}, 6.5 \times 10^{-1}, 7 \times 10^{-1}, 7.5 \times 10^{-1}$
6	Highly likely	The occurrence likelihood of accident is very likely	$7 \times 10^{-1}, 7.5 \times 10^{-1}, 8 \times 10^{-1}, 8.5 \times 10^{-1}$
7	Definite	The accident occurs given the occurrence of the failure event.	$8 \times 10^{-1}, 8.5 \times 10^{-1}, 1, 1$

TABLE 25.6

Definitions of Qualitative Descriptors of RL

Index	Qualitative Descriptors	Description	Parameters of MFs (Trapezoid)
1	Low	Risk is acceptable	0, 0, 1, 2
2	Possible	Risk is tolerable but should be further reduced if it is cost-effective to do so	1, 2, 4, 5
3	Substantial	Risk must be reduced if it is reasonably practicable to do so	4, 5, 7, 8
4	High	Risk must be reduced to safe in exceptional circumstances	7, 8, 10, 10

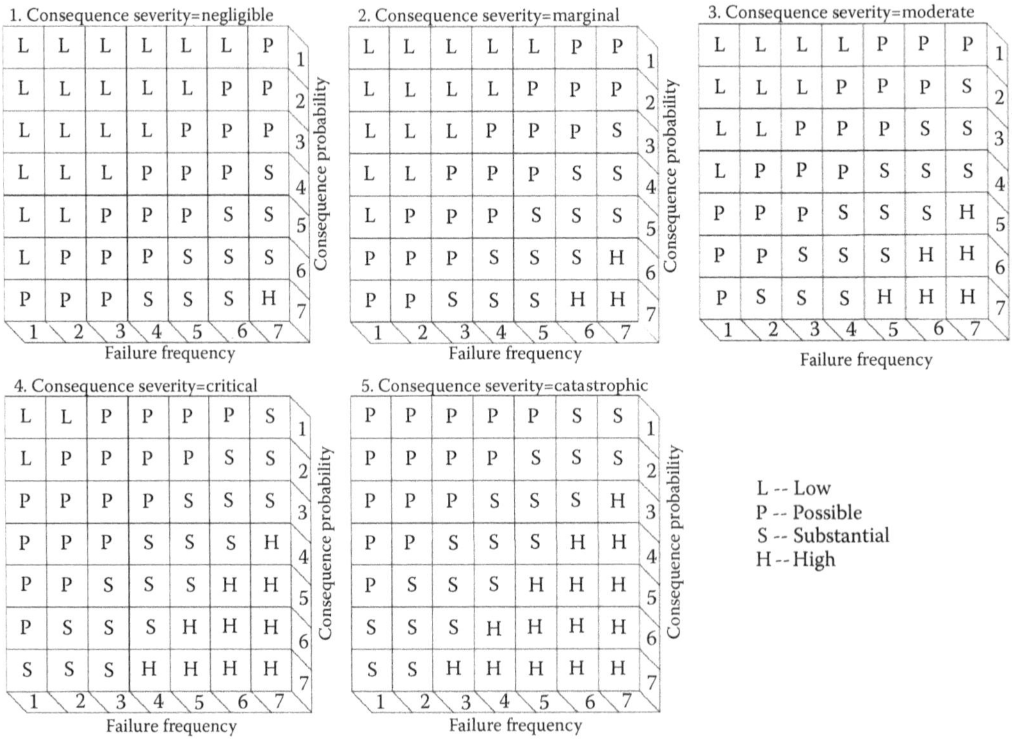

1. Consequence severity=negligible (Consequence probability rows 1–7, Failure frequency columns 1–7)

	1	2	3	4	5	6	7
1	L	L	L	L	L	L	P
2	L	L	L	L	L	P	P
3	L	L	L	L	P	P	P
4	L	L	L	P	P	P	S
5	L	L	P	P	P	S	S
6	L	P	P	P	S	S	S
7	P	P	P	S	S	S	H

2. Consequence severity=marginal

	1	2	3	4	5	6	7
1	L	L	L	L	L	P	P
2	L	L	L	L	P	P	P
3	L	L	L	P	P	P	S
4	L	L	P	P	P	S	S
5	L	P	P	P	S	S	S
6	P	P	P	S	S	S	H
7	P	P	S	S	S	H	H

3. Consequence severity=moderate

	1	2	3	4	5	6	7
1	L	L	L	L	P	P	P
2	L	L	L	P	P	P	S
3	L	L	P	P	P	S	S
4	L	P	P	P	S	S	S
5	P	P	P	S	S	S	H
6	P	P	S	S	S	H	H
7	P	S	S	S	H	H	H

4. Consequence severity=critical

	1	2	3	4	5	6	7
1	L	L	P	P	P	P	S
2	L	P	P	P	P	S	S
3	P	P	P	P	S	S	S
4	P	P	P	S	S	S	H
5	P	P	S	S	S	H	H
6	P	S	S	S	H	H	H
7	S	S	S	H	H	H	H

5. Consequence severity=catastrophic

	1	2	3	4	5	6	7
1	P	P	P	P	P	S	S
2	P	P	P	P	S	S	S
3	P	P	P	S	S	S	H
4	P	P	S	S	S	H	H
5	P	S	S	S	H	H	H
6	S	S	S	H	H	H	H
7	S	S	H	H	H	H	H

L -- Low
P -- Possible
S -- Substantial
H -- High

FIGURE 25.6
Fuzzy rulebase matrices.

TABLE 25.7

EIs for Five Experts

Expert	Years of Experience	EI
E1	10	2.10×10^{-1}
E2	10	2.40×10^{-1}
E3	10	2.10×10^{-1}
E4	8	1.80×10^{-1}
E5	5	1.60×10^{-1}

25.6.3 FRA Risk Estimation

As the data of FF, CP, and CS in this case are directly related to subhazard group, the assessment is carried out from subhazard group level, then up to hazard group level, and finally to depot level. The risk data of subhazard groups are then processed following the procedures described earlier.

25.6.3.1 Step 1: Inputting of FFs, CSs, and CPs

As the data obtained are in numerical format, these crisp values of FF, CP, and CS of subhazard groups are captured in an input template file. For example, the FF, CP, and CS of the subhazard group "derailment (typical outcome)" are 3.33×10^{-2}, 99%, and 0.005, respectively, as shown in Table 25.8.

25.6.3.2 Step 2: Conversion of Inputs into UFNs

These crisp values are converted into corresponding UFNs according to Table 25.1. The converted UFNs A_{FF}, A_{CP}, and A_{CS} of the FF, CP, and CS of the subhazard group "derailment (typical outcome)" are

$$A_{FF} = \left\{ 3.33 \times 10^{-2}, 3.33 \times 10^{-2}, 3.33 \times 10^{-2}, 3.33 \times 10^{-2} \right\},$$
$$A_{CP} = \left\{ 0.99, 0.99, 0.99, 0.99 \right\},$$
$$A_{CS} = \left\{ 0.005, 0.005, 0.005, 0.005 \right\}.$$

25.6.3.3 Step 3: Aggregation of UFNs

As the data are obtained from the historical record, there is no need to aggregate experts' judgments, and the process directly moves to step 4.

25.6.3.4 Step 4: Transfer of UFNs into Fuzzy Sets

UFNs are then transferred into fuzzy sets according to sets 2–6. The fuzzy sets $\tilde{A}_{FF}$, $\tilde{A}_{CP}$, and $\tilde{A}_{CS}$ of the FF, CP, and CS of the subhazard group derailment (typical outcome) are

$$\tilde{A}_{FF} = \left\{ \left(u, \mu_{A_{FF}}(u) \right) \middle| u \in [0,100], \ \mu_{A_{FF}}(u) = \begin{cases} 1 & u = 3.33 \times 10^{-2} \\ 0 & \text{otherwise} \end{cases} \right\},$$

$$\tilde{A}_{CP} = \left\{ \left(v, \mu_{A_{CP}}(v) \right) \middle| v \in [0,1], \ \mu_{A_{CP}}(v) = \begin{cases} 1 & v = 0.99 \\ 0 & \text{otherwise} \end{cases} \right\},$$

$$\tilde{A}_{CS} = \left\{ \left(w, \mu_{A_{CS}}(w) \right) \middle| w \in [0,5], \ \mu_{A_{CS}}(w) = \begin{cases} 1 & w = 0.005 \\ 0 & \text{otherwise} \end{cases} \right\}.$$

TABLE 25.8

Shunting at Hammersmith Depot

Operation	Hazard Groups	Index	Subhazard Groups	Proposed Model				
				Failure Frequency	Consequence Probability (%)	Consequence Severity	Risk Score	Risk Category
Shunting at Hammersmith depot	Derailment	1	Derailment (typical outcome)	3.33×10^{-2}	99.00	5.00×10^{-3}	3.00	Possible: 100%
		2	Derailment (worst-case scenario)	3.33×10^{-2}	1.00	1.25×10^{-1}	0.81	Low: 100%
	Collision	3	Collision between trains (worst-case scenario)	3.00×10^{-2}	20.00	6.25×10^{-1}	2.52	Possible: 100%
		4	Collision between trains (typical outcome)	3.00×10^{-2}	80.00	2.50×10^{-2}	3.00	Possible: 100%
		5	Collision hazard (typical outcome)	1.20×10^{-1}	99.00	5.00×10^{-3}	4.04	Possible: 96% Substantial: 4%
		6	Collision hazard (worst-case scenario)	1.20×10^{-1}	1.00	1.25×10^{-1}	0.82	Low: 100%
	Train fire	7	Train fire (typical outcome)	1.50×10^{-2}	100.00	5.00×10^{-3}	3.00	Possible: 100%
	Electrocution	8	Electrocution (typical outcome)	5.71×10^{-2}	75.00	6.25×10^{-1}	6.00	Substantial: 100%
		9	Electrocution (best-case scenario)	5.71×10^{-2}	25.00	1.25×10^{-1}	0.80	Low: 100%
	Slips/trips	10	Slips/trips (minor injury)	2.00×10^{-1}	80.00	5.00×10^{-3}	3.00	Possible: 100%
		11	Slips/trips (major injury)	2.00×10^{-1}	15.00	1.25×10^{-1}	0.80	Low: 100%
		12	Slips/trips (fatality)	2.00×10^{-1}	5.00	6.25×10^{-1}	3.00	Possible: 100%
	Falls from height	13	Falls from height (minor injury)	1.43×10^{-2}	15.00	5.00×10^{-3}	0.75	Low: 100%
		14	Falls from height (major injury)	1.43×10^{-2}	80.00	1.25×10^{-1}	3.00	Possible: 100%
		15	Falls from height (fatality)	1.43×10^{-2}	5.00	6.25×10^{-1}	0.75	Low: 100%
	Train strikes person	16	Train strikes person (major injury)	1.00×10^{-1}	50.00	1.25×10^{-1}	3.00	Possible: 100%
		17	Train strikes person (fatality)	1.00×10^{-1}	50.00	6.25×10^{-1}	3.54	Possible: 100%

25.6.3.5 Step 5: Fuzzy Inference Process

There are 245 rules in the rulebase. They are subjectively defined based on expert experience and engineering judgment. The fire strength of each rule with input fuzzy sets can be calculated by Equation 25.8. Then, the fuzzy implication is applied to obtain the conclusion fuzzy sets of fired rules. The truncated MF of the conclusion fuzzy set of each rule can be obtained by Equation 25.9. Finally, all f the conclusion fuzzy sets are aggregated by Equation 25.10 to form a single fuzzy set which represents the fuzzy output of RL. For example, the fuzzy sets $\tilde{A}_{FF}$, $\tilde{A}_{CP}$, and $\tilde{A}_{CS}$ are the input fuzzy sets and are fired with all the rules in the rulebase. However, there are only two rules with nonzero fire strength:

- R_{28}: If FF = *Rare* and CP = *Definite* and CS = *Negligible*, then RL = *Possible*.
- R_{29}: If FF = *Infrequent* and CP = *Definite* and CS = *Negligible*, then RL= *Possible*.

When input fuzzy sets are fired with rule 28, the MFs of qualitative descriptors of rule 28 are obtained according to Equations 25.2 and 25.7):

$$\mu_{B_{Rare}}(u) = \begin{cases} (u-0.002)/0.004, & u \in [0.002, 0.006], \\ 1, & u \in [0.006, 0.015], \\ (u-0.04)/0.025, & u \in [0.015, 0.04], \\ 0, & \text{otherwise} \end{cases},$$

$$\mu_{B_{Definite}}(v) = \begin{cases} (v-0.8)/0.05, & v \in [0.8, 0.85], \\ 1 & v \in [0.85, 1], \\ 0, & \text{otherwise} \end{cases},$$

$$\mu_{B_{Negligible}}(w) = \begin{cases} 1, & w \in [0, 0.009], \\ (w-0.2)/0.011, & v \in [0.009, 0.02], \\ 0, & \text{otherwise} \end{cases},$$

$$\mu_{B_{Possible}}(x) = \begin{cases} x-1, & x \in [1,2], \\ 1, & x \in [2,4], \\ 5-x, & x \in [4,5], \\ 0, & \text{otherwise} \end{cases}.$$

The fire strength of rule 28 α_{28} is calculated by Equation 25.8:

$$\alpha_{28} = \min\left[\max\left(\mu_{A_{FF}}(u) \wedge \mu_{B_{Rare}}(u)\right), \max\left(\mu_{A_{CP}}(v) \wedge \mu_{B_{Definite}}(v)\right), \right.$$
$$\left. \max\left(\mu_{A_{CS}}(w) \wedge \mu_{B_{Negligible}}(w)\right) \right]$$
$$= \min(0.25, 1, 1) = 0.25.$$

The MF conclusion fuzzy set of rule 28 $\mu'_{B^{28}_{RL}}(x)$ is calculated by Equation 25.9:

$$\mu'_{B^{28}_{RL}}(x) = \alpha_{28} \wedge \mu_{B_{Possible}}(x) = \begin{cases} x-1, & x \in [1,2], \\ 0.25, & x \in [2,4], \\ 5-x, & x \in [4,5], \\ 0, & \text{otherwise} \end{cases}.$$

The MF conclusion fuzzy set of rule 29 $\mu'_{B^{29}_{RL}}(x)$ is also calculated:

$$\mu'_{B^{29}_{RL}}(x) = \alpha_{29} \wedge \mu_{B_{Possible}}(x) = \begin{cases} x-1, & x \in [1,2], \\ 0.25, & x \in [2,4], \\ 5-x, & x \in [4,5], \\ 0, & \text{otherwise} \end{cases}, \quad \alpha_{29} = 0.75.$$

The MF of final output RL is obtained by Equation 25.10:

$$\mu'_{B_{RL}}(x) = \overset{245}{\underset{i=1}{V}} \mu'_{B_{RL}}(x) = \mu'_{B^1_{RL}} \vee \mu'_{B^2_{RL}} \vee, \cdots, \vee \mu'_{B^{28}_{RL}} \vee \mu'_{B^{29}_{RL}} \vee, \cdots, \vee \mu'_{B^{245}_{RL}},$$

$$= \mu'_{B^{28}_{RL}} \vee \mu'_{B^{29}_{RL}} = \begin{cases} x-1, & x \in [1,2], \\ 0.75, & x \in [2,4], \\ 5-x, & x \in [4,5], \\ 0, & \text{otherwise} \end{cases},$$

25.6.3.6 Step 6: Defuzzification

The crisp value of RL can be defuzzified from the fuzzy set of the output RL by Equation 25.11. The RL of the subhazard group derailment (typical outcome) is finally obtained:

$$RL = \frac{\sum_{j=1}^{10} \mu'_{B_{RL}}(x_j) \cdot x_j}{\sum_{j=1}^{10} \mu'_{B_{RL}}(x_j)} = \frac{0.75 \times 2 + 0.75 \times 3 + 0.75 \times 4}{0.75 + 0.75 + 0.75} = 3,$$

where the number of quantization levels is set to 10, which is appropriate to defuzzify the output fuzzy set.

By following the six steps earlier, all the RLs of subhazard groups are calculated, and the result is shown in Table 25.8. As the subhazard groups are of equal importance to their hazard groups, there is no need to perform fuzzy-AHP analysis to obtain the RL of hazard groups, so the RLs of hazard groups are obtained by aggregating fuzzy sets of the RLs of the subhazard groups and then performing a defuzzification operation based on the aggregated fuzzy sets.

25.6.4 Fuzzy-AHP Risk Estimation

In order to assess the RL at the railway depot level, the relative importance of the contribution of hazard groups to the RL of shunting at Hammersmith depot is considered and estimated by fuzzy-AHP, which is quantified as WFs in the developed risk model. Thus, the WFs of hazard groups in this phase are calculated firstly and then synthesized with the RLs of hazard groups to finally determine the RL at railway depot level. The process is demonstrated as follows.

25.6.4.1 Step 1: Establishment of Estimation Scheme

The judgments were made on the basis of the estimation scheme, which all the experts agreed with.

25.6.4.2 Step 2: Pairwise Comparison of Factors in Risk Tree

Experts' judgments about the relative importance between hazard groups are shown in Table 25.9. Thanks to improved fuzzy-AHP, there are only six comparisons that have been made. Experts can use linguistic terms, numerical numbers, ranges, and fuzzy numbers to present their opinions. For example, in comparison 1 in Table 25.9, experts agree that DHG is more important than CHG. However, they have different opinions on the degree of importance, and E1, E2, E3, E4, and E5 chose "WI"; "BWS"; "4, 6"; "4, 5"; and "BWS," respectively.

25.6.4.3 Step 3: Conversion of Inputs into UFNs

The judgments are then converted into UFNs according to Table 25.1, and the converted UFNs are listed in Table 25.9.

25.6.4.4 Step 4: Aggregation of UFNs

The converted UFNs are then aggregated with respect to EIs in Table 25.7 by Equation 25.3. For example, the aggregated UFN $A_{1,2}$ of comparison 1 in Table 25.9 is obtained by Equation 25.3:

TABLE 25.9

Expert Judgments

Comparison	E1		E2		E3		E4		E5		Aggregated UFNs
	Judgment	UFNs	Judgment	UFNs	Judgment	UFNs	Judgment	UFNs	Judgment	UFNs	
DHG vs. CHG ($A_{1,2}$)	WI	1,1,1,2	BWS	3,4,4,5	4, 6	4,5,5,6	4, 5	4,4.5,4.5,6	BWS	3,4,4,5	2.94,4.00,4.00,5.00
CHG vs. TfHG ($A_{2,3}$)	WI	2,3,3,4	BWS	3,4,4,5	BWS	3,4,4,5	SI	4,5,5,6	BWS	3,4,4,5	2.38,3.18,3.18,3.99
TfHG vs. EHG ($A_{3,4}$)	EQ	1,1,1,2	EQ	1,1,1,2	EQ	1,1,1,2	EQ	1,1,1,2	EQ	1,1,1,2	1.00,1.00,1.00,2.00
EHG vs. SHG ($A_{4,5}$)	BWS	3,4,4,5	BWS	3,4,4,5	WI	2,3,3,4	BWS	3,4,4,5	BEW	1,2,2,3	2.47,3.47,3.47,4.47
SHG vs. FHG ($A_{5,6}$)	BWS	3,4,4,5	BWS	3,4,4,5	SI	4,5,5,6	WI	2,3,3,4	BWS	3,4,4,5	3.03,4.03,4.03,5.03
TsHG vs. FHG ($A_{7,6}$)	BSV	5,6,6,7	SI	4,5,5,6	BSV	5,6,6,7	BSV	5,6,6,7	SI	4,5,5,6	4.55,5.56,5.56,6.67

$$A_{1,2} = \left\{ a_{1,2}, b_{1,2}, c_{1,2}, d_{1,2} \right\} = \left\{ \frac{\sum\limits_{k=1}^{5} a_k EI_k}{\sum\limits_{k=1}^{5} EI_k}, \frac{\sum\limits_{k=1}^{5} b_k EI_k}{\sum\limits_{k=1}^{5} EI_k}, \right.$$

$$\left. \frac{\sum\limits_{k=1}^{5} c_k EI_k}{\sum\limits_{k=1}^{5} EI_k}, \frac{\sum\limits_{k=1}^{5} d_k EI_k}{\sum\limits_{k=1}^{5} EI_k} \right\} = \{2.94, 4, 4, 5\}$$

$$= \left\{ \frac{1 \times 0.21 + 3 \times 0.24 + \ldots + 3 \times 0.16}{0.21 + 0.24 + \ldots + 0.16}, \ldots, \right.$$

$$\left. \frac{2 \times 0.21 + 5 \times 0.24 + \ldots + 5 \times 0.16}{0.21 + 0.24 + \ldots + 0.16} \right\} = \left\{ 2.94, 4, 4, 5 \right\},$$

where a_k, b_k, c_k, and d_k are the four parameters in a converted UFN. All the aggregated UNFs are shown in Table 25.9.

25.6.4.5 Step 5: Calculation of Fuzzy WFs from Comparison Matrix

Then, the entire comparison matrix for the hazard group can be developed. The comparison matrix M is then established with these aggregated UNFs according to Equation 25.12:

$$M = [A_{i,j}] = \begin{bmatrix} A_{1,1} & A_{1,2} & \cdots & A_{1,7} \\ A_{2,1} & A_{2,2} & \cdots & A_{2,7} \\ \vdots & \vdots & \ddots & \vdots \\ A_{7,1} & A_{7,2} & \cdots & A_{7,7} \end{bmatrix},$$

$$i, j = 1, 2, \ldots, 7 = \begin{bmatrix} 1,1,1,1 & A_{1,2} & \cdots & A_{1,7} \\ A_{2,1} & 1,1,1,1 & \cdots & A_{2,7} \\ \vdots & \vdots & \ddots & \vdots \\ A_{7,1} & A_{7,2} & \cdots & 1,1,1,1 \end{bmatrix}.$$

By using propositions 3 and 4 as described in Section 25.4, the entire comparison matrix can be obtained, as shown in Table 25.10. However, for example, value $A_{1,6}$ is not in the interval [1/9,9]; the following transformation functions must be applied according to Equation 25.20;

$$f(x_a) = x_a^{1/\log_9^{897.12}}, f(x_b) = x_b^{1/\log_9^{897.12}}, f(x_c) = x_c^{1/\log_9^{897.12}}, f(x_d) = x_d^{1/\log_9^{897.12}}.$$

TABLE 25.10

Pairwise Comparison Matrix M Established for Hammersmith Depot

$A_{i,j}$	1	2	3	4	5	6	7
1	1.00,1.00,1.00,1.00	2.94,4.00,4.00,5.00	7.00,12.72,12.72, 19.95	7.00,12.72,12.72, 39.90	17.28,44.14,44.14, 178.35	52.37,177.88, 177.88,897.12	7.85,32.00,32.00, 197.19
2	0.20,0.25,0.25,0.34	1.00,1.00,1.00,1.00	2.38,3.18,3.18,3.99	2.38,3.18,3.18,7.98	5.88,11.03,11.03, 35.67	17.81,44.47,44.47, 179.42	2.67,8.00,8.00,39.44
3	0.05,0.08,0.08,0.14	0.25,0.31,0.31,0.42	1.00,1.00,1.00,1.00	1,1,1,2	2.47,3.47,3.47,8.94	7.48,13.98,13.99, 44.97	1.12,2.52,2.52,9.88
4	0.03,0.08,0.08,0.14	0.13,0.31,0.31,0.42	0.50,1.00,1.00, 1.00	1.00,1.00,1.00,1.00	2.47,3.47,3.47,4.47	7.48,13.98,13.98, 22.48	1.12,2.52,2.52,4.94
5	0.01,0.02,0.02,0.06	0.03,0.09,0.09,0.17	0.11,0.29,0.29,0.40	0.22,0.29,0.29,0.40	1.00,1.00,1.00,1.00	3.03,4.03,4.03,5.03	0.45,0.72,0.72,1.11
6	1.11×10^{-3}, 5.62×10^{-3}, 5.62×10^{-3},0.02	5.57×10^{-3},0.02, 0.02,0.06	0.02,0.07,0.07,0.13	0.04,0.07,0.07,0.13	0.20,0.25,0.25,0.33	1.00,1.00,1.00,1.00	0.15,0.18,0.18,0.22
7	5.07×10^{-3},0.03, 0.03,0.13	0.03,0.12,0.12,0.37	0.10,0.40,0.40,0.89	0.20,0.40,0.40,0.89	0.90,1.38,1.38,2.20	4.55,5.56,5.56,6.67	1.00,1.00,1.00,1.00

Table 25.11 lists the transferred entries that are used as the completed comparison matrix *M* for the Fuzzy-AHP process.

The UFN geometric mean $\bar{A}_i$ of the *i*th row in the comparison matrix is calculated by Equation 25.13:

$$\bar{A}_1 = \left\{\bar{a}_1, \bar{b}_1, \bar{c}_1, \bar{d}_1\right\} = \left\{\sqrt[7]{\prod_{j=1}^{7} a_{i,j}}, \sqrt[7]{\prod_{j=1}^{7} b_{i,j}}, \sqrt[7]{\prod_{j=1}^{7} c_{i,j}}, \sqrt[7]{\prod_{j=1}^{7} d_{i,j}}\right\}$$

$$= \left\{\sqrt[7]{1 \times 1.41 \times \ldots \times 1.95}, \ldots, \sqrt[7]{1 \times 1.61 \times \ldots \times 5.51},\right\}$$

$$= \{1.894, 2.394, 2.394, 3.254\},$$

$$\bar{A}_2 = \{1.305, 1.529, 1.529, 1.982\},$$

$$\bar{A}_3 = \{0.940, 1.052, 1.052, 1.329\},$$

$$\bar{A}_4 = \{0.854, 1.052, 1.052, 1.170\},$$

$$\bar{A}_5 = \{0.571, 0.704, 0.704, 0.804\},$$

$$\bar{A}_6 = \{0.355, 0.449, 0.449, 0.537\},$$

$$\bar{A}_7 = \{0.590, 0.781, 0.781, 0.973\}$$

Then fuzzy WFs W_i of hazard groups are calculated with $\bar{A}_i$ by Equation 25.14:

$$W_1 = \left\{a_1, b_1, c_1, d_1\right\} = \left\{\frac{\bar{a}_1}{\sum_{i=1}^{7} \bar{d}_i}, \frac{\bar{b}_1}{\sum_{i=1}^{7} \bar{c}_i}, \frac{\bar{c}_1}{\sum_{i=1}^{7} \bar{b}_i}, \frac{\bar{d}_1}{\sum_{i=1}^{7} \bar{a}_i}\right\}$$

$$= \left\{\frac{1.305}{1.305 + 0.904 + \ldots + 0.590}, \ldots, \frac{1.982}{1.982 + 1.329 + \ldots + 0.973}\right\}$$

$$= \{0.190, 0.901, 0.901, 0.500\},$$

$$W_2 = \{0.130, 0.192, 0.192, 0.305\},$$

$$W_3 = \{0.094, 0.132, 0.132, 0.204\},$$

$$W_4 = \{0.085, 0.132, 0.132, 0.180\},$$

$$W_5 = \{0.057, 0.088, 0.088, 0.124\},$$

$$W_6 = \{0.035, 0.056, 0.056, 0.082\},$$

$$W_7 = \{0.059, 0.098, 0.098, 0.150\}.$$

TABLE 25.11

Final Pairwise Comparison Matrix M Established for Hammersmith Depot

A_{ij}	1	2	3	4	5	6	7
1	1.00,1.00,1.00,1.00	1.42,1.57,1.57,1.68	1.88,2.27,2.27,2.63	1.88,2.27,2.27,3.29	2.51,3.40,3.40,5.34	3.59,5.34,5.34,9.00	1.95,3.06,3.06,5.52
2	0.59,0.64,0.64,0.71	1.00,1.00,1.00,1.00	1.32,1.45,1.45,1.56	1.32,1.45,1.45,1.96	1.77,2.17,2.17,3.17	2.54,3.41,3.41,5.35	1.37,1.96,1.96,3.28
3	0.38,0.44,0.44,0.53	0.64,0.69,0.69,0.76	1.00,1.00,1.00,1.00	1.00,1.00,1.00,1.25	1.34,1.49,1.49,2.03	1.92,2.35,2.35,3.42	1.04,1.35,1.35,2.10
4	0.30,0.44,0.44,0.53	0.51,0.69,0.69,0.76	0.80,1.00,1.00,1.00	1.00,1.00,1.00,1.00	1.34,1.49,1.49,1.62	1.92,2.35,2.35,2.73	1.04,1.35,1.35,1.68
5	0.19,0.29,0.29,0.40	0.32,0.46,0.46,0.56	0.49,0.67,0.67,0.75	0.62,0.67,0.67,0.75	1.00,1.00,1.00,1.00	1.43,1.57,1.57,1.69	0.77,0.90,0.90,1.03
6	0.11,0.19,0.19,0.28	0.19,0.29,0.29,0.39	0.29,0.43,0.43,0.52	0.37,0.43,0.43,0.52	0.59,0.64,0.64,0.70	1.00,1.00,1.00,1.00	0.54,0.57,0.57,0.61
7	0.18,0.33,0.33,0.51	0.30,0.51,0.51,0.73	0.48,0.74,0.74,0.96	0.60,0.74,0.74,0.96	0.97,1.11,1.11,1.29	1.63,1.74,1.74,1.85	1.00,1.00,1.00,1.00

25.6.4.6 Step 6: Defuzzification and Normalization

The crisp value w_i' of fuzzy WF W_i can be calculated by Equation 25.15:

$$w_1' = \frac{a_1 + 2(b_1 + c_1) + d_1}{6} = \frac{0.190 + 2 \times (0.301 + 0.301) + 0.5}{6} = 0.315,$$

$$w_2' = 0.201, \ w_3' = 0.138, \ w_4' = 0.132, \ w_5' = 0.090, \ w_6' = 0.057, \ w_7' = 0.100.$$

The final WF of hazardous groups WF_i is obtained by Equation 25.16:

$$WF_1 = \frac{w_1'}{\sum_{i=1}^{7} w_1'} = \frac{0.315}{0.315 + 0.201 + \ldots + 0.100} = 0.31,$$

$$WF_2 = 0.19, \ WF_3 = 0.13, \ WF_4 = 0.13, \ WF_5 = 0.09, \ WF_6 = 0.06, \ WF_7 = 0.10$$

25.6.4.7 Step 7: Synthesis of RLs and WFs

Once the WFs of hazard groups are obtained, the RL at railway depot level RL_{Depot} can be derived from the synthesis of the WFs and RLs of the hazard groups using Equation 25.17:

$$RL_{Depot} = \sum_{i=1}^{7} RL_i \cdot WF_i = 2.31 \times 0.31 + 3.30 \times 0.19 + \ldots + 3.54 \times 0.10 = 2.99,$$

which indicates that the overall RL of shunting at Hammersmith depot is 2.99 belonging to "possible" with a belief of 100%.

25.6.5 Risk Response Phase

The overall RL of shunting at Hammersmith depot is 2.99, belonging to the risk category "possible" with a belief of 100%. This requires risk reduction measures to reduce the overall RL of depot to ALARP. Seven hazard groups affect the overall RL estimation at the Hammersmith depot (Table 25.12). It should be noted that each hazard group contributes a different weight value to the overall RL of the depot. It can be seen from Table 25.13 that the major contributions are from the hazard groups "derailment," "collision," and "electrocution," which contributed 24%, 21%, and 19%, respectively, to the overall RL of shunting at Hammersmith depot. Each hazard group consists of a number of hazardous events. For example, in this case, there are six main hazardous events in the "derailment" hazard group, which are track-related faults, signal-related faults, rolling stock faults, structural failures, falling objects from trains, and human errors, which result in derailment. Based on the accident and incident reports and statistics, the majority of derailment risk (92%) is put down to human errors such as overspending and incorrect routing. Therefore, in order to reduce the RLs of derailment, staff training should be provided to shunters, signalers, and drivers; at the same time, speed should be limited at the depot, liaison between Metronet Rail shunters and signalers should be improved, and reference manual procedures should be provided.

TABLE 25.12

RLs of Hazard Group in Hammersmith Depot

Operation	Index	Hazard Groups	Risk Score	Risk Category
Shunting at Hammersmith depot (3.29, Possible: 100%	1	Derailment	2.31	Possible: 100%
	2	Collision	3.30	Possible: 100%
	3	Train fire	3.00	Possible: 100%
	4	Electrocution	4.47	Possible: 53% Substantial: 47%
	5	Slips/trips	2.40	Possible: 100%
	6	Falls from height	2.17	Possible: 100%
	7	Train strikes person	3.54	Possible: 100%

TABLE 25.13

Hazard Groups' Risk Contribution Ranking for Hammersmith Depot

Operation	Index	Hazard Groups	WF	Contribution (%)
Shunting at Hammersmith Depot	1	Derailment	3.1×10^{-1}	24.00
	2	Collision	1.9×10^{-1}	21.00
	4	Electrocution	1.3×10^{-1}	19.00
	3	Train fire	1.3×10^{-3}	13.00
	7	Train strikes person	1.0×10^{-3}	11.00
	5	Slips/trips	9.0×10^{-2}	7.00
	6	Falls from height	6.0×10^{-2}	4.00

The hazard groups "train fire," "train strikes person," "slips/trips," and "falls from height" contribute less than the preceding hazard groups with 13%, 11%, 7%, and 4%, respectively. Although these hazard groups have relatively a minor contribution to the overall RL of shunting at Hammersmith depot, the control measures are still carried out to reduce those hazardous events whose RLs fall in the transition region, i.e., "possible" and "substantial." For example, the hazard group of "train fire" contributes 10% to the overall RL of depot. As the major fire-related hazardous events lead to system failure or personal injury and health hazards include arcing and mechanical failure, the suggested control measures are to provide maintenance and inspection of fleet and track assets regularly.

25.7 Summary

Traditionally, risk assessment techniques currently used in the railway industry have adopted a probabilistic approach, which heavily rely on the availability and accuracy of data; sometimes, they are unable to adequately deal with incomplete or uncertain data. This chapter presents a case study on the risk assessment of shunting at Hammersmith depot using the proposed risk assessment model based on FRA and improved fuzzy-AHP. The outcomes of risk assessment are the RLs of hazardous events, hazard groups, and a railway system and corresponding risk categories as well as risk contributions. It will provide railway risk analysts, managers, and engineers with useful information to improve safety management and set safety standards.

References

An, M., J. Wang, and T. Ruxton, 2000a. Risk analysis of offshore installation using approximate reasoning in concept design stage. *Proceedings of the ESREL 2000 and SRA-Europe Annual Conference. Edinburgh.*

An, M., J. Wang, and T. Ruxton, 2000b. The development of a fuzzy rule base for risk analysis of offshore engineering products using approximate reasoning approaches. *Proceedings of Engineering Design Conference (EDC 2000). London.*

An, M., W. Liu, and A. Stirling, 2006. Fuzzy-reasoning-based approach to qualitative railway risk assessment. *Proceedings of the Institution of Mechanical Engineers, Part F: Journal of Rail and Rapid Transit,* **220**(2): pp. 153–167.

An, M., S. Huang, and C. J. Baker, 2007. Railway risk assessment—The FRA and FAHP approaches: A case study of shunting at Waterloo depot. *Proceedings of Institute of Mechanical Engineers, Part F: Journal of Rail and Rapid Transit,* **221**: pp. 1–19.

An, M., Y. Chen, and C. J. Baker, 2008. Development of an intelligent system for railway risk analysis. *Proceeding of 3rd International Conference on System Safety.* Birmingham, pp. 850–852.

An, M., Y. Chen, and C. J. Baker, 2011. A fuzzy reasoning and fuzzy-analytical hierarchy process based approach to the process of railway risk information: A railway risk management system. *International Journal of Information Science,* **181**: pp. 3946–3966.

Bojadziev, G., and M. Bojacziev, 1997. *Fuzzy Logic for Business, Finance, and Management.* Singapore: World Scientific.

Chen, Y., M. An, S. Huang, and C. J. Baker, 2007. Application of FRA and FAHP approaches to railway maintenance safety risk assessment process. *Proceedings of the International Railway Engineering Conference 2007, London.*

Chen, Y., and M. An, 2011. A modified fuzzy-AHP methodology in railway decision making process. *Proceedings of the International Railway Engineering Conference* (CD Format), ISBN 0-947644-45-16. Birmingham.

Herrera, F., E. Herrera-Viedma, and F. Chiclana, 2001. Multiperson decision-making based on multiplicative preference relations. *European Journal of Operational Research,* **129**: pp. 372–385.

Herrera, F., S. Alonso, F. Chiclana, and E. Herrera-Viedma, 2009. Computing with words in decision making: Foundations, trends and prospects. *Fuzzy Optimization and Decision Making,* **8**(4): pp. 337–364.

Huang, S., M. An, and C. Baker, 2006. Application of FRA and FAHP approaches to risk analysis. *World Journal of Engineering,* **21**(2), pp. 226–238.

Lee, H. K., 2005. *First Course on Fuzzy Theory and Applications.* Berlin: Springer.

LUL, 2001. London Underground Limited Quantified Risk Assessment Update 2001. LUL, London, 2001.

Metronet, 2005. *Framework for the Assessment of HS&E Risks.* Second Metronet SSL Interim Report, London.

Murtaza, M. B., 2003. Fuzzy-AHP application to country risk assessment. *American Business Review,* **21**(2): pp. 109–116.

Railway Safety, 2002. *Guidance on the Preparation of Risk Assessments within Railway Safety Cases.* Railway Group Guidance Note—GE/GN8561, 1, June 2002, London.

Saaty, T. L., 1980. *Analytical Hierarchy Process.* New York: McGraw-Hill.

Vaidya, O. S., and S. Kumar, 2006. Analytical hierarchy process: An overview of applications. *European Journal of Operational Research,* **169**(1): pp. 1–19.

Zadeh, L. A., 1965. Fuzzy sets. *Information and Control,* **8**: pp. 338–353.

26

Independent Safety Assessment Process and Methodology

Peter Wigger

CONTENTS

26.1 Introduction

To date, the railway application standards EN 50126 ff series [1]—being transferred to the International Electrotechnical Commission level—have reached a mature state and are implemented in practically every new rail technology project. These standards apply not only to heavy rail systems but also to light rail and urban mass transportation including metro and people mover systems. The standard 50126 requires a "safety plan," covering—beyond others—planning for independent safety assessment.

It is recommended to use an independent safety assessor at least for the overall railway system/urban guided transport system comprehensive aspects as well as for the safety-related subsystem signaling and train control. If the safety case concept of EN 50129, after matching it to individual technologies, is applied on a total railway system/urban guided transport system basis including all subsystems, then an appropriately distributed safety level for the whole system can be achieved. This implies that independent safety assessment of the safety evidence of all system functions and all subsystems including control center, depot and operation, and maintenance aspects shall be performed. The initial introduction and involvement of an independent safety assessor should take place in the early life cycle approach stages.

The chapter at hand describes a typical independent safety assessment methodology and works out the benefits thereof.

26.2 Definition of Independent Safety Assessment

The independent safety assessment services should be performed following the principles and processes described in the Comité Européen de Normalisation Électrotechnique (CENELEC) railway application standards. EN 50129 defines *assessment* as the process of analysis to determine whether the design authority and the validator have achieved a product that meets the specified requirements and to form a judgment as to whether the product is fit for its intended purpose. CENELEC independent safety assessment is based on the following:

- Application of life cycle/V-model as per EN 5012x
- Safety case(s) structured as per EN 50129
- Methods of quality management as per ISO 9001 ff

26.3 Independent Safety Assessor's Role and Responsibility

The independent safety assessor's role is to verify that the required level of safety and quality is achieved by all actors concerned to ensure a safe implementation of the project to the required safety standards.

This means that the supplier carries the overall responsibility for the detailed verification and validation (V&V) activities and, thus, the evidence of safety within his/her contractual scope, whereas the independent safety assessment will focus on the judgment whether the supplier's verification and validation and safety management organization has applied appropriate processes and techniques in line with the requirements of the standards.

It is the goal of an independent safety assessment to put the safety authority into a position to grant the allowance for operations (licensing) after the finalization of all necessary evaluations, examinations, analyses, inspections, tests, etc. The license can be issued for single modes or submodes of operation.

At the discretion of the safety authority, an independent safety assessor may be part of the supplier's organization or operator's organization, but, in such cases, the assessor should be authorized by the safety authority, be totally independent from the project team, and report directly to the safety authority.

26.4 Independent Safety Assessment Methodology

Referring to the independent safety assessor's role and responsibility as described earlier, the independent safety methodology focuses on the judgment on the supplier's organization and processes, in particular, on the supplier's V&V activities as well as on the functional and technical safety of the system/function/subsystem/item under assessment.

The independent safety methodology basically distinguishes between the *aspects* to be assessed and the *techniques* applied therefore.

The aspects of independent safety assessment are as follows:

- Assessment of operator's system criteria
  - Overall safety target and acceptance criteria
  - Preliminary hazard and risk analysis
  - Overall system safety requirements
- Assessment of supplier's quality management system
  - Quality plan, quality policy, quality procedures
  - Quality management organization, responsibility, authority, and communication
  - Codes and standards to be applied
  - Design and development, configuration management
  - Requirements implementation and traceability, interface management control
  - Control of production, identification, and traceability
  - Monitoring and measuring devices
  - Control of nonconformity of product
  - Competence and training
  - Servicing and maintenance requirements
  - Corrective and preventive action
  - Internal and external audits
- Assessment of supplier's safety management system
  - Safety management organization
  - V&V organization
  - Independence to safety integrity level
- Assessment of supplier's safety process implementation
  - Safety plan
  - Safety procedures
  - Safety analysis tools
  - Safety requirements specification process
  - Safety requirements traceability process
  - Test plans and specifications
  - Test equipment and tools
  - Test specifications and reports
  - V&V plan
  - V&V process
  - V&V reports
  - Safety case plan, safety case
  - The supplier's organization capacity in hazard and risk analyses
  - Identification of hazards and respective risk analysis, hazard log
  - Specification of safety requirements and safety integrity levels

- Specification of hazards mitigation measures at the design stage
- Proper implementation of hazards mitigation measures at implementation stage
- System validation/system tests and hazard closure
- Assessment of functional and technical safety
  - Preliminary hazard analysis, risk analysis, hazard log
  - Safety architecture, safety integrity allocation
  - Technical documentation, analyses
  - Separation of safety relevant and non-safety-relevant functions
  - System and subsystem analysis and refinement
  - Requirement specifications and traceability
  - Test specifications, test plans and test reports, acceptance criteria
- Assessment of related safety cases
  - Application of cross acceptance to the extent possible, if applicable
  - Safety-related application conditions

Independent safety assessment principle techniques are applied as appropriate:

- Preparation of an assessment plan
- Audits/interviews of suppliers quality and safety management organization/ processes
- Review of documentation such as safety plans, safety concepts, hazard and risk documentation, requirements, design and test specifications, test reports, and safety cases for completeness, validity, unambiguity, comprehensibility, and consistency
- Test witnessing/site inspection throughout different phases, factory and site inspection, commissioning tests, trial run tests, etc.
- Inspection of safety cases/evidence documentation
- Usage of cross acceptance to the extent possible
- Raise observations and findings
- Prepare assessment reports and certifications
- Interactive, solution-oriented communication
- Recommendations and lessons learnt to the benefit of the project
- Project accompanying safety certification approach

The preceding described independent assessment methodology is proven by project examples to be suitable suitable for the following:

- Combination of classical and new technologies
- Application for classical systems with driver as well as for modern driverless operation
- Most complex systems with multiple networked software-driven systems

- Application of generic product and generic application approvals as base for specific applications
- Application of cross acceptance to the extent possible

26.5 Independent Safety Assessment throughout the Life Cycle

The independent safety assessment is based on the application of the life cycle/V-model approach of the railway application standards, applicable for both the railway sector and urban guided transport sector.

Figure 26.1 shows the typical involvement of the independent safety assessor in relation to the life cycle phases. At the discretion of the operator and/or the safety authority, the independent safety assessor can be involved in all life cycle approach phases.

26.5.1 Intermediate Assessment and Approval Milestones

The more complex the system the more detailed has the assessment and approval process to be planned. The life cycle model (also in conjunction with other standards/regulations)* may form the basis to define an acceptance/authority approval process based on authority approval milestones derived from the life cycle approach.

In order to allow for a project accompanying safety approval and continuous evaluation, whether all relevant safety activities and the corresponding submittals are in place, it is recommended to introduce interim safety approval milestones after certain life cycle approach phases.

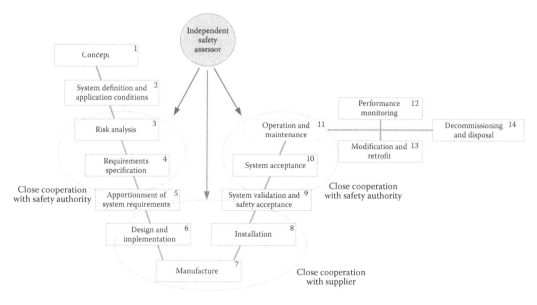

FIGURE 26.1
Independent safety assessor involvement.

---

* Other (project-specific) specific standards, rules, and regulations may apply in addition, for example, the new standard EN 50657 for software on Rolling Stock.

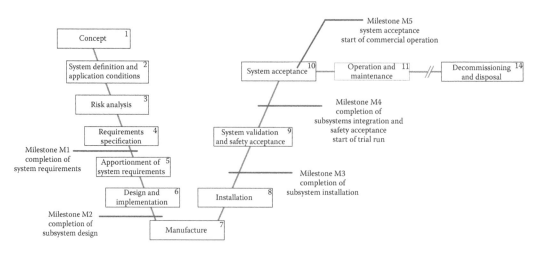

FIGURE 26.2
Safety approval milestones—examples.

The following project milestones/safety approval milestones as shown in Figure 26.2 are recommended. They should be matched with respective supplier's project planning. The final definition of milestones is topic of mutually agreement between the actors affected, namely, the operator and the supplier as well as the independent safety assessor and the safety authority.

Intermediate approval milestones can be applied as go-ahead gates, allowing the supplier to enter the next stage only after the respective intermediate approval milestone has been achieved to the satisfaction of the independent safety assessor and the safety authority. This avoids, for example, the start of detailed design (based on potentially incomplete, inconsistent, or invalid requirements) prior to having achieved acceptance of the functional and safety requirements.

26.5.2 Handling of Subsystems

In case the life cycle approach is applied to a total railway system/urban guided transport scheme that includes the integration of various subsystems such as rolling stock, permanent way, signaling and communication, and stations the handling of subsystem requires a further refinement. This also covers renewal/modernization projects where new subsystems such as signaling or rolling stock are integrated into existing total railway or urban guided transport systems, also affecting operational aspects in case of change in the grade of automation.

Therefore, for complex systems, it is highly recommended to refine the general life cycle approach model in order to address overall system aspects as well as subsystem specific issues. Some of the equipment being used could also be developed as part of the project, but this activity would also have its own V life cycle approach that would integrate with the V life cycle approach of the overall project. This situation is illustrated in Figure 26.3.

This means, in practice, that the design and development processes as well as the related independent safety assessment for the individual subsystems run in parallel.

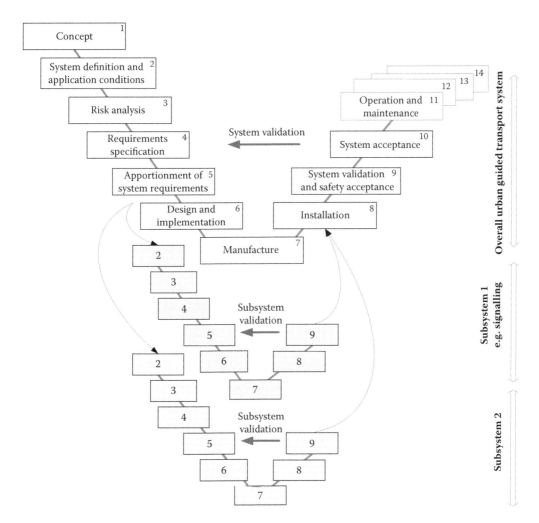

FIGURE 26.3
Application of the life cycle approach for a system and its subsystems.

The subsystems are developed, manufactured, tested, and assessed in the factory (according to the respective subsystem life cycle approach) before finally being integrated into the overall railway/urban guided transport system.

The handling of subsystems depends on whether the subsystems are part of an overall railway/urban guided transport system process or whether they are considered stand alone, for example, a signaling/train control system in case of a signaling system renewal project or in the case of buying new rolling stock for an existing railway/urban guided transport system.

In the first case (subsystem part of a [new] overall system), the subsystem may run its own life cycle as an "underlying life cycle approach"; this should apply at least for the signaling/train control subsystem. In the second case (stand-alone subsystem), the subsystem may run its own life cycle approach as a "stand-alone life cycle approach"; this should apply at least for the signaling/train control subsystem. In both cases, basically, the same life cycle process applies as described for the overall railway/urban guided transport system, of course, limited and focused on the sub-system boundaries.

26.5.3 Life Cycle Approach Phase-Related Roles and Responsibilities

Figure 26.4 shows the life cycle approach phase-related roles and responsibilities and related interfaces, in particular, the involvement of an independent safety assessor and the safety authority.

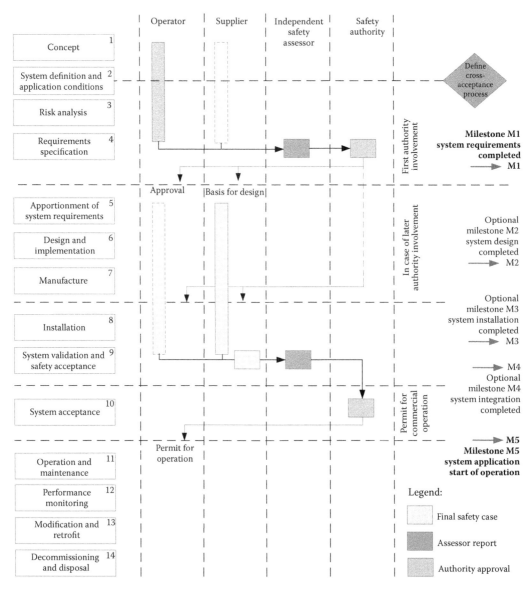

FIGURE 26.4
Life cycle approach-related roles, responsibilities, and interfaces.

26.5.4 Integration of Subsystems into System Safety Case

Overall, the system life cycle approach phases 1–5 provide the input and requirements for the individual subsystems, whereas each subsystem itself (once having completed the subsystem design and development process) comes with potential application conditions and operation and maintenance requirements to be observed on system level forming the input for life cycle approach phases 8–10.

On this basis, the overall life cycle approach (as well as the independent safety assessment process) can focus on a clear and precise set of safety documentation and, afterward, on thorough checks that the defined targets have been achieved. Figure 26.5 shows the principle hierarchy of safety documentation, following a top–down approach at the beginning of the project (life cycle approach phases 1–5) and a final bottom–up approach for the collection of the subsystem safety evidence (life cycle approach phases 8–10) and forming an integrated system level safety case.

26.5.5 Supplier's V&V and Related Assessment Activities

V&V applies on both the railway system/urban guided transport system level (ref. EN 50126 [1]) and subsystems, at least for the signaling and train control subsystem (ref. EN 50128 [2] and EN 50129 [3]). This means that validation and safety acceptance applies first for the signaling and train control subsystem for the sole subsystem functions and second for the whole system with respect to functional integration and interfaces including operational aspects. Figure 26.6 shows the main actors for the system validation and safety acceptance phase (life cycle phase 9).

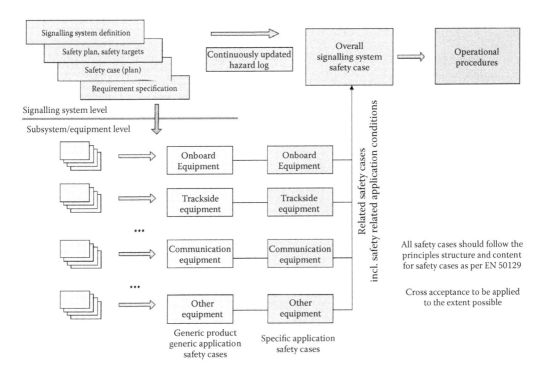

FIGURE 26.5
Principle hierarchy of documentation.

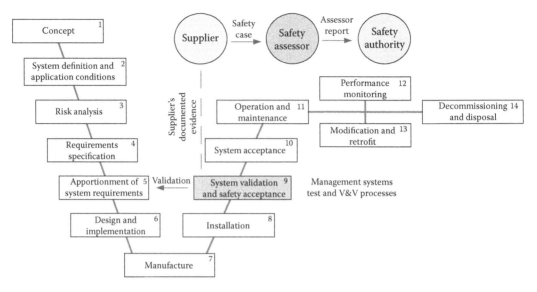

FIGURE 26.6
Supplier's V&V and assessment activities.

26.6 Conclusion and Benefits of Independent Safety Assessment

Considering the development in the railway and urban guided transport sector (see also MODSafe)* during the last decade, the following can be concluded:

- Systems become more and more complex and software driven, requiring adequate safety means.
- Due to increasing complexity, traditional methods of proof become more problematic.
- Trends in standardization show increasing requirements on independent safety assessment.
- Involvement of independent safety assessment becomes increasingly important.
- Approval authorities more frequently require the involvement of an independent safety assessor.
- Authorities, operators, and even suppliers may not be able to ensure full scale qualification.

---

* MODSafe—Modular urban transport safety and security analysis—is one of the latest projects in the European Transport sector under the Seventh Framework Programme (FP7) for Research and Technological Development of the European Union, performed from 2008 to 2012. The purpose of the MODSafe project was the development of a safety analysis and model reference for future urban guided transport projects. Even if the rail safety landscape in urban guided transport is highly diversified, the sector will benefit from some kind of harmonization. TÜV Rheinland was the project coordinator on behalf of the European Commission and furthermore responsible for the life cycle work package—both in the person of Peter Wigger, the author of the chapter at hand. The raw material for the chapter at hand has been taken from the MODSafe deliverables, respectively, from TÜV Rheinland in-house sources. Refer to http://www.modsafe.eu for details.

In conclusion, the application of independent safety assessment is recommended since it comes along with the following benefits for the project and involved actors:

- Independence of the assessor ensures opinions being free from project constraints.
- Independent safety assessor qualification and competence ensures views on the relevant aspects.
- There is a reduction of project risk due to competent third-party opinion.
- Independent safety assessor may contribute with lessons learnt from previous projects.
- Involvement of different independent safety assessor team members ensures a variety of views.
- Positive assessment report (and authority approval) result in marketing and sales benefits.
- Positive assessment report (and authority approval) may allow to reduce liability insurance.
- The cross-acceptance approach will significantly reduce project risks.
- Assessor organizations may be accredited as inspection bodies, ensuring criteria compliance.
- Assessor organizations may be accredited as certification bodies, ensuring criteria compliance.
- Operators are given increased confidence independent from potential supplier constraints.
- Responsible safety authorities can limit their resources and activities to the legal responsibility.
- Participation of an independent safety assessor is a win–win situation for all parties involved.

References

1. CENELEC (Comité Européen de Normalisation Électrotechnique). EN 50126: *Railway Applications—Specification and Demonstration of Reliability, Availability, Maintainability and Safety (RAMS)*. CENELEC, Brussels.
2. CENELEC. EN 50128: *Railway Applications—Communication, Signalling and Processing Systems—Software for Railway Control and Protection Systems*. CENELEC, Brussels.
3. CENELEC. EN 50129: *Railway Applications—Communication, Signalling and Processing Systems—Safety-Related Electronic Systems for Signalling*. CENELEC, Brussels.

27

Application of the Interface and Functional Failure Mode Effects and Criticality Analysis (IFF-MECA) for RAM and Safety Assessment of Rail Electrification

Qamar Mahboob

CONTENTS

27.1 Introduction and Background

Railway-related standards require from the railway suppliers and operators to demonstrate the reliability, availability, and maintainability (RAM) and safety in projects (EN 50126 [CEN 1999]). Reliability, availability, maintainability, and safety (RAMS) studies should (a) give answers to the questions associated to the failures in interfaces (INs), functions, and components/subsystems (SSs); (b) consider system and SS level hazards; and (c) show the evaluation of the results in reference to the contracts and standards compliances. Existing methods have limitations in the combined handling of three areas mentioned earlier. Existing RAMS analysis methods (such as failure modes and effects analysis [FMEA], hazard log management, IN hazard analysis, fault tree analysis [FTA]) mainly take into account the system evaluation by considering the subsystem and component level failures of a system (Zio 2007; Aven 2012; Braband 2001; Jahanian and Mahboob 2016). For example, one big limitation of the classical FMEA is that it does not consider the combination of component failures in a system. A single component is considered at a time, and it is assumed that other components in the system are working correctly (IEC 60812, 2006-01 [IEC 2006]). In usual engineering systems, there can be more than one failing component at a time. Additionally, existing analysis approaches only consider portion of system failures. For additional readings on the FMEA, we refer to the study by Carlson (2017) and references therein.

The new approach presented herein is called interface and function failure mode effects and criticality analysis (IFF-MECA) and provides a framework to handle the system properties, functions, INs, components, and combination of all these together with external events. In the modern engineering systems, there is considerable amount of no identified defects caused by neglecting INs, properties, and function-related failures in the systems.

The structure of the study explodes in case the system under analysis is large and complex, which is usual in modern engineering systems (Mahboob 2016). The complexity in systems arises due to the microprocessors, computers, and inter- and intradependencies in the system and its INs. Demonstrating acceptable safety of a technology requires combined, careful, and precise handling of the information on complex and large systems such as rail electrification systems (Chen et al. 2007; Mahboob 2014). In order to demonstrate acceptable safety of a technology, considerations on INs and functions of one system with other systems are important; because RAMS-related requirements can be exported and or imported to other systems. Usually, system development is based on historical experiences, and system correctly performs its intended functions. However, a wrong method or system model for the purpose of RAMS analysis can lead to over- or underestimation of the system RAMS (Mahboob et al. 2012). Consequently, complications toward, e.g., demonstrating acceptable safety and high related costs can occur. Classical methods have following main limitations:

- Mainly limited to the component level considerations (such as FMEA/failure mode, effects and criticality analysis)
- Functions and INs (also required for safety integrity level [SIL] quantification) are not dealt with
- No combinations of failures and no handling of dependence
- SIL-related issues are not dealt with
- Bottom–up approaches, but contractual requirements require top–down demonstrations
- Limitations toward demonstration of "functional tendering documents"
- Extra delivery of functions leading to expensive customer solutions
- Transparency and visualization problems leading to long time of acceptance

This chapter will present the application of the IFF-MECA to a large and complex rail electrification system as an example of any engineering application. The new methodology offers a simple, flexible, and concise visualization of the assessment and demonstration. It offers top–down analysis with focus on system functions and INs. One of the major advantages of the IFF-MECA is that the realization of function, which is usually dependent on hardware, and the incorporation of software can be done at the early stages of the product design and development. In other words, this approach offers flexibility to replace function realization via hardware to software and vice versa. Additionally, IFF-MECA supports decision makers even with limited knowledge in the area of RAMS to understand and then handle the related requirements and constraints adequately.

27.2 IFF-MECA

The IFF-MECA provides a framework to handle the system functions, INs, components, and combination of all these together with external events. This approach is presented by Mahboob et al. (2016, 2017) where the application of the IFF-MECA is presented for a small system called invertor and a large system called rail electrification system, respectively. The IFF part of IFF-MECA analyzes INs, functions, and SS/component-related failure causes in a system, whereas the MECA part of IFF-MECA is used for RAMS effects in a system. The general description of the system; its primary functions (PFs) and secondary functions (SFs); INs; and components supporting primary, secondary, and safety-related functions are identified. The system descriptions provide basis on how a system functions in order to obtain its main purpose(s). Functions are basically decomposed into PFs and SFs. The PF of the system is the major function that a system must deliver and is contractually binding, e.g., for a project delivery. SF is a subfunction of system, which is based on SS performances and provides support to the PF. Failure of a SF can possibly lead to the failure of PF as well. INs among other systems can be hardware or physical, information or communication, material, energy, or the combinations of all of them. An additional advantage of using PF and SF is to differentiate and deal with primary failures and secondary failures that are often part of contractual RAMS requirements in projects. Furthermore, function, IN, and component-based analysis differentiates between and facilitates the identification of (a) safety functions, (b) safety-related function, and (c) non-safety-related functions, meaning functions with and without direct effect on safety in case of failure of one system or component. The IFF-MECA moreover handles the quantification and classification of their integrity requirements by automatically allocating SILs based on frequency probabilities of failure occurrences.

The objective is to introduce a method that can provide traceable, consistent, transparent, easy to handle, and logically structured information related to large and complex engineering system (includes SSs, components, functions, INs, external influences, hazards) and associated RAMS topics. The ultimate goal is to obtain acceptance toward the RAMS compliances and documentation with the lowest possible effort in different project phases by meeting the contractual and standards-related requirements. Many of these can be achieved by applying the IFF-MECA. The development, maintenance, and administration of overall RAMS documentation can be simplified even with improved quality if the IFF-MECA is applied. Importantly, the IFF-MECA saves resources and can considerably contribute toward the independent safety assessor acceptance. The use of the IFF-MECA methodology is shown through its real-world application to the rail electrification system.

A classical approach such as FMEA is limited to the component-level failure modes analyses. The proposed IFF-MECA methodology offers combined handling of components, INs, and function-related failures in a clear and flexible way and thus provides a novel approach for the RAMS assessment.

Function-based tendering documents for railway projects are in market. Please refer to the study by Doelling and Schmieder (2016) and the references therein. These tendering documents mainly contain functional requirements rather than detailed technical specifications and associated parameters. The main purpose of the functional tendering is to explore for an economical and sustainable solution that meets all functional requirements at a better price and performance ratio—no more and no less. Such tenders define the required performance in accordance with the stated objectives. For example, to construct a rail system with a maximum operating speed of 350 km/h and availability of 99.95% (every year) for a 200 km track length is one of the main objectives of functional tendering. To achieve this main objective, functional tendering documents provide comprehensive requirements in reference to operator specifications, standards, and laws such as in RAMS-related standard EN 50126. IFF-MECA is useful in analyzing functional tendering documents that require and order innovative solutions, which are able to essentially reduce investments and life cycle costs. The main features of the IFF–MECA are given in the following:

- Top–down approach better suited for higher-level RAMS demonstration, also combines bottom–up ways
- Includes components, INs, functions, external events, hazards, and RAMS assessment criteria
- Standardized wording (e.g., from project or product design) from the beginning to the end of the project
- Comprehensive, complete, and concise product database is behind the approach
- Creates standard documents with few clicks
- Realization of functions via HW (Hardware) and SW (Software) (in case of functional tendering), thus leading to more cost optimal solution
- Dependencies can be handled and are traceable
- Suitable for both RAMS assessments in projects
- Provides support to the applicable railway RAMS standards

27.3 System under Analysis

The IFF-MECA approach is applied to a DC 750 V traction power supply system for a metro system, which is responsible for distributing sufficient traction power to trains within the contractual limits of scope.

Redundancies in the supply networks are provided to ensure the availability of the traction power to rolling stock. For example, TSS_1 has redundancies in terms of incoming and outgoing switchgears, transformer, rectifier, and cable ring (CR). Additionally, the load parameters of an individual traction supply station (TSS) (and its components) are designed in a way that it should be able to supply the required traction power to the operating trains in case the neighboring TSS (herein TSS_2) completely fails to perform its intended function. In other words, in all cases, the design of the traction power supply provides possibilities of feeding power to any of the traction power zones (TPZs) from more than one TSS. For example, TPZ_2 can be fed by TSS_2 given that TSS_3, which is a regular supply source for TPZ_2, has been failed due to any reason. Please refer to Figure 27.1 for the graphical representation of the system and Table 27.1 for brief description of the associated components. The system receives its power via the 33 kV utility feed cables and the bulk supply substation (BSS) from public grid. The bulk supply station mainly consists of two bus bars, which are connected to the public grid via circuit breakers (CBs) on the input side (CB_1, CB_2) and to the 33 kV cables on the output side (CB_4, CB_5, CB_6, CB_7). The two bus bars can also be connected if necessary to disconnect the bulk station from the public grid without losing the traction power. Additionally, there is one protection device (P_1, ..., P_7) attached to each of the CBs to monitor its processing parameters, e.g., the voltage, and to initialize a trip in case of exceeding permissible limits.

The bulk supply station is subsequently connected to the TSS via the CR section which is an arrangement of power cables in the form of a section ring configuration providing

TABLE 27.1

SSs and Components in the Single Diagram for DC 750 V Traction Power Supply for a Metro System

ID	Item	Brief Functional Description
UF	Utility feed cables	Electrical connections between BSS and main grid station for high-voltage transmission such as 33 kV
BSS	Bulks supply station	Receiver station/intake station of the local utility provided for connection of the rail electrification to the public (utility) grid
BB	Bus bar	Electrical rigid conductor within a switchgear
CB	Circuit breaker	Electrical switch that is able to cut off a short circuit current
P	Protection device	Electronic device monitoring process parameters, e.g., voltage and initiating trip in case of exceeding permissible limits
CR	Cable ring section	Arrangement of power cables in the form of section ring configuration, to introduce redundant power connection. For example, CB4, P4, C1, CB8, and P8 form one CR section. One CR is formed by combining CR1, CR2, CR3, CR4, and CR5
C	Cable	An electrical cable composed of one or more wires, called conductors, and is used to provide electrical connections between two fixed connection points
TSS	Substation	Subordinate electrical station for transformation and distribution of electrical power
T	Transformer	Electrical device that transforms electrical energy between two or more circuits through electromagnetic induction
R	Rectifier	Electrical device that converts electrical power from AC to DC
TPZ	Traction power zone	A segment in the TR system that can be powered on and off separately
TR	Third rail	Distribution of electrical traction current along the line to the current collectors of the train
REL	Rail electrification	SS responsible for distribution of sufficient traction power to the trains within the contractual limits of scope

redundant power connections. Thus, one CR is formed by combining two cables (C) connecting one bulk station from the BSS with two different traction substations (CR$_1$, CR$_5$) and three cables connecting two TSSs (CR$_2$, CR$_3$, CR$_4$).

The TSS itself is a subordinate electrical station for the transformation and distribution of the electrical power to the third rail (TR). Similar to the bulk station, one TSS contains two incoming CBs, each provided with a protection device and a connecting bus bar. The adjacent conversion area is redundantly incorporated: One is composed of a CB (e.g., CB$_{24}$) with its protection device (P$_{24}$), a transformer (e.g., T$_3$), a rectifier (e.g., R$_3$), and a disconnector. The transformer transforms electrical energy between two or more circuits through electromagnetic induction, whereas the adjacent rectifier converts it from AC to DC. The function of the disconnector is similar to the CB, but the disconnector is not able to cut equally off high currents.

Both conversion areas within one TSS are then connected via a second bus bar, followed by two parallel conductors, which connect the TSS to the TR. Each of these parallel conductors includes a CB and a disconnector and can be connected via a load breaker. The TR distributes the electrical traction current along the line to the current collectors of the train. It is divided into several TPZs that can be powered on and off separately.

In reference to the whole system, one BSS feeds seven TSSs.

The supply of traction power and the train traction power conversion are considered as safety functions (Figure 27.1).

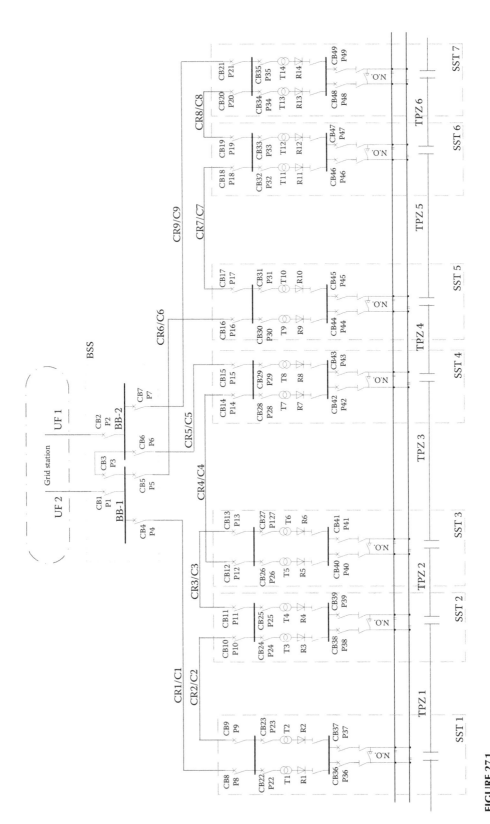

FIGURE 27.1
SLD for described TPS. (Adapted from Q. Mahboob et al., *Risk, Reliability and Safety: Innovating Theory and Practice*, 2330–2334, CRC Press, Boca Raton, FL, 2016.)

27.4 IFF-MECA Application

This section presents an application based on the system explained in the last section. The results (obtained from Microsoft tool implemented with Visual Basic) are summarized in Table 27.2. The headers are explained in the following:

- Sub-SSs
  - States the name of the sub-SS.
- PF
  - A list of PFs is provided for each sub-SS here. PFs are major functions that the system must deliver. PFs are in general contractually binding.
- SF
  - SFs of the sub-SS are presented here. SFs, in general, support to the PFs, and failure in SF can possibly lead to the failure in PF too.
- External INs
  - Identifies SSs, components within the system that are the cause(s) of a non-availability of the primary or SF. Here external events are also considered.
- Component/component description
  - All components of one sub-SS are listed here in a sensible order. Additionally, a description is provided for each of them.
- Mean time between failures (MTBF)
  - Specifies an MTBF value for each of the components.
- Failure modes
  - Provides the possible types of failure for each component. This is, in general, "loss of function" (LO), "uncommanded close" (UC), "error" (ER), and/or "not applicable" in case that a definitive classification is not possible.
- Frequency of failures
  - Frequency of functional failure is taken from the FTA, which is not possible to discuss here. For details on the FTA please refer to the book by the National Aeronautics and Space Administration (2002) and the references therein. The FTA is carried out to logically structure and model the failure combinations for each function. In this way, the failure data associated with each failure are combined and simulated in the FTA.
- Frequency of wrong side failures
  - Wrong side failures can lead to some human safety consequences (e.g., for maintenance personal). For example, wrong open and close positioning of switches (via remote or onsite operation) during maintenance actions could lead to personal harms. The percentage of wrong side failures is given in this column for each PF. It is important to mention that wrong side failures here portion functional failures coming from the FTA.
- Number of possible hazards
  - In case of wrong side failures, a number of hazardous situations can arise.

TABLE 27.2

Overview of the IFF-MECA for a Railway Traction Power Supply System

Sub-SSs (According to Single Line Diagram [SLD])	PF	REL— External INs	SF	Components (Partial)	Failure Modes	Functional Failure	Frequency (of Wrong Side) Failures	Number of Possible Hazards
Passage post (PP) (Figure 27.2a)	(PF02) Distributing electrical power on 33 kV level to SST and LPS via 33 kV CR arrangement	SEC grid station	SF02A receiving power from public grid (SEC)	High-voltage (66 kV) switchgear panel (Incoming)	LO, UC, ER, NA	1.84×10^{-9}	5.52×10^{-10}	31
		COM	SF02B electrical separation from public grid (SEC)	HV switchgear incoming CB	LO, UC, ER, NA			
		HVAC	SF02C coupling of the 33 kV bus bar inside switchgear (on demand)	110 V CB	LO, UC, ER, NA			
		PP station control	SF02D electrical protection of 33 kV switchgear and connected cables through tripping	Bus coupler	LO, UC, ER, NA			
				MV switchgear outgoing CB	LO, UC, ER, NA			
CR and connections	(PF03) transmission of electrical power on 33 kV level between PP and SST		SF03A electrical insulation of the conductor	MV Cable BulkSt to TSS	LO, UC, ER, NA	1.77×10^{-7}	1.77×10^{-7}	8
				MV Cable TSS to TSS	LO, UC, ER, NA			
TSS (Figure 27.2b)	(PF04) receiving 33 kV from PP via 33 kV cable	HVAC	SF05A electrical separation on demand	110 V CB	LO, UC, ER, NA	5.83×10^{-7}	5.83×10^{-9}	38
	(PF05) supplying electrical power to the traction transformer	COM	SF09A protects connected equipment through tripping	Bus coupler	LO, UC, ER, NA			
	(PF06) transformation of 33 kV to 585 V	Control system (station control system)	SF09A transformation of 585 V to 33 kV (at dedicated stations only)	Medium voltage (20 kV) switchgear panel (outgoing)	LO, UC, ER, NA			
	(PF07) AC DC conversion from the traction transformer and converts it to nominal 750 V DC		SF10 DC AC inversion (at dedicated stations only)	MV switchgear outgoing CB	LO, UC, ER, NA			
	(PF08) distribution of 750 V DC traction power to the TR system		SF08A provides electrical separation of the DC switchgear	MV switchgear outgoing protection	LO, UC, ER, NA			
	(PF09) protects the DC connection cables through tripping		SF08B provides electrical bypass for DC traction power on demand	Protection relay	LO, UC, ER, NA			
	(PF10) connection point of return current collection		SF08C transforms 33 kV to 2 × 585 V	110V CB	LO, UC, ER, NA			
	(PF11) station control		SF11A Protection from internal and external overvoltage and short circuit	MV auxiliary transformer	LO, UC, ER, NA			
Zone insulation post (Figure 27.2c)	(PF13) Separation of TPZs	(Natural environment)		Single-pole load break switch with stored energy mechanism	LO, UC, ER, NA	5.71×10^{-7}	5.71×10^{-8}	29
	(PF14) separation of the line and the depot area	(Station control panel of the next station control)						

(Continued)

Exposure Probability of Human	Severity Level	Expected Risk	Risk Level	Safety Effect	Safety Level	Safety-Related Function	Hazards	Hazard Description	Failure Rate	SIL
9.13×10^{-4}	Critical	1.56×10^{-11}	Negligible	Yes	Acceptable	(S1.1) supply of traction power	H19	Evacuation in tunnel due to loss of traction power	$1.00 \times 10^{-5} \leq \lambda < 1.00 \times 10^{-4}$	No SIL required
5.71×10^{-4}	Marginal	8.08×10^{-10}	Negligible	No	Acceptable	(S1.1) supply of traction power	H19	Evacuation in tunnel due to loss of traction power	$1.00 \times 10^{-5} \leq \lambda < 1.00 \times 10^{-4}$	No SIL required
5.02×10^{-3}	Critical	1.11×10^{-9}	Tolerable	Yes	Acceptable	(S1.1) supply of traction power	H19	Evacuation in tunnel due to loss of traction power	$1.00 \times 10^{-5} \leq \lambda < 1.00 \times 10^{-4}$	No SIL required
1.54×10^{-3}	Critical	2.55×10^{-9}	Tolerable	Yes	Acceptable	(No safety related)				

(Continued)

TABLE 27.2 (CONTINUED)

Overview of the IFF-MECA for a Railway Traction Power Supply System

Sub-SSs (According to Single Line Diagram [SLD])	PF	REL— External IN	SF	Components (Partial)	Failure Modes	Functional Failure	Frequency (of Wrong Side) Failures	Number of Possible Hazards
TR (Figure 27.2d)	(PF17) transmission of traction power to the trains	RST	SF17A providing rigid electrical contact	Cables and wires, cable sheath, cable screen, cable shield, cable sealing ends, cable joints	LO, UC, ER, NA	3.20×10^{-6}	3.20×10^{-7}	5
Return current collection (Figure 27.2e)	(PF18) collection of return current from train and transmitting it to the SST	Running rails		Return current DC cables	LO, UC, ER, NA	1.77×10^{-7}	1.77×10^{-7}	2
				Return conductors	LO, UC, ER, NA			

Note: COM, communication; HV, high voltage; HVAC, heating, ventilation and air conditioning; LPS, low voltage power supply; MV, medium voltage; RST, rolling stock; SEC, supply electric company; SST, traction substation.

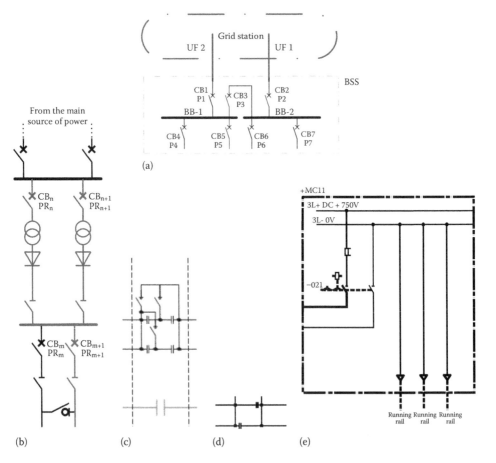

FIGURE 27.2
Section in SLD: (a) passage post, (b) TSS, (c) zone insulation post, (d) TR, and (e) return current collection.

Exposure Probability of Human	Severity Level	Expected Risk	Risk Level	Safety Effect	Safety Level	Safety-Related Function	Hazards	Hazard Description	Failure Rate	SIL
5.71×10^{-3}	Critical	9.13×10^{-9}	Tolerable	Yes	Acceptable					
5.71×10^{-3}	Marginal	2.02×10^{-9}	Negligible	No	Acceptable					

- Exposure probability
  - Exposure probability means the exposure of the person to the hazardous situations. It mainly considers expected maintenance actions and provides a key figure that reflects maintenance time and number of staff. One preventive and one corrective maintenance actions per year per relevant SF is considered. It is assumed that each maintenance action takes 1 hour.
- Severity level
  - Defines the possible severity that results from the failure. Severity levels are according to Table 27.3, based on EN 50126-1:1999 (CEN 1999).
- Expected risk
  - As it is known that the expectation for the severity in railway electrification projects is one victim, the formula for calculating the risk can be simplified: Expected Risk = Wrong side failure frequency × Number of apportioned possible hazards × Exposure probability:

$$R = F \cdot N \cdot P = h^{-1} \tag{27.1}$$

- Risk level
  - Risk level is quantified based on risk evaluation criteria (in Figure 27.3) and frequency description given in Table 27.4, also adopted from (EN 50126-1:1999

TABLE 27.3

Severity Level Description

Severity Level	Consequence to Persons or Environment
Catastrophic	Fatalities and/or multiple severe injuries
Critical	Single fatality and/or severe injury
Marginal	Minor injury
Insignificant	Possible minor injury

Range frequency	Frequency of hazardous event	Severity of hazardous event			
		Insignificant	Marginal	Critical	Catastrophic
$f > 1 \times 10^{-5}$	Frequent	R2	R1	R1	R1
$1 \times 10^{-6} < f \le 1 \times 10^{-5}$	Probable	R3	R2	R1	R1
$1 \times 10^{-7} < f \le 1 \times 10^{-6}$	Occasional	R3	R2	R2	R1
$1 \times 10^{-8} < f \le 1 \times 10^{-7}$	Remote	R4	R3	R2	R2
$1 \times 10^{-9} < f \le 1 \times 10^{-8}$	Improbable	R4	R4	R3	R3
$f \le 1 \times 10^{-9}$	Incredible	R4	R4	R4	R4

The risk categories are R1—intolerable, R2—Undesible, R3—Tolerable, R4—Negligible. For the definitions of frequencies, severities and risk categories, refer to consotium safety procedures.

FIGURE 27.3
Risk matrix.

TABLE 27.4

Frequency Description

Frequency Level	Description
Frequent	Likely to occur frequently. The hazard will be continually experienced.
Probable	Will occur several times. The hazard can be expected to occur often.
Occasional	Likely to occur several times. The hazard can be expected to occur several times.
Remote	Likely to occur sometime in the system life cycle. The hazard can be reasonably expected to occur.
Improbable	Unlikely to occur but possible. It can be assumed that the hazard may exceptionally occur.
Incredible	Extremely unlikely to occur. It can be assumed that the hazard may not occur.

[CEN 1999]). For example, R1, …, R4 are selected based on the frequency of functional failure and the severity of functional failure.

- Safety effect
  - Information about whether a wrong side failure or nonavailability of the system has safety impact is presented here. A conservative approach is applied: meaning that all relevant functional failures somehow can lead to safety effects, given that there are wrong side failures. In this way, safety functions can be identified and analyzed from the view of safety level and their integrity requirements, if required.

TABLE 27.5

SIL Allocation

Tolerable Hazard Rate (THR) per Hour	Severity Level	SIL	Consequence to Service
$1 \times 10^{-9} \leq f < 1 \times 10^{-8}$	Catastrophic	SIL 4	Loss of all the system
$1 \times 10^{-8} \leq f < 1 \times 10^{-7}$	Critical	SIL 3	Loss of a major system
$1 \times 10^{-7} \leq f < 1 \times 10^{-6}$	Marginal	SIL 2	Severe system(s) damage
$1 \times 10^{-6} \leq f < 1 \times 10^{-5}$	Insignificant	SIL 1	Minor system damage
$f \geq 1 \times 10^{-5}$	No safety related	No SIL required	–

- Level
  - According to the railway RAMS-related standards, a railway technology or system is regarded as safe when the risk associated with it is controlled to an acceptable level. As low as reasonably possible is the usual risk acceptance criteria in the rail industry (EN 50126 [CEN 1999]; Melchers 2001). A safety function is therefore a function that in case of failure of such function, the associated risk is beyond that acceptable level. The safety level is acceptable if the risk level is negligible or tolerable as visualized in IFF-MECA (see Risk level).
- Safety-related function
  - Safety-related functions are functions carrying responsibility for safety. Here all functions that affect the safety requirements of this sub-SS are listed.
- Hazards/hazard description
  - Provides a list of relevant hazard(s) corresponding to the given safety-related function.
- Failure rate
  - The failure rate is calculated as

$$\lambda = \frac{f_a}{P(X|FD \cap \text{trigger}) \cdot f_d \cdot \tau}, \tag{27.2}$$

  where f_a is the frequency of occurrence of the worst case accident associated with the referred hazard in a defined period, f_d is the frequency target associated with the reference hazard, τ is the interval at which proper performance of the analyzed safety function is checked, and $P(X|FD \cap \text{trigger})$ is the probability of occurrence of accident A, given that the safety function has failed and the trigger has occurred.
- SIL
  - The SIL classification is determined based on the failure rate interval (tolerable hazard rate [THR]). Please refer to Table 27.5 for further information on the SIL allocation.
- Document ID
  - This column provides a list of documents or references related to the sub-SS or its functions.

27.5 Conclusion and Outlook

A functional, IN, and component failure-based approach called IFF-MECA is developed and used for the RAMS assessment of an example but real system. RAM parameters are calculated and were made available for comparison and decision making in reference to the widely accepted RAMS-related standards. It is argued that the functions of the system and its INs, together with components and their failure modes, must be considered for comprehensive RAMS studies, and this is possible using the IFF-MECA methodology presented herein. Due to the combined handling of system components, INs, and functions using IFF-MECA, a novel approach for the RAMS assessment is achieved. One of the major advantages of this top–down approach is that the realization of not only the functions can be analyzed from the hardware point of view but also software can replace hardware for the same function(s) in the analysis. In this way, future systems can possibly be improved by replacing hardware with software and vice versa. One-time developed methodology is suitable to apply for similar future projects with lowest efforts, high consistency, traceability, transparency, and completeness. This approach is suitable for analyzing *functional tendering documents* for technical systems that offers great potential of innovation with reduced overall life cycle costs. The application of the IFF-MECA methodology and its usefulness to other similar RAMS-related problems should be investigated further. For example, systems involving functions, complex INs (e.g., INs related to heat, information, and energy flows), and component-based systems, such as smart energy networks (include software and digital INs), can be analyzed using the IFF-MECA.

References

Aven, T., 2012. Foundational issues in risk assessment and risk management. *Risk Analysis*, 32(10), pp. 1647–1655.

Braband, J., 2001. A practical guide to safety analysis methods. *Signal+Draht*, 93(9), pp. 41–44.

Carlson, C., 2017. *FMEA Corner*. Accessed on May 5, 2017 from http://www.weibull.com/hotwire /fmea_corner.htm.

CENELEC (Comité Européen de Normalisation Électrotechnique). EN 50126. *Railway Applications— Specification and Demonstration of Reliability, Availability, Maintainability and Safety (RAMS)*. CENELEC, Brussels.

Chen, S., Ho, T., and Mao, B., 2007. Reliability evaluations of railway power supplies by fault-tree analysis. *IET Electric Power Applications*, 1(2), pp. 161–172.

Doelling, A., and Schmieder, A., 2016. Innovative railway energy supply systems—Incentives by functional tendering. *Elektrische Bahnen*, 114(12), pp. 686–700.

International Electrotechnical Commission, 2006. IEC 60812, 2006-01: *Analysis Techniques for System Reliability—Procedure for Failure Mode and Effects Analysis (FMEA)*, Geneva: International Electrotechnical Commission.

Jahanian, H., and Mahboob, Q., 2016. SIL determination as a utility-based decision process. *Process Safety and Environmental Protection*, 102, pp. 757–767.

Mahboob, Q., 2014. *A Bayesian Network Methodology for Railway Risk, Safety and Decision Support*, PhD Thesis, Dresden: Technische Universität Dresden.

Mahboob, Q., 2016. Identification of reliability critical Items in large and complex rail electrification systems. In *Smart Grid as a Solution for Renewable and Efficient Energy*, Hershey, PA: IGI Global, pp. 226–248.

Mahboob, Q., Bernd, A., and Stephan, Z., 2016. Interfaces, functions and components based failure analysis for railway RAMS assessment. In *Risk, Reliability and Safety: Innovating Theory and Practice*, Boca Raton, FL: CRC Press, pp. 2330–2334.

Mahboob, Q., Bernd, A., and Stephan, Z., 2017. IFF-MECA: Interfaces and function failure-modes effect and criticality analysis. In *Safety and Reliability—Theory and Application*, Boca Raton, FL: CRC Press.

Mahboob, Q., Schoene, E., Kunze, E., and Trinckauf, J., 2012. *Application of Importance Measures to Transport Industry: Computation Using Bayesian Networks and Fault Tree Analysis*, Chengdu: IEEE Xplore.

Melchers, R., 2001. On the ALARP approach to risk management. *Reliability Engineering and System Safety*, 71(2), pp. 201–208.

National Aeronautics and Space Administration, 2002. *Fault Tree Handbook with Aerospace Applications*. Washington, DC: NASA Office of Safety and Mission Assurance.

Zio, E., 2007. *Introduction to the Basics of Reliability and Risk Analysis: Series on Quality, Reliability, and Engineering Statistics*, Vol. 13, Singapore: World Scientific.

28

RAMS as Integrated Part of the Engineering Process and the Application for Railway Rolling Stock

Georg Edlbacher and Simone Finkeldei

CONTENTS

28.1 Introduction

Reliability, availability, maintainability, and safety (RAMS) became of growing importance over the past decades. Compared to other functions in engineering, RAMS is a relatively new field of expertise, and as such, it had to find recognition within engineering first, before it could be fully established and deployed in the product development process.

In the past, train operators have been state owned very often, and operators were holding both rolling stock and infrastructure with no market competition on their networks. Those companies were in fact required to act economically, nevertheless, life cycle costs (LCCs) were not in the focus as the governments still financially supported them, and priorities were usually set to the promotion of public transport, network coverage, and provision of adequate rolling stock.

Liberalization attempts led to a split into infrastructure manager and rolling stock operator with different train operators aiming to compete against each other on one and the same network. This caused rolling stock operators to establish total cost of ownership considerations for being able to provide their service to lowest achievable costs. Through this reliability, availability and LCC came more and more important in bids. Nowadays, LCC can reach up to 30–40% of the bid evaluation besides compliance with technical and commercial requirements.

Safety considerations have been implemented early in the railway history to prevent accidents. In earlier times, safety was ensured by the implementation of product standards and guidelines. Over the past decades and being promoted by several train accidents, safety assessments have been formalized, and safety processes including definition of risk acceptance criteria have been defined which go beyond compliance of the product to technical standards. Proof of safety is required toward the rolling stock operator and to ensure that safety risks are controlled and minimized for passengers and personnel. Safety also has to be demonstrated toward the authorities so that the train can be homologated. A further aspect of safety demonstration is to validate design and manufacturing for product liability reasons. This supports manufacturers in case of accidents so that although they can be made responsible for harm and damages in court cases, criminal proceedings can most probably be avoided if no negligence can be demonstrated.

This chapter provides an overview of the RAMS process as part of the engineering process and the required qualifications and skills of the RAMS engineer so that he/she is capable of driving RAMS analyses and influencing the system development to ensure that the product is optimized from the RAM/LCC and safety perspective.

28.2 RAMS Engineering along the System Life Cycle

RAMS activities in a project are divided into different phases as defined and described by the V representation of the system life cycle in Figure 28.1 in accordance with the EN 50126 standard (CEN 1999). However, the RAMS process is not a sequence of stand-alone process phases by its own but is embedded in the system life cycle and is as such integral part of the product development process. This means that the V representation of the life

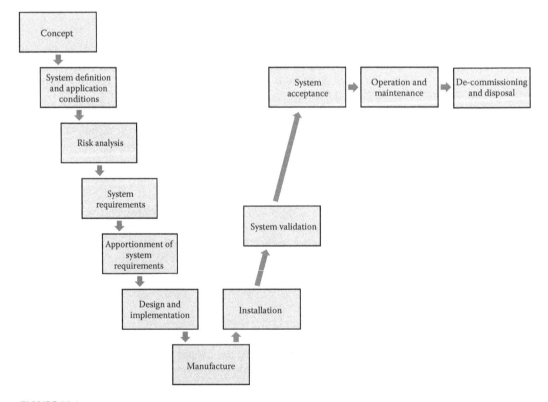

FIGURE 28.1
V representation of the system life cycle. (Based on CEN, EN 50126:1999-09. Railway Applications—The Specification and Demonstration of Reliability, Availability, Maintainability and Safety (RAMS), CEN, Brussels, 1999.)

cycle is to be understood as representation for the whole engineering, manufacturing, and commissioning process with RAMS being a part of it.

Consequently, the RAMS engineer is an active member within the engineering team and contributes to the project tasks in line with the defined project phases and project schedule.

Already in the *concept phase* (see Figure 28.1), which might be a generic product development, a bid or a project start-up phase, the RAMS engineer is involved in product decisions by comparing different options through analytic comparison of RAM/LCC parameters or by using available field data of comparable projects. Concept and design decisions in this phase should not only be taken by retail price arguments, but should also consider the full life cycle of the product as from customer's point of view the LCC of the asset is more relevant than the lowest purchase price. It is therefore essential to find the optimal balance between acquisition costs and operational/maintenance costs for being able to provide the most competitive offer with regard to the total sum of those costs. For rolling stock equipment, the most important after-sales costs are operating costs (e.g., labor costs, energy costs, track access charges), maintenance costs, and scrapping costs.

The minimization of maintenance costs is normally linked to three parameters: Frequency of the maintenance tasks, task times, and material costs. During the concept phase, the RAMS engineer supports the product manager in finding the best compromise between purchase costs and maintenance costs. The RAMS engineer uses techniques such as maintainability analysis to compare task times and availability. He/she aims to extend maintenance intervals and to reduce failure rates by arguing for specific materials, redundant system architectures, or suppliers being well known for their innovative and high-quality products. Some of those targets are in contradiction to each other, e.g., specific materials or dedicated suppliers being required for long maintenance intervals or system architectures built for high reliability might not only cause higher product costs but could also affect the maintenance costs themselves.

The determination of optimized total costs of a product by simultaneous consideration of market needs is normally an iterative process with close cooperation between product manager, RAMS engineer, existing customers, and suppliers.

In addition to maintenance cost considerations, the RAMS engineer shall consider reliability, availability, and maintainability (RAM) requirements from potential or dedicated customers, and he/she shall ensure that these are respected in the product design.

The concept phase is also important from the safety point of view. In this stage, possible safety implications should be analyzed by review of safety incidents, previously achieved safety performance and requirements from safety standards such as EN 50126 (CEN 1999), EN 50128 (CEN 2011), and EN 50129 (CEN 2003). A first preliminary risk analysis should be undertaken for the desired functions of the product to evaluate if those functions are safety relevant and to define safety targets accordingly. In the case of safety relevance, the system architecture is normally influenced by those safety targets to a large extent. Safety decisions that have to be taken again can have an impact on maintenance costs and sales price, so that a further boundary condition needs to be considered in the total cost of ownership optimization.

The tasks in the concept phase are closely linked to the ones in the *system definition and application conditions phase* (see Figure 28.1). In this stage, the concept shall be further developed and elaborated in a detailed system description. The proposed design solution shall reflect customer requirements, operational and environmental conditions, mission profile, and infrastructure constraints. With regard to RAMS, also basic maintenance strategies and conditions have to be considered. The RAMS engineer establishes the first issues of a

RAM plan and a safety plan, both describing RAMS and project organization and, additionally, the RAMS process. These documents shall be in line with the project schedule and therefore have to be agreed with project management. Both RAM and safety plan keep records of the basic (mostly external) RAMS requirements that have to be considered in the product development process and describe the methods that will be used to fulfill those requirements. The safety plan also defines risk acceptance criteria that have been agreed with customer and/or authorities.

Part of the project ramp-up is to analyze and identify potential risks. For the overall project, those are primarily technical or commercial risks, whereas the RAMS engineer concentrates on risks of not being able to fulfill the basic RAMS requirements and on safety risks that are caused by the product or by functions of the product. Risks of failing to comply with RAMS targets have to be communicated to the project management as early as possible so that adequate mitigation actions can be taken to minimize those project risks. With regard to product safety, the RAMS engineer performs a preliminary hazard analysis and a risk analysis. The purposes of those analyses are as follows:

- To identify any contributors to any credible hazard
- To define safety requirements to control those hazards
- To define safety-related functional requirements
- To determine the required quality of the proposed control measures

This can be specified by a tolerable hazard rate or by the safety integrity level of a safety function.

After the determination of the principle system architecture and its functions and achieving a common understanding of the operational environment and potential risks, all requirements need to be systematically collected in the system design specification. Safety requirements are collected in a safety requirements specification, but safety requirements that have an impact on the design will also be progressed further into the system design specification.

The next task is to apportion the system design requirements to subsystem and component requirements and to define acceptance criteria accordingly. This shall be done in line with the apportionment of subsystem and component RAM and safety requirements which shall again be integral part of the system requirements specification.

The requirements defined are the basis for the design and the development of the product. The system engineer and his/her design team have to ensure that all requirements are adhered to, and they have to demonstrate the compliance to those requirements by provision of appropriate evidence documentation. The RAMS engineer specifically proposes evidence measures and documents that he/she will need to provide satisfactory evidence of adherence to RAM and safety requirements toward internal directives and guidelines, toward the end customer or toward independent assessors (if required) and authorities. It is good practice to establish regular (e.g., weekly) meetings with the design team to discuss all aspects of possible design solutions, risks, and objections with all technical experts. It is essential that the RAMS engineer plays an active part in the design process and in those design reviews and that he/she contributes to decisions to be taken considering the RAM and safety aspects he/she is responsible for. The RAMS engineer analyses proposed design solutions by means of failure mode, effects and criticality analysis (FMECA), fault tree analysis, and reliability block diagrams, and he/she justifies RAM and safety-related design decisions. In this phase, the RAMS engineer also establishes the hazard log, where he/she tracks the progress in complying with safety requirements and where he/she collects the verification evidence documentation.

Examples of analysis methods and evidence documentation to be used for safety demonstration are the following:

- Adherence to relevant standards and design principles
- Structural calculations
- Dynamic simulations
- Component type tests
- Functional type tests
- System tests
- Vehicle tests
- Environmental condition type tests
- Electromagnetic compatibility (EMC) tests, fire certificates
- Safety data sheets for fluids
- Adherence to architecture and hardware requirements according to standard EN 50129 (CEN 2003)
- Software process and documentation according to standard EN 50128 (CEN 2011) and EN 50657 (CEN 2017) (including software tests and software/hardware integration tests)
- FMECA to prove that single point failures do not provide an unacceptable risk
- Fault tree analysis to prove that tolerable hazard rates for safety functions can be reached and to prove that requirements with regard to redundancies and common cause failures are fulfilled

If the ongoing discussion or if evidence documentation shows that proposed design solutions are in contradiction to safety regulations and rules or to safety requirements, the RAMS engineer has to report those to the project team. In cases where the responsible designer refuses to adapt the design to safety demands due to commercial or other reasons, this situation needs to be escalated to a project-independent safety authority within the company, who then has to decide accordingly.

Along with the design process, safety-related application conditions (SRACs) shall be established. In the case of a subsystem developer, these do normally include safety requirements that need to be adhered to by the system integrator. These can be environmental or operational restrictions for use, functional and technical design requirements, diagnostic requirements, or advises to include certain safety aspects into the higher-level system documentation. On the train system level, SRACs normally contain requirements and restriction for use to be adhered to by the operator. The responsible RAMS engineer of the subsystem shall verify together with the RAMS engineer and the design team of the superior system that all requirements are fully understood and incorporated.

At the end of the design phase and in the manufacturing and installation phase, the RAMS engineer has to be in close contact with the technical publications team to ensure that requirements for maintaining and operating the train or subsystems are correctly described in the maintenance and operation manual and in the training documentation. Such requirements are specified maintenance activities, maintenance intervals, and any additional regulation to perform those activities. The operator needs clear instructions for intended use in fault free conditions and for trouble shooting in case of any credible fault. Operation instructions have to be agreed with the train operator to ensure correct integration of the train into the existing infrastructure and the operational environment.

The RAMS engineer should also perform an operating and support hazard analysis to consider safety risks resulting from maintaining and operating rolling stock assets. The outcome of this analysis can consist of design control measures, safety measures to be taken by the staff, specific task descriptions, or warnings. These shall be included in the maintenance and operation manuals and shall be part of the training to be provided to operator and maintainer.

A further task of the RAMS engineer is to accompany the manufacturing and installation of the system to make use of any findings with regard to maintainability or to verify task times that have been estimated in the LCC prediction before. He/she does also collect manufacturing quality assurance documentation to demonstrate as part of the safety documentation that the product is manufactured in line with quality requirements.

In this phase, the failure reporting and corrective action system (FRACAS) process shall also be established, as during installation and system testing first incidents might already occur that have to be analyzed with regard to their RAM and safety impact. Those issues need to be transferred to the engineering so that design or manufacturing mitigations can be implemented accordingly.

After vehicle testing and system acceptance, the train will go into operation. In this phase, the FRACAS process will be continued, and the train undergoes a RAM and safety validation. With regard to safety, the train shall prove in daily operation that the safety targets are met and that safety functions and barriers are fully operational as intended. It is important that all changes and modifications made after approvals are analyzed toward their safety relevance by using the same safety process and safety criteria as in the design phase. If required, the safety documentation and any supplementary documentation shall be updated accordingly.

The operation and maintenance phase and the performance monitoring phase are also used to validate RAM/LCC parameters which have been committed toward the train operator. These can be reliability (split into different failure classes), availability, task times, maintenance intervals, LCC, and others. This validation is normally performed through regular meetings between train operator and train manufacturer. The scope of those meetings is to analyze the collected field data, to discuss the findings in detail, and to determine if individual data records are included into the validation or not. Datasets, which are very often excluded, relate to failures not in the responsibility of the manufacturer or are of minor quality so that a clear conclusion cannot be taken. In case of not fulfilling committed RAM/LCC parameters and depending on contractual agreements, the manufacturer can either be required to pay penalties or to rework the product so that it does fulfill the contractual commitments.

Besides this interaction with the train operator due to contractual obligations, it is highly recommended to establish a close cooperation with the maintainer to obtain a more detailed product knowledge from practical point of view. This knowledge can be used to verify and update theoretical predictions made in the design phase, which will help provide more accurate RAM and LCC calculations in other bids, so that potential project risks with regard to RAM/LCC commitments can be reduced. Field data can also be used to reveal design weaknesses and optimization needs, which can be used to improve RAM/LCC parameters of the product so that these have a positive impact on competiveness. Of specific interest is the collection of failure rates and replacement or refurbishment intervals of wear parts (e.g., brake pads, wheels, or filters), measurement of exchange times, results from maintenance interval extension programs, and recordings of material prices. A collaboration with a train maintainer can also be used to explicitly examine specific components in recurrent intervals to investigate changes of material characteristics, which can be the base to determine maintenance intervals.

Project and RAMS activities are very often iterative processes, especially in case of design novelties and innovative or complex systems and functions. The repetition of individual process steps can be caused by design solutions being proposed that are not capable of fulfilling the relevant requirements. A close cooperation between design team and RAMS engineer is therefore mandatory to ensure that RAMS requirements are understood and adhered to. The RAMS engineer is also required to perform RAMS analyses beforehand and parallel to the product development so that gaps between requirements, RAMS findings, and design decisions are captured as early as possible to prevent the need for redesign or to minimize the time between an already taken decision and its necessary revision.

The project management must therefore be aware of the function and role of RAMS management and must ensure that RAMS is fully integrated in the project team in line with the individual project phases.

28.3 RAMS Analysis Depending on the Role in the Supply Chain

The size of the RAMS organization, task allocation, and depth of RAMS analysis and documentation are normally dependent on the scope of supply and the RAM and safety impact of the component, subsystem, or system. While rolling stock manufacturers developed full RAMS organizations with a clear allocation of RAM, safety and engineering functions, and a well-defined separation between the roles accordingly, smaller component suppliers only might have a reduced or even no particular RAMS organization. In those companies, often, the responsible design engineer himself/herself or an application engineer carries out RAMS tasks. In this case, the RAMS responsible is normally not a RAMS expert and therefore needs support and guidance from the superior system RAMS engineer.

The allocation of RAMS requirements and evidence documentation requested also needs to respect the scope of supply of the component supplier. It is inappropriate to pass over all high-level system requirements to all subsystems and components in an undifferentiated manner, but they need to be broken down and need to be specific so that component suppliers can understand and fulfill those requirements.

Suppliers of simple components or sub-systems shall also not be asked for the complete set of documents that are required on vehicle level as some analyses do not provide additional benefit at all. As an example, fault tree analysis as a method to analyze multiple failure conditions is essential to assess a complex function such as the door closing demand, but is inadequate for use as a method for the supplier of the pure mechanical door leaf. Of course, subsuppliers shall accordingly provide information so that adequate RAM/LCC and safety analysis can be carried out on system level; nevertheless, this information can also be provided by specific input to a high-level analysis rather than asking the component supplier to provide a full RAM or safety analysis which serves as a subanalysis of the system analysis only and possibly causing gaps on the subsystem/system interface.

The adaptation of RAMS requirements and deliverables request does normally cause higher efforts in the start-up phase as clear considerations are needed for every subsystem and subcomponent, but those efforts will be awarded once the input to the system documentation is provided back as this input then is normally technology based rather than being of a more formal kind.

28.4 Teamwork of RAMS and Design Engineers

Project managers and RAMS engineers potentially pursue different targets. They sometimes also have different opinions about RAM/LCC or safety needs. The project manager directs his/her focus on technical aspects of the design, on compliance with customer requirements, on delivery in time, and on commercial and contractual issues, specifically on cost reductions with regard to manufacturing of the product. The focus of RAMS is on service performance, especially reliability, availability, and LCC and on minimizing the safety risks for operator, passengers, maintenance staff, and the environment around the rolling stock asset. The requirements for RAMS can cause conflicts between design and RAMS as they have the potential for cost impacts and delays in the project schedule. RAMS requirements and findings can even be in contradiction to train operator requirements and wishes. Such conflicts can be caused by safety analysis resulting, e.g., in the need to design a redundant system or to implement an additional supervision system, which has not been foreseen in the product concept, in the bid phase, or even in the project phase so far, resulting in additional engineering hours and a potential project delay. A conflict with train operator requirements is also possible in case he/she asks for a certain maintenance interval, which cannot be accepted from safety perspective in case the requested maintenance interval has the potential to create unacceptable safety risks. In case of such conflicts, the RAMS engineer shall raise those as internal project risks as early as possible so that adequate countermeasures can be agreed. Also, negotiations with the train operator are required at an early state for not running into the risk of nonacceptance of the final design by the operator.

Within the design team, RAMS is sometimes seen as an additional function besides engineering, which does not support the development process but adds further requirements, which need to be adhered to. If the product development team is not aware of the necessity for RAMS, it is up to the RAMS engineer to provide explanations and trainings to the team, to "educate" the team so that the function is fully understood.

These efforts are supported by helping others understand the way of thinking a RAMS engineer does, which is specifically valid for product safety aspects. As an example, the RAMS engineer must analyze all possible and credible failure modes of every component and must discuss effects on the system and on humans, he/she then has to assess the severity and initial frequency of the event and to define and to document mitigation actions. If not accordingly prepared, the responsible design engineer may sometimes believe that his/her system does not fail at all; thus, he/she does not understand why a RAMS engineer wants to discuss several failures and their consequences. It occurred that the responsible design engineer feels questioned or even criticized for his/her design. It is up to the RAMS engineer to explain to him/her that the safety investigations are, of course, required to find any safety weaknesses but that, normally, the design is automatically providing safe solutions if the designer follows established standards and processes and especially when the design engineer is well experienced. The design engineer, for example, needs to understand that the possible risk of fracture of a structural component can be mitigated by strength and fatigue calculations with adequate safety factors applied and that he/she automatically ensures a safe system design when conducting calculations according to available standards.

The RAMS engineer shall also explain that design, manufacturing, and maintenance measures can never ensure absolute safety but that the mitigation measures taken are foreseen to either minimize the frequency of occurrence of possible failures or to reduce the severity of the consequence resulting from the failure to ensure an acceptable risk.

It can be a possible convincing argument toward the design engineer that his/her belief in his/her design is caused by the measures he/she has already taken and that additional measures do only need to be taken because of the systematic safety analysis, which also takes the mitigation measures into account that have already been deployed.

RAMS engineering and project engineering should not be opposite to each other, but both functions shall aim for a close cooperation from start-up of a project on. This minimizes the risk of iterative design changes and repetitive design steps. Both functions can profit from each other in case the processes are correctly established and in case both are willing to work as a team.

28.5 Skills of the RAMS Engineer

The RAMS engineer has to play an active role in the engineering process, and he/she has to drive the RAMS process in close cooperation with the design team for being able to correctly analyze the product and to influence the design in a way that optimized reliability, and LCC can be achieved and that the product is safe.

While failing RAM/LCC requirements normally bears the risk of missing contractual obligations, an unsafe system can easily lead to injuries and mortalities. The high responsibility demands therefore require a dedicated technical education, certain skills and knowledge, and ideally some years of experience from a RAMS engineer.

Technical knowledge and detailed technical understanding of the product is the base for being able to produce correct RAMS analyses. The RAMS engineer is not necessarily required to be a technical specialist, but at least, he/she must have fundamental knowledge to a level that he/she can challenge the engineering team so that specialists can accordingly provide input or that they initiate design chances in case of any issues revealed. Not having appropriate product knowledge causes the RAMS engineer to be a collector of information only, compiling inputs to a formal RAM analysis or safety case, but prohibits him/her to make analyses of required depth and quality. This situation leads to parallelism in the design and RAMS process but not to an interlinking between both processes and mutual influences. The result of such a case can either be inadequate or lack of RAMS requirements or the design team following up design solutions based on their own understanding of RAMS requirements, which might not be in line with certain decisions that would have to be taken to optimize the product. In no case shall the role of the RAMS engineer be understood as an administrative task, with the only purpose of formally supervising the adherence to RAMS process and the delivery of documentation without examination and full understanding of the content of the document.

The role of a RAMS engineer shall also not be understood as pure office job, but it is essential to be present at manufacturing and in the customer's workshop. This shall be used to verify RAM and safety assumptions, to discuss certain maintenance aspects with staff that performs the maintenance, to benefit from maintenance methods and strategies that are successfully applied by maintainers, and to gain a deeper product knowledge in general.

Besides technical knowledge, the RAMS engineer is also required to understand relevant RAM and safety standards. He/she must have knowledge of RAMS analysis methods and tools, and he/she must be familiar with applying these methods. This includes the willingness to keep knowledge on standards, processes, and homologation requirements up to date. RAMS analyses and tools (such as FMECA) shall be used to support the RAM and safety process and shall not only be seen as formal deliverables only. The analyses shall therefore focus on technical content rather than on too formalistic approaches, and it is advised to describe technical aspects in an appropriate level of detail so that addressees of the analyses who are not directly involved in the project are able to understand all aspects of those documents. The RAMS engineer shall, in addition, be able to explain to all involved people the RAM and safety process, the motivation and need to follow these processes, reason and source of RAMS requirements, reasons why certain decisions have to be taken, and method and contents of RAMS analyses. Technical experts or project managers are not required to use RAMS analysis tools, but they need to understand the methods and tools to be able to provide most beneficial input and to understand their RAM and safety responsibility.

As mentioned before, successful RAMS project activities can only be carried out in close cooperation with engineering. RAMS engineers therefore have to be team players and willing and capable to interact with different people. RAMS engineers have to play an active role and have to be able to convince people. At the same time, they have to be open for arguments and for explanations they receive back. RAMS requires excellent communication skills as challenging negotiations with project team, management, internal and external customers, assessors, and authorities are daily business. These discussions very often include contractual aspects and so, sometimes, require intuition and the ability to find agreements without compromising relevant RAMS requirement or the safety of the product.

Summing up the requirements for a RAMS engineer, he/she must have fundamental technology, product, and RAMS knowledge, with additional communication skills, ability to act actively and as a team player, and willing to compromise but convincing others with the ability to justify specific RAMS topics.

References

CEN (European Committee for Standardization). 1999. EN 50126:1999-09: *Railway Applications—The Specification and Demonstration of Reliability, Availability, Maintainability and Safety (RAMS)*. CEN, Brussels. (Refer to IEC 62278 for the international standard.)

CEN. 2011. EN 50128:2011-06: *Railway Applications—Communication, Signalling and Processing Systems—Software for Railway Control and Protection Systems*. CEN, Brussels. (Refer to IEC 62279 for the international standard.)

CEN. 2003. EN 50129:2003-02: *Railway Applications—Communication, Signalling and Processing Systems—Safety-related Electronic Systems for Signalling*. CEN, Brussels. (Refer to IEC 62425 for the international standard.)

CEN. 2017. EN 50657:2017-11: *Railway Applications—Rolling Stock Applications—Software on Board of Rolling Stock, Excluding Railway Control and Protection Applications*. CEN, Brussels.

29

Safety and Security Assurance in Complex Technological Train Control System

Datian Zhou, Ali Hessami, and Xiaofei Yao

CONTENTS

29.1 Model-Based Approach to Identifying Hazards for Modern Train Control Systems

29.1.1 Hazard Identification of Computer-Based Train Control Systems

Since the rather deterministic days of hardwired relay logic, the proliferation of modern computing and communications technologies into the safety-critical railway signaling and control has posed a new challenge in the understanding and assurance of systems emergent properties, specifically safety. This transition has transformed the approach to design and implementation of control systems from electrical circuits to communicating and controlling functions implemented in software and firmware. A typical train control system is composed of many supervisory and control functions, and the concurrency, inter-dependency, and criticality pose a hugely complex dilemma to modern system designers in assuring overall system resilience and safety. In such a setting, it is necessary to systematically search and identify the undesirable system states at the earlier phases of the life cycle to save on effort and reengineering. A model-based approach to system representation, analysis, and safety/resilience assurance provides a proactive and potent tool in the face of pervasive complexity in modern train control systems.

29.1.1.1 Context of the Standards Related to Modern Train Control Systems

Alongside EN 50129, as one of 137 technical specifications for interoperability, the European Committee for Standardization or the European Committee for Electrotechnical Standardization (CENELEC) standards are required for achieving the interoperability of the trans-European railway network by the Directive 2008/57/EC. EN 50126 was produced to introduce the application of a systematic RAMS management process in the railway sector. The system life cycle model combines 12 stages, including concept, system definition and operational context, risk analysis and evaluation, specification of system requirements, architecture and apportionment of system requirements, design and implement, manufacture, integration, system validation, system acceptance, operation maintenance and performance monitoring, and decommissioning, as shown in Figure 29.1.

It can be systematically applied throughout all phases of the realization of a railway system/application, to develop railway-specific RAMS requirements, and to achieve compliance with these requirements. However, the process defined by EN 50126 assumes that the practitioners have business-level policies addressing quality, performance, and safety, which does not normally apply to people outside the European context. To the practitioner who is not sufficiently competent in the thorough and systematic application of EN 50129 in developing a complete system, a feasible way is integrate "well-developed" components of others into their system. Capability Maturity Model Integration (CMMI) is a process level improvement training and appraisal program. CMMI was developed by the CMMI project, which aimed to improve the usability of maturity models by integrating many different models into one framework. The integrity process advocated by CENELEC standards is analogous to what CMMI proposed to integrate engineering systems from systems engineering capability model, software engineering from Software CMM v2C, and integrated process and product development from the integrated product development CMM v0.98 in a single model that incorporates both the staged and continuous representations. Integrating some

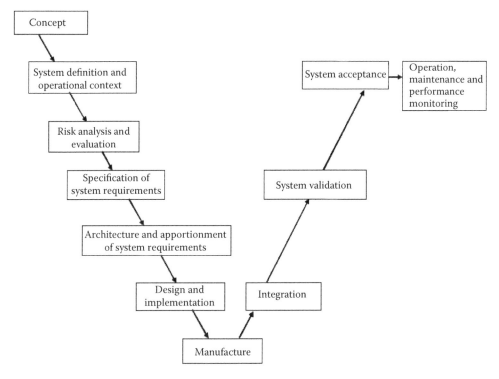

FIGURE 29.1
Life cycle for developing a safety-critical signaling system.

packages of outsourced software into the whole software system requires that the organization is mature enough to solve the problem that not all the parts were fully under control in a single existing framework challenged by the tremendous scale of software development.

European countries have always pursued a long-term plan of constructing a high-speed rail network. There is more than 11,000 km in operation and 3,800 km in schedule as the sum total of high-speed rail mileage in Europe. In Asia, the network of high-speed railways reached 24,000 km. When the projects in progress are completed, the figure will almost be doubled at 42,700 km. Modern train control systems will be deployed as the most cutting-edge technology-intensive backbone of these high-speed railway systems. Facing the requirement of the long life cycle of service accompanied with several intended upgrades and the rapid advancement of technologies, integrating more subsystems or components from the dedicated component suppliers, is likely to be the norm. This complex and large-scale task might be carried out by the iteration of purchase–overhaul–creation, as shown in Figure 29.2.

However, this strategy has some issues, which involves, with the hazard identification, the risk management, etc., for integrating a developed multiple-disciplinary cross-cultural train control system and some other systems into local signaling system engineering. The iterative process of improving the safety of a system, identifying the potential hazard, and controlling the safety risk is still an evolving discipline in the face of the emerging technologies and consequent challenges.

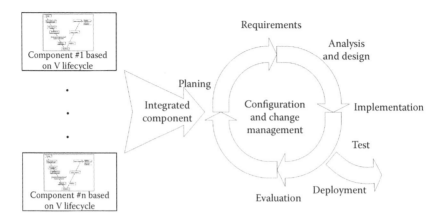

FIGURE 29.2
Iteration of rejuvenating modern train control system.

29.1.1.2 Hazard Identification

As a widespread method for identifying hazard, hazard and operability study (HAZOP) was well defined in IEC 61882. It is characterized by some paired conceptions, such as structuring and grouping, environment and operability, guide word, and experience-based methods. HAZOP originated in the background of the process industries. Without a powerful modeling method to set forth the design intent of a complex system, it is hard to conduct a completed hazard identification and reduce the possibility of overlooking some hazards or operability problems well investigated. Therefore, HAZOP has some inherent methodological bias to define the design intent for the chemical industry, which is not appropriate or sufficient for a complex computer-based control system, such as the train control system. Using the model-based approach increases the ability to describe the design intent for a complex control system while facilitating the traceability between a hazard and a model element.

29.1.2 Model-Based Hazard Identification

To achieve effective hazard identification for a complex train control system by HAZOP, some challenges should be faced. The first one relates to the text-based specification. The natural language has never been regarded as an exact approach to building an explicit system specification. Somehow, a term usually has different meanings to different people in different cultures. Also, different people can perceive different understandings from the same specification. Several fragments of description on the same topic distributed at different parts of more than one specification do pose an obstacle to build a whole picture to most readers. Often some cryptic graphics cannot give an intelligible understanding, but mislead the viewer.

Another issue with HAZOP is the requirement for the participant to focus on the deviations between the design intent and the designed and deployed system. In the HAZOP standard—IEC 61882—the definition characteristic is naturally oriented from the process industries and represented by some qualitative and quantitative properties of an element such as pressure, temperature, and voltage. For a complex train control system, detailed and specific reference to system parts and functions is essential.

Moreover, HAZOP is a team collaboration work making it easy to share, discuss, argue, modify, control version, release, and make an agreement that the described information is what the system is. So that the group can study whether the design intent is proper to combined with the guide words, generating a valuable deviation to trigger an abnormal condition or state of the system to check the solution of the design for all the abnormal scenarios.

Considering the challenges mentioned, the new model-based method should have some features to easily understand, support multiple different viewpoints, build inter-model connection, be complete and traceable, and estimate the workload. A combined approach is proposed for the model the train control system, including reference model (REM), state transition model (STM), functional hierarchical model (FHM), and sequence model (SEM). It explicitly facilitates the explicit specification of a complex train control system by these four kinds of different models. The soundness of the multiple models can be achieved by adding four pairs of inter-model constraints, such as REM versus FHM, STM versus SEM, REM versus SEM, and FHM versus SEM. An inbuilt individual identity of each element in the models helps trace any hazard back to the model of the train control system. After acknowledging the combined system model, it is possible to make the HAZOP study plan and assess the volume of effort.

29.1.2.1 Notations of Model Representations

For the convenience of modeling train control systems, several notations are used. Principally, most notations were created before. Some are common for each kind of model. Others are only used in a specific model.

29.1.2.1.1 REM

There are five notations that can be used in REM. The first one is the boundary, as shown in Figure 29.3.

The boundary is used to identify whether a component is part of a system of interest, and an interface is connected outside the system to be analyzed. Then, the component is represented by a rectangle, as shown in Figure 29.4.

FIGURE 29.3
Boundary notation.

FIGURE 29.4
Component notation.

The component is representing a part of a system. It can also be regarded as subsystem. For the convenience of defining the inter-model constraints lately, the component is abbreviated as N-REM-COM. The interface is used to define a connection between a subsystem and somewhere outside the system, as shown in Figure 29.5. An interface could be single input, single output, and bidirectional input–output.

If an interface has just a logical definition, it would be a functional interface specification (FIS). And if an interface is needed to be defined both logically and electronically, it would be form fit functional interface specification (FFFIS). The last notation of REM is the note, which is also a common notation to FHM, STM, and SEM, as shown in Figure 29.6.

The note notation is very important and necessary to record the information of the author and some other critical awareness, which cannot be addressed elsewhere.

29.1.2.1.2 STM

STM is used to describe the system states from the outside viewpoint, which are not offered by any of the subsystems independently. There should be one high-level STM for a given system. The high-level system functions can be grouped by several states in an STM. The start state notation is a filled cycle, as shown in Figure 29.7.

The start state is the default in a STM and represents the initial state of the system, which indicates that the system will provide some system functions after transiting from the start

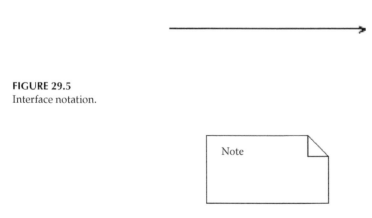

FIGURE 29.5
Interface notation.

FIGURE 29.6
Note notation.

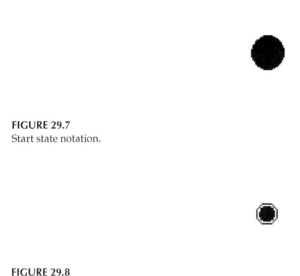

FIGURE 29.7
Start state notation.

FIGURE 29.8
End state notation.

state. There should be an exclusive start state in an STM. Another default state is the end state, which is a filled cycle orbited by a hollow cycle, as shown in Figure 29.8.

The end state represents the finishing of the work of the system, which indicates that the system will not provide any more system functions. The next one is a specific state notation, depicted by a round rectangle divided by a bar, as shown in Figure 29.9.

A state is named by the modeler and could be bundled with some system functions to show that these functions can only be provided in the specific state, as shown in Figure 29.9. For the convenience of defining the inter-model constraints, a system function incorporated in a specific state is abbreviated by N-STM-STATE-NAME-SEMi, where i can be any number increased from 1 to N incrementally. N means there are N system functions bundled with the specific state named STATE-NAME. There could be several specific states in an STM, except the start state and the end state. The transition notation is used to connect one state and another state or itself, as shown in Figure 29.10.

The transition should connect two points. Its name can be specified by the modeler.

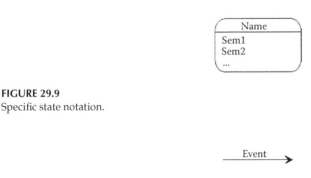

FIGURE 29.9
Specific state notation.

FIGURE 29.10
Event guard transition notation.

29.1.2.1.3 FHM

To a given system, there could be several FHMs usually. FHM builds a function tree from the top level to the low level representing the construction of the function of a subsystem, which can only be effectively analyzed at a detailed level. The component is the only top-level notation in a FHM. The component represents a subsystem defined in the REM of the system, as shown in Figure 29.11.

For the convenience of defining the inter-model constraints, the component is abbreviated by N-FHM-COMP-NAME. The detailed function notation represents a function contained in a component, as shown in Figure 29.11. A component can consist of more than one detailed function. And a detailed function can comprise more than one detailed function.

For the convenience of defining the inter-model constraints, the detailed function is abbreviated by N-FHM-FUNC-NAME. A detailed function belonging to a component can be connected by a connector called the consist notation, as shown in Figure 29.12.

A component or a detailed function can comprise any number of detailed functions or others detailed functions as a designer wants.

An FHM tree could be built in any level as the analyzer needs.

29.1.2.1.4 SEM

SEM is used to describe how the relevant subsystems cooperate together to achieve a function in a system state defined in the STM of the system. Usually, a system could have several SEMs. For the convenience of defining the intermodel constraints, the SEM is abbreviated by N-SEM-SEM-NAME. The component notation is shaped as a rectangle, as shown in Figure 29.13.

For the convenience of defining the inter-model constraints, the component is abbreviated to N-SEM-COMP. The component notation represents a subsystem defined in the REM of the system. The component contributes a specific function in system level. The behavior could include forwarding, exchanging, or conducting something. The nature of

FIGURE 29.11
Function notation.

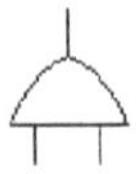

FIGURE 29.12
Consist notation.

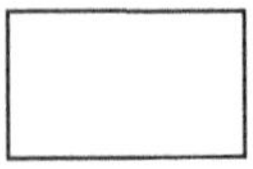

FIGURE 29.13
Component notation.

FIGURE 29.14
Control notation.

FIGURE 29.15
Data-type message notation.

what is forwarded by a component could be one of three types. The first type is the control notation representing that the sender requires the receiver conduct some control functions by sending command information, as shown in Figure 29.14.

The second type is the data, which represents that the receiver uses the data to do some calculations, as shown in Figure 29.15.

The last type is the status notation, representing that a receiver is informed of the status of a sender, as shown in Figure 29.16.

Besides the three kinds of forwarding, a component conducting something is defined by the activity notation, which is shaped as a hollow bar put vertically, as shown in Figure 29.17. For the convenience of defining the inter-model constraints, the activity is abbreviated by N-SEM-COMP-NAME-ACTIVITY.

The activity notation represents that a component is conducting an activity, which is defined in the FHM of the component. The destined or source component is represented by the delegate notation, which is like a dashed line, shown in Figure 29.18.

The delegate notation should connect to a component. From the side of the delegate connecting point with a component to another side, it represents the chronology of time sequence.

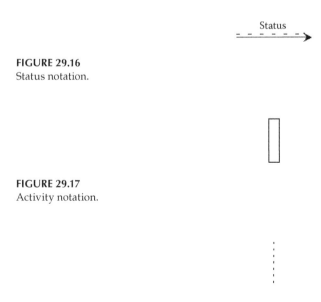

FIGURE 29.16
Status notation.

FIGURE 29.17
Activity notation.

FIGURE 29.18
Representative notation.

29.1.2.1.5 Demonstration of Modeling

For simplicity, a single example is demonstrated to show the multiple-model method.

29.1.2.1.5.1 REM The model system to be analyzed is composed of four subsystems named A, B, C, and D. The REM of the case is shown in Figure 29.19.

There are two external interfaces connected to C and D from the outside of the boundary of the system and three internal interfaces connected to B from A, D, and C.

There are three single directive interfaces from A to B, from D to B, from outside to D, and two dual-directive interfaces between B and C, and between C and outside.

There are three FFFIS interfaces connected to B from C, to C from outside, and to D from outside, and two FIS interfaces connected to B from A and to B from D.

29.1.2.1.5.2 STM The top-level functions of the system are divided into five states, including the start state, the end state, S1 state, S2 state, and S3 state. The STM of the case is shown in Figure 29.20.

As shown in Table 29.1, there are seven transitions involving the five states. If the system is in the start state and T1 occurs, the system will transit from the start state to the S1 state. If the system is in the S1 state and T4 occurs, the system will be kept in the S1 state or transited from the S1 state to itself.

The system has five top-level functions F1.1/1.2/1.3/2.1/3.1. The three functions F1.1/1.2/1.3 can be provided only when the system is in the S1 state. The system can conduct function F2.1 only in the S2 state and function F3.1 only in the S3 state.

29.1.2.1.5.3 FHM There could be several FHMs according to the REM. One of the cases is shown in Figure 29.21.

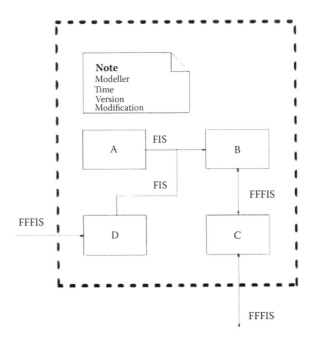

FIGURE 29.19
Demonstration of REM.

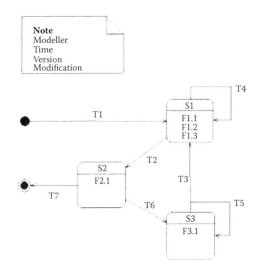

FIGURE 29.20
Demonstration of STM.

TABLE 29.1

Transitions in the Demonstrated STM

Original State	Transition	New State
Start	T1	S1
S1	T2	S2
S3	T3	S1
S1	T4	S1
S3	T5	S3
S2	T6	S3
S2	T7	End

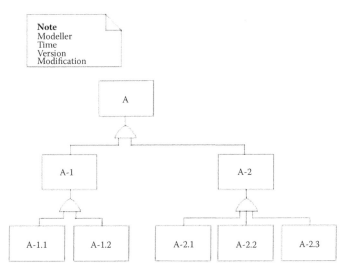

FIGURE 29.21
Demonstration of FHM.

The function of component A consists of two independent low-level functions A-1 and A-2. In order to analyze the function of component A adequately, a further level is delineated. A-1 function consists of two independent detailed functions A-1.1 and A-1.2. The function of A-2 consists of three independent detailed functions A-2.1, A-2.2, and A-2.3.

29.1.2.1.5.4 SEM   There could be several SEMs according to the STM. One of the cases is shown in Figure 29.22.

This SEM is showing how the four components/subsystems cooperate together to achieve the top-level function of the system.

First, component B sends a control command #1 to component D. According to the 1# control command, D conducts some function and sends a data-type message 2# back to component B. The 2# message spurs B to perform some function. After receiving a data-type message 3#, B goes about some function and sends another control command 4# to component C. Depending on the control command 4#, after conducting some function, C sends status information numbered 5# back to component B. Then, with the information 5# status information, B conducts some function. At last, until deriving a data-type message 6# from C, B does some function.

In this manner, the function of the given system is accomplished by the four components interactively.

29.1.2.2 Constraints of Model Representations

The multiple-model method exhibits many advantages and characteristics against the single model method. First, the multiple-model can provide more useful aspects since different experts have different perspectives about the system. Sometimes it is difficult to reach an agreement, because they think from different viewpoints. It is not cost-effective to put

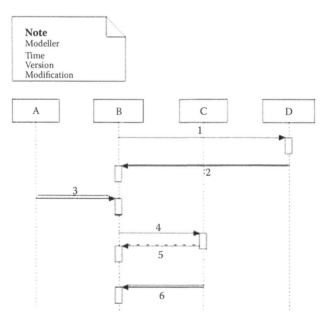

FIGURE 29.22
Demonstration of SEM.

FIGURE 29.23
Four types of inter-model constraints.

much effort into this kind of argument. Secondly, the multiple model can represent the static and dynamic characteristics of the system. Some models can easily describe the static system, such as the structure. Others have the advantage of showing the dynamic interactions among the different subsystems. Thirdly, each model of the multiple-model method has fewer notations than a sophistic model, so that it could model a system as explicitly and easily understanding as possible.

However, the multiple-model approach has some weaknesses which must be overcome to make sure that the model-based HAZOP delivers useful outcomes. The main point is about the isolation. Each model of the multiple-model method is inaccessible to others. Without the necessary connections, a person can hardly simulate the total system thoroughly in the mind. Moreover, the isolated multiple-model can suffer from conflicts. Some contents in different isolated models could be agonized to make the HAZOP ineffective. Moreover, the multiple independent models are incomplete to be studied further.

The proposed inter-model constraints include the rules connecting four pairs of models, REM versus FHM, STM versus SEM, REM versus SEM, and FHM versus SEM, as shown in Figure 29.23.

29.1.3 Case of Model-Based Approach to Train Control System Safety Assurance

In June 2012, a crucial project named Research on the Key Technologies of Safety and Risk for Railway Signalling System was founded by the then Chinese Ministry Of Railway. One of three missions of the work focused on identifying the hazard for Chinese train control system level 3 (CTCS-3). CTCS-3 is a modern conceptual designation for Chinese high-speed railway, which allows the trains equipped with certified CTCS-3 onboard equipment operated in the track deployed with CTCS-3 ground equipments and crossing the track deployed with CTCS-3 ground equipments supplied by different manufactures. CTCS-3 is realized by three separated suppliers. Until the end of 2012, China had 7735 km high-speed railway equipped with CTCS-3 system in service, and there had been 39 released specifications on CTCS-3 system within a span of 5 years, which posed an enormous and incredible challenge to the development team. The completeness of hazard identifying for CTCS-3 system was the top issue indeed. Several kinds of hazard identification methods had been methodologically contemplated, including preliminary hazard analysis, subsystem hazard analysis, system hazard analysis, operating and support hazard analysis, fault tree analysis, event tree analysis, failure mode and effects analysis, fault hazard analysis, functional hazard analysis, Petri net analysis, hazard and operability analysis, and cause–consequence analysis. After a tough and detailed decision making, none of the existed methods was considered suitable, but HAZOP was chosen to be modified as the best-suited method. Another key issue is the possibility of overlooking some hazards, which could dramatically reduce the effectiveness and make the cost unacceptable. Usually tracing the hazard from the system is a sound approach. However, the system is too complex and large in scale to model and keep the trace from system to a hazard and ensure that it is not overlooked. Naturally, to tackle the issue, the method of model-based Hazard Identification was proposed.

29.1.3.1 System Representation and Modeling

29.1.3.1.1 REM

The reference model in Figure 29.24 represents a basic structure of the CTCS-3 onboard system, and describes all components/modules of the CTCS-3 onboard system and the interfaces between them.

29.1.3.1.2 STM

The state model of CTCS-3 onboard system in Figure 29.25 describes different states or operational scenarios of an operating system from the state of power on to power off and the transitions between all the states. All operational states of the CTCS-3 onboard system are presented in STM, which are divided into two categories: normal type and abnormal type. The normal types reflect the behaviors of the onboard system when the train is moving or stopped, respectively, including normal run, shunting, emergency status, driver responsible run, enter C3, and end C3, as well as shut down, power on, and start of mission. The abnormal types describe the unusual status of modules of the onboard system and how the system will deal with these unusual states.

Moreover, each type has several states, which describe a complete series of the operational behaviors of the CTCS-3 onboard system. Unique names or numbers are assigned to the states because each state corresponds to a SEM, which is more specific and detailed with interactions among different components.

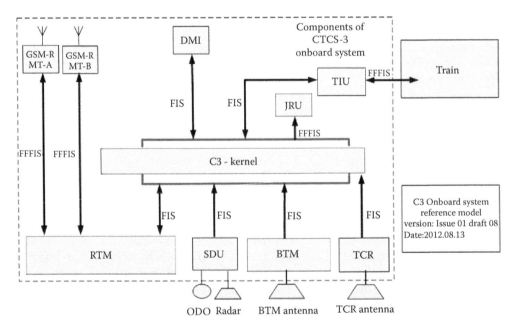

FIGURE 29.24
REM of CTCS-3 onboard system.

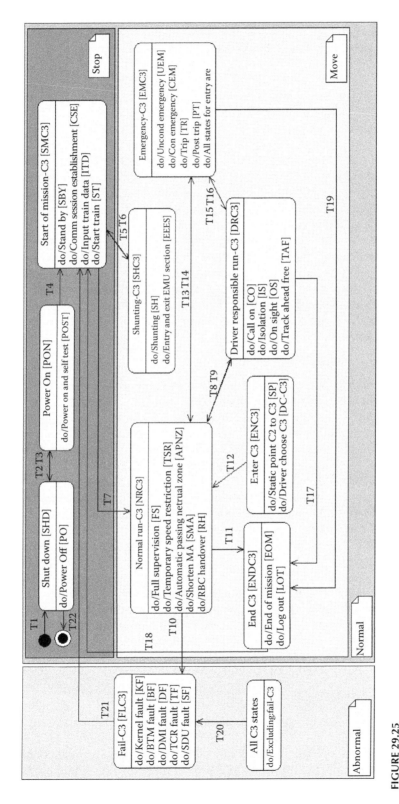

FIGURE 29.25
STM of CTCS-3 onboard system.

The transitions in the STM describe the transforming relationships between the states. For example, all normal type states may transit to the fail states, where will transit to the state of start of mission only when the system is recovered from the failure.

29.1.3.1.3 FHM

The function hierarchical model is developed to present the functions of a component/ subsystem of CTCS-3 onboard system. It can be used to examine whether all the designed functions are achieved in operations or not and identify the hazards when executing these functions.

Each component within the system boundary, as defined in the reference model, plays its own role to guarantee the safe and reliable operation of the onboard system. A method of constructing a hierarchy of functions is suggested to explore all the functions of C3-Kernel completely and comprehensively. Functions in lower levels are much more detailed and specific.

Fox example, the function hierarchical model of the C3-Kernel is as shown in Figure 29.26. The C3-Kernel is the core unit of the whole onboard system, which receives all kinds of information from other subsystems, processes the information, and outputs commands to realize safe train control. The primary function of the C3-Kernel is protecting the train from exceeding speed limitations and safe distance. The detailed functions of the C3-Kernel can be summarized as function of processing data, train locating, and auxiliary function, as well as safe control of speed and moving authority, door protection, and supervision function.

29.1.3.1.4 SEM

In the sequence model, all the objects are determined according to the structure of the onboard system. In other words, each object corresponds to a component of the CTCS-3 onboard system in the reference model. Furthermore, a sequence model depicts the process how the system transits into a state, which is defined in STM. The sequence model shows the detailed processes of how a function of the onboard system is achieved or how an event happens. As the preceding descriptions, the sequence model of the temporary speed restriction (TSR) is shown in Figure 29.27.

The TSR is defined in order to enable a separate category of speed restriction for track infrastructure, which can be used in working areas and so on. TSR orders should be set, commanded, and revoked by the TSR server (TSRS), which is deployed in centralized traffic control (CTC). The planned TSR orders of the whole line are set by a dispatcher through the TSR terminal. Once acknowledged by CTC, the TSR order will be stored by TSRS and transmitted to the radio block center (RBC) and train control center when it is implementable after the activation of the signalman. Then the RBC sends the TSR order to the corresponding train on the condition of exact and continued radio communication with the train. The onboard system will implement this order and brake the train to the restrictive speed once the train receives the TSR order, until the train passes the TSR region or the TSR order has been revoked by the trackside system.

29.1.3.2 Constraint Checking

To a given multiple model of system, the constraint checking can be started at anytime, shown in Figure 29.28.

In this case, the result of checking constraints could be seen in the output dock. It can be seen that there three pieces of information. The first piece of text announces that the constraint between the SEM named COBO-SEM-NRC3-FS and the STM named COBO-STM

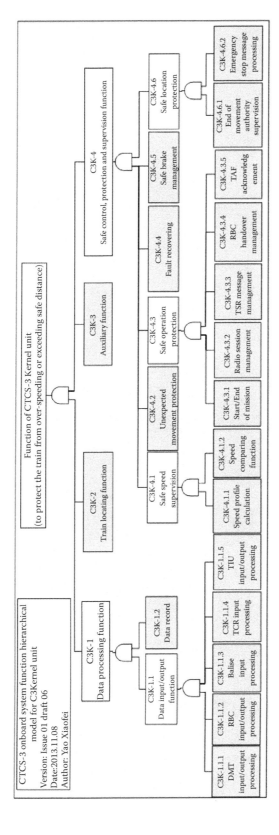

FIGURE 29.26
FHM of C3-Kernel.

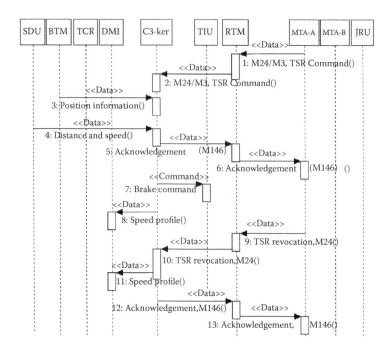

FIGURE 29.27
SEM of TSR.

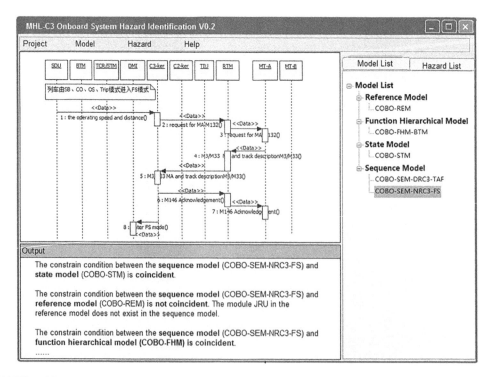

FIGURE 29.28
Constraint checking output in the demonstrated project.

passes the check. The second piece of text issues a warning that a constraint between the SEM named COBO-SEM-NRC3-FS and the REM named COBO-REM fails to pass the check. The third piece of text announces that the constraint between the SEM named COBO-SEM-NRC3-FS and the FHM named COBO-FHM passed the check. All the information reveals that the relationships of COBO-SEM-NRC3-FS VS COBO-STM and COBO-SEM-NRC3 VS COBO-FHM are coincident, and the relationship of COBO-SEM-NRC3-FS VS COBO-REM is not coincident because of the subsystem judge record unit in the COBO-REM not existing in the COBO-SEM-NRC3-FS.

29.1.3.3 Hazard Identification

For a given multiple model of system, the hazard identification can be performed in the process as Figure 29.29.

1. From a model, select an element to be examined by HAZOP study.
2. Choose one of nine guide words to be assembled with the selected element.
3. Combine the guide word with the selected element to generate an abnormal scenario representing the deviation context according to the design intent.
4. Agreeing with the specialists whether the abnormal scenario is a credible hazard.

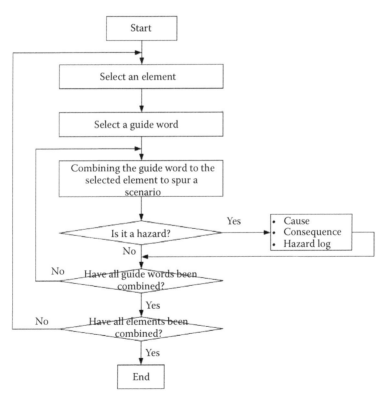

FIGURE 29.29
Procedure of hazard identification.

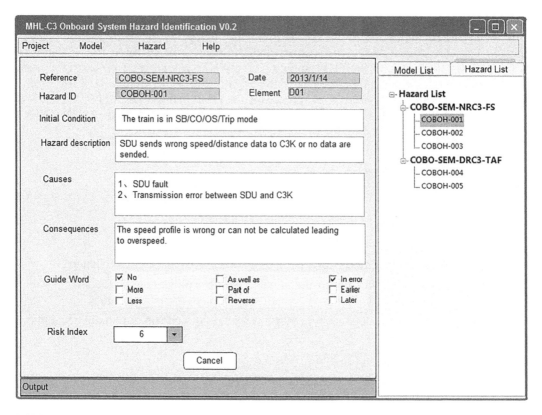

FIGURE 29.30
Instance of hazard in the demonstrated project.

5. If it is an agreed hazard, the causes and consequences should be found and logged. Otherwise, neglect it. Then, go to step 2 and choose another guide word until the rest of the guide words are used once.

6. To a given element, if all the guide words have already been opted, go to step 1 and choose another guide word from the candidate of the rest elements of the model.

7. If all the candidates of the elements have already been analyzed, end the process.

8. A hazard case is shown in Figure 29.30.

There is some information to be logged for a recognized hazard, including the model name in the blank of reference, the unique identity number of hazard, the unique element identity number of a element, date, initial condition, hazard description, causes, consequences, and guide words.

29.1.3.4 Traceability Checking

Generally, an unconnected hazard can play a few roles in the HAZOP of the system. It is easy to generate a chaotic the hazard log. Judging the completeness of the hazard log is impossible. It is hard to find out whether some similar hazards are duplicated and can be represented by a single hazard. Overlooking and overweighting some hazards are difficult

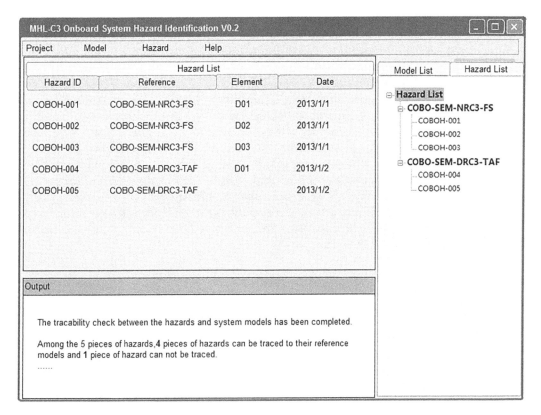

FIGURE 29.31
Traceability checking in the demonstrated project.

to avoid, so that the confidence and trust is reduced. These issues can be tackled by building traceability. Indeed, a traceable hazard can help the reader to understand the context described in a model. And with the background information, it is easy to distinguish the tiny difference between two similar hazards. Coupled with the traced model, a hazard can be easily understood exactly.

To a given multiple model of system, the traceability checking can be conducted at any time, shown as Figure 29.31.

The result shows that four of five hazards have the trace relationship to the elements of models including the elements of three COBO-SEM-NRC3-FS model and the element of a COBO-SEM-DRC3-TAF model. It can also be found that one of five hazards is an orphan without a trace to any element, because it might be an external hazard which is not generated by the procedure of presented model-based HAZOP method.

29.1.4 Conclusion

As a software-intensive sophisticated safety-critical control system, train control systems should be carefully analyzed for identifying hazards in order to keep the modern railway system operating safely during all the stages of the life cycle. Traditional HAZOP can play limited role in the middle stage of the life cycle, especially to a train control system in service on a large scale. Using a case of CTCS-3, model-based HAZOP has verified the effectiveness of dealing with the issues of understanding the textual specification of the system, checking the inter-model constraints, and tracing the relationship between a recognized hazard and a model. A prototype software suite is developed, and the first version has been certified by the State Intellectual Property Office of the People's Republic of China.

There are also some further valuable works worthy to be carried out in the future. The first one is about adding more model types into the multiple-model approach. The current four-model approaches cannot obviously satisfy various practitioners' modeling requirements. A Petri net is being considered as an add-in. The next issue is expanding the inter-model constraint set. Defining and checking more inter-model constraints could dramatically improve the correctness and coverage of the multiple-model approach. The hazard identification quality could be improved as well as detailed in Section 29.2. Additionally, to make the software suite more open and mature, it is necessary to systematically define the abstract grammar and the semantics for the models, the constraints, etc. Moreover, it is valuable to reduce the hard and extensive team effort, by introducing the automatic identification for the hazard.

29.2 Safety and Security Assurance: The Way Forward

29.2.1 System-Level Perspective

This sub-section revisits the principles of safety assurance in modern computer and communications-based complex systems and explores the synergies and differences between safety and the new emerging concerns over security in such systems. The strengths and deficits of the current methodologies and techniques are reviewed to highlight the areas that require further innovation and development. A system-level framework is also proposed for a comprehensive, thorough, and integrated assurance of safety and security while highlighting the future challenges in assurance.

29.2.2 Background

Modern railways exploit many advanced electronic, computing, and communications innovations in order to enhance performance, reliability, availability, safety, and capacity. A typical train control system is composed of many supervisory and control functions, and their concurrency, interdependency, and criticality pose a hugely complex dilemma to modern system designers and integrators alike in assuring overall system resilience, performance, and safety. The incorporation of advanced technologies in the railways has not only brought tangible benefits in terms of efficiency, speed, safety, and performance but also helped increase the complexity and vulnerability of the network and the operations due to inherent attributes of the modern computing, networking, and communications technologies involved. The rush for the exploitation of the opportunities needs to be balanced by the identification, evaluation, and assessment of potential threats and hazards/risks. This is often the aspect that lags behind the desire to exploit the benefits of the innovative technologies and can result in major savings on effort and high costs of re-engineering during later stages of development and even deployment.

29.2.3 Safety and Security

Aided by transportation, communications, and computer technologies, the modern global village presents many opportunities and benefits to humankind from rise in international trade and creation of new industries and professions to redistribution of wealth. However, alongside these emerging opportunities, innovations, scientific discoveries, and burgeoning complexities in products, processes and indeed human relationships pose a challenge to humankind, threatening health, safety, security, and the natural habitat. The prudent exploitation of the opportunities and credible assessment/management of potential adversities in the face of these rapidly evolving drivers demands a more systematic and potent approach to tackling these global challenges. To arrive at a credible and inclusive integrated framework for tackling system safety and security concerns in modern complex systems, it is instructive to revisit the principal premise for safety and security.

29.2.3.1 Safety

Safety is synonymous with freedom from unacceptable levels of harm to people and is a highly desirable property of products, systems, processes, or services. However, in view of ever-increasing complexity, development, and change, safety is often difficult if not sometimes impossible to entirely predict, manage, and guarantee. At the same time, rising social awareness and the more stringent legal requirements almost globally demand higher levels of safety performance from the duty holders arising from their products, processes, systems, and services.

Safety problems are technically and legally characterized by unintended yet harmful incidents and accidents which, apart from acts of nature, are mainly traceable to human shortcomings in concept, design, development, deployment, or maintenance of products and services. Safety as a principal and desirable attribute of products, processes, systems [1–4], and undertakings is regulated, and health, safety, and welfare of people are under legal protection in most developed countries.

29.2.3.2 Security

Security is synonymous with freedom from unacceptable levels of harm to people or damage to business operation/property or the natural habitat. Unlike safety, security problems are largely characterized by underlying malicious intent (threat) which, aligned with structural or operational vulnerabilities, cause incidents and accidents with significant potential to cause harm and consequent loss [5]. In the absence of the malicious intent, vulnerabilities can still accidentally pose a risk to life and property, but these are not generally regarded as security, issues. However, complexity, rapid change, global geopolitics, social and protest movements and even novelty pose increasing threat to the security of products, processes, and systems requiring an enhanced degree of assurance. Security, as yet, is not generally regulated, and a duty of care regarding freedom from intentional harm to people and property from third parties is largely left to the discretion of the duty holders. In the context of complex systems, much of the largely recent security concern relates to the unauthorized access to and misuse of data or intellectual property, malicious takeover of control systems, fraud, and espionage on presence of malicious agents on the computer networks and the web. This is often referred to as cyber-security [6,7].

29.2.4 Hazard Identification Process and Challenges

The history of human-made accidents is a stark reminder of the extent of lack of preparedness for major unexpected events that entail large-scale harm to people, detriments to the society and enterprises, as well as damage to the natural habitat. The human and commercial losses associated with the *Titanic* disaster (1912) or the catastrophic environmental impact of the Chernobyl nuclear accident (1986) illustrate the necessity to look beyond capitalizing the benefits arising from promising opportunities and additionally strive to identify and manage the potential adversities with the same rigor. This is a formidable task in the face of constant innovation, discoveries, and burgeoning complexities in products, services, systems, and undertakings. In the face of the pervasive challenges, there is a need for developing more efficient and effective approaches to identifying precursors to accidents and proactively devising potent mitigating measures to reduce the likelihood or the scale of harm and losses should incidents and accidents happen.

One fundamental shortcoming in this endeavor has been the overt emphasis on learning from accidents at the expense of more systematic approaches to prediction and prevention through identifying and tackling the precursors. The principal flaw in the learning from accidents is the myriad of paths that could potentially lead to the same accident, and each path may entail different scenarios and consequent scale of losses. Learning from a given accident scenario at best reveals one of many paths and offers no guarantees for the avoidance of future occurrences. However, the adversarial approach to accident investigation is largely driven by searching for the causes and legal liability for a culprit. Once this is achieved, the interest in establishing the root causes with a view for learning and prevention wanes; hence, learning from accidents is seldom a fruitful or viable approach to prevention. The precursors or hazards, however, have a more manageable causation chain in comparison and have a wider temporal distance to an accident, which implies that there is more potential to detect a hazardous state and be able to take a preventative or mitigatory action.

In this spirit, the circumstances with a potential to lead to loss, i.e., harm to people, financial/material detriment, or environmental damage are associated with most activities and undertakings to varying degrees of impact. While it is relatively straightforward to identify these within the context of familiar day-to-day tasks and experiences, more complex products, processes, services, and undertakings generally pose a more arduous if not insurmountable challenge in this respect. Rapid development and widespread exploitation of cost-saving or performance-enhancing technologies generally exacerbate the situation and increase the scope for larger potential losses in the event of unforeseen or unprotected faults, errors, and failures.

The structured comprehensive identification of hazardous circumstances arising from threats or unintended failure of products, services, systems, processes, or human error/action is fundamental to any safety and security process. It is, however, even more pertinent to large-scale or complex undertakings with a potential to lead to significant losses in the event of hazardous occurrences. In the absence of a systematic and robust hazard identification phase, all the subsequent analyses and risk control actions amount to no more than an exercise in vain, creating an illusion of safety/security and a false sense of confidence. This is particularly pertinent to circumstances where, due to a poor process, a number of significant hazards remain unidentified hence dormant within the system posing intrinsic vulnerabilities.

The determination of the domain/sphere of influence of a product, process, service, or undertaking is another by-product of the systematic hazard identification process. This is essential in establishing the scope of the subsequent assessment and should be employed in preference to the classical approach based on the physical boundaries of the subject under consideration. The radiation of electromagnetic interference typifies instances where the domain of influence extends well beyond the physical boundaries of a poorly designed and constructed electronic system.

The systematic identification of hazardous circumstances in a complex product, process, system, or undertaking entails two key stages at the outset:

- Empirical phase
- Creative phase

However, in view of the extensive resource requirements, the approach described here is more appropriate to products, services, processes, and undertakings that are likely to lead to significant losses or harm due to their scope, scale, or novelty.

29.2.4.1 Empirical Phase

Traditionally, the knowledge and experience of the past adversities in a given context, in the form of checklists, do's and don'ts, have been applied to the determination of the potential hazardous circumstances in products, processes, missions, and undertakings. This approach is seldom adequate in isolation especially when there are novelties or significant changes in the functionality, technology, complexity, composition, environment (time /space) or the mode of operation of the matter under consideration. It is essential therefore to compile and maintain a checklist of hazardous circumstances pertinent to specific products, processes, or undertakings, in order to facilitate a simple first cut identification of the likely problem spots and where possible, avoid the errors, failures, and losses of the past.

However, there are constraints and limitations in employing the knowledge and experience of the past as the only compass to navigating the future free from failures and adversities in the face of novelty, innovation, change, complexity, and new environments. Where the product, process, system, or undertaking lend themselves to a more detailed scrutiny, failure mode and effects analysis (FMEA) for equipment/systems [8] and its human-related counterparts, action error analysis and task analysis, may be applied in order to identify the particular component/subsystems or human failures conducive to hazardous circumstances. These, however, require a detailed knowledge of the failure modes of the components and subsystems, including human actions and the likely errors in different operating conditions.

The application of checklists, FMEA, action error analysis, and task analysis are generally not resource intensive and may be carried out by suitably competent individuals and appropriately recorded for further analysis. These are broadly empirical and based on the body of knowledge and experience arising from the events and observations of the past. The hazards identified through the application of these techniques generally constitute a sub-set of the total potential hazard apace that should be further explored with the aid of the complementary creative techniques especially in circumstances and environments where the potential magnitude of harm and loss from likely accidents are regarded as highly undesirable or intolerable.

29.2.4.2 Creative Phase

The systematic and creative techniques have an established pedigree in the analysis and resolution of complex problems and undertakings. These generally capitalize on cognitive diversity through a team-based multiple-disciplinary approach, composed of members with diverse and complementary knowledge and expertise. Furthermore, in view of their reliance on lateral perception, divergent thinking, and imaginative creative faculties, the structured and systematic variants of these techniques generally share a number of key characteristics, namely, the following:

- Planning and process management
- Competent study panel selection and briefing
- Hierarchical decomposition and graphical representation of the problem domain
- High-level probing of the key elements of the system and coarse determination of the critical subsystems and interfaces
- Comprehensive, step-by-step probing of the subsystems and interfaces with a more meticulous and focused scrutiny of the critical areas
- Deployment of the tools and technologies, visualization, and simulations to promote creative thinking
- Identification and structured recording of the hazardous circumstances and vulnerable states including causes, consequences, and potential mitigation and control measures
- Expert driven qualitative ranking of the identified hazards and vulnerabilities employing appropriate frequency/consequence matrices
- Adoption of a systems framework for the advanced quantitative analysis of the significant/critical hazards/vulnerabilities with a view to develop a total risk profile for the critical product, process, system, or undertaking
- Maintenance, update, and management of the hazard/vulnerability and risk records throughout the life of the product, process, or undertaking

The hazards and threat/vulnerabilities identified through the empirical processes must be reviewed at appropriate stage(s) during the creative phase and recorded together with the other attributes alongside the newly identified items in a structured log for further processing and management. The empirical phase is sometimes employed as a completeness/coverage test or means of detailed probing of specific hazards and failures, subsequent to the creative identification phase. Whichever the temporal order, the empirical and creative phases should be applied in a consistent and complementary manner to reinforce and increase confidence in the hazard portfolio.

The two-phase complementary process enhances the integrity and coverage of the potential hazard and threat/vulnerability space, increasing the effectiveness and confidence in the safety and security process and outcomes. It has to be born in mind, however, that the hazard identification exhibits an essentially non-linear gain in that a creative identification of a single significant hazard may outweigh the contribution of a large number of less severe items. In this spirit, it is the quality and not the quantity of the identified hazards that is of the essence as the success criteria for the resources invested in the process. The methodologies that generate an unrealistically large number of mostly trivial hazards are wasteful of resource, misleading, and unproductive and should be avoided wherever possible. Furthermore, the subsequent analytical treatment of hazards as detailed in this chapter should be applied on a prioritized basis, beginning with the highest-ranking critical hazards.

29.2.4.3 Current Best Practice, Challenges, and Enhancements

Despite the significant advancement in the application of the empirical and creative processes elaborated earlier, the rapid pace of technological developments and scientific advancement and emerging innovations, shorter product and system life cycles and market-driven demand for higher efficiency and performance have rendered these approaches ineffectual if not misleading in achieving sufficient levels of assurance. While the advancing scientific and technological know-how offer opportunities for improved performance and value, they also pose new emergent risks that pose the likelihood for new forms of accident causation and harm that defy our empirical deterministic approaches to systematic assurance in safety and security of products, systems, and services. This rapid transformation inevitably tips the balance from reliance on empiricism toward more creative and scientific processes since there is virtually insufficient time for gaining adequate understanding and confidence in the properties of complex technological systems before these are upgraded, enhanced, or rendered obsolete in favor of even more promising alternatives.

The current practice in searching for hazards in complex products, systems, or undertakings [9] is largely manual, effort intensive, and, consequently, costly. In a rapidly evolving global economic and social setting with competitive commercial and cost reduction pressures, the duty holders from public to private sector tend to shy away from these lengthy processes with no apparent or immediate gain toward more resource efficient solutions that save on cost and effort in assurance. There is a need for more efficient, rational, and cost-sensitive processes to counter this trend and achieve an acceptable level of safety and security assurance in the face of such trends.

The current best practice [9] typified by sector-specific safety and security assurance standards [10–20] suffers from many shortcomings that tend to make compliance part of the problem in the sense of high costs and resource-intensive nature of requisite activities. The key shortcomings in the current best practice standards and codes of practice is composed of, among many, a suite of conceptual and practical dysfunctions, chiefly,

- An inordinate set of linear and un-prioritized mandatory and optional activities demanded as normative that entails a huge effort in expertise and time scale to achieve with no credible justification on value or impact.

- Lack of risk-based resource allocation and appreciation of structural and behavioral complexity differences in various subsystems of a complex product, process, or undertaking. This is rather paradoxical in the face of claims by these standards to be advancing a risk-based approach!

- Lack of an intrinsic evaluation and assessment framework to facilitate process management, optimal resource allocation, and demonstration of an acceptable level of compliance.

- Lack of sensitivity to effort and consequent costs inherent in achieving acceptable or desirable compliance and assurance.

- Inflexibility in terms of giving a real-time perspective on the degree of assurance achieved at any stage of the life cycle especially during the design and development phases.

- Missing if not highly inadequate process metrics and quantification of the interim and final assurance achieved in a large and complex project.
- Insufficient if not unacceptable level of assurance modularization in the form of process and deployment components and libraries arising from process and implementation learning.
- Insistence on complete compliance with all mandatory/normative requirements, lacking the impact assessment and cost justification in the face of formidable cost consciousness by the public and private enterprises globally.
- Insufficient or complete lack of recognition of the nature of the assurance activities as essentially a form of "knowledge management" that requires collaborative multiple-disciplinary effort and advanced tools for integration and management of the assurance process.
- Lack of systemic guidance on the three key assurance dimensions composed of people, process, and product/system especially the requisite human resource competences for handling various aspects of implementation and deployment.
- Finally, lack of guidance on how to proceed on achieving compliance in a multiple-disciplinary and multiple-standard setting for a given product, process, or system/undertaking to avoid inefficiencies, overlaps, as well as deficits. This is practically the most striking deficiency in the domain of safety and mission critical standards compliance and certification.

An advanced and potent new approach to the systems-based safety and security assurance needs to address the preceding dysfunctions as well as facilitate and incorporate a number of essential quality factors that enhance process integrity and confidence in the outcomes, namely, the following:

- Ensuring diversity and competence of the participating experts in the safety and security assurance processes and activities
- Adopting and deploying a risk-based process in terms of structure, coverage of the domain, problem domain representation in terms of physical, and functional and quality of management
- Hygiene factors such as workshop duration, space, comfort, expert engagement, pace, and pause/rest periods
- Examining local and global/system-level occurrences and failures in the course of searching for hazardous states and scenarios [8]
- Enhanced focus on the assurance of the inherent design intent before focusing on deviations from the intent
- Taking into account dependency and concurrency during the creative search as well as common modes of failures and their causative factors
- Factoring in the impact of the host environment on the behavior, performance, and resilience of the overall system of interest
- Building on current knowledge and of the past employing past records and knowledge repositories in a structured and integrative manner
- Adopting a creative mind-set, such as exploring beyond the obvious including lateral thinking

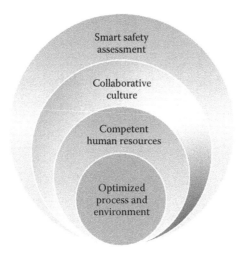

FIGURE 29.32
Essential success factors in SSA philosophy.

29.2.5 Smart Safety Assessment

To this end, an advanced evaluation and assessment-focused approach for assurance is called for. The smart safety assessment (SSA) solution that is the outcome of the recent research by the author [21] at the Beijing Jiaotong University. In response to many dysfunctions and shortfalls of the current standardization and assurance processes in vogue globally today, SSA embodies a suite of advanced systemic features and illustrates the next-generation class of assurance paradigms that are desperately needed to credibly face the challenges cited earlier as shown in Figure 29.32.

Devising a theoretical framework for assessment, the SSA methodology is founded on continual evaluation and assessment that assists in focusing on the development/deployment of critical high-risk areas of a product, process, or system as shown in Figure 29.33.

SSA is designed to empower the duty holders to responsively deliver optimal value from resources invested for the assurance of safety and security. The SSA architecture is illustrated in Figure 29.34.

29.2.6 Holistic System Security Assurance

Much of the assurance rules, codes of practice, standards, and activities have so far been focused on attaining acceptably safe performance in complex products, undertakings, and systems. With the emergence of security as a new concern largely due to the vulnerabilities that the incorporation of advanced computing and communications technologies bring [22–25], there is a need for an equally potent and systematic approach to the assurance of secure performance in products, systems, and processes. The comprehensive system-level assurance of security involves consideration of threats and vulnerabilities within three complementary contexts as shown in Figure 29.35:

1. Cyber-system (including communications, data, information)
2. Organizational and governance (including personnel considerations, supply chain, enterprise culture, procedures, and services)
3. Physical (including infrastructure, facilities, machinery, and fixed and mobile installations/assets)

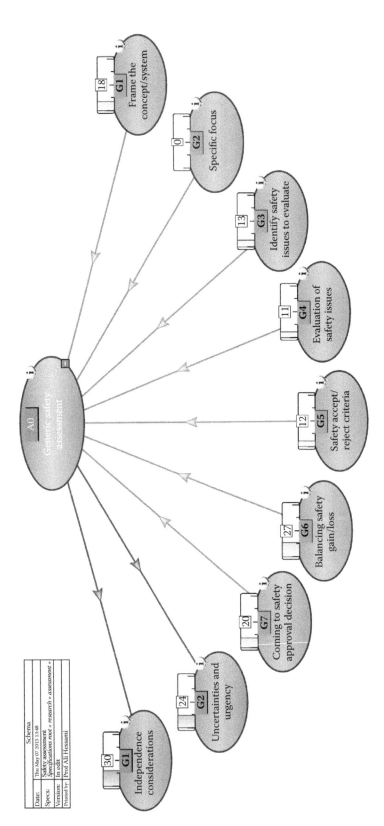

FIGURE 29.33
Schema for generic safety assessment.

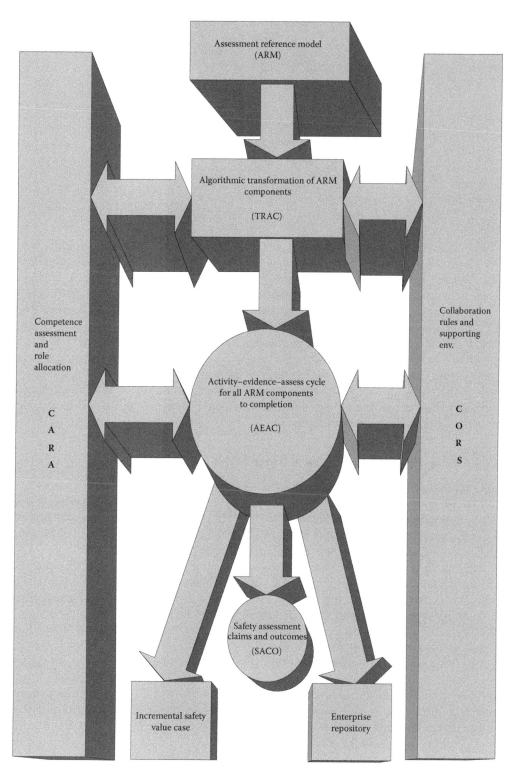

FIGURE 29.34
Illustrative SSA architecture.

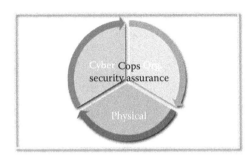

FIGURE 29.35
COPS total systematic security assurance framework.

These collectively constitute a cyber, organizational, physical security (COPS) framework for total security assurance [26]. The COPS framework ensures a consistent, efficient,
and coherent treatment of security concerns at system level and will assist in ensuring a
joined-up approach to total security assurance across the three domains much akin to systems safety practices. The COPS framework caters for the overlaps and synergies between
the vulnerability to malicious intent of the three prongs of cyber (technology dimension),
organization (human dimension), and physical assets. This is largely driven by the interdependence between the three domains and the necessity to address security, such as
freedom from unacceptable harm/loss arising from malicious intent, through a comprehensive integrated approach. The three-pronged integrated approach in COPS framework
is also cognizant of the diversity in processes and protection mechanisms that may prove
necessary when assuring each dimension.

29.2.6.1 Emerging European Railways' Response to Cybersecurity

The European rail operators in all rail networks including conventional, high-speed, and
mass transit have obligations under the domestic and the European Union (EU) law to
protect/maintain the safety of their operations. Their failure in taking reasonable care to
do so may make them accountable for the resulting harm to people or some of the operational and commercial losses incurred by third parties [27].

While this duty of care is largely focused on the general safety that arises from random
and unintended failures, a new concern on security of services and operations has emerged
lately. This is driven by the vulnerability of transportation networks to malicious attacks
due to the open access and public nature of multiple-modal transportation network. In this
spirit, security has now gained prominence alongside safety as an extension or addition to
the duty of care related to potential harm to people arising from inadequate provision of
security focused prevention and protection measures. Apart from industry best practice as
a response to the new threat, a new suite of national and EU-wide regulations is currently
in preparation. The UK Department for Transport has a general requirement for Security
Informed Safety Case in Railways [28], which is indicative of the emerging requirements
and necessitates consideration of system security alongside safety matters in all railway
development projects and the end state railway.

The consideration of security matters in large programs and complex undertakings
is gaining prominence due to the pervading nature of computing, communications,
and Internet technologies as enablers if not core infrastructures for service delivery.

While security is not new to transportation as a public service, new vulnerabilities arising from deployment of modern programmable and communications technologies in rail transportation emerge that demand a new perspective on total security assurance alongside the traditional focus on safety. In this spirit, safety and security in products, processes, systems, and undertakings should be collectively assured via an integrated systematic framework. This is lacking in the current European approach to the security assurance that is largely focused on the information technology and control/command (signaling) subsystems.

29.2.7 Integrated Safety and Security Assurance

In principle, safety and security assurance processes have significant synergies and overlap that can result in a more effective and efficient approach if these important dimensions of performance assurance are integrated. Adopting a hazard and threat-based approach to risk assessment and management generates a more systematic framework for coping with varieties of existing and emerging risks. A defect–error–failure sequence is proposed to address the processes leading to the realization of hazardous states in a product, process, or system. The consideration of the post-hazard horizon in this approach involves identifying the potential escalation scenarios, the defenses against accidents, the range of accidents that arise due to the failure of defenses, and optimal response and recovery regimes for each major accident scenario. The principal difference in this paradigm for security-related threats and vulnerabilities is that fault–error–failure is replaced by threat/vulnerability consideration in a product, process, or system. The integrated framework for safety and security assurance [29] is composed of the following seven principles:

1. Prediction and proactivity
2. Prevention
3. Containment and protection
4. Preparedness and response
5. Recovery and restoration
6. Organization and learning
7. Continual enhancement

These principles collectively address the total risk landscape and are interrelated in a systemic fashion. The seven principles fall into three broad categories:

The first principle, proactivity, is mainly concerned with establishing an environment and a baseline for the product, process, service, system, or organization in terms of its desirable properties and performance. It represents an antithesis to reactivity in facing the potential of accidents. In this spirit, proactivity is fundamental to the achievement and improvement of performance since it emphasizes that plans and resources must be devised, secured, and applied in advance of incidents, threats, and accidents to enable the duty holders to eliminate, control, or mitigate the risks.

The second group is composed of prevention, protection, response, and recovery, which are mainly associated with the causation and escalation of accidents and the optimal preparedness in responding to these and emergencies with a view to minimize losses.

The third and final group of two principles relates to the significant role that the human organization, communications, responsibilities, competencies, certification, regulation, and corporate memory/learning play in the attainment and improvement of overall performance.

This includes a drive for continual enhancement based on an audit, measurement, and feedback loop to ensure a set of common indicators are continually monitored to empower the duty holders to take effective remedial and improvement actions as appropriate.

The seven fundamental principles collectively constitute a systematic and systemic framework for assurance of overall safety and security performance in the face of threats and risks.

29.2.7.1 Principle Scope and Intent

29.2.7.1.1 Prediction and Proactivity

Setting policy and strategy, identifying all stakeholders and interfaces, hazard/threat identification, planning, resourcing, and data collection. Modeling, assessing baseline risks, identifying key performance indicators, and implementing policy. Developing safety, security, sustainabilty cases, and relevant management manuals.

29.2.7.1.2 Prevention

All measures, processes, activities, and actions including maintenance aimed at eliminating or reducing the likelihood/frequency of threats/hazardous states with a potential to cause harm and loss.

29.2.7.1.3 Containment and Protection

All measures, processes, activities, and actions aimed at reducing the likelihood/frequency or severity of potential accidents arising from the hazardous states or security breaches.

29.2.7.1.4 Preparedness and Response

All plans, measures, processes, activities, and resources relevant to managing and responding to degraded, failure modes, and emergencies; investigation of the causes; collection; maintenance; sharing of records.

29.2.7.1.5 Recovery and Restoration

All plans, measures, processes, activities, and resources relevant to recovery from planned and unplanned disturbances, degraded and failure modes, and emergencies toward full resumption of production/service including the criteria and organization for authorizing the system back into service post-disruptions and emergencies.

29.2.7.1.6 Organization and Learning

Structuring, communications, training, certification, competencies, roles and responsibilities, and validation for human organization, as well as ensuring lessons are learned from incidents and accidents, and key points recorded, shared, and implemented.

29.2.7.1.7 Continual Enhancement

All processes associated with setting and reviewing targets, measuring/assessing, processing, auditing, reviewing, monitoring, regulating, and sustaining/improving performance including decision aids and criteria. The systemic assurance framework of seven principles, when collectively applied, can systematically underpin the attainment, maintenance (principles I–VI), and improvement (principle VII) of overall safety and security performance. A framework founded on systemic principles is more credible and universally applicable than specific context-related suite of actions, processes, or methodologies.

29.2.8 Conclusions

The advancing technology and pervasive complexity [30–32] and pace of change in modern complex products, processes, and systems such as those employed in railway control and command applications pose significant challenges to the assurance of desirable system performance properties such as safety and security [33–38]. Most modern systems incorporate advanced computing, networking, and communications technologies in support of supervisory and control functions that tend to add much higher degrees of complexity due to software intensive subsystems, enhanced functionality, and large-scale distributed nature of systems. Added to these is the cost consciousness that is pervasive across the public as well as private enterprises. These tend to render the classical approaches inadequate to face the challenges of value, efficiency, shorter life cycles, pervasive and rising complexity, and faster delivery. A whole new potent and capable approach to assurance of desirable properties of complex systems is called for in response to the emerging social, legal, technological, and market requirements.

29.3 Way Forward and Recommendations

The challenges highlighted demand a new paradigm, processes, and suite of advanced supporting tools and methodologies in the safety and security assurance domains and beyond. This by necessity also calls for new modes of stakeholder engagement and collaboration. The quality (fitness for purpose) and adequacy of the expert knowledge likewise needs assurance through focus on a systematic and preferably quantitative approach to the competence management for the human resources.

Other aspects of the new paradigm relate to the diversity and complementarity of the problem and system representation covering physical, functional, layout, temporal, and structural forms including the use of simulations, visualization graphics, virtual reality, dynamics, and behavioral modeling alongside the classical static perspectives.

A higher degree of process automation employing a suite of validated tools and libraries of assurance modules will further enhance efficiency and coverage of the problem space of concern and create a knowledge repository for reuse and continual enhancement. The use of formal specifications and proof fall into this category of tools and processes.

We have explored the challenges, existing practices, and emerging requirements in the systematic and credible assurance of safety and security in the safety-critical railway supervisory and control context. This is an evolving scene that is in need of major paradigm shift and process enhancements. The frameworks illustrated demonstrate the desirable characteristics of this new paradigm. All practitioners have a role and professional duty to contribute to this advance.

References

1. BSI (British Standards Institution) (2001) BS EN 50128:2001: *Railway Applications— Communications, Signalling and Processing Systems: Software for Railway Control and Protection Systems*. BSI, London.
2. BSI (2011) BS EN 50128:2011: *Railway Applications—Communications, Signalling and Processing systems. Software for Railway Control and Protection Systems*. BSI, London.
3. BSI (2003) BS EN 50129:2003: *Railway Applications—Communication, Signalling and Processing Systems: Safety related Electronic Systems for Signalling*. BSI, London.
4. Hessami, A. G. (1999). Safety assurance: A systems paradigm, hazard prevention, *Journal of System Safety Society*, Volume 35 No. 3, pp. 8–13.
5. Hessami, A. G. (May 2004) A systems framework for safety and security—The holistic paradigm, *Systems Engineering Journal*, Volume 7, Issue 2, pp. 99–112..
6. BSI (2014) BS ISO/IEC 27000:2014: *Information Technology—Security Techniques—Information Security Management Systems—Overview and Vocabulary*. BSI, London.
7. ISO (International Organization for Standardization) (2013) ISO 27001:2013: *Information technology— Security Techniques—Information Security Management Systems—Requirements*. ISO, Geneva.
8. BSI (2016) BS EN 60812:2016: *Analysis Techniques for System Reliability—Procedure for Failure Mode and Effects Analysis (FMEA)*. BSI, London.
9. BSI (2016) BS EN 61882:2016: *Hazard and Operability Studies (HAZOP studies)—Application Guide*. BSI, London.
10. *Cybersecurity for Industrial Control Systems: Classification Method and Key Measures*. http://www .ssi.gouv.fr/uploads/2014/01/industrial_security_WG_Classification_Method.pdf, National Cybersecurity Agency of France, Paris, France.
11. *Cybersecurity for Industrial Control Systems: Detailed Measures*. http://www.ssi.gouv.fr/uploads /2014/01/industrial_security_WG_detailed_measures.pdf, National Cybersecurity Agency of France, Paris, France.
12. IEC (International Electrotechnical Commission) (2010) IEC 61508-1 ed2.0 (2010-04): *Functional Safety of Electrical/Electronic/Programmable Electronic Safety-Related Systems—Parts 1–7*. IEC, Geneva.
13. CENELEC (European Committee for Electrotechnical Standardization) (2007) TR50451: *Systematic Application of Safety Integrity Requirements-Railway Applications*. CENELEC, Brussels.
14. IEC (2009) IEC TS 62443-1-1:2009: *Industrial Communication Networks—Network and System Security—Part 1-1: Terminology, Concepts and Models*. IEC, Geneva.
15. IEC (2010) IEC 62443-2-1:2010: *Industrial Communication Networks—Network and System Security—Part 2-1: Establishing an Industrial Automation and Control System Security Program*. IEC, Geneva.
16. IEC (2015) IEC TR 62443-2-3:2015: *Security for Industrial Automation and Control Systems—Part 2-3: Patch Management in the IACS Environment*. IEC, Geneva.
17. IEC (2015) IEC 62443-2-4:2015: *Security for Industrial Automation and Control Systems—Part 2-4: Security Program Requirements for IACS Service Providers*. IEC, Geneva.
18. IEC (2008) IEC PAS 62443-3:2008: *Security for Industrial Process Measurement and Control— Network and System Security*. IEC, Geneva.
19. IEC (2009) IEC TR 62443-3-1:2009: *Industrial Communication Networks—Network and System Security—Part 3-1: Security Technologies for Industrial Automation and Control Systems*. IEC, Geneva.
20. IEC (2013) IEC 62443-3-3:2013: *Industrial Communication Networks—Network and System Security—Part 3-3: System Security Requirements and Security Levels*. IEC, Geneva.
21. Hessami, A. G. Smart safety assessment, SSA, *Proceedings of the Safety Critical Club (SSS'16) Conference, Brighton, February 2–4, 2016*.

22. Homeland Security (2011) *Common Cybersecurity Vulnerabilities in Industrial Control Systems*, Homeland Security, Washington, DC, https://ics-cert.us-cert.gov/sites/default/files/recommended_practices/DHS_Common_Cybersecurity_Vulnerabilities_ICS_2010.pdf.

23. NIST (National Institute of Standards and Technology) (2012) *Guide for Conducting Information Security Risk Assessments*, NIST, Gaithersburg, MD, NIST 800-30, September 2012.

24. NIST (2014) *Framework for Improving Critical Infrastructure Cyber-security*, NIST, Gaithersburg, MD, NIST version 1.0, February 2014.

25. *Manage Industrial Control Systems Lifecycle Centre for the Protection of National Infrastructure*, V1.0, 2015, https://www.cpni.gov.uk/, CPNI (Centre for the Protection of National Infrastructure), MI5, Thames House, London, UK.

26. *HMG Security Policy Framework 3*, Cabinet Office, National security and intelligence, and Government Security Profession, UK.

27. *What is the EU Doing to Improve Security and Safety of Transport in the EU?* http://ec.europa.eu/transport/themes/security/index_en.htm, Directorate General for Mobility and Transport, European Commission.

28. *Light Rail Security-Recommended Best Practice* (2014), Department for Transport, Great Minster House, 33 Horseferry Road, UK.

29. Guedes Soares, C. (editor) (2010) *Safety and Reliability of Industrial Products, Systems and Structures*, CRC Press, Boca Raton, FL, pp. 21–31.

30. Karcanias, N., and Hessami, A. G. (2010) Complexity and the notion of systems of systems: Part (I): General systems and complexity, *Proceedings of the 2010 World Automation Congress International Symposium on Intelligent Automation and Control (ISIAC), Kobe, September 19–23, 2010.*

31. Maier, M. W. (1998) Architecting principles for system of systems, *Systems Engineering*, Volume 1, No. 4, pp. 267–284.

32. Karcanias, N., and Hessami, A. G. (2010) Complexity and the notion of systems of systems: Part (II): Defining the notion of systems of systems, *Proceedings of the 2010 World Automation Congress International Symposium on Intelligent Automation and Control (ISIAC), Kobe, September 19–23, 2010.*

33. BSI (2002) BS ISO/IEC 15288:2002: *Systems Engineering: System Lifecycle Processes*. BSI, London.

34. Bowler, D. (1981) *General Systems Thinking*, North Holland, New York.

35. DeLaurentis, D. (2007) *System of Systems Definition and Vocabulary*, School of Aeronautics and Astronautics, Purdue University, West Lafayette, IN.

36. Kaposi, A., and Myers, M. (2001) *Systems for All*, Imperial College Press, UK.

37. Karcanias, N., and Hessami, A. G. (2010) System of systems and emergence: Part (I): Principles and framework, *Proceedings of the 2010 World Automation Congress International Symposium on Intelligent Automation and Control (ISIAC), Kobe, September 19–23, 2010.*

38. Karcanias, N., and Hessami, A. G. (2010) System of systems and emergence: Part (II): Synergetic effects and emergence, *Proceedings of the 2010 World Automation Congress International Symposium on Intelligent Automation and Control (ISIAC), Kobe, September 19–23, 2010.*

39. Section 2: Désignation des opérateurs d'importance vitale, des délégués pour la défense et la sécurité et des points d'importance vitale http://www.legifrance.gouv.fr/affichCode.do?idSectionTA=LEGISCTA000006182855&cidTexte=LEGITEXT000006071307&dateTexte=20080505. 1998.07.17, Lionel Jospin.

40. Code de l'environnement—Article L511-1http://www.legifrance.gouv.fr/affichCodeArticle.do?cidTexte=LEGITEXT000006074220&idArticle=LEGIARTI000006834227&dateTexte=&categorieLien=cid. 2011.01.20, Council of State, France.

41. BSI (2009) BS ISO 31000:2009: *Risk Management—Principles and Guidelines*. BSI, London.

42. EU 95/46/EG—*Data Protection Directive*. http://ec.europa.eu/justice/policies/privacy/docs/95-46-ce/dir1995-46_part1_en.pdf, 1995.12, Official Journal of the European Communities, No L 281/32.

43. *Quadro Strategico Nazionale per la Sicurezza dello Spazio Cibernetico* (2014). http://www.sicurezzanazionale.gov.it/sisr.nsf/wp-content/uploads/2014/02/quadro-strategico-nazionale-cyber.pdf. 2013.12, Presidency of the Council of Ministers, Italy.

30

Application of Risk Analysis Methods for Railway Level Crossing Problems

Eric J. Schöne and Qamar Mahboob

CONTENTS

30.1 Introduction

This chapter introduces the two main fields of application of risk analysis for level crossings and gives a short overview of working process:

- To select appropriate protection technologies
- To determine technical safety targets

For a better understanding of special requirements of level crossings, Section 30.2 shows the basic features of this intersection between railways and roads. It presents the most important risks and gives essential definitions and distinguishing features of safety technologies.

For the selection of appropriate protection technologies (especially the type of protection), we need to pay special attention to the risks resulting from human error of the road users. For this, we need to model the environmental conditions of rail traffic and human behavior. Section 30.3 describes this application scenario.

When determining the technical safety targets, we more strongly focus on how the protection systems work and which characteristics they can have. Risks from technical defects are described in Section 30.4.

Please note that risk analysis in this chapter is intentionally shown at a severely simplified level to reduce complexity and to focus on specific problems of level crossings. Section 30.5 gives conclusion and outlook to possible extensions.

30.2 Basic Features of Level Crossing Protection

30.2.1 Specific Features of the Intersection

Level crossings are intersections where roads or paths cross a railway line at the same level. They pose special hazard points, as this is where two transport systems with fundamentally different properties meet.

On the one hand, in road traffic, vehicles can achieve relatively short braking distances as the adhesion between rubber wheels and the road surface is high. In addition, they can avoid obstacles on short notice. Safety is prevalently based on traffic rules and therefore strongly depends on the behavior of the road users. Road users are allowed to go only at a speed that enables them to stop within the distance they can see. The authorities try to influence the road users' behavior, particularly through road design. Technical measures aim at passive safety to reduce accident severity (e.g., seat belts and air bags).

On the other hand, rail traffic has fundamentally different properties. The adhesion between steel wheels and steel rails is very small, and therefore, braking distances are significantly longer. Due to the track guiding, the vehicles cannot avoid obstacles quickly; they can change direction only on designated points. The safety mainly depends on technical systems which secure the route. Train drivers cannot see as far as their braking distance, but adjust braking and acceleration chiefly according to the signals. The main focus for rail traffic is therefore active safety to avoid accidents.

This means that on level crossings, two different safety philosophies clash; their characteristics have to be taken into consideration when designing these intersections: The behavior of the road users should not significantly influence the safety of the rail traffic, but at the same time, the properties of the rail traffic must not impair the safety of the road users inappropriately.

The dangerous effects of human error can be completely eliminated only by avoiding or removing level crossings. In all other cases, errors of the road users have to be taken into consideration und factored into the design of the level crossing. In particular, this is achieved by selecting and implementing the appropriate protection measures.

30.2.2 Classification and Severity of Risks

The most important undesirable event on a level crossing is the collision between a railway vehicle and a road user, as here, the most severe damage can be expected. Statistics from different countries show that on average, one road user is killed in every fourth collision (Schöne 2013).

There are other undesirable events which have significantly less severe effects. Nevertheless, we need these events to define technical safety requirements. These include collisions of road users with safety equipment and collisions of road users with other road users.

About 80–90% of collisions with railway vehicles are caused by the behavior of the road users. In the remaining cases, errors of railway staff and other influences (such as vehicle defects) played an important role. Regarding the behavior of the road users, we can find significant differences between intentional and unintentional errors (Griffioen 2004; RSSB 2011); these differences should be kept in mind when developing and selecting safety measures.

Collisions are essentially caused in two areas of the level crossing, which are shown in Figure 30.1 (Schöne 2013):

- In the approach and decision zone
- In the clearance zone

In the approach and decision zone, collisions can be caused by road users overlooking working safety equipment unintentionally or ignoring it intentionally. In the clearance zone, the road traffic environment is the main factor (e.g., if traffic regulations are unsuitable or if road vehicles cannot pass each other due to narrow streets and have to stop in the danger zone). To identify appropriate safety measures, we have to differentiate between these zones.

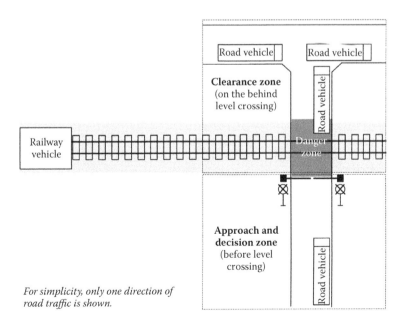

FIGURE 30.1
Causation areas for risks on level crossings.

30.2.3 Safety Technologies

Level crossing protection aims at avoiding collisions between road users and railway vehicles. Here, two approaches exist (Theeg et al. 2009):

- Passive railway crossings without technical equipment, which need to be protected by the road users themselves (through hearing and seeing)
- Active railway crossings, where technical equipment warns the user if railway vehicles approach

This section will further focus on active railway crossings. Active railway crossings can be further divided according to the parameters shown in Figure 30.2.

The *type of protection* denotes the types of technical equipment that inform the road users about approaching railway vehicles (i.e., how the crossing looks from the road). Typical types of protection include the following:

- Light signals
- Light signals with half barriers
- Light signals with full barriers

The *type of activation* describes how the technical protection equipment is activated when the railway vehicle is approaching. Some common technologies for this purpose include the following:

- Detection of railway vehicles by sensors (automatic)
- Integration into routes in the interlocking system (automatic or semiautomatic)
- Activation by railway staff using a control device (manual)

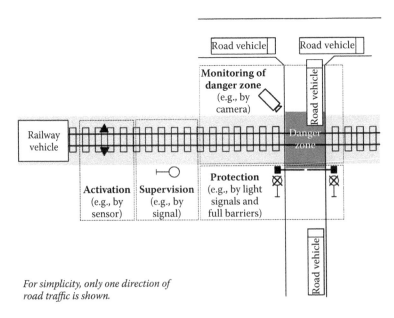

FIGURE 30.2
Distinguishing features of active level crossings.

The same options apply for switching off the technical protection after the railway vehicle has left the crossing.

The *type of supervision* refers to the type of verification and evaluation of the working status aimed at finding malfunctions in the level crossing protection equipment, so that the railways can react to it. This can include the following:

- Special supervision signals on the railway line
- Dependence on existing main signals
- Remote supervision from a neighboring station

The *type of monitoring of the danger zone* refers to solutions that verify in certain, regulated cases if the danger zone is free of trapped road users. Some typical technologies for this purpose include the following:

- Monitoring of the danger zone through sensors (automatic)
- Indirect visual confirmation by operators, e.g., via a camera system (manual)
- Direct visual confirmation by operators (manual)

In practice, the listed protection, activation, monitoring, and notification types should always be considered in correlation. It is neither possible nor useful to combine all types; e.g., when using the automatic monitoring of the danger zone, it should also be supervised using signals to make sure that railway vehicles receive a stop signal when road vehicles are still on the crossing.

30.3 Selection of Appropriate Protection Measures

30.3.1 Introduction and Procedure

Depending on the legal regulations of the country, it is usually state authorities (railway infrastructure companies, road traffic authorities, or similar) who choose the appropriate protection measures. Also, the decisions of the authorities can differ from country to country. To use the available financial means efficiently, it is recommended to make decisions based on risk-based approaches. Based on the system definition (see Section 30.3.2), this task is subdivided into four stages:

- Hazard identification (Section 30.3.3)
- Risk analysis (Section 30.3.4)
- Risk assessment (Section 30.3.5)
- Risk control (Section 30.3.6)

During the risk analysis stage, the parameters are determined and quantified to calculate risks in various conditions (for either one particular level crossing or a group of level crossings with similar properties). Then, risk assessment must answer the question if the existing risks are within justifiable limits. This calls for defining a risk acceptance criterion. If the existing risks are beyond the acceptable risk limits, we have to choose appropriate measures to reduce the risks during the risk control stage. These measures also have to be quantified.

These stages can also be perceived as an iterative process, which is repeated until risk assessment yields an acceptable risk. This procedure can be used both for new level crossings and for the regular review of existing crossings.

30.3.2 System Definition and Assumptions

The scope of the system definition depends on the necessary depth of risk analysis. In all cases, we need to model the properties of the rail traffic, the road traffic, and the level crossing equipment. Figure 30.3 shows a model that is based on data that are widely available for level crossings.

The system definition shown implies knowledge about influencing factors and interdependencies of level crossing safety, which is available in the relevant literature (e.g., the studies by RSSB [2007], SELCAT Working Group [2011], or ALCAM Working Group [2011]).

For our example, we shall make the following simplified assumptions:

- We will use a single-track railway line.
- We will consider only one direction of road traffic.
- As road users, we will consider passenger vehicles only.
- We will disregard the crossing angle of road and railway line.
- We will consider only the types of protection mentioned in Section 30.2.3.
- No risks occur in the clearance zone behind the crossing.
- We will not distinguish between intentional and unintentional errors of the road users.

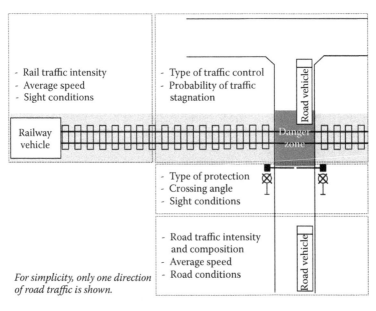

FIGURE 30.3
Example model for risk analysis.

- We will disregard human errors of railway staff.
- We will disregard technical failures of the safety equipment.
- The severity of damage includes only personal injuries of road users.

With today's state-of-the-art technologies, the actions of railway staff (e.g., of the train driver) are usually monitored by additional precautionary measures (e.g., train protection systems). We disregard technical failures because they are less likely than human error of the road users by several decimal powers. We limit the severity of damage to personal injuries of the road users because they represent 95–99% of all personal injuries on level crossings (Schöne 2013); furthermore, personal injuries can also be used as an indicator for material damage.

30.3.3 Hazard Identification

As described in Section 30.2.2, the most severe undesirable event is the collision between a railway vehicle and a road user. As the prevention of this event is the main task of protection measures at level crossings, we have to consider it as the top hazard in our risk analysis.

30.3.4 Risk Analysis

We start our analysis with the usual mathematical definition of risk (the product of frequency F of an undesirable event and the expected severity of damage S):

$$R = F \times S. \tag{30.1}$$

For level crossings, we can subdivide the frequency of collisions into *exposure E* (how often a railway vehicle and a road vehicle approach at the same time), *hazard H* (probability that the protection measures will not be effective), and *unavoidability U* (accident probability if there is a hazard) (Schöne 2013):

$$F = E \times H \times U. \tag{30.2}$$

Thus, we can calculate the following mathematical relation:

$$R = E \times H \times U \times S. \tag{30.3}$$

For our example, this level of detail is sufficient. If we need a more detailed analysis, we would have to add the risks we disregarded in the clearance zone into the equation; we would also have to distinguish between intentional and unintentional behaviors (see Section 30.2.2). We would also have to differentiate between the different types of road users (pedestrians, cyclists, lorries, etc.), as they have different properties and behave differently. For more detailed analysis, we would also have to model all directions of road traffic and all traffic relations before and after the crossing.

The example considers the collective risk on one single crossing. However, we can easily convert this into individual risks of the road users using the crossing frequencies.

30.3.4.1 Exposure

Exposure is defined as the number of collisions per time unit, which would theoretically happen if road and rail traffic would cross the crossing independently of each other, without protection measures and without the possibility to prevent the collision. We use the unit "theoretical strikings per time unit" for exposure.

It is commonly accepted that traffic intensity on the road and on the railway line influences the probability of an encounter. Many risk models are based on the product of both traffic intensities (Austin and Carson 2002; Oh et al. 2006).

However, if we consider this more closely, we can see that the influence of the road traffic intensity is not linear, but is saturated and reversed due to traffic jam effects: Because the level crossing protection device is activated a certain amount of time before the railway vehicle approaches; the road traffic is "dammed up" in front of the crossing, behind the first waiting vehicle. This causes a decline in the probability of the next car reaching the danger zone before the railway vehicle approaches. Stott (1987) described this phenomenon for the first time for British crossings; Heavisides and Barker (2008) further elaborated on the topic.

Figure 30.4 compares the conventional linear approach with Stott's approach. If this effect is neglected, the exposure—and with it, the risk on level crossings with high road traffic intensity—is overestimated. The precise values of the graph are influenced by further factors—especially by the approach warning time.

If a more detailed model is needed, one can also include the geometrical conditions of the crossing and the speed of road and railway vehicles, as these, among other things, determine how long they remain in the danger zone.

30.3.4.2 Hazard

In risk analysis, the hazard is defined as the probability with which a situation will occur at a certain exposure, which differs from the planned sequences, which is dangerous, and

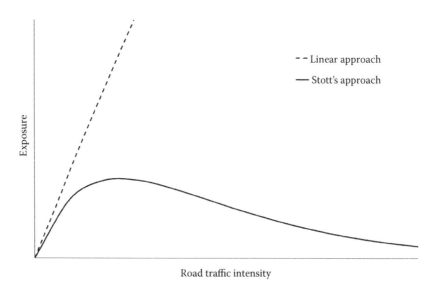

FIGURE 30.4
Impacts of traffic jam effects on exposure.

TABLE 30.1

Hazard for Certain Types of Protection

Type of Protection	Average Hazard Probability
Passive	10^{-1}
Light signals	10^{-2}
Light signals with half barriers	10^{-3}
Light signals with full barriers	10^{-5}

which can lead to an accident. Therefore, we choose "hazards per theoretical striking" as the unit.

The "planned sequence" is the rule-compliant reaction of the road users to safety equipment. E.g., for active rail crossings, this means that road users will stop when the light signals are turned on; for passive crossings, they will stop when a railway vehicle approaches. A premise for both cases is that the road users pay appropriate attention.

Numerous studies show that the type of protection influences the probability of human error (Berg and Oppenlander 1969; Heilmann 1984; Raslear 1996; Rösiger 2006). Table 30.1 shows the average hazard probabilities we can assume for motorized road users. As we do not distinguish between intentional and unintentional errors (see Section 30.3.2), the values are roughly simplified, however, in a correct relation and order of magnitude.

To increase the precision of the analysis, we can model further parameters that influence the results. Example parameters are the waiting time when the safety equipment is turned on (when the patience limit is exceeded, this can lead to intentional human error) or the road design, which influences the recognizability and visibility of the safety equipment.

30.3.4.3 Unavoidability

Not every hazard leads to an accident: often, an emergency reaction of the participants can avert the accident, if they are able to recognize the hazardous situation on time. For this, we use "accidents per hazard."

There are no confirmed values because we can hardly do empirical studies on unavoidability. However, the fundamental relevance for the risk is unchallenged. To quantify unavoidability, we have to resort to estimations.

It is inherent to the system that in the prevalent number of cases, the train driver cannot avert a collision. This leaves us with the possible actions of the road users: notably braking, accelerating, and evasive maneuvers. But this is possible only if road users recognize the imminent accident and react accordingly. Therefore, visibility conditions play an important role.

Table 30.2 shows estimated values for unavoidability under certain conditions; these values are following an approach of Bepperling (2008), which is based on error probabilities of

TABLE 30.2

Unavoidability for Certain Types of Protection

Visibility Conditions for Road Users	Unavoidability
No visibility, avoiding not possible	1
Medium visibility, avoiding rarely possible	10^{-1}
Good visibility, avoiding often possible	10^{-2}

Hinzen (1993). We define the visibility conditions as average values for all traffic directions and lines of vision. This parameter refers to the combination between sight conditions and speed of road and railway vehicles: low speed and bad sight can lead to the same value for visibility conditions as high speed and good sight, similar to calculation method for passive level crossings.

For a more precise model, we would have to distinguish between the type of human error: intentional violations (e.g., driving around half barriers, usually at lower speed) depend on the visibility directly in front of the level crossing, whereas for unintentional errors (e.g., overlooking the safety equipment, usually without braking), we need to take the visibility in the braking distance in front of the level crossing into account.

30.3.4.4 Severity of Damage

In this case, the severity of damage means the expected personal injuries of the road users during an accident. As a unit, we use "weighted personal injuries per accident"; one weighted personal injury is an equivalent of the following:

- 1 fatality
- 10 seriously injured persons
- 100 slightly injured persons

To quantify this parameter, we have a high number of confirmed values, as it can be directly deduced from accident statistics. One deduction is that the speed of the railway vehicle has a strong influence on the severity of damage. Except very low (under 30 km/h) and very high (over 160 km/h) speed, which has weak relevance for level crossings, the dependency is approximately linear. International studies (Cooper and Ragland 2007; Beard and Melo 2010; RAIB 2011) allow us to calculate a value of 0.003 weighted personal injuries per kilometer per hour railway speed (on average for all road users). As we focus in our example on passenger vehicles, we can deduce the following quantity equation (Schöne 2013):

$$S = 0.0024 \times v_{\text{railway}}. \tag{30.4}$$

In this equation, v is the average speed of railway vehicles in kilometers per hour. We can identify similar formulas for other road users, with pedestrians showing the highest severity of damage when considering the same railway speed.

30.3.5 Risk Assessment

Even with significant time and effort, we cannot fully exclude risk. Therefore, the acceptance criterion has to be larger than zero. Thus, we can formulate the following condition:

$$0 < R \leq R_{\text{accepted}}. \tag{30.5}$$

During risk analysis, we focused only on collective risks. We will keep this level for the risk assessment and will therefore use the sum of damage per level crossing and period under consideration.

It is rather hard to choose an appropriate threshold, as this process has to take into account technical, legal, ethical, and sociopolitical issues. Also, we cannot use technical acceptance criteria such as Risk Acceptance Criteria for Technical System (RAC-TS) or Minimum Endogenous Mortality (MEM) (Braband 2004) in this case because they represent functional safety requirements for technical systems; in our case, we focus on human behavior, which includes a considerable amount of intentional violations, where road users intentionally put themselves into danger. As an alternative, we could use the following approaches:

- Using average and maximum values of previous accident statistics, with the condition that the existing risk should not be exceeded in the future
- Benefit–cost analysis including economic criteria, e.g., using the life quality index (see the book by Mahboob et al. [2014])
- Calculating theoretical scenarios where the most adverse combination of all parameters with the most effective safety equipment is accepted

As an example, we will show the first approach with statistics from Germany. During the years 2009–2013, we determined an average of 0.002 weighted personal injuries per level crossing and year for all level crossings in the country (Schöne 2016). If we consider 1 year, we directly get the accepted risk:

$$R_{\text{accepted}} = 0.002 \text{ weighted personal injuries per level crossing and year.}$$

30.3.6 Risk Control

Risk control is necessary if the existing or expected risks on a level crossing exceed the set risk threshold. Essentially, risk can be reduced by adjusting the parameters determined in Section 30.3.4, e.g., by removing the level crossing or by selecting more effective safety equipment. In these cases, we can directly calculate the effects by changing the adjusted input parameters. If the local or financial conditions do not allow for such a change of the parameters, other measures to reduce risks become necessary.

The relevant literature includes numerous studies about the effectiveness of additional protection measures. Overviews can be found, for example, in the study by Saccomanno et al. (2009) or Washington and Oh (2006). Examples of especially effective measures include the following:

- Constructional division of lanes to prevent driving around half barriers (reduction of intentional errors by up to 80%)
- Installation of cameras to monitor red light violations (reduction of intentional errors by up to 70%)

To implement these measures into the procedure, we need to take into account the base values and conditions of the studies—often, these do not apply to all road users or do not exclusively pertain to hazard as a risk factor. If necessary, these values need to be converted. Here, we need a hazard reduction factor (a probability between 0 and 1), which we will multiply by factor H (hazard).

As we did not distinguish between intentional and unintentional behaviors in our risk analysis, it is complicated to implement additional measures. Therefore, we will limit risk control to adjusting the type of protection.

TABLE 30.3

Risk Analysis and Risk Assessment for Example Scenarios

Parameter	Scenario 1	Scenario 2a	Scenario 2b
E (theoretical strikings per year)	72	144	144
H (hazards per theoretical striking)	0.01	0.01	0.001
U (accidents per hazard)	0.01	0.01	0.01
S (weighted personal injuries per accident)	0.19	0.19	0.19
R (weighted personal injuries per year)	0.0014	0.0027	0.00027
$R \leq R_{accepted}$	Yes	No	Yes

30.3.7 Example

Our example level crossing will have the following properties:

- 100 passenger vehicles per day
- Light signals as safety equipment
- Good visibility conditions
- Average speed of 80 km/h for railway vehicles

The calculations refer to a period of 1 year and are shown in Table 30.3. Scenario 1 is based on a railway line traffic intensity of 40 trains per day; scenario 2a has an intensity of 80 trains per day. The parameters are calculated according to the explanations in Section 30.3.4; because of its complexity, we do not show the calculation of the exposure (for further details on this topic, see the study by Schöne [2016]).

While in scenario 1, the threshold for the risk acceptance criterion is met, the higher number of trains in scenario 2 leads to a risk which is unacceptable. Therefore, risk control measures become necessary. In our example, we can reduce the risk by changing the type of protection. Using half barriers reduces the hazard by one decimal power, which results in an acceptable risk (scenario 2b).

30.4 Determining Technical Safety Targets

30.4.1 Introduction and Procedure

The railway infrastructure company has to determine the necessary safety targets to allow manufacturers to identify, develop, and approve new technologies for the protection of level crossings. The targets should be abstract enough to ensure that they are independent of a particular technical implementation, allowing manufacturers to develop various solutions. We can achieve this by setting safety targets for functional units such as activation and protection. How detailed these targets need to be depends among other things on interactions between the level crossing protection and other technical systems for which the railway infrastructure company is responsible.

We will show the basic procedure in the following sections using a highly simplified example. In the first step, we need to define and outline the system (see Section 30.4.2). This can result in several working models, for which we have to identify possible hazards

(Section 30.4.3). Then, we need to establish a global safety target for the system in question (Section 30.4.4). On the basis of this global target, we need to deduce safety targets for the individual functional units (Section 30.4.5). With the results of the risk analysis, manufacturers can then assign safety integrity levels and draw conclusions for the development of their systems accordingly.

The approach was first described by Braband and Lennartz (1999) and exemplarily applied on level crossing problems by Braband and Lennartz (2000). It provided the basis for development of European standard EN 50129 (CEN 2003).

30.4.2 System Definition and Assumptions

Usually, the system definition can be deduced from the functional requirements of the railway infrastructure company for level crossings. Our example will be based on the system setting shown in Section 30.2.3; we will consider only level crossings where the danger zone is not monitored. This gives us the following functional units:

- Activation
- Supervision
- Protection
- Deactivation

Our analysis will consider only level crossings that are switched on and off by the detection of railway vehicles, that are monitored by supervision signals or remote supervision, and that are protected by light signals. By looking at two different supervision technologies, we get two different system versions for which we want to establish safety targets (Table 30.4).

If there is no malfunction on the technical side, for system version 1, events unfold in the following manner:

- The railway vehicle approaches the level crossing.
- The activation sensor detects the railway vehicle approaching the level crossing.
- The road signals are activated.
- The supervision signal is activated.
- The railway vehicle enters the level crossing.
- The deactivation sensor detects that the railway vehicle is leaving the level crossing.
- The supervision signal is deactivated.
- The road signals are deactivated.

TABLE 30.4

Examined Technologies in Two System Versions

Functional Unit	System Version 1	System Version 2
Activation	Activation sensor	Activation sensor
Supervision	Supervision signals	Remote supervision
Protection	Road signals	Road signals
Deactivation	Deactivation sensor	Deactivation sensor

Apart from the activation and deactivation of the supervision signals, the sequence for system version 2 is identical. In this scenario, remote supervision allows for the detection of technical malfunctions in order to repair the equipment; however, it is usually not able to stop an approaching railway vehicle. This means we can consider system version 2 as version 1 with virtual, continuously activated supervision signals.

Our analysis considers technical malfunctions only. We do not take human behavior into consideration. We assume that road users respect the safety equipment. This analysis also does not take into consideration if the train driver respects the supervision signals.

Another premise we make concerns safety-related events. For reasons of simplification, we consider only the collision of road user and railway vehicle as an "undesirable event." In real-life applications, we would also have to take into account other events with potential damage.

We also assume that each failure of the technical safety equipment leads to a hazard. This rather pessimistic assumption is justified, as it usually takes longer to detect the technical malfunction (e.g., by inspection) than for another road user to cross the level crossing.

30.4.3 Hazard Identification

Based on the system definition and the assumptions, we consider the dangerous failure of the technical safety equipment as the "top event." This malfunction can be caused by different failures of the individual functional units; on the other hand, not all failure modes cause a top event. Table 30.5 shows a simplified failure mode and effects analysis to identify relevant failures. The failures refer to undetected technical malfunctions of functional units (e.g., if an untimely lighting of the supervision signals remains undetected). The system development (based on the set safety targets) aims at monitoring these functions.

From the system definition and the failure mode and effects analysis, we can deduce fault trees for quantitative surveys. Figures 30.5 and 30.6 show these for both system versions.

TABLE 30.5

Simplified Failure Mode and Effects Analysis

Functional Unit	Failure Mode	Hazard	
		System Version 1	System Version 2
Activation	1a) Railway vehicle not detected	Only if failure 2b occurs at the same time	Yes
	1b) Nonexistent railway vehicle detected	No	No
Supervision	2a) Supervision signals do not light after activation	No	N/A (remote supervision)
	2b) Supervision signals light without activation	Only if failure 1a occurs at the same time	N/A (remote supervision)
Protection	3a) Road signals do not light after activation	Yes	Yes
	3b) Road signals light without activation	No	No
Deactivation	4a) Railway vehicle not detected	No	No
	4b) Nonexistent railway vehicle detected	Yes	Yes

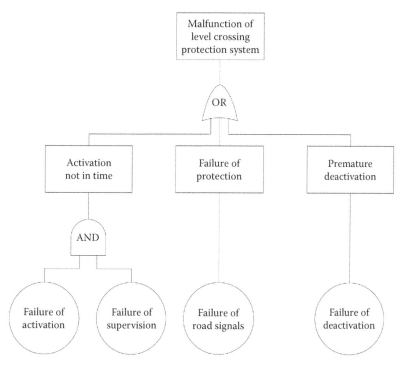

FIGURE 30.5
Fault tree for system version 1.

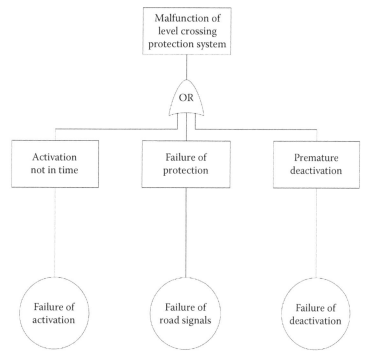

FIGURE 30.6
Fault tree for system version 2.

TABLE 30.6

Risk Matrix on the Basis of EN 50126 (1999)

Frequency of Occurrence of a Hazard (per Hour)		Risk Acceptance Levels			
Frequent	$10^{-5} < \text{HR} \leq 10^{-4}$	Unacceptable	Unacceptable	Unacceptable	Unacceptable
Probable	$10^{-6} < \text{HR} \leq 10^{-5}$	Unacceptable	Unacceptable	Unacceptable	Unacceptable
Occasional	$10^{-7} < \text{HR} \leq 10^{-6}$	Acceptable	Unacceptable	Unacceptable	Unacceptable
Rare	$10^{-8} < \text{HR} \leq 10^{-7}$	Acceptable	Acceptable	Unacceptable	Unacceptable
Improbable	$10^{-9} < \text{HR} \leq 10^{-8}$	Acceptable	Acceptable	Acceptable	Unacceptable
Incredible	$\text{HR} \leq 10^{-9}$	Acceptable	Acceptable	Acceptable	Acceptable
		Insignificant	Marginal	Critical	Catastrophic
		Severity of an accident (caused by a hazard)			

30.4.4 Global Safety Target

Unlike the risk assessment shown in Section 35.3.5, the technical safety targets only refer to risks caused by technical failures. Risk acceptance criteria, which can be applied to level crossings, already exist. For example, the CSM (Common Safety Methods) directive (2009) contains the following requirement: "For technical systems where a failure has a credible direct potential for a catastrophic consequence, the associated risk does not have to be reduced further if the rate of that failure is less than or equal to 10^{-9} per operating hour." Thus, we can calibrate the risk matrix by using a combination of a catastrophic severity of damage and a hazard rate of 10^{-9} per hour and grade the other combinations depending on this value (Table 30.6).

While we can assume catastrophic severity of damage (meaning, for calculation purposes: several casualties), for example, in case of a point failure for a high-speed line, we can expect critical severity (meaning, for calculation purposes, several seriously injured persons) for collisions on level crossings (based on empirical data). In the risk matrix, we can see that the tolerable hazard rate for this severity is 10^{-8} per hour.

Before we break down the hazard rate into the functional units, we have to verify if there are any barriers between hazard and accident, which could prevent the hazardous situation. Unlike in the analysis in Section 30.3.4 (which refers to the nonobservance of properly activated level crossing protection), we now confront the road users with safety equipment which is switched off from their point of view. For this, the assumed probability that an approaching railway vehicle can be recognized by chance and the accident can be avoided is considerably smaller.

But as avoidability is possible even in this case, we will use it for increasing the tolerable hazard rate. A factor of about one decimal power seems appropriate; due to logarithmic grading, we get a value of $\sqrt{10}$ (approximately 3). That means that on average, about every third hazard leads to an accident. Therefore, we can set the tolerable hazard rate for the top event "hazardous malfunction of the technical safety equipment" to a value of 3×10^{-8} per hour.

30.4.5 Safety Targets of Functional Units

The last step is to apply the set global safety target to the fault trees defined in Section 30.4.3 (Figures 30.5 and 30.6). On the second level of the fault tree, the tolerable hazard rate of 3×10^{-8} per hour is split onto the three OR-related events, resulting in 1×10^{-8} per hour per event. For system version 1, we have two further AND-related events on level 3.

TABLE 30.7

Deduced Safety Targets for Functional Units

Functional Unit (Number of Units)	Failure Mode	Tolerable Hazard Rate Malfunctions per Hour and Functional Unit	
		System Version 1	**System Version 2**
Activation (2)	Railway vehicle not detected	5×10^{-5} per direction	5×10^{-9} per direction
Supervision (2)	Supervision signals light without activation	5×10^{-5} per direction	N/A (remote supervision)
Protection (2)	Road signals do not light after activation	5×10^{-9} per direction	5×10^{-9} per direction
Deactivation (1)	Nonexistent railway vehicle detected	1×10^{-8}	1×10^{-8}

We assume that both events contribute evenly to the risk; this means that we get a value of 1×10^{-4} per hour for each functional unit. For system version 1, we can directly copy the values from the second level to the functional units on the third level.

We also need to take into account the number of functional units per level crossing. For simplification purposes, we will assume the following:

- Activation: two units (single-track railway line with two traffic directions)
- Supervision: two units (single-track railway line with two traffic directions)
- Protection: two units (road with two traffic directions)
- Deactivation: one unit (for both railway line directions together)

For the final calculation of safety targets, we have to divide the values calculated using the fault trees by the number of individual functional units per level crossing. The higher the number of functional units is, the higher are the safety requirements for an individual element. Table 30.7 shows the results of the calculations.

Due to the separate activation supervision in system version 1, the requirements for the individual functional units are lower than those for system version 2. But as the safety targets for protection and deactivation are the same in both versions, the manufacturer will most likely apply the same high requirements to central units such as the control unit during the development process.

30.5 Conclusions and Outlook

The chapter showed two different application scenarios of risk analysis for level crossings: on the one hand, the selection of safety measures based on conditions such traffic intensity, on the other hand, determining technical safety targets for different types of protection systems.

However, risk assessment for level crossings in Germany is currently limited to a one-time determination of safety targets by the railway operator, which has to be fulfilled by the signaling industry. This analysis was carried out in a similar way as described in Section 30.4.

Recently, a new table-based approach for this task was developed and is currently being tested (Meissner 2016). It offers a method to define safety targets for any railway signaling system, thus also for level crossings. Whereas the scope of procedure described in

Section 30.4 is primarily assigning hazard rates to functional units, the focus of the table-based approach is more the complete system under consideration of different operating conditions.

No quantitative risk assessment is done for the selection of protection measures at level crossings in Germany. Rather, these decisions are carried out based on legislative rules without scientific basis. In the interest of efficient allocation of resources, it is strongly recommended to complement these rules with quantitative approach.

We can identify different extensions of the presented approaches. The definition of acceptable risk could be extended to F–N diagrams instead of using the simplified acceptable annual weighted personal injuries. To integrate macroeconomic perspective, also the approach of life quality index could be useful. For more information concerning these aspects, see the corresponding chapters of this book. The extension of consequences to also economic and environmental aspects is possible too, e.g., for railway lines or road with frequent transport of dangerous goods, which can lead to extraordinary risks.

Finally, a useful future challenge is to bring together the two presented tasks—selecting protection measures and determining technical safety targets—in one procedure under joint consideration of human errors and technical failures.

References

ALCAM Working Group (Editor) (2011): *The Australian Level Crossing Assessment Model—ALCAM in Detail.* http://www.transport.nsw.gov.au/sites/default/file/levelcrossings/ALCAM_In _Detail-NSW.pdf, June 9, 2011.

Austin, R. D., and Carson, J. L. (2002): An alternative accident prediction model for highway-rail interfaces. *Accident Analysis and Prevention* 34, pp. 31–42.

Beard, M., and Melo, P. (2010): Level crossing risk tool for Portuguese railways. *11th World Level Crossing Symposium, Tokyo.*

Bepperling, S.-L. (2008). *Validierung eines semi-quantitativen Ansatzes zur Risikobeurteilung in der Eisenbahntechnik.* PhD Thesis, Technische Universität Braunschweig, Braunschweig.

Berg, W. D., and Oppenlander, J. C. (1969): Accident analysis at railroad-highway grade crossings in urban areas. *Accident Analysis and Prevention* 1, pp. 129–141.

Braband, J. (2004): Risikoakzeptanzkriterien und-bewertungsmethoden—Ein systematischer Vergleich. *Signal und Draht 96* 4 pp. 6–9.

Braband, J., and Lennartz, K. (1999): Systematisches Verfahren zur Festlegung von Sicherheitszielen für Anwendungen der Eisenbahnsignaltechnik. *Signal und Draht 91*, 9, pp. 5–10.

Braband, J., and Lennartz, K. (2000): Risikoorientierte Aufteilung von Sicherheitsanforderungen— Ein Beispiel. *Signal und Draht 92* 1, 2, pp. 5–10.

Cooper, D. L., and Ragland, D. R. (2007): *Driver Behavior at Rail Crossings: Cost-Effective Improvements to Increase Driver Safety at Public At-Grade Rail-Highway Crossings in California.* California PATH Program, Institute of Transportation Studies, University of California, Berkeley, CA.

CSM-Verordnung (2009). Verordnung (EG) Nr. 352/2009 der Kommission vom 24. April 2009 über die Festlegung einer gemeinsamen Sicherheitsmethode für die Evaluierung und Bewertung von Risiken gemäß Artikel 6 Absatz 3 Buchstabe a der Richtlinie 2004/49/EG des Europäischen Parlaments und des Rates.

CEN (European Committee for Standardization) (1999) EN 50126-1: *Railway Applications—The Specification and Demonstration of Reliability, Availability, Maintainability and Safety (RAMS)— Part 1: Generic RAMS Process.* CEN, Brussels.

CEN (2003) EN 50129: *Railway Applications—Communication, Signalling and Processing Systems—Safety Related Electronic Systems for Signalling.* CEN, Brussels.

Griffioen, E. (2004): Improving level crossing using findings from human behaviour studies. *8th International Level Crossing Symposium, Sheffield.*

Heavisides, J., and Barker, J. (2008): Impact of traffic flow on level crossing risk. *10th World Level Crossing Symposium, Paris.*

Heilmann, W. (1984): *Grundlagen und Verfahren zur Abschätzung der Sicherheit an Bahnübergängen.* PhD Thesis, Technische Hochschule Darmstadt, Darmstadt.

Hinzen, A. (1993). *Der Einfluß des menschlichen Fehlers auf die Sicherheit der Eisenbahn.* Dissertation, Rheinisch-Westfälische Technische Hochschule Aachen, Aachen.

Mahboob, Q., Schöne, E., Maschek, U., and Trinckauf, J. (2014). *Investment into Human Risks in Railways and Decision Optimization. Human and Ecological Risk Assessment.*

Meissner, G. (2016): Experience in applying the DIN VDE 0831-103 draft standard in commercial projects. *Signalling + Data Communication 108* 9, pp. 67–72.

Oh, J., Washington, S. P., and Nam, D. (2006): Accident prediction model für railway-highway interfaces. *Accident Analysis and Prevention 38,* pp. 346–356.

RAIB (Rail Accident Investigation Branch) (Editor) (2011): *Rail Accident Report—Investigation into the Safety of Automatic Open Level Crossings on Network Rail's Managed Infrastructure.* Report 12/2011. Derby.

Raslear, T. G. (1996): Driver behavior at rail-highway grade crossings: A signal detection theory analysis. In: Carroll, A. A., and Helser. J. L. (Editors): *Safety of Highway-Railroad Grade Crossings,* Volume II—Appendices. Washington, DC.

Rösiger, T. (2006): *Wirkungsüberprüfung eines dynamischen Rückmeldesystems auf das Fahrerverhalten am technisch nicht gesicherten BÜ.* Diploma Thesis, Technische Universität Dresden, Dresden.

RSSB (Rail Safety and Standards Board) (Editor) (2007): *Use of Risk Models an Risk Assessment for Level Crossings by Other Railways.* RSSB, Cambridge, UK. http://www.rssb.co.uk /SiteCollectionDocuments/pdf/reports/research/T524_rpt_final.pdf, 24.06.2010.

RSSB (Editor) (2011): *Road-Rail Interface Special Topic Report.* RSSB, Cambridge, UK. http://www.rssb .co.uk/SiteCollectionDocuments/pdf/reports/road-rail_interface_str_full.pdf, 09.08.2011.

Saccomanno, F. F., Park, P. Y., and Fu, L. (2009): Estimating countermeasure effects for reducing collisions at highway-railway grade crossings. *Accident Analysis and Prevention 39,* pp. 406–416.

SELCAT Working Group (Editor) (2011): D3—Report on Risk Modelling Techniques for Level Crossing Risk and System Safety Evaluation. http://www.iva.ing.tu-bs.de/levelcrossing /selcat/lcDocuments/866-866-26_SELCAT-D3.pdf, 04.06.2011.

Schöne, E. (2013): *Ein risikobasiertes Verfahren zur Sicherheitsbeurteilung von Bahnübergängen.* PhD Thesis, Technische Universität Dresden, Dresden.

Schöne, E. (2016). Bahnübergänge: zwischen Restrisiko und Sicherheitslücke. *Eisenbahntechnische Rundschau 65* 3, pp. 12–17.

Stott, P. F. (1987). *Automatic Open Level Crossings—A Review of Safety.* London.

Theeg, G., Svalov, D., and Schöne, E. (2009): Level crossings. In: Theeg, G., and Vlasenko, S. (Editors): *Railway Signalling and Interlocking—International Compendium.* Hamburg.

Washington, S., and Oh, J. (2006). Bayesian methodology incorporating expert judgment for ranking countermeasure effectiveness under uncertainty: Example applied to at grade railroad crossings in Korea. *Accident Analysis and Prevention 38,* pp. 234–247.

31

Human Reliability and RAMS Management

Malcolm Terry Guy Harris

CONTENTS

31.1 Introduction and Historical Perspective into Human Reliability

Human reliability represents a field of human factors and ergonomics that is concerned with the reliability of humans within an industrial context. There are several factors that can affect human performance and may significantly influence the reliability and safety levels of complex technical systems.

It has been widely recognized that human error is a significant contributor, or at least in part, to accidents in the operation of complex railway systems. The UK Health and Safety Executive (HSE) findings from UK Rail Accidents reports have highlighted that human factors play a critical role in the causation of the accident in Clapham Junction (Hidden, 1989), Southall (HSC, 2000), Hatfield (HSC, 2000), and Ladbroke Grove (HSC, 2001a, 2001b; Cullen, 2001).

The growing trends in the rail industry identified by the UK Rail Safety and Standards Board (RSSB) (2008) are as follows:

- Technical systems are becoming more far reaching and complex, which places greater importance in terms of considering their effects on a wider group of working people and on the organization as a whole.
- The nature of working is placing greater demands on people.
- There is a growing interest by organizations in regarding employees and technology as valuable investments.

These trends place a greater focus on the relationship between the individual, equipment, and the working environment, which all require a balance to work efficiently.

This chapter considers the influence of the individual, equipment, and working environment on rail organizations and management. It references current research work and ideas in the field of human performance.

31.2 Human Reliability and Human Error Definitions

Human reliability refers to the reliability of humans in an industrial context, such as manufacturing, transportation, and medical. Human reliability focuses on the ability of humans to perform specific tasks, which can be affected by many factors, such as age, physical health, attitude, state of mind and ability for certain common mistakes, errors, and cognitive biases.

Historically early accident models tended to consider the main causes of accidents either by technical failures or by human errors (Reason, 1991). With technological advances improving the reliability of equipment and systems, failures associated with human interaction were identified as the main causes of accidents. In the 1960s, the estimated contribution of human error failures to system failure was around 20%. By the 1990s, this estimate had increased to about 80% (Hollnagel, 1993). Reason stressed that the significant increase was not due to the increased fallibility of people but instead was the result of significant advances in system technology where human interfaces are common. These changes placed the concept of human error into a more prominent position in accident investigations (Reason, 2008).

The railway industry has reported that human error is attributed to up to 75% of fatal railway accidents in Europe between 1990 and 2009 (Evans, 2011), which has led to developing controls and measures to help reduce them.

There have been several definitions of human error by various researchers of human reliability, including those by Reason (1990), Hansen (2006), Senders and Moray (1991), and HSE (1999). Table 31.1 shows a list of typical definitions used to describe human error from various industries.

The estimation of human error is calculated through the analysis of potential human error consequences when applied to specific tasks. Each task is assigned a human error probability, and the summation of these failures from tasks on a system is then collated to provide an overall failure estimation for that system. Section 31.4 illustrates a simple example using the human error probabilities for a specific operation based on the *Red Book* (Committee for the Prevention of Disasters 1997).

TABLE 31.1

Human Error Definitions

Reference	Definition
Hansen (2006)	Human error is used to describe the outcome or consequence of human action, the causal factor of an accident, and as an action itself.
Senders and Moray (1991)	Human error means that something has been done that was "not intended by the actor; not desired by a set of rules or an external observer; or that led the task or system outside its acceptable limits." "It is a deviation from intention, expectation, or desirability."
Reason (1990)	*Error* is taken as a generic term that encompasses all those occasions in which a planned sequence of mental and physical activities fails to achieve the intended outcome and when these failures cannot be attributed to the intervention of some chance agency.
HSE (1999)	Human error is an action or decision which was not intended, which involved a deviation from an accepted standard, and which led to an undesirable outcome.

Source: F. D. Hansen, *Journal of Air Transportation*, 11, 61, 2006; J. W. Senders and Moray, N. P., *Human Error: Cause, Prediction, and Reduction*, Lawrence Erlbaum Associates, Mahwah, NJ, 1991.

31.3 Classification of Human Failures

A clearer understanding of the different sources of errors can provide engineers with a means for determining the different actions needed to prevent them. The classification of human failures provides a means to understanding how different human failures have different causes and remedial actions when assessing human failure causation in incident and accident investigations. Kletz (2005) suggests the following classification of human error:

- *Mistakes*—These are human errors due to not knowing what action is to be done or, worse, thought they knew what was required but did not.

- *Violations*—These are human errors due to a deliberate decision not to follow instructions, perhaps to make the task easier but often because a better way to perform the task has not been identified.

- *Mismatches*—This type of human error arises because the task was beyond the physical or mental ability of the person asked to perform it and is perhaps beyond a reasonable person's ability.

- *Slip or lapses of attention*—This is the last type of human error. This is considered the most common cause of human errors by frontline operators, maintenance workers, vehicle drivers, but not designers, or managers as these latter groups have time to review their decisions.

Kletz states that two possible easy ways of reducing errors involves changing people's behavior or changing the environment where they work to minimize the opportunities for error. Figure 31.1 illustrates an example of human failure taxonomy based on the research conducted by the Energy Institute Human and Organisational Factors Working Group (King, 2010). This can be applied to all industries and organizations to provide a transparent classification of human failures.

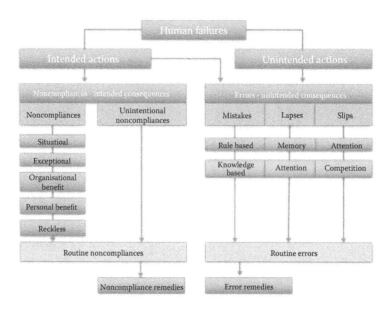

FIGURE 31.1
Overview of the human failure taxonomy model. (Based on S. King, *Petroleum Review*, 2010.)

The UK RSSB guidance on human factors was developed for designers, suppliers, managers, supervisors, trainers, and health and safety staff who work in the railway industry. This guide presents techniques and knowledge on human performance (RSSB, 2008). Figure 31.2 illustrates four key questions on human performance at work taken from the guidance document.

The RSSB guide provides several reasons behind these queries on human performance including but limited to the following areas:

1. The three types of performance that can lead to human errors are the following:
   - Skill-based performance—These are routinely performed highly practiced activities with little conscious effort, for example, this could include the replacing of a rail clip, closing an equipment panel, or setting a specific route on a panel.

FIGURE 31.2
Key questions on human performance. (From RSSB, *Understanding Human Factors—A Guide for the Railway Industry*, RSSB, London, 2008.)

- Rule-based performance—These activities require more mental involvement with an application of previously learned rules to specific situations and tasks that the person has been trained for, for instance, evacuating a station platform or planning a route in a signal box are examples of such rule-based activities.
- Knowledge-based performance —These activities have the highest mental involvement that is often applied to novel situations, for example, when attending an accident investigation scene or counseling a staff member.

2. There are three types of rule-based performance that people make in error when applying known rules as follows:

- Misapplying a good rule—This situation involves applying a rule that is not appropriate. Generally, this is a rule that is frequently applied and seems to fit the situation well.
- Applying a bad rule—For this case, the task is completed but with an unwanted consequence.
- Failing to apply a good rule—In this case, the rule that is applicable and valid for the situation is ignored.

3. The four types of skilled-based performance error that people make are generally considered as slips or lapses:

- Familiarity slips—Where the frequent action overtakes a similar but infrequent action by a person, for instance, dialing in a familiar number than entering the less frequent but similar one.
- Similarity slips—Where the action is a frequent action performed by a person so that the action is conducted on the wrong object, for instance, activating a switch on a panel because of its proximity to similar switches.
- Memory lapses—Whereby a routine sequence of steps or a goal is omitted from a sequence of actions, for example, when a person forgets the reason for entering a room.
- Association slips—This arises when the brain makes a faulty connection between two ideas, for instance, where an external stimulus provokes an action, for instance, a driver reacts to an alarm as if another one was being activated.

4. Knowledge-based performance that is prone to error generally arises from a lack of knowledge, uncertainty, lack of concentration, or a misapplication of knowledge, particularly in novel situations. Two examples of this type of error are as follows:

- Availability bias—This involves favoring a course of action because it is one that comes readily to mind. For instance, by performing a procedure that is straightforward to implement but is technically flawed.
- Confirmation bias—This type of bias can arise through overconfidence or when making shortcuts to reduce the complexity of a task. In this situation, the person looks for information that confirms their belief about the situation, while ignoring or filtering out anything else which disagrees.

In both cases, the error bias can lead the person to make a faulty judgment of the situation and thereby prepare and execute a faulty plan to achieve the task.

The RSSB guidance on human factors presents a comprehensive guide to the causes and techniques applied to assess human error and provide useful recommendations for further reading.

31.3.1 How Can Organizations Reduce Errors?

For organizations, error management should be specifically tailored to the context of the organization under consideration. The key for organizations is to develop a working environment where people can make mistakes but without being penalized for the consequences. Five main areas that need to be considered by organizations in combination with the human reliability techniques described in this chapter and in RSSB (2008) include the following:

- Design is applied to ensure that people cannot make certain types of mistakes, for instance the use of signal interlocking or to assist the user to review their decision before implementing it, for example a dialogue box within a human machine interface within a train cab that prompts you to confirm your selection.
- Training is used to ensure persons are adequately rehearsed in their skill sets and knowledge for dealing with situations such that the likelihood of making mistakes is minimized.
- Staffing is used to ensure there is right balance of people placed in the right jobs.
- Culture is developed by the organization through its leadership, management, and collective teamwork to ensure a supportive, healthy environment is developed, free of blame.
- Working conditions are developed with the objective of minimizing mistakes that lead to consequences of stress, poor motivation, and morale.

31.4 Practical Issues in Human Reliability

Human reliability analysis (HRA) is composed of methods used to model and quantify human errors within a comprehensive analysis. The HRA method is widely recognized to possess several merits, including the following:

- A systematic comprehensive analysis of factors that can influence human performance
- A sustainable active approach whereby the outcomes result in recommendations for improvement
- A recognized method to support safety cases with specific attention towards critical tasks analysis
- Having direct applications outside the nuclear industry; most generic tools can be applied to any sector to provide an insight to risk assessment

Conversely, there are several practical issues with use of the HRA methods when applied to industrial processes. Researchers in this field have identified several disadvantages in applying HRA methods including the following:

- Some HRA methods are not fully validated.
- The techniques, being specialized, requires input from a technical human factor specialist for some methods.
- The method being considered is too time consuming, given the level of risk associated with human error for a task.

31.5 Human Reliability Techniques and Tools

Historically, the use of human reliability methods was initially developed in the aerospace and nuclear industries and was later adapted to the oil and gas, process, and other high-risk industries. Early human reliability studies in the 1960s were generally focused on the influences of human activity on equipment failures. In 1972, the Institute of Electrical and Electronic Engineers published a report on human reliability.

The HRA methods can be classified into the following groups:

- First-generation tools
- Expert judgment methods
- Second-generation tools

The earlier first-generation human reliability methods, such as the technique for human error rate prediction (THERP) and human error assessment and reduction technique (HEART) were mainly focused on human error failure and human operational error, from 1920 to 1990, to help predict and quantify the likelihood of human errors. This tool enabled the user to decompose a task into component parts and consider the impact from modifying factors, such as time pressure, stress, and equipment design. The outcome from these tools focused on the skill level and rule-based level of the human action to determine a nominal human error probability (HEP).

The development of expert judgment methods such as absolute probability judgment and paired comparison were developed around the same time as the first-generation methods (HSE, 2009). These methods were proven popular in non-safety-critical environments.

The 1990s and ongoing saw the development of second-generation human reliability methods, such as cognitive reliability and error analysis method (CREAM) and a technique for human event analysis (ATHEANA); these tools considered the context and errors of commission in human error prediction. These methods are yet to be empirically validated (HSE 2009, Cooper et al., 1996).

The RSSB (2012) developed the railway action reliability assessment method for quantifying human error probabilities as part of the risk assessments.

TABLE 31.2

HRA Tools in the Railway Industry

Method	Description	Reference
ATHEANA	A technique for human error analysis	Nureg (2000)
CAHR	Connectionism assessment of human reliability	Strater (2000)
CESR	Commission errors search and assessment	Reer (2004)
CREAM	Cognitive reliability and error analysis method	Hollnagel (1998)
MERMOS	Assessment method for the performance of safety operation	LeBot (2004)
THERP	Technique for human error rate prediction	William (1985)
HEART	Human error assessment and reduction technique	Williams (1985, 1986)

Part 3 of the RSSB guidance on human factors presents a list of other human factor techniques and outlines the advantages and limitations of these techniques in their application in the rail industry (RSSB, 2008), such as the following:

- A fault tree is a top–down technique to identify system failures and causes and to estimate their probabilities. The analyst/designer develops a treelike diagram to represent hardware failures and human errors. The contributing causes are then linked together by AND/OR relationships, and each contributing cause is further analyzed in terms of its own contributing causes. The analyst must take care to keep the analysis going until the underlying root causes are properly identified and recorded.

- Human error HAZOP uses a structured approach that is keyword driven to analyze errors and potential mitigation solutions. The team agrees on a set of guide words, such as "sooner than," "later than," and "not done," which they then use to evaluate operations that have been systematically described beforehand—often as a hierarchical structure using a technique such as hierarchical task analysis (HTA). This requires a selected team of specialist individuals to participate and support this activity, which is often resource intensive.

- Murphy diagrams are similar in approach to fault trees to analyze errors or failures by their causes by operators or teams. This approach starts with a comprehensive description of the task; the analyst classifies each task step into one of several decision-making categories and breaks this down into successes and failures.

- Systematic human error reduction and prediction approach) is a human error technique that enables tasks to be analyzed and potential solutions to errors to the presented in a structured manner to the designer.

31.5.1 Application of HRA on a Rail Operator Case

31.5.1.1 Context

In the event of an emergency, the rail operator can switch a synchronization device to divert the main feed to an emergency supply. The operator activates the synchronization device of bus bar (B-1) first by pressing a button by a button on a panel. After activation of synchronization of bus bar (B-1), the operator then waits until the position indicator of the circuit breaker (S-1) to indicate a closed position on the panel and the position indicator of circuit breaker (S-2) to indicate an open on the panel. After the successful completion of the restoration of bus bar (B-1), the operator will continue with the restoring the emergency power supply to bus bar (B-2) by the activation of the synchronization device of bus bar (B-2) on the panel. If the restoration of bus bar (B-1) is not successful, the operator has the stop the emergency activation process and will have to seek assistance from the maintenance operators. Failure of the synchronization device implies that the power supply to the emergency bus bar is stopped.

A written procedure is available to perform an automatic synchronization of both bus bars. However, in practice, this procedure is not used because it is normally a simple step-by-step task that is performed from memory.

What is the probability of failure of the power supply to both emergency bus bars?

31.5.1.2 Task Analysis

The first step of the analysis is to identify the human actions and equipment failures that can lead to the unwanted failure of the power supply to both emergency bus bars. After the review of the procedures and the design documents and site visit, the following tasks could be identified:

- Step 1: Start automatic synchronization of emergency bus bar (B-1) by activating the synchronization device of emergency bus bar (B-1).
- Step 2: The operator must wait until the synchronization process of emergency bus bar (B-1) is completed.
- Step 3: After completion of the synchronization of emergency bus bar (B-1), the operator must check the positions of the circuit breakers (S-1) and (S-2). The circuit breaker (S-1) must be in the closed position and circuit breaker (S-2) must be in the open position.
- Step 4: If the synchronization process is not successfully completed, the operator must stop the synchronization procedure and ask for assistance from the maintenance operators.
- Step 5: After the successful completion of the synchronization process of emergency bus bar (B-1), the operator must start the automatic synchronization device of emergency bus bar (B-2).
- Step 6: The operator must wait until the synchronization of emergency bus bar (B-2) is completed and must check whether the synchronization was successful or not.

31.5.1.3 Identification and Classification of Potential Human Errors

For this example, only errors of omission are considered. The following potential errors of omission are therefore identified:

1. A: No execution of the synchronization process.
2. B: The operator does not wait until the synchronization of the emergency bus bar (B-1) is completed.
3. C: The operator does not check the position of the circuit breakers (S-1) and (S-2) after completion of the synchronization process of emergency bus bar (B-1).
4. D: The operator does not stop the synchronization process after failure of the synchronization of the emergency bus bar (B-1).

It is noted that other potential errors of omission can be identified but are not important in relation to failure of both emergency bus bars (B-1) and (B-2).

31.5.1.4 Human Error Event Tree

Considering all potential errors identified, a human error event tree for improper restoration of both bus bars can be constructed; see Figure 31.3. All potential human errors are represented by capital letters. Technical failure of the synchronization device is represented by capital Greek letter eta Σ.

The inspection of the HRA event tree shows that three failure sequences (F2, F3, and F4) can lead to unavailability of both emergency bus bars:

- A: No execution of procedure.
- B: Operator does not wait.
- C: Operator does not perform the check.
- D: Operator does not stop.
- $\Sigma1$: Technical failure of synchronization emergency bus bar (B-1).
- $\Sigma2$: Technical failure of synchronization emergency bus bar (B-2).

To generate a quantitative result, the analyst/designer must estimate the probability of each failure or error included in the HRA event tree. Data for all the human failures and errors are available in tables in the handbook by Swain and Guttman (1983). The analyst/designer must modify these data as necessary to account for specific characteristics of the work situation, such as stress levels and equipment design features.

Experience shows that the failure probability of the automatic synchronization device is equal to 0.1 per demand. This figure is applicable for failures, which lead to loss of the power supply to the bus bar in consideration.

Table 31.3 summarizes the data used in this problem. From Table 20-11 in the handbook by Swain and Guttman (1983), it can be concluded that misinterpretation of the status lamp indicators is negligible.

After completing Table 31.3, the analyst/designer calculates the total probability of failure of both emergency bus bars given a demand for synchronization. The probabilities of a specific path is calculated by multiplying the probabilities of each success and failure branch in the path. Table 31.4 summarizes the calculations of the HRA results.

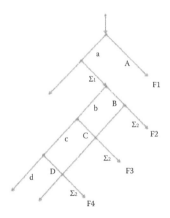

FIGURE 31.3
HRA event tree human performance. (From Committee for the Prevention of Disasters, *Methods for Determining and Processing Probabilities* ('Red Book'), SDU, The Hague, 1997.)

TABLE 31.3

Events Included in the HRA Event Tree

Failure Symbol	Failure Description	Estimated HEP	Data Source
A	No execution of procedure	0.01	
Σ_1	Failure of synchronization device emergency bus-bar (B-1)	0.1	Field data
B	Operator does not wait	0.05	
Σ_2	Failure of synchronization device emergency bus bar (B-2)	0.1	Field data
C	Operator fails to perform a check	0.05	
D	Operator fails to stop	0.05	

Source: Committee for the Prevention of Disasters, *Methods for Determining and Processing Probabilities ('Red Book')*, SDU, The Hague, 1997.

TABLE 31.4

HRA Event Tree Results

Sequence	Probability of Failure Given a Demand
$F2 = a\ \Sigma_1 B\ \Sigma_2$	0.0005
$F3 = a\ \Sigma_1\ b\ C\ \Sigma_2$	0.00047
$F4 = a\ \Sigma_1\ b\ c\ D\ \Sigma_2$	0.00045
$F_{total} = F2 + F3 + F4$	0.001

The probability of failure of both emergency bus bars (F_{total}), given a demand for synchronization, is equal to 0.001. Improvements can be achieved by training of the operators to use the written procedures.

31.6 Human Reliability Process and Integration into a Reliability, Availability, and Maintainability and Safety Process

Companies that undertake a safety management process toward managing railway hazards need to conduct system safety analyses and quantitative and qualitative risk analysis as part of their continuous improvement program development. Those organizations that undertake human reliability analyses are focused on developing an integrated process to reduce human error within their reliability, availability, and maintainability and safety activities, such as operational or maintenance management, which would typically follow a systematic process.

Table 31.5 outlines the key steps involved in a generic HRA based on the work of Kirwan (1994).

31.6.1 Brief Description of the Process

A brief description of the eight-staged process is described as follows:

Step 1: Preparation and problem definition: This step is used to understand the relative impact from different failures upon system safety. It is also required to priorities risk reduction efforts by implementing adequate controls with the aim

TABLE 31.5

Generic HRA Process

Step	Process Stage
1	Preparation and problem definition
2	Task analysis
3	Failure identification
4	Human failure modeling
5	Human failure quantification
6	Impact assessment
7	Failure reduction
8	Review

Source: B. A. Kirwan, *A Guide to Practical Human Reliability Assessment*. Taylor & Francis, London, 1994.

of developing (HEPs). These are defined as the number of failures on demand divided by the number of demands. The use of a checklist should be used to prepare and define a comprehensive problem.

Step 2: Task analysis: This involves defining the specific tasks undertaken by the system. This often involves splitting a task into the subtasks that are necessary to carry it out. This needs to be fully defined to provide a more successful understanding of the identified potential failures.

Step 3: Failure identification: This involves describing how the human failure will impact the validity of the HRA. This is best undertaken using a series of guide words that are applied to the steps identified in the task analysis. For instance, the guide words "action omitted," "action too late," or "action in wrong order" all provide a prompt to describe the observable failure.

The use of a checklist during this step provides a more comprehensive and systematic approach towards the identification of failures.

Step 4: Human failure modeling: In this step, an HRA event tree is developed using the findings from the task analysis and from the failure identification review.

Step 5: Human failure quantification: This involves the identification and quantification of the failures of concern from the HRA event tree based on generic data sources. The outcome from this step will include the estimation of the overall failure probability. For example, this could include failure of a driver stopping the train within a designated zone or an operator required to perform an abnormal operational task.

Step 6: Impact assessment: This step identifies the dominant human failure mechanism based on an assessment/review of the HRA event tree. The dominant case can be identified as the one that leads to the greatest overall human failure probability.

Step 7: Failure reduction: This step identifies the number of possible actions that could be implemented to reduce the outcome of the hazardous accident event, in terms of failure event frequency.

Step 8: Review: This includes developing conclusions from the findings and recommendations for reducing human error. This may include providing the project team with an insight into the problems undertaking the task. It should also extend

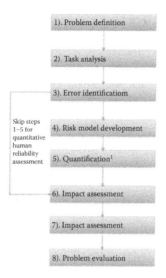

FIGURE 31.4
Generic human reliability process. (From RSSB, *Development of Human Error Quantification Technique for Use in the Railway Industry*, Research Brief T270, RSSB, London, 2012.)

to providing recommendations such as reliving the issue at hand or focusing on improving the lines of communication between those involved in undertaking the task and those developing changes to the procedure for instance. Figure 31.4* outlines the key steps involved in a generic HRA process.

- Advantages
  - The method is relatively straightforward to learn and does not require a significant level of training.
  - The method is auditable as it can be recorded on a standard error pro forma.
  - The taxonomy provides a prompt for analysts/designers for potential errors.
  - The technique can be applied to a variety of domains, for instance, a control room, drivers console, or maintenance environment.
  - The method is ideal for small groups of tasks.
- Disadvantages
  - For larger lists of tasks, it can take longer to complete and the size of the document can become large.
  - Extra work is involved if the HTA element is not available.

---

* RSSB outlines a generic quantification technique for assessing task analysis; refer to Table 31.2. This step requires applying a quantification technique to assess work tasks. Work conducted by RSSB (2012) outlines a generic quantification technique for assessing specific tasks, which include the following:
  • Identifying the specific tasks required for quantification
  • Comparing the specific task with a list of generic tasks which are listed in the technique
  • Applying the generic probability to the specific task
  • Evaluating the negative performance shaping factors and amend the probability based on the outcome

It should be noted that while it is not necessary to be a human factors specialist to apply this technique, it is always very useful if not essential to gain advice from such specialist in interpreting and applying the findings from the results of the analysis, for example, through interventions in training, system, or job design.

The generic process would require providing the required technical resourcing within the organization. To provide a valued assessment of the department or requirements of a company in developing and implementing a successful HRA processes several operational activities should be considered.

Chapter 10 on human factors describes this process in further detail in relation to the EN 50126 standard.

31.7 Trends in Human Reliability Research for Rail Applications

Human reliability research in the rail industry can extend within the following potential areas of future research:

- Use of HRA evaluation techniques
- Use of Alternative approaches

31.7.1 Use of HRA Techniques in the Rail Industry

The known practical limitations associated with HRA tools that are described in this chapter highlight the need for research to adaptable techniques that can consider the complexities of modern industrial systems. Most HRA tools are deterministic in their approach towards managing human failure, for instance, they consider the identification and management of cause and effect relationships. It has been recognized that these tools tend to be dynamic but highly complex in their interpretation, which often results in difficulty in the confidence level from the results. Recent work by Schwencke et al. (2012) focuses on the human's ability to adapt to and recover from hazardous situations (resilience) instead of human errors. Their approach uses performance-shaping factors in a proactive, goal-orientated way to model human ability to cope with unfamiliar situation. This work draws on the findings from the SMSmod project, a German funded research program for developing a new human reliability assessment method. The project has focused on the importance of performing shaping factors based on literature, reviews from existing methods, and expert evaluations which have been applied to the ergonomic evaluation of technical systems to improve the human reliability by adapting the designs of these systems.

31.7.2 Use of Alternative HRA Techniques

The shortfalls of using the probabilistic approach as an effective tool for addressing the issue of uncertainty within HRA and system reliability has been widely reported (Bell and Holroyd, 2009) and, specifically, where just applying a mean and variance or confidence levels is insufficient to address the problem. The emerging use of a Bayesian approach in HRA models has been applied to estimate the human error probability (HEP) for human failure events. The HEP provides the conditional probability of a human failure event given context shaped by a set of performance shaping factors or performance influencing factors, which are formed into discrete levels.

Groth and Swiler (2012) describe the use of Bayesian networks (BNs) for predicting human error probabilities when set within performance shaping factors and, hence, HRA. The standardized plant analysis risk-HRA (SPAR-H) [3] BN method was developed to estimate HEPs for use in the SPAR-H models to assess human reliability within 70 nuclear power plant facilities.

31.8 Conclusion

This chapter presents an outline on human error as a subset of human factors. It describes the classification of human failures and explores some of the practical issues associated with the HRA method in assessing human errors.

A simple example of the human error method is applied to illustrate how this approach can be used for the identifying and quantifying errors by operators and maintenance personnel.

Some emerging trends in human error research provide the analyst/designer with new practices and approaches toward understanding the complexities of human error parameters and their interpretation to design.

References

Bell, J., and Holroyd, J. (2009). *Review of Human Reliability Assessment Methods*. Harpur Hill, UK: Health and Safety Laboratory.

Barriere, M., Bley, D., Cooper, S., Forester, J., Kolaczkowski, A., Lukas, W., Parry, G. et al. (2000). *Technical Basis and Implementation Guidelines for a Technique for Human Event Analysis (ATHEANA)*, Rev. 1. NUREG-1624, Washington DC, US: Nuclear Regulatory Commission.

Committee for the Prevention of Disasters. CPR12E. *Methods for Determining and Processing Probabilities ('Red Book')* NRG, Second edition. The Hague: 1997.

Cooper, S., Ramey-Smith, A., Wreathall, J., Parry, G., Bley, D., Luckas, W., Taylor, J., and Barriere, M. (1996). *A Technique for Human Error Analysis (ATHEANA)—Technical Basis and Methodology Description*, NUREG/CR-6350. Washington, DC: Nuclear Regulatory Commission.

Cullen, W. D. (2001). *The Ladbroke Grove Rail Inquiry* (Part 1 Report). Norwich, UK: HSE Books.

Evans, A. W. (2011, January). Fatal train accidents on Europe's railways: 1980–2009. *Accident, Analysis and Prevention*, vol. 43, no. 1, pp. 391–401.

Groth, K. M., and Swiler, L. P. (2012, January). Use of a SPAR-H Bayesian network for predicting human error probabilities with missing observations. In *Proceedings of the International Conference on Probabilistic Safety Assessment and Management (PSAM 11)* (Helsinki, Finland, June 25–29, 2012).

Hansen, F. D. (2006). Human error: A concept analysis. *Journal of Air Transportation*, vol. 11, no. 3, pp. 61.

Hidden, A. (1989), *Investigation of the Clapham Junction Railway Accident*, Department of Transport.

Hollnagel, E. (1993). *Human Reliability Analysis: Context and Control*. London: Academic Press.

Hollnagel, E. (1998). *Cognitive Reliability and Error Analysis Method—CREAM*. Amsterdam: Elsevier.

HSC (Health and Safety Commission) (2000). *The Southall Rail Accident Report*. Norwich: HSE Books.

HSC (2001a). *The Ladbroke Grove Rail Inquiry. Part 2 Report.* Norwich: HSE Books.

HSC (2001b). *The Southall and Ladbroke Grove Joint Inquiry into Train Protection Warning Systems.* Norwich: HSE Books.

HSE (Health and Safety Executive) (2009). *Review of Human Reliability Methods,* RR679 HSL. Norwich: HSE Books.

HSE (1999) *Reducing Error and Influencing Behaviour HS(G)48),* Second edition. Norwich: HSE Books.

King, S. (2010). Human failure types, *Petroleum Review.*

Kirwan, B. (1994). *A Guide to Practical Human Reliability Assessment.* London: Taylor & Francis.

Kletz, T. (2005). Preventing human error: "What could I do better?" *Safety & Reliability,* vol. 25, No. 2.

LeBot, P. (2004). Human reliability data, human error and accident models—Illustration through the Three Mile Island accident analysis. *Reliability Engineering and System Safety,* vol. 82, no. 2.

Reason, J. (1990). *Human Error.* Cambridge, UK: Cambridge University Press.

Reason, J. T. (1991). Too little and too late: A commentary on accident and incident reporting systems. In T. W. Van der Schaaf, D. A. Lucas, and A. R. Hale (Eds.), *Near Miss Reporting as a Safety Tool* (pp. 926). Oxford, UK: Butterworth–Heinemann.

Reason, J. T. (2008). *The Human Contribution: Unsafe Acts, Accidents and Heroic Recoveries.* Aldershot, UK: Ashgate.

Reer, B. (2004). The CESA method and its application in a plant-specific pilot study on errors of commission. *Reliability Engineering and System Safety,* vol. 82, No. 2.

RSSB (Rail Safety and Standards Board) (2008, June). *Understanding Human Factors—A Guide for the Railway Industry.* London: RSSB.

RSSB (2012, July). *Development of Human Error Quantification Technique for Use in the Railway Industry,* Research Brief T270.

Schwencke, D., Lindner, T., Milius, B., Arenius, M., Sträter, O., and Lemmer, K. (2012). *A New Method for Human Reliability Assessment in Railway Transport,* PSAM 11/ESREL 2012.

Senders, J. W., and Moray, N. P. (1991). *Human Error: Cause, Prediction, and Reduction.* Mahwah, NJ: Lawrence Erlbaum Associates.

Strater, O. (2000) *Evaluation of Human Reliability on the Basis of Operational Experience.* GRS-170. Koln: Gesellschaft fur Anlagen und Reaktorsicherheit.

Swain, A. D., and Guttmann, H. E. (1983). *Handbook of Human Reliability Analysis with Emphasis on Nuclear Power Plant Applications.* NUREG/CR-1278, Washington, DC: Nuclear Regulatory Commission.

Williams, J. C. (1985). HEART—A proposed method for achieving high reliability in process operation by means of human engineering technology, in *Proceedings of a Symposium on the Achievement of Reliability in Operating Plant, Safety and Reliability Society,* September 16, 1985.

Williams, J. C. (1986) HEART—A proposed method for assessing and reducing human error. In *9th Advances in Reliability Technology Symposium,* University of Bradford, Bradford.

32

Generic Approval Process for Public Transport Systems

Peter Wigger

CONTENTS

32.1 Introduction

In Europe, trams, light rail, and metro systems—in general, public transport systems—are handled rather differently when it comes to acceptance and approval. Different rules and regulations and requirements as well as different methods, roles, and responsibilities make a comparison of the approval processes of the various European Member States hardly possible.

Currently, there are no standardized procedures for acceptance, approval, and operation of public transport systems in Europe. There are no standardized safety requirements since each European Member State uses its own rules for the assessment of compliance with safety requirements. Even in Germany, one will find differences due to different federal responsibilities. The current public transport system applications more and more made use of the European standards EN 50126 ff [1].

Many representatives of the transport sector support the development of European standards in order to make systems and processes comparable and to enable mutual acceptance through a common binding requirement basis. Years ago, the European Commission presented a draft directive for the public transport sector modeled on the directives on the harmonization of the high-speed and conventional rail transport; but at the time, this draft was rejected. Critics of the directive argue that public transport systems would not operate across borders and that there is no need for standardization. Proponents of the directive—in particular, from the industry—ask the question why, for example, a tram system once approved in another European Member State almost has to go through the complete approval once again and request mutual acceptance (so-called cross acceptance) in order to save time and money.

The European Commission favors this harmonization approach and has initiated the MODSafe project for this purpose in the year 2008. This project analyzes the differences of the approval processes for public transport systems in Europe and drafted a proposal for a simplified and generic process. The results available since 2012 shall be considered by the standardization bodies for the current revision of rules and regulations. This chapter describes a generic process for the acceptance and approval of public transit systems on the basis of the MODSafe results [2].*

32.2 Definition of Terms

The *owner* of a public transport system is the owner of the system, for example, a city or a municipality, or a private organization.

The *operator* is the organization which carries out the operation of the public transport system with its own personnel and related regulations. Owner and operator can be identical; the owner can, however, also outsource the operation.

The *assessor* is a qualified person or organization with recognition of the approval authority for an independent evaluation of compliance with the safety requirements and any other requirements such as availability.

The *approval authority* is the state organization, which issues the legitimization/the permit for the operation/the approval of the public transport system and oversees the operation.

Acceptance refers to the declaration of the customer (for example, the owner or the operator of a public transport system) for contractual and functional acceptance according to the specific criteria (for example, acceptance of a modernization or a new system).

Safety acceptance means the acceptance of the proof of compliance with the safety requirements and the acceptance of the safety functions by an expert and/or by the approval authority. Also, the owner and/or operator can declare for a safety acceptance.

Approval is the legitimization to operate/the issued permit to operate/the operating license issued to the owner and/or to the operator by the responsible approval authority.

---

* MODSafe—Modular Urban Transport Safety and Security Analysis—is one of the latest projects in the European Transport sector under the Seventh Framework Programme (FP7) for Research and Technological Development of the European Union, performed from 2008 to 2012. The purpose of the MODSafe project was the development of a safety analysis and model reference for future urban guided transport projects. Even if the rail safety landscape in urban guided transport is highly diversified, the sector will benefit from some kind of harmonization. TÜV Rheinland was the project coordinator on behalf of the European Commission and furthermore responsible for the life cycle work package—in person of Peter Wigger, the author of the chapter at hand. The raw material for the chapter has been taken from the MODSafe deliverables from TÜV Rheinland in-house sources. Refer to http://www.modsafe.eu for details [2].

32.3 Methodology for the Definition of a Generic Approval Process

For the purpose of defining a generic approval process, an overview of the processes in Europe was created in MODSafe [2] first, and then the processes were analyzed for comparable generic activities to extract the similarities and differences, which ultimately form the basis for a generic approval process.

Looking at the public transport systems in Europe, it quickly becomes clear that on the one hand, Member States such as Germany and France have a long history of these systems, while other Member States only have one or two or even no public transport systems. For the preparation of an overview of the European approval processes, standardized questionnaires related to all Member States were generated first, with whose help the following key aspects were queried:

- Legal basis and regulations and required standards
- Processes, roles, and responsibilities
- Methods of risk assessment and specification of requirements
- Procedural and technical requirements
- Type of compliance evidence and required documentation
- Acceptance criteria, obligations, and liability
- Specific procedures and national peculiarities

For the analysis of the similarities and differences, selected approval processes were analyzed in more detail, including the processes of the countries of Germany, France, Great Britain, Hungary, Denmark, and Sweden, on the one hand, because of their history, and on the other hand, due to new trends. The selected approval processes were presented in the form of flowcharts, which have gradually been standardized, i.e., there was a distinction between the actors of the processes (*who*), the activities (*what*), the time of the activity (*when*), the manner of execution (*how*), and finally the basis therefore (*why*), so who is doing what, when, how, and why. There were not only many similarities with hints on generic elements, but also many differences and overlaps and gaps and thus difficulties of comparison. The activities (*what*) were the key focus of the further considerations, regardless of who performs them, as well as when and why and how this activity will be carried out.

32.4 Similarities of the Approval Processes in Europe

First, similarities in the hierarchy of legislation, regulations, and standards are striking. Figure 32.1 shows the so-called legislation pyramid, emanating from a law or a royal (Belgium) or state act (decree); specific degrees and regulations follow for the public transport systems and finally eventually standards, technical rules, and application guidelines. Also, project-related requirements of the owner or operator can be added.

In principle, these three hierarchical levels can be found in all European approval processes. In Germany, for example, the passenger transport act, the regulation for the construction and operation of tramways, and application-specific/project-related standards

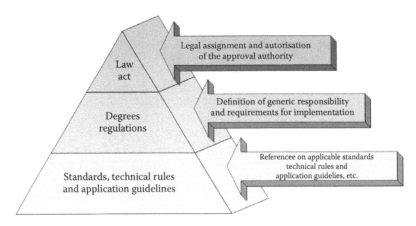

FIGURE 32.1
Legislation pyramid.

and regulations apply, often including rules and regulations of the association of German transport companies such as the technical rule concerning the approval of signaling and train control systems and increasingly the CENELEC railway application standards EN 50126 ff [1] and (increasingly) more technical European and (decreasingly) less national standards.

Common to all approval processes is the state legitimization to issues a permit to operate. Mostly this is issued by the state-assigned approval authority; occasionally, the legitimization is also awarded by the owner of the public transport system. In the degrees and regulations as well as in the standards and rules, requirements for the owner and the operator can be found in great difference (the owner may be the operator or he/she can outsource the operation to an own organization) as well as for the system supplier/suppliers of subsystems (such as vehicles, train control, etc.). Independent assessors are, if at all, required in standards and regulations.

In general, the process participants and their responsibility and legitimization (*who*, *what*, and *why*) can be identified from there, despite differences in detail in scope and way of conducting and the time of activities.

One more aspect in common is the life cycle of the public transport system. While historically, each new system or any extension or modernization of a public transport system naturally such as any technical construction or procurement project, followed an engineering process (for example, concept development, specification of technical requirements, development, installation, test, and commissioning), the life cycle concept of the CENELEC railway application standard EN 50126 [1] is increasingly applied.

This standard was originally developed for railway applications, but it increasingly finds application in public transport projects as well. Although these life cycles are not new in principle, the phased approach with activities and prove in each phase brings a clear structure and therewith a clear advantage to introduce and implement standardization and comparability/cross acceptance in public transport.

Common to all approval processes is the step-by-step approach to issuing the operating license, however, in strongly varying degrees regarding what type and scope activity, the number of project phases, and the time of activities of the process participants. Naturally, the owner of the public transport system stands at the beginning with the idea of his/her project, while the approval authority with their permit to operate stand at the end.

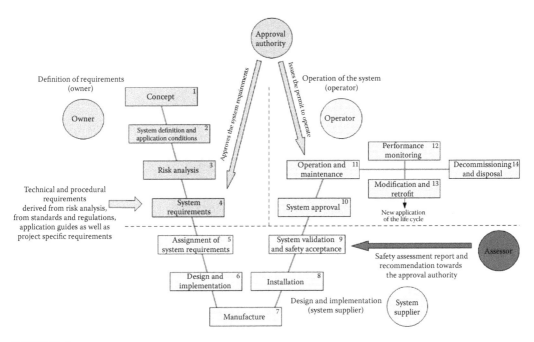

FIGURE 32.2
Life cycle.

In general, the process participants and their principle activities can be identified thereof (*who*, *what*, *when*, and *how*), despite differences in detail in scope, level of deepness, way of conducting, and the time of activities.

On the way to the definition of a generic approval process, the life cycle (see Figure 32.2), with its 14 stages, applies, and the generic aspects common to all approval processes are assigned to the respective life cycle phases. The owner of the system bears the main responsibility for the phases 1–4; the system supplier, for the phases 5–9; and the operator, for phases 10–14.

The assessor is working at the same time along the progress of the life cycle, i.e., the results of each phase are presented to the assessor after completion of that phase (and not only at the end of the project for the safety acceptance).

The approval authority is ideally integrated also at an early stage; in addition to the final permit to operate, an early approval of the system requirements makes sense as well.

32.5 Elementary Activity Modules

After the analysis of similarities and differences of selected approval processes in Europe as well as on the basis of a uniform definition of acceptance and approval and an identification of the major participants in these processes, generic activities, the so-called elementary activity modules, can be defined now. The approval processes were depicted in a comparable way as functional flowchart as the basis for determining these elementary activity modules.

Elementary activity modules are generic, comparable activities in the different approval processes, which are however performed with a different degree of independence and possibly with a different level of detail and by different participants at different times in the life cycle. The elementary activity modules were hierarchically sorted as follows:

- Overall system
  - Specification of system requirements
  - Assessment of system requirements
  - Evidence of compliance of system requirements
  - Assessment of the evidence of compliance of system requirements
  - Acceptance and approval
- Function
  - Specification of functional requirements
  - Assessment of functional requirements
  - Evidence of compliance of functional requirements
  - Assessment of the evidence of compliance of functional requirements
- Safety
  - Specification of safety requirements
  - Assessment of safety requirements
  - Evidence of compliance of safety requirements
  - Assessment of the evidence of compliance of safety requirements

The elementary activity modules of a generic approval process are now arranged according to the life cycle phases and the preceding hierarchy as shown in Figure 32.3. This is based on the assumption that a specification of a transport system takes place top–down, i.e., starting from the concept and definition of the system (life cycle phases 1 and 2) and a risk analysis (life cycle stage 3); the specification of the requirements follows in form of system requirements, functional requirements, and safety requirements (life cycle stage 4).

On the next level, the breakdown to technical subsystems of the public transport system follows such as rolling stock, guideway, train control system, and power supply with an underlying life cycle for each subsystem and a similar distinction in subsystem requirements, functional requirements for the subsystem, and safety requirements for the subsystem. A further refinement is, of course, possible. In general, the system, functional, and safety requirements for the overall system (similar to follow for each subsystem and all further underlying components) are assessed (assessment of the requirements) regarding the criteria validity, correctness, completeness, appropriateness, etc.

After the development and installation of the system/of the subsystems (life cycle phases 6–8), the proof of fulfillment of the requirements follows with respect to system, functional, and safety requirements. The evidence can be provided analytically or by testing. Evidence/tests must be carried out in a bottom–up manner, i.e., initially for the underlying components (in the factory) and next for the various subsystems (initially in the factory, later on site); and, finally, the evidence/tests follow for the integrated overall system/public transport system.

In general, the evidence of compliance with the system-functional, and safety requirements for the overall system (in analogue before for the underlying components and every

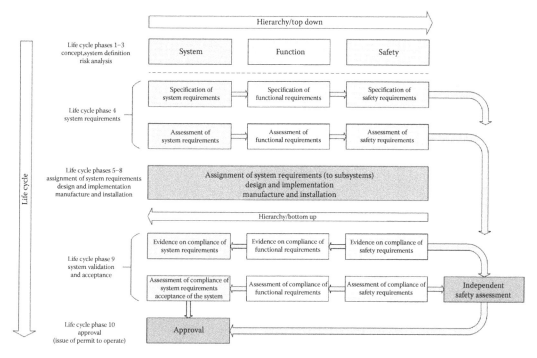

FIGURE 32.3
System of elementary activity modules.

further subsystem) is assessed (assessment of evidence of compliance) regarding the criteria correctness, completeness, fulfillment of the specification, application conditions, etc.

In addition, as a further elementary activity, one module finds the often used independent safety assessment as well as the final formal approval act (the issuance of the permit to operate).

32.6 Generic Approval Process

The main participants of an approval process are as follows:

- The operator of the public transport systems
- The system supplier (of the system or a subsystem)
- The independent assessor/the independent assessment organization (optional)
- The approval authority
- (Optional) Consultants and project management organizations

The linking of the elementary activity modules to the participants of the approval process is shown in Figure 32.4 as a generic approval process. The sequence is derived from the timing throughout the life cycle of the system.

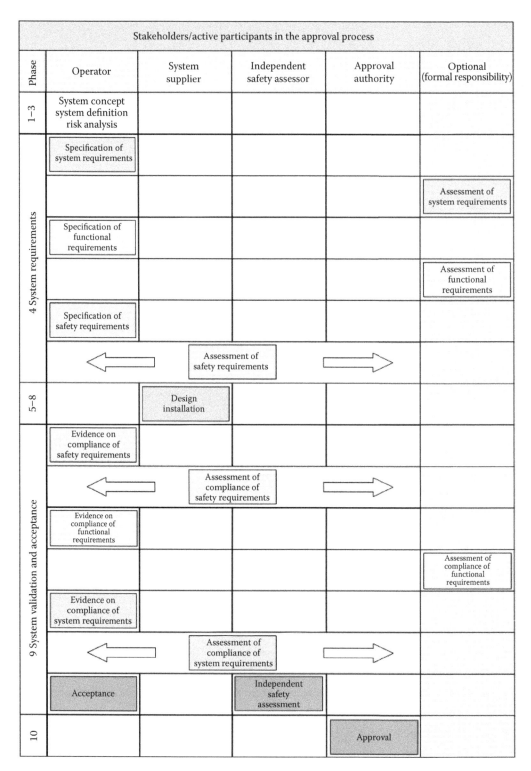

FIGURE 32.4
Generic approval process.

The elementary activity modules are assigned to the different participants of the approval process. The responsibility of a participant for the performance of the elementary activity modules may be distinguished as follows:

- Legal responsibility (as a result of a law or a degree)
- Responsibility out of a contract (for example, between operator and supplier)
- Responsibility due to generally accepted rules of technology

Beyond the performance by the independent safety assessor, the elementary activity modules "assessment of safety requirements" and "assessment of evidence of compliance of safety requirements" are also performed by the supplier in the context with his/her internal test and validation process. In addition, these activities can at least partly be performed by the operator (often with focus on operational aspects during the system trial run) as well as by the approval authority (in case the authority does not solely rely on the independent safety assessor but performs own evaluations with own expertise). This constellation is shown in Figure 32.4 with arrows to the right and left. The same applies for the "assessment of the evidence of compliance of system requirements," here often with focus on the operator on performance criteria for the overall public transport system.

32.7 Cross Acceptance: Experience from Praxis

Praxis shows that the approval process for public transport systems is still not uniformly treated. Cross acceptance offers a solution of assisting and fastening the approval process by the reuse of available equipment/products/subsystems already preapproved/precertified as generic products or generic applications by previous assessor and/or safety authority approvals. Figure 32.5 (ref MODSafe [2]) shows the principle process of the application of cross acceptance.

The decision process should be applied in the early life cycle approach stages on public transport system level and/or at the beginning of the design stage in case of applying for subsystems such as reuse of preapproved signaling and train control equipment. The acceptance criteria for cross acceptance can be derived from CLC/TR 50506-1:2007 [3].

32.8 Summary and Outlook

This chapter describes a generic process for the acceptance and approval of public transport systems on the basis of the MODSafe results [2]. Starting from generic activities (so-called elementary activity modules) of the involved participants, roles, and responsibilities, a process has been developed consisting of generic activities, which can be a project specifically adapted as needed. This process furthermore forms the basis for the acceptance (cross acceptance) of already issued (part) approvals of public transport systems and its subsystems and components.

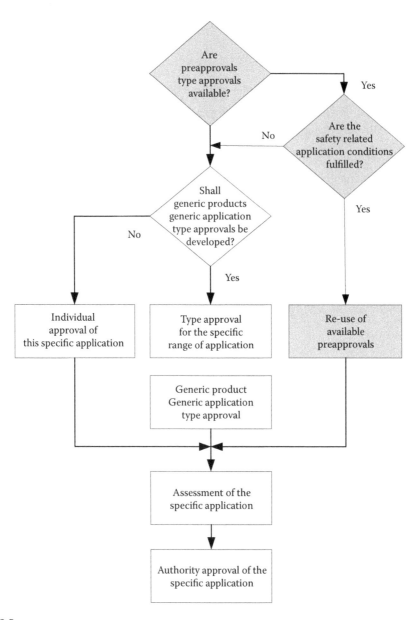

FIGURE 32.5
Cross acceptance.

By using the elementary activity modules, a common language is established, which will ease the understanding and comparison of different processes and increase the application of the envisaged cross acceptance. The established generic process for the acceptance and approval of public transport systems is recommended as the base and allows for project-specific adoptions, be it for simplification of less complex processes for trams or be it for addition of further aspects for more complex processes such as for driverless underground metro systems. This approach can be further simplified with the application of cross acceptance.

References

1. CENELEC (Comité Européen de Normalisation Électrotechnique). EN 50126: *Railway Applications—Specification and Demonstration of Reliability, Availability, Maintainability and Safety (RAMS)*. CENELEC, Brussels.
2. TÜV Rheinland. http://www.modsafe.eu.
3. CLC/TR 50506-1:2007: Railway Applications—Communication, Signalling and Processing systems—Application Guide for EN 50129—Part 1: Cross-Acceptance. CENELEC, Brussels.

33

Importance of Safety Culture for RAMS Management

Malcolm Terry Guy Harris

CONTENTS

33.1 Safety Culture Overview

The modernist approach toward changing health and safety (H&S) systems of organizations is focused on improving the core values, beliefs, and attitudes of the company and to develop positive group behavior between employees toward H&S aspects. Generally, companies refer to a "safety culture" when describing the behavior of employees to comply with rules or acts of safety. However, the concept of safety culture and the interaction with management styles have been more significant in historical cases, where management have applied a natural, unconscious bias leaning toward production over safety or may have had a tendency to focus on short-term targets or exhibit a highly reactive behavioral pattern.

There have been several major historical accidents in various industries that have been associated with poor safety culture. These have been widely reported within the oil and gas, nuclear, and space exploration industries. Table 33.1 lists some of the more significant recent disasters by industry associated with poor safety culture. Table 33.2 shows a list

TABLE 33.1

Major Accidents by Industry

Industry	Accident	Date	No. of Fatalities
Nuclear	Chernobyl disaster, Ukraine	April 26,1986	31[a]
Oil and gas	Piper Alpha disaster, UK	July 6, 1988	167[b]
Space exploration	Space Shuttle Challenger, US	January 28, 1986	7[c]
Space exploration	Space Shuttle Columbia, US	February 1, 2003	7[d]

[a] The reported death toll as an immediate result of the accident included staff and emergency workers (IAEA 2005).
[b] The report *Public Inquiry into the Piper Alpha Disaster* (short: Cullen Report) was released in November 1990. A second part of the report made 106 recommendations for changes to North Sea safety procedures and culture within the Industry (Cullen 1990).
[c] The Rogers Commission, which investigated the Space Shuttle Challenger disaster at the Kennedy Space Center in Florida on January 28, 1986, made several recommendations that addressed organizational, communications, and safety oversight issues and also determined the following:
  • National Aeronautics and Space Administration (NASA) solid rocket booster experts had expressed safety concerns about the Challenger launch.
  • NASA's culture prevented these concerns from reaching the top decision makers.
  • An environment of overconfidence was exhibited within NASA due to past successes.
  • Project pressures to maintain the launch schedules may have prompted flawed decision-making.
[d] The Columbia Accident Investigation Board (CIBA), which investigated the destruction of the Space Shuttle Columbia upon reentry on February 1, 2003, made 29 recommendations to NASA to improve the safety of future shuttle flight. CIBA also determined the following:
  • NASA had not learned from the lessons of the Challenger launch.
  • Communications problems still existed within the NASA project teams.
  • Abnormal events were not reviewed in sufficient detail.
  • Hazard and risk assessment that were performed were still insufficient in detail.
  • Project schedule still dominated over safety concerns.

TABLE 33.2

Fatal Accidents in the UK Railway Industry

Industry	Accident	Date	No. of Fatalities	No. of Injuries
UK Railway	Clapham Junction	December 12, 1988	35	415
	Southall	September 19, 1997	7	139
	Hatfield	October 17, 2000	4	70
	Ladbroke Grove	October 5, 2001	31	520

of the major UK rail disasters that have occurred over the last 30 years where poor safety culture has been a contributor.

In the UK railway industry, the earliest reference to safety culture and human factors interaction was from the findings reported in the railway accident inquiries. Catastrophic rail accidents in the United Kingdom, such as at Clapham Junction (Hidden, 1989), Southall (HSC, 2000, 2001b), Hatfield (ORR, 2006), and Ladbroke Grove (HSC, 2001a, 2001b) all highlighted the critical role both human factors and safety culture play in the causation paths of an accident (Whittingham, 2004).

This chapter considers the influence of safety culture on rail organizations and management. It draws on previous research work and ideas in the field of safety culture.

33.1.1 Historical Train Failures

A comparison of major rail accident statistics in the United Kingdom and the European Union (EU) rail industries is given in Figure 33.1. The figure shows the comparative incidents

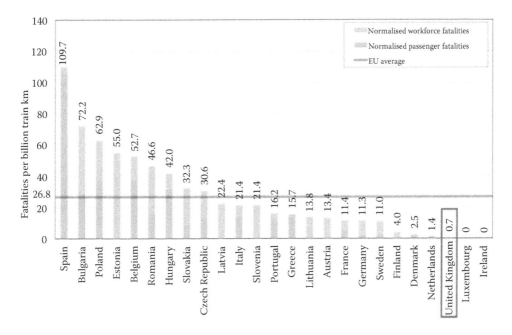

FIGURE 33.1
Comparison of passenger and workforce fatality rates across the European Union railway industry, 2010–2014. (From RSSB, *Annual Safety Performance Report, 2015–2016*, RSSB, London, 2016.)

of fatalities between 2010 and 2014 based on the Rail Safety and Standards Board (RSSB) datasets (RSSB, 2016). Figure 33.1 shows a general downward trend for both the UK and EU rail industries compared with data from passenger and workforce fatality rates across the EU railways from 2007 to 2011 as reported in the 2013 H&S report (ORR, 2013). However, these statistics can be significantly altered by the existence of a single major accident.

It has been widely recognized that a significant proportion of historical rail accidents and near misses have been attributed to poor safety culture.

33.2 Safety Culture and Organizational Maturity

The concept of a safety culture within an organization focuses on the organization's core values, beliefs, attitudes, and group behavior toward developing HS&E aspects.

A substantial amount of literature exists on the safety culture and safety climate of an organization, which has been derived mainly from researchers in various high-risk industries. These industries have historically experienced high accident rates and have been the subject of reviews, such as the nuclear (INSAG, 1991), chemical (Lees, 1980), and offshore oil and gas (Cullen, 1990).

33.2.1 Safety Culture versus Safety Climate: Differences

While several similarities exist between safety culture and safety climate for an organization, the term *safety culture* is more encompassing while safety climate tends to reflect a

TABLE 33.3

Definitions in Safety Culture and Safety Climate

Reference	Definition
International Atomic Energy Agency (1991)	Safety culture is the assembly of characteristics and attitudes in organizations and individuals which establishes that as an overriding priority, nuclear plant safety issues receive the attention warranted by their significance.
ACSNI (1993)	"The safety culture of an organization is the product of individual and group values, attitudes, perceptions, competencies, and patterns of behaviour that determine the commitment to, and the style and proficiency of, an organisation's health and safety management."
Pidgeon (1992)	Safety cultures reflect the attitudes, beliefs, perceptions, and values that employees share in relation to safety (safety culture).
Cullen (1990)	Safety culture is a corporate atmosphere or culture in which safety is understood to be, and is accepted as, the number one priority
Turner et al. (1989)	Safety culture is the set of beliefs, norms attitudes, roles, and social and technical practices that are concerned with minimizing the exposure of employees, managers, customer, and members of the public to conditions considered dangerous or injurious.
Williamson et al. (1997)	Safety climate is a summary concept describing the safety ethics in an organization or a workplace. This is reflected in employees' beliefs on safety and is believed to predict the way employees behave with respect to safety in that workplace.

temporal position within the organization. Cox and Flins' (1998) view is that safety culture is seen as an indicator of the culture of the organization as perceived by the employees at a specific point in time.

Table 33.3 shows a list of typical definitions used to describe safety culture and safety climate from various industries.

Companies also possess an organization culture, which represents collective values, beliefs, and principals held by the employees. This type of culture is a product of several factors such as the history of the company, market, type of employees, and the style of management. Handy's (1999) review of Harrison's work on culture and associated links to organizational structures within companies identified four specific cultures. His work on role, power, task, and person culture identified that these have specific organizational structures and systems appropriate to the organization culture. Handy identified four types of activities, which characterize each part of the organization and are shown in Handy's organizational diversity model (Figure 33.2).

Schein's work on organizational culture defines three levels, by which the cultural phenomenon is visible to observe:

1. Visible artifacts
2. Values, beliefs, behavioral norms, and operational rules
3. Underlying assumptions

Organizational culture has been defined by several researchers of corporate culture, including Hofstede and Hofstede (2005), Schein (1990), and Handy (1999) who have all provided comprehensive definitions.

These researchers all widely acknowledge the importance of organization culture and specifically the subset of safety culture as being critical to the success or failure of an organization.

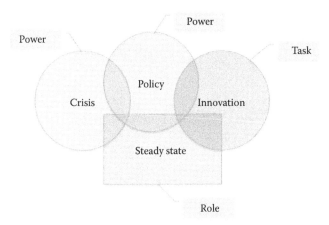

FIGURE 33.2
Organizational diversity model. (Based on C. Handy, *Understanding Organizations*, 4th Edition, Penguin Books, London, 1999.)

33.3 Safety Culture Maturity

In most industries, such as the oil and gas, energy, and software, the concept of capability maturity models has been developed and adopted to facilitate the safety culture process. The primary aim of these capability maturity models is to assist organizations in understanding the level of maturity of their safety culture. In this way, organizations can plan core activities to develop a "positive" safety culture that will be used to improve safety performance.

The safety culture maturity model concept relies on the fact that safety processes are formed from a set of activities, methods, practices, and transformations that employees can use to develop and maintain safety culture. As an organization matures, the safety culture process becomes more defined and better utilized within the organization as employees implement the tools and drive the process more consistently. Research conducted by the Institution of Occupational Safety and Health (IOSH) (2004) on organizations led to the published IOSH guide *Promoting a Positive Culture—Health and Safety Culture*.

The guide proposes that an action plan is required and developed by companies to focus on the following:

1. Organizational changes within the workplace
2. Training
3. Behavioral safety
4. A survey carried out to assist organizations target resources effectively.

The IOSH guidance recommends that a reporting culture needs to be initially created to identify all incidents, near misses, and concerns and learn from them, through the use of a safety climate survey. A review of behavior modification is then followed which is likened to the culture of the organization. The paper suggests that a maturity model for culture can be applied to companies to assist the selection and implementation of a desirable behavior modification of a workforce that is appropriate to that organization. The safety culture maturity model is illustrated in Figure 33.3.

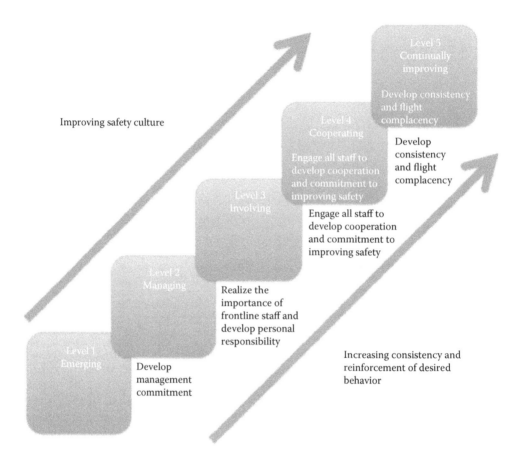

FIGURE 33.3
Safety culture maturity model. (Based on IOSH, *Promoting a Positive Culture—A Guide to Health and Safety Culture*, IOSH, Wigston, UK, 2004.)

The key features of the five mature levels are shown in Table 33.4, which starts from a bottom-up direction and has been adapted for the rail sector. These features supporting the maturity model can be managed through the organization's safety management system (SMS).

Health, Safety and Environment (HSE) work under the Hearts and Minds program that was formerly developed in the petrochemical industry led by Shell has been applied to other industries (Hudson, 2007). The program uses nine organizational aspects to facilitate the safety improvement process.

The process encourages organizations to facilitate group discussions and engage all members of the organization through the change process.

The Hearts and Minds tools have been applied to the oil and gas industry and evidence has shown that the application of these tools are effective, particularly when used in the context of a sound HSE management system. This system requires cooperation by employees with attention to personal responsibilities and consequences to become effective (Holstvoogd et al., 2006).

The RSSB recently published a guide/strategy position document "Leading Health and Safety on Britain's Railway." This document illustrates the maturity of collaboration across 12 risk areas to reduce H&S risks (RSSB, 2016). This work looks at a systemwide view of the UK rail industry and provides an approach for a collaborative effort to reduce H&S risk.

TABLE 33.4

Definitions in Safety Culture and Safety Climate

Maturity Level	Key Features
Level 1: emerging	• Safety is not perceived as a key business risk. • Safety department is perceived as primarily responsible for safety. • Most accidents are unavoidable.
Level 2: managing	• Safety is perceived as a key business risk requiring management, time, and resource to prevent accidents. • Safety is mainly focused on adherence to rules, procedures, and engineering controls. • Accidents are perceived as preventable. • Accident rates are approximately near to the industry sector average, but more serious accidents tend to occur. • Incentives are focused on loss time incidents.
Level 3: involving	• Accident rates are low compared to the industry sector average. • Management recognize the accident sequence indicating that a wide range of factors can lead to accidents and that root causes are likely to be linked to management decisions.
Level 4: cooperating	• Majority of employees believe safety is important. • Employees feel valued and treated fairly and are recognized within the organization. • The organization takes proactive measures to reduce accidents within the workplace.
Level 5: continually improving	• The organization experiences a sustained period (years) without a reported accident or high incident, without feeling complacent. • The organization applies a variety of safety performance indicators to monitor safety but is not driven by performance; there is confidence in the safety process. • The organization invests significant effort toward promoting the H&S of employees.

33.4 Conflict in Safety Culture

Handy's (1999) work on organizations identified the following symptoms of organizational conflict, which can impact a safety culture:

- Poor communication both vertically and laterally within the organizational structure
- Intergroup hostility
- Interpersonal friction between individuals
- Escalation of arbitration
- Proliferation of rules and regulations
- Low morale of the group expressed by frustration at efficiency

Other areas of conflict involve the universal conflict between commercial pressures and safety, which is well documented within the rail industry.

While the causes of conflict can lead to frustrating arguments or escalating competition, which are considered as intermediate causes, the underlying cause can be categorized into two issues:

- Goals and ideologies
- Territory

A strategy to manage conflict would involve applying the following:

- Turning the conflict into fruitful competition or provide an open purposeful argument between groups
- Apply a control to the conflict through arbitration or through rules and procedures, for example

Hofstede's (2005) work on organizational culture describes key differences between feminine societies, masculine societies, education, and consumer behavior. He highlights how the masculinity–femininity dimension affects the handling of industrial conflicts within the workplace. It suggests that masculine cultures exhibit a tendency to resolve conflicts through aggression with a "let the best man win" approach while a feminine culture will tend to manage conflict by compromise and negotiation. These will have different outcomes within an organization in terms of risk and reward and will need to be managed.

33.5 Organizational and Safety Indicators

The measurement of safety performance has traditionally been based on safety outcomes, such as lag indicators. Measurement is important within organizations as it is regarded to be an accepted part of the "plan–do–check–act" management process activity and forms a critical function for a H&S management system.

Generally, there has been increasing concern in the use of lag indicators that have resulted in the movement toward leading indicators for safety (Flin, 1998; Flin et al., 2000). In contrast, the use of lead indicators is associated with proactive activities, which assist toward managing risk (Grabowski et al., 2007).

Table 33.5 shows selected examples of leading and lagging safety performance indicators for the railway industry. Table 33.6 illustrates selected examples of leading and lagging safety culture indicators in the railway industry.

33.6 Requirements for a Positive Safety Culture for RAMS Projects

One principal conclusion that was drawn from three separate public rail inquiries took a fundamental approach on the generic issues impacting safety in the UK railway industry (HSC, 2000, 2001a,b). The inquiry reports lead to 295 recommendations that were setting a "necessary and challenging criteria to change the state of the railways" (HSC, 2005).

Under the Railways and Other Guided Transport (Safety) Regulations (2006), a key requirement for duty holders is to develop and maintain an SMS, hold a valid safety certificate or safety authorization, and cooperate and collaborate with other relevant parties to ensure the safety of the railway.

The introduction of the common safety methods places additional requirements for the duty holder in terms of safety levels, achievements of safety targets, and compliance with other safety requirements under EU law. This requires duty holders to risk evaluate and

TABLE 33.5

Safety Performance Indicators in the Railway Industry

Control	Measure	Key Performance Indicator	
		Leading Indicator	Lagging Indicator
Track inspection/ maintenance	Work planning/ cost	• Percentage of preplanned preventative maintenance program activities completed within a defined timescale	• Maintenance cost
	Failures		• Number of unexpected broken rails or irregular buckled track
Rolling stock inspection/ maintenance	Work identification/ downtime	• Percentage of maintenance work orders with human-hour estimates within 10% of actual over the specified time period • Percentage of maintenance work orders closed within 3 days, over the specified time period	• Number of scheduled maintenance related downtime
Signaling infrastructure maintenance	Failures	• Number of potential problems identified before a broken rail occurs (in terms of mean time between failures) through ultrasonic testing	• Number of wrong-sided equipment failures
Operational procedures	Work scheduling	• Number of pretrain daily inspections which identified the handbrake positioned in an energized (on) state and was then rectified	• Number of hot axles damaged • Number of times a hot axle changeout does not occur due to unclear or incorrect operational instruction

assess their activities and to cooperate to assess and control risk at shared interfaces with other parties. Furthermore, any material changes to the system will require the risk associated with the changes to be assessed to ensure that the safety levels are maintained or exceeded.

A more recent development in the UK legislation has been the emergence of Corporate Manslaughter and Corporate Homicide Act 2007. This legislation, which came into force in 2008, is unique in that it is the first time that companies and corporations can be found liable of corporate manslaughter if serious failures have resulted in a breach of a duty of care. This placed a greater importance on safety culture and the interaction with human factors for railway organizations within the legal compliance domain.

A key requirement under the regulation is that companies and organizations should keep their H&S management systems under review, in particular, the way in which their activities are managed and organized by senior management.

In Europe, the principal focus on Health and Safety law is based on insurance as opposed to authority regulation, which leads to a system of defined compensation and rehabilitation programs as opposed to legal enforcement, prosecution, and civil claims for compensation.

TABLE 33.6

Safety Culture Indicators in the Railway Industry

		Key Performance Indicator	
Control	**Measure**	**Leading Indicator**	**Lagging Indicator**
Track inspection/ rolling stock maintenance/	Leadership	• Percentage of preplanned preventative maintenance program activities completed within a defined timescale • Percentage of preplanned leadership program completed by supervisors per quarter	• Maintenance cost
	Communication	• Percentage of preplanned safety briefings completed a in a defined timescale	• Number of unexpected broken rails or irregular buckled track
	Involvement of staff	• Percentage of observation and feedback programs completed by workers, supervisors, and managers within a defined timescale	• Cost of observation and feedback programs completed by workers and supervisors
	Learning culture	• Percentage of recommendations from incidents implemented and communication to the employees within a defined timescale	
	Prevailing culture	• Percentage of failures investigated and followed up by the safety team and supplier A within a defined timescale	• Number of failures due to supplier A

The following key aspects are considered to be typical indicators of a positive safety culture:

- Management commitment through active monitoring.
- Effective communications between all levels of employees.
- Visible management: Participation by managers on H&S committee meetings or through conducting safety tours. In the United Kingdom, Lord Cullen (2001) suggested that rail companies should decide the amount of time their leaders spent in the field. However, best practice indicated that senior execs should spend 1 hour per week; middle managers, 1 hour per day, and first line managers should spend at least 30% of their time in the field.
- Mangers getting actively involved in reactive monitoring: This includes investigating accidents that resulted in injury or near miss occurrences or have resulted in an occupational ill health case.

The application of a positive safety culture environment for the reliability, availability, and maintainability (RAM) process requires a framework within the RAM organization with the following key characteristics:

- A strong senior management commitment to safety
- An organization design that applies a feedback system for continuous organizational learning

- The use of a flexible and realistic handling mechanism for addressing unsafe practices throughout the business
- Consideration for hazards associated with the organization's activities to be shared by the workforce

The following sections describe safety culture methods and tools that organizations can use and illustrates by an example of how this process can be applied to a rail group.

33.7 Safety Culture Methods and Tools

There are a variety of methods and tools that address specific aspects of safety culture. These methods and tools have been based on current best practices and research performed by HSE (2005).

A review of safety culture tools and methods used in the rail industry was conducted by the Keil Centre for RSSB (2003). Table 33.7 lists the main tools and methods used to address safety culture and safety climate within rail organizations.

TABLE 33.7

Safety Climate, Safety Culture Tools in the Railway Industry

Method/Tool	Description	Climate or Culture Tool	Reference
Checklist	Safety Culture Checklist[a]	Culture/climate	Reason (2008)
Questionnaire/survey	H&S Climate Survey Tool (CST)[b]	Climate	HSE (1999)
Questionnaire	The Keil Centre Safety Culture Maturity® Model[c]	Culture	
Questionnaire	Occupational Psychology Centre Safety Culture Questionnaire (SAFECQ)	Culture	HSE (2005)
Survey	Loughborough University Safety Climate Measurement Toolkit	Climate, culture	HSE (2005)
	British Safety Council Safety Standard and Journey Guide	Culture	
	University of St. Andrews Safety Culture Tool	Culture	HSE (2005)
Questionnaire	Aberdeen University OSQ99	Climate	HSE (2005)
Survey	Marsh Fleet Safety Culture Survey	Climate, culture	
Questionnaire	Quest Safety Climate Questionnaire	Climate	HSE (2005)
Questionnaire/survey	Hearts and Minds Toolkit	Culture	Holstvoogd et al. (2006)
Questionnaire/survey	Rail Safety and Standards Board (RSSB) Safety Culture Tool I[d]	Culture	RSSB (2012a,b), HSE (2005),
Questionnaire	SafeCulture	Culture	UIC (2004)

[a] The Railway Safety Management Systems Guide – Transports Canada 11/2010. Safety Culture Checklist – Transports Canada 11/2010 (http://www.tc.gc.ca/media/documents/railsafety/sms_checklist.pdf). The checklist provides an evaluation of safety culture/climate and uses safety culture assessment toolkits.
[b] The CST is based on HSE, HSG65 (HSC 1997) and HSG48 (HSE, 1999).
[c] The Keil Centre Safety Culture Maturity Model.
[d] The RSSB Safety Culture Toolkit provides rail organizations with a self-assessment safety culture package, guidance for members for improvement and an option to share good practices on all aspects of safety culture (HSE, 2005).

33.8 Integration of Safety Culture into RAMS Management

Organizations intending to develop or improve their safety culture will need to follow a systematic approach toward understanding safety culture and the factors that influence human performance with the goal of improving the safety indicator for the business.

This safety culture performance indicator would focus on several manageable key safety themes associated with maintenance activities that have the potential of minimizing human error and providing a positive safety management approach via an SMS.

The following key elements of the safety culture process would need to be applied:

- Identify the critical needs and issues of the organization, for instance, poor management of change, complacency of risks, and poor internal communication/flow of information/involvement of staff.

- Define the safety culture model including the key drivers and tools for achieving a successful outcome.

- Assess the maturity of the company through site surveys, interviews with employees, questionnaires; see Section 33.7 for safety culture methods and tools.

- Develop action plans.

- Implement actions.

- Measure the safety performance to assess any gaps requiring improvement. The use of the SMS of the company should be applied to monitor the process.

Figures 33.4 through 33.7 illustrate an example of a behavior framework model for developing a positive safety behavior across the organization toward a risk management element of the safety culture based on the work of Hayes et al. (2008). Table 33.8 presents a detailed overview of the safety culture framework model.

This example applies to all employees, supervisors, and managers within the organization and shows how the key themes, derived from a safety culture assessment, can be developed as a minimum set of safety rules.

FIGURE 33.4
Overview of the safety culture framework model. (Based on Hayes et al., 2008.)

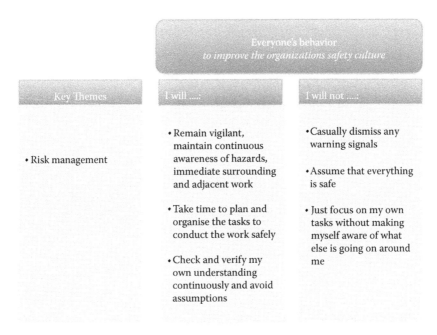

FIGURE 33.5
Safety culture framework model: employee's risk management example. (Based on Hayes et al., 2008.)

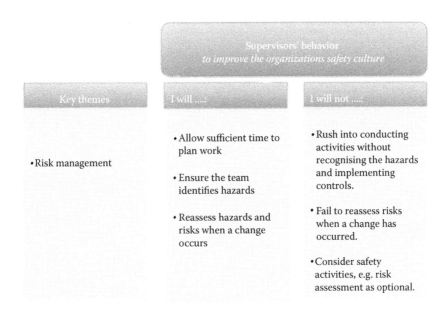

FIGURE 33.6
Safety culture framework model: supervisors' risk management example. (Based on Hayes et al., 2008.)

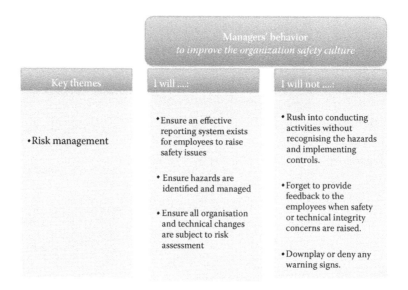

FIGURE 33.7
Safety culture framework model: manager's risk management example. (Based on Hayes et al., 2008.)

TABLE 33.8

Detailed Overview of the Safety Culture Framework Model for a RAM Department

Theme	All Employees	Supervisors	Managers
Company standards	Follow safety rules	Ensure product compliance	Set high standards
		• Perform regular site visits	• Perform regular management site visits
Company communication	Speak openly	Team encouragement	Open communication
	• Safety meetings • Observation and feedback program	• Observation and feedback program	• Observation and feedback program
Risk management	Risk awareness	Promote risk awareness	Challenge risk
	• Hazard reporting program	• Incident investigation training and toolkit • Hazard reporting program	• Incident investigation training and toolkit • Regular incident review meeting
Involvement	Get involved	Involve the team	Involve the workforce
	• Observation and feedback program		
All themes	• Office building induction • Site specific induction • Performance/ development plans	• Personal safety development plan	• Personal safety development plan

These rules could form part of the wider general management practices for a maintenance business unit, for instance. These practices could then be integrated within the rail organizations SMS, which will address the four behavioral themes across three typical occupational groups.

33.9 Safety Culture Research

Research into safety culture in the rail industry can extend within the following potential areas of future research:

- Examining blame culture across national cultures on rail safety
- Industrywide safety climate and safety culture measurement
- Use of alternative evaluation techniques

33.9.1 Examining Blame Culture across National Cultures on Rail Safety

The findings from the major accidents from various industries including the UK rail industry have indicated that a blame culture has been one of the main contributors to poor safety culture. Whittingham (2004) describes these organizations as those that assign blame to the individual for the error rather than the defective system, organization, or management process.

This research should look at the impact from a blame culture on safety behavior, specifically where employees are actively engaged in safety reporting and decision-making. A wider research into blame culture across national cultures could be examined to understand the influences on organizations within the rail industry.

33.9.2 Industrywide Safety Climate and Safety Culture Measurement

The numbers of recommendations made from UK rail accident inquiries have identified deficiencies due to poor safety culture to be a persistent problem in the rail industry. The need to provide a nationwide evaluation of safety culture/safety climate measurement on an industrywide basis is an area for future consideration. A similar recommendation by Glendon (2008) was concluded for the Australian rail industry. The limitations in the use of these tools, in terms of information to provide an understanding of the organizations safety culture, needs to be considered when developing a safety culture position. A balanced approach that gathers information on both perceptions, attitudes, and behaviors with a focus on social subsystems and from retrospective accident investigations/analysis reports needs to be put into place. This will provide a broader and more comprehensive understanding of the situational issues and provide a means to validate the safety climate/culture an organization at several levels. This should be further examined to reinforce information and findings from studies that identified problems using accidents as a measurement variable due to the rarity of the incident.

33.9.3 Evaluation of Alternative Techniques for Studying Complex Interrelationships

Historical studies on safety climate/safety culture have focused on quantitative evaluation of the effects of safety climate/safety culture on the rail safety outcomes often using a single quantitative method.

Future experimentation should explore the application of alternative validation methods, such as the triangulation method, also referred to as the mixed method, to provide an effective technique for evaluating several research methods to study safety culture/safety climate.

Hofstede (2001) applied the triangulation method in organizational culture research.

Glendon (2008) used the triangulation method as a complimentary technique in safety climate research.

Harris (2011) used the triangulation method in his work on the safety culture of small and medium enterprise (SME) organizations operating in the UK oil and gas industry.

This approach provides a means for solving complex decision-type problems or validating mixed research methods where a degree of uncertainty exists. The approach should be further explored to study the causal relationships within safety culture for complex railway systems. The relationships can then be identified and validated using a broader mixed-methods approach.

References

ACSNI (Advisory Committee on the Safety of Nuclear Installations) (1993). *Study Group on Human Factors*, Third Report: Organising for Safety. London: Her Majesty's Stationery Office.

Cox, S., and Flin, R. (1998). Safety culture—Special issue. *Work & Stress*, 12, pp. 187–306.

Cullen, W. D. (1990). *The Public Inquiry into the Piper Alpha Disaster*. London: Her Majesty's Stationery Office.

Cullen, W. D. (2001). *The Ladbroke Grove Rail Inquiry*, Part 1 Report. Norwich: HSE Books, Her Majesty's Stationery Office.

Flin, R. (1998). Safety condition monitoring: Lessons from man-made disasters. *Journal of Contingencies and Crisis Management*, 6(2), 88–92.

Flin, R., Mearns, K., O'Connor, P., and Bryden, R. (2000). Measuring safety climate: Identifying the common features. *Safety Science*, 34(1–3), 177–192.

Glendon, A. I. (2008). Safety culture and safety climate: How far have we come and where could we be heading? *Journal of Occupational Health and Safety, Australia and New Zealand*, 24(3), 249–271.

Grabowski, M., Ayyalasomayajula, P., Merrick, J., Harrald, J. R., and Roberts, K. (2007). Leading indicators of safety in virtual organizations. *Safety Science*, 45(10), 1013–1043.

Handy, C. (1999). *Understanding Organizations*, 4th Edition, London: Penguin Books, Reprinted edition.

Harris, M. T. G. (2011). *The Effectiveness of Health & Safety and Sustainability Performance Management within UK Oil and Gas Small and Medium Enterprises*. Chifley Business School, PDM Unit 301.

Hidden, A. (1989). *Investigation of the Clapham Junction Railway Accident*. London: Department of Transport.

Hayes, A., Lardner, R., and Novatsis, E. K. (2008, January 1). *Our Safety Culture: Our Behaviour Is The Key*. Society of Petroleum Engineers. Summary Guide to Safety Climate Tools: Prepared by MaTSU. Offshore Technology Report 063.

HSC (Health and Safety Commission) (1997). *Successful Health and Safety Management*. HS(G) 65. London: HSE Books.

HSC (2000). *The Southall Rail Accident Report*. London: Professor J. Uff, HSE Books.

HSC (2001a). *The Ladbroke Grove Rail Inquiry*. Part 2 Report. London: The Rt Hon Lord Cullen, HSE Books.

HSC (2001b). *The Southall and Ladbroke Grove Joint Inquiry into Train Protection Warning Systems*. London: Professor J. Uff and the Rt Hon Lord Cullen, HSE Books.

HSC (2005). *The Rail Public Inquiries, HSC report on overall progress as of March 2005 on the remaining recommendations from the Rail Public Inquiries*. The Southall Rail Accident Inquiry Report; The Joint Inquiry into Train Protection Systems; The Ladbroke Grove Rail Inquiry Part 1 Report; The Ladbroke Grove Rail Inquiry Part 2 Report. London: The Rt Hon Lord Collen, HSE Books.

HSE (Health and Safety Executive) (1999). *Reducing Error and Influencing Behaviour*. HS(G) 48. London: HSE Books.

HSE (2005). *A Review of Safety Culture and Safety Climate Literature for the Development of the Safety Culture Inspection Toolkit*, Research Report 367. Liverpool, UK: HSE.

Holstvoogd, R. et al. (2006). *Hearts and Minds Programmes the Road Map to Improved HSE Culture*. IChemE Symposium Series No. 151.

Hudson, P. R. W. (2007). Implementing a safety culture in a major multi-national. *Safety Science*, 45(6), 697–722.

Hofstede, G. H. (2001). *Culture's Consequences: Comparing Values, Behaviors, Institutions, and Organizations across Nations*. Thousand Oaks, CA: Sage.

Hofstede, G., and Hofstede, G. J. (2005). *Cultures and Organizations: Software of the Mind*. London: McGraw-Hill.

IAEA (1991). *Safety Cultures* (Safety Series No. 75-INSAG-4), A Report by the International Nuclear Safety Advisory Group (INSAG), IAEA, Vienna.

IAEA (2005, May). *Frequently Asked Chernobyl Questions*. Vienna: Division of Public Information, International Atomic Energy Agency. Archived from the original on February 23, 2011.

IOSH (Institution of Occupational Safety and Health) (2004). *Promoting a Positive Culture—A Guide to Health and Safety Culture*. Wigston, UK: IOSH.

Lees, F. P. (1980), *Loss Prevention in the Process Industries*, Oxford, UK: Butterworth-Heinemann, 1980.

(ORR) Office of Rail Regulation (2006). *Train Derailment at Hatfield: A Final Report by the Independent Investigation Board*, ORR, London.

(ORR) (2013). *Health and Safety Report*. ORR, London.

Pidgeon, N., Hood, C., and Jones, D. (1992). Risk perception. In *Risk: Analysis, Perception and Management, Report of a Royal Society Study Group*: The Royal Society, London, pp. 89–134.

RSSB (Rail Safety and Standards Board) (2003, May). *Measurement of Safety Culture in the Rail Industry*. Research Programme. London: RSSB.

RSSB (2012a). *Safety Culture Toolkit*. Retrieved December 26, 2012, from http://www.safetyculture toolkit.rssb.co.uk/home.aspx.

RSSB (2012b). *Sample Safety Culture Questionnaire*. London: RSSB.

RSSB (2016). *Annual Safety Performance Report, 2015–2016*. London: RSSB.

Schein, E. (1990). Organisational Culture. *American Psychologist*, 45, 109–119.

Turner, B.A., Pidgeon, N., Blockley, D., and Toft, B. (1989). Safety culture: Its importance in future risk management. Position paper for the Second World Bank Workshop on Safety Control and Risk Management, Karlstad.

UIC (International Union of Railways) (2004). *SafeCulture: A Method for Assessing Organizational Safety at Interfaces*. Paris: UIC.

Whittingham, R. (2004). *The Blame Machine: Why Human Error Causes Accidents*. Oxford, UK: Elsevier Butterworth–Heineman.

Williamson A. M., Feyer, A. M., Cairns, D., and Biancotti D. (1997). The development of a measure of safety climate: The role of safety perceptions and attitudes. *Safety Science*, 25(1–3), 15–27.

34

Railway Security Policy and Administration in the United States: Reacting to the Terrorist Threat after September 11, 2001

Jeremy F. Plant, Gary A. Gordon, and Richard R. Young

CONTENTS

34.1 Introduction: US Rail System Is Complex and Unique

Attempting to arrive at a general approach to protecting railway systems that would, by definition, include the United States may not be a simple endeavor. In this chapter, we shall attempt to present the argument that the US railway network is both complex and unique. Being both complex and unique also suggests that this network has vulnerabilities as well as resilient features that demand somewhat different approaches both to assessing risks and for mitigating them. We readily admit that this chapter about the United States is somewhat of an outlier in a book that largely focuses on Europe.

As the reader winds his/her way through this chapter, it will be instructive to keep in mind that European systems have a greater emphasis on moving passengers with their frequent and punctual services. Conversely, the US railway network has evolved to be the envy of many others with its ability to move large volumes of freight, often as unit trains, but increasingly as intermodal services for higher value goods.

34.2 Risk Assessment for the US Railway System

Risk assessment is at the heart of any effort to secure rail systems against natural and human-made catastrophic events. Risk assessment is usually considered as a four-step process: identification of critical assets to be assessed; inventory security measures in place or needed to secure the assets; review of threats and weighting in terms of probability and salience; and finally, assessment of vulnerabilities in light of weaknesses and security gaps (Gordon 2011, p. 4).

In the United States, risk assessment for homeland security has gained considerable importance since the terrorist attacks of September 11, 2001, and subsequent threats to critical infrastructure, including rail systems, in the years following. The focus of the federal government has been on aviation security, with only 3% of the budget of the lead transportation security agency, the Transportation Security Administration (TSA) of the Department of Homeland Security (DHS), devoted to surface transportation security efforts. The result has been a highly collaborative and decentralized security approach that has evolved over the past 15 years in partnership with various stakeholders. Peter Neffenger (2016, p. 2), the administrator of the TSA, characterized this approach in April 2016 in this way:

> Direct responsibility to secure surface transportation systems falls primarily on the system owners and operators. TSA's role in surface transportation security is focused on program oversight, system assessments, operator compliance with voluntary industry standards, collaborative law enforcement and security operations, and regulations. TSA could not accomplish its essential counterterrorism security mission without our partners voluntarily adopting security improvements and sharing best practices.

Can such an approach develop a systematic and comprehensive method of assessing risk to the US rail system? Can a collaborative approach as outlined by the TSA utilizing voluntary acceptance of standards and practices accurately point out threats and vulnerabilities and respond to incidents in a way that protects and secures rail infrastructure and operations? These are the critical tests that the United States faces as it attempts to secure the vast and highly complex system of freight, intercity passenger, and commuter rail operations against terrorist threats, human-generated accidents, and natural disasters.

In this chapter, we describe the rail network in the United States, review the efforts made to assess risk and develop programs to secure rail infrastructure and trains against human-made and natural threats, and suggest why the United States has chosen the approach it has taken to railway security given the factors that characterize not only the US rail system but also the political and economic environment in which it operates. In many respects, the rail system in the US (and to a similar degree in Canada) is substantially different from rail systems in other developed nations. We begin by highlighting the most salient characteristics of the US rail system and suggest factors that help explain the approach to rail security taken since September 11, 2001. We then review policy and program development based on the key aspects of that approach: risk-based threat and vulnerability assessments (TVAs) undertaken in a cooperative and collaborative partnered and regulatory approach. We conclude by considering the costs and benefits of such an approach compared to an ideal quantitatively based reliability, availability, maintainability, and safety system.

34.3 US Rail System: Expansive, Organizationally Complex, and Critical to the Nation's Economy

In contrast to the practices in most of the rest of the world, US railroads have been largely an activity of the private sector. It has only been during the past 50 years that government-controlled railroads appeared, those being for passenger operations. Even Conrail, a government-conceived and government-financed freight carrier, was still a private entity even though the US government was a major stockholder. Today, it is the intercity and commuter passenger railroads that are owned and/or operated by either government units or government-controlled operating authorities.

The US railways emerged in the first half of the nineteenth century largely through private initiative, supported in part by indirect government subsidies such as land grants or partial public funding. Both passenger and freight operations were in the control of private corporations, at first largely free from government regulation. Beginning in the 1880s and evolving over many decades, the federal government assumed regulatory control over the railroads, eschewing direct operational control except for an unhappy experiment in nationalization during World War I.* Regulations involved both economic factors such as freight and passenger rates; mergers; abandonments; and antitrust actions and safety factors, such as regular inspections of equipment and rules on speed limits and crew time of service. Rail regulations were and continue to be largely the responsibility of the federal government and not state or local governments, based on the doctrine of interstate commerce.

In the early 1980s, economic regulations were largely eliminated, allowing market forces to play the dominant role in setting rates and competition and allowing mergers. Railroads then generally became free to negotiate competitive rates to charge shippers, abandon unprofitable branch lines, and curtail service where there was insufficient business volume. This last point is significant because it is via the acquisition of branch lines where government continues to play a role in rail freight activities. In no instance does the governmental unit operate freight services per se, but rather contracts this to private sector organizations, which depending on the arrangement may or may not also be responsible for maintaining the line (Staggers 1980).

Around the same time, significant growth in intermodal freight operations and the growth of the coal industry after interruptions in oil imports in the 1970s led to unprecedented profitability for the dominant freight carriers, namely, the class I railroads that remain after decades of merger consolidation. Intercity passenger operations were transferred from the private freight railroads to a government-supported corporation, the National Rail Passenger Corporation (Amtrak) in 1971 and commuter operations a little more than a decade thereafter. The federal government supervised a merger of unprofitable railroads in the northeastern United States in 1976 with the caveat that the new operation, the Consolidated Rail Corporation (Conrail), achieve profitability. By the early to mid-1980s, it too had shared in the financial success that was being enjoyed by other major private sector systems. By 1999, when Conrail was divided between the two other major railroads in the eastern half of the country, CSX and Norfolk Southern, the system as it now appears was essentially complete: seven major class I railroads, two each east and west of the Mississippi River plus a smaller north–south system, the Kansas City Southern,

---

* The US Railway Administration was established in 1917 to aid the US war efforts. Government action was the last resort after the ill-fated railroad war board. A private sector initiative ran into antitrust conflicts.

connecting the Midwest and Mexico; and the two Canadian railroads, Canadian National and Canadian Pacific, each of which operates substantial trackage within the United States.

What has resulted is a system that appears bifurcated between freight and passenger operations, but which, in reality, shares much of the overall trackage and many of the densest and most critical and vulnerable rail corridors between freight and passenger operations. Commuter passenger operations are owned and/or operated by public agencies, usually regional or state transportation organizations. Freight operations are divided between the main trunk lines, operated by the large Class I railroads, and over 500 regional and short line railroads that provide local and connector service to communities across the nation and feed traffic to the major railroads at interchange points.

From the standpoint of rail security, it is important to also note that the increasing business of railroads in intermodal freight movement requires them to work with maritime, port, and trucking organizations to ensure the secure movement of containerized freight that moves from one mode of transport to another. Similarly, freight security requires collaboration between the freight railroads and critical industries such as the chemical industry, agricultural companies, and energy producers. A significant portion of the freight rolling stock used for such movements is owned and/or operated by the industries and not the railroads, and much of it is loaded or stored on sites secured by the industries.*

As if the myriad of freight railroad companies plus Amtrak and the regional commuter rail entities did not make for a sufficiently complex network with its inherent security concerns, there is the matter of shared trackage as well as shared rights-of-way. First, freight railroads will contractually acquire the right to move trains over the rails of another freight railroad. But the practice does not end there because Amtrak and the commuter lines not only operate over the tracks of freight railroads, but there are also freight trains that operate over the tracks of the passenger railroads. A good example of this is the Amtrak line from Philadelphia to Harrisburg, Pennsylvania, which is owned by Amtrak, but sees frequent Norfolk Southern freight trains. The situation is the reverse just west of Harrisburg, where the rails are owned by Norfolk Southern, and Amtrak is the tenant railroad.

Stemming from the prior ownership of the lines, even when there are no trackage rights, it is often common for freight and passenger railroads to operate parallel tracks in the same right-of-way. Moreover, there are also instances where light rail (similar to German S-Bahn and U-Bahn operations) passenger lines may operate parallel lines.

Add to this organizational complexity the vast scale and expanse of the US rail network and it immediately becomes apparent that the issues are also significantly different from what is found in most other parts of the world. Although mergers and abandonments of redundant trackage have reduced the overall mileage of the network, it nevertheless remains huge.† In most areas, at least two major freight railroads maintain competitive trackage between major nodes (e.g., Los Angeles/Chicago or New York/Chicago). As a test for security programs, some of the densest lines profoundly vary in geography and access; two of the most heavily trafficked corridors are the passenger-based Northeast Corridor between Washington, New York, and Boston, operating in the most densely populated region in the country, and the freight-oriented Transcontinental Line of the Burlington Northern Santa Fe, which crosses the almost uninhabited Mojave Desert and remote locations in the Great Plains. Both see as many as 100 plus trains per day.

---

* Recent Association of American Railroad (AAR) statistics show approximately 500,000 railcars controlled by the railroads with nearly 800,000 controlled by industry (AAR 2016).

† The US freight railroads operate nearly 140,000 mi of road (note that this differs from miles of track because the former excludes trackage rights or the contractual use of the line of one railroad by a second railroad, which would result in some double counting (AAR 2016).

Prior to September 11, 2001, security focused on trespass, theft, and vandalism rather than securing rail systems against terrorism. Safety regulations were the responsibility of the Federal Railroad Administration (FRA), a component of the US Department of Transportation. Railroads have long employed their own police forces to guard against trespass and vandalism and to provide visible security at passenger terminals and major freight nodes. Railroad police forces tended to be small in size but professionally trained and authorized by their respective state legislatures to enjoy the same law enforcement powers as their public service counterparts.* However, a system for sharing information, developing sophisticated TVAs, and other security measures had to be developed almost overnight in the wake of the terrorist threat to US transportation exposed in the 9/11 attacks. In the following section, we outline the development of security protocols for railways and stress the voluntary and collaborative nature of these efforts, raising the question of how systematic and analytical such measures can be in the context of limited overall control by government and limited resources to prevent and respond to catastrophic events.

34.4 Potential Threats and Vulnerabilities of US Rail Assets

With only empirical evidence of attacks on railroads coming from events outside the United States, and with such examples limited to intercity and passenger operations, what are the potential threats and vulnerabilities of the US freight and passenger assets? How can planners anticipate risk factors and come up with a systematic and comprehensive approach to largely assess hypothetical situations?

The most obvious type of threat is the attack on passenger operations: commuter trains, urban stations, other vulnerable infrastructure critical to passenger operations (e.g., the Hudson River tunnels between Manhattan and New Jersey, crowded commuter stations along the Northeast Corridor). In such scenarios, the threat can come from active shooters inside trains or in terminals; improvised explosive devices (IEDs) on the right of-way or in trains; radioactive dispersal devices commonly known as "dirty bombs"; or biological, chemical, or poisonous gases released in confined spaces, such as belowground stations. Trains may be the targets in some situations as in the case of freight trains, the means of conveying toxic or other weapons of mass destruction (WMDs).

Vulnerabilities related to the rails include at least the following:

- The free movement of individuals around stations and boarding and deboarding of trains. This is especially true of dense commuter rail operations, where it is infeasible to institute time-consuming screening of passengers or restricted access to terminal facilities. As a system, terrorists may choose to board a commuter train at a remote commuter station where there may be little to no security, yet since they are then "in the system" such persons can move relatively freely to more heavily used terminals, their access, thereby, going largely unnoticed.

- The possibility of WMDs being left on trains, in unattended luggage, briefcases, and other personal effects.

---

* The aggregate of railroad-employed police officers currently stands between 2500 and 3000 nationwide.

- The close connection or proximity of passenger rail and freight rail trains or infrastructure, to include bridges and tunnels, where sabotage of freight rail assets can cause catastrophic harm to passengers and surrounding areas.
- Interference with the signaling or communications systems of the railway, either by cyberattacks or physical attacks, leading to collisions or derailments of trains in or out of terminals.
- The hijacking of passenger or freight trains by terrorists to cause catastrophic damage.

Compared to threats to freight operations, passenger rail is an inviting target for terrorists for the graphic images of dead and injured passengers in the aftermath of explosions or derailments or of mass shootings on trains or in terminals. Unlike air crashes, which usually result in total fatalities and little in the way of graphic imagery of pain and suffering, rail incidents are played on the media to full effect and provide terrorists and terrorist organizations with a message that resonates with followers' intent on subverting a society's sense of security. Attacks on trains and terminals in urban settings also cause substantial disruption of travel and economic costs as the normal routine of life is disturbed.

While many of these vulnerabilities are shared with freight railroads, there are additional vulnerabilities of freight operations. Freight trains can be extremely vulnerable to tampering at intermodal nodes, on unattended sidings and yards, and while traversing remote and largely unmonitored locations, particularly at slower speeds and at traffic control points known as "interlockings." Terrorists may see freight trains as an ideal means of transporting WMDs or becoming a WMD itself, either in intermodal shipments where a dirty bomb or biological weapon is already in place or by affixing an explosive device at a convenient location. Freight trains also carry the bulk of hazardous materials in the United States, often passing through urban settings, making them vulnerable to attack by derailment or by an explosion releasing toxic materials. Several non-terrorist-induced accidents have shown the catastrophic effects of such accidents. Even in situations where toxic materials are not released, derailments may cause extensive damage to surrounding infrastructure through fire or collision with adjacent buildings as was the case in 2013 at Lac Megantic, Quebec, when a unit train of Bakken crude oil slipped its brakes and rolled into a small wayside village killing 47 persons and injuring scores more. A major portion of the town was destroyed by the ensuing fire; however, much of the remaining structures had to be demolished due to contamination (Canadian Transport Safety Board 2014). As catastrophic as the event was, the new regulations that were imposed in its aftermath (e.g., hazmat train routing regulations and new tank car standards for those carrying flammable materials) easily run into the hundreds of millions of US dollars.

Freight railroads in the United States also carry the bulk of agricultural products such as animal feed, corn syrup for industry, and grain for export or domestic use. This creates possibilities for produce for human use or for animal feed to be contaminated and to enter the food system with great potential for harm.

The role of freight railroads in intermodal freight movement is one of the clearest areas of potential risk. While port security has been probably the second highest priority of transportation security in the United States after airline security, it still cannot guarantee that containers that are conveyed by ship to port and then to the rail network are adequately monitored. The requirements for efficient and swift movement of such cargo are the key for rail being an integral part of the supply chain and may weigh against more intrusive and time-consuming checks of intermodal movements. Delays, whatever their reason, result in the need of importers to maintain higher levels of inventory, a consequence that underscores the economic damage that terrorist activity, or the threat of terrorist activity, can have on a transportation network.

The freight rail system, moreover, is too large and too complex to be monitored by persons or remote electronic means or both in tandem. Railroad police forces are understaffed to provide real-time surveillance of even a fraction of the rail system considering its expanse. Drone technology is being introduced and may be a means of reducing the vulnerabilities inherent in freight (and passenger) rail operations. Some years ago, BNSF-initiated a program known as Citizens for Rail Security that began with registering rail hobbyists* and providing them with ready access to the BNSF police in the event that they see anything that might look out of place. Initially, many of the other railroads scoffed at the idea, but after being in place for several years and demonstrating clear value, it is now being replicated by most of the industry (Plant and Young 2007).

Lacking actual terrorist-induced threats to the US rail system, homeland security professionals have been forced to rely upon attacks on passenger rail assets in other national settings; vulnerabilities exposed by non-terror-related human error; and natural disasters such as Gulf Coast hurricanes, floods, and forest fires. Despite these inherent limitations, and despite limited resources to devote to rail security, significant efforts have been made to develop TVAs. It is to the development of such risk-based assessment we turn our attention in the next section of this chapter.

Finally, there are the cyberthreats. Modern rail systems are remotely controlled with trains being switched among various routes and their respective signals being managed by extensive computer systems. Many of these controls are supervisory control and data acquisition (SCADA)† that were likely installed before anyone was aware of the ease with which they could be hacked. In a similar fashion, the power provided to electrified railroads, whether third rail or catenary, is controlled by similar means and is consequently just as vulnerable.

During 2016, the United Sates has seen passenger train accidents caused by locomotives suddenly accelerating, derailing their trains, and causing death and destruction. While we are not calling terrorism the cause, the ability for terrorists to hack into locomotive computer control systems could potentially cause similar events elsewhere.

34.5 Development of a Risk-Based Security Approach in the United States

It is difficult to overstate the importance of the attacks of September 11, 2001, in New York City, Washington, and Pennsylvania on subsequent public policy development in homeland security. The United States had not experienced any successful terrorist attacks approaching the magnitude of these, nor had it established any comprehensive framework of policy and administrative capacity to deal with the prevention of, and the recovery from, acts of terrorism against specific targets such as the rail system. Only one recorded attack on a passenger train, in Arizona in 1995, was attributed to

---

* Also known as *railfans*, these individuals are often photographers, are very knowledgeable about railroad operating practices, and are often compared to neighborhood watch programs where residents are watching for unusual activities in their respective locales.
† *SCADA* in the railroad environment is a system for remote monitoring and control that operates with coded signals over communication channels (using typically one communication channel per remote station).

terrorism and that happened to be from a domestic and not a foreign terrorist organization. Terrorist attacks on railways in other national settings were not seen as salient to the US situation, as they were attributed to long-standing international disputes, as in the case of India and Pakistan, or to internal political conflicts, as in Mozambique and Russia.

While the September 11 attacks utilized air transport as the means of attack, it was unclear in the immediate aftermath if further attacks on the nation's transportation system, including rail, might be imminent. The geographic focus of the 9/11 attacks was on the heavily populated northeast, also the region most reliant upon rail passenger service, both intercity and commuter operations. As a result, rail services were temporarily suspended, and rail operators and government immediately began to develop the means of sharing information on possible threats and vulnerabilities.

To respond to what became increasingly known as "the new normal," the post–September 11, 2001, reality of the probability of terrorist attacks, especially on critical infrastructure, the United States considered two alternative policy and administrative approaches to deal with the problem. One was to charge existing government departments and agencies to consider terrorist threats within the purview of their existing programs and activities, with coordination provided by a homeland security coordinator in the executive office of the presidency, reporting directly to the president on imminent threats and proactive approaches to prevent catastrophic events. The second was to create a new DHS to bring under one administrative umbrella a plethora of extant agencies and programs and to directly administer the labor-intensive screening of passengers at airports in place of the privately run security operation in place on September 11.* While it is not possible to provide a detailed account of the reasons why the latter was chosen, it became an immediate challenge for the new department to simultaneously create a functioning entity integrating 22 separate administrative organizations and develop ways of analyzing threats and vulnerabilities to all forms of critical infrastructure in a large and economically developed nation.

Transportation modes, especially air transport, were seen as the most likely target of future terrorist attacks after 9/11. Congress and the White House quickly responded by forming the TSA through the Aviation and Transportation Security Act of 2001, signed into law on November 19, 2001. Within a month, the TSA was up and running within the US Department of Transportation. After the formation of the DHS in November 2002, the TSA was moved to the new department on March 9, 2003, where it remains today.

The TSA is charged by law to protect the nation's entire transportation system, including all modes of transportation and pipelines. However, rail security was given far less attention and resources than that given to air, reflecting the continuing impact of air attacks on September 11. However, the attacks on commuter trains in Madrid in 2004 and the London subway system in 2005 helped bring attention to the possible risks and vulnerabilities of passenger rail systems in the United States. In addition to these actual attacks, a number of models and scenarios had been developed to suggest the impact of attacks on rail infrastructure, both freight and passenger rail, and on ports where rail plays an important role in intermodal freight movements (Wilson et al. 2007).

The timing of these new factors coincided with a commitment by the, then, newly appointed DHS secretary Michael Chertoff to make risk assessment the foundation of DHS activities across the range of department programs and operations. Risk assessment

---

* Protection of air travel was the likely obvious choice because all three of the September 11 attacks were on civilian aircraft, and aviation is a much higher profile mode of transportation in the eyes of the public.

was both a way of making activities move from a reactive to a proactive approach and a way of developing a sense within DHS of a departmental mission and philosophy and a break from the idea that each component agency had its own way of operating and a specific historical mission apart from that of the broader department.

Intelligence gathering has increasingly been recognized as a proactive approach to managing vulnerability. Using the prevent–protect–mitigate–respond–recover model of Federal Emergency Management Agency (FEMA) (2016), it is far more cost effective, with the term *cost* applied in its broadest sense of the word, to prevent, protect, and mitigate than to respond and recover. Understanding the threats and their corresponding vulnerabilities becomes the domain of intelligence gathering, which is a complex and messy business. Some years ago, Amtrak established an intelligence unit staffed with former personnel from the Central Intelligence Agency and Office of Naval Investigation and charged it not only with understanding vulnerability, but also with determining first responder assets that would be available in proximity to its lines. In the wake of the September 11 attacks, it became readily apparent that there were many data points that if connected, could have provided advance warning to the events. Often this is described as being able to "connect the dots."

A more robust approach, which operates in real time that has been gaining adherence as well as participation by both the passenger and the freight railroads, is the fusion center. First established as an information clearinghouse staffed by representatives from a broad spectrum of stakeholders, but often with a state police agency as an anchor unit, these have forged working relationships, whereby otherwise disparate data points are linked to better understand threats and their respective vulnerabilities. The concept has expanded beyond public sector law enforcement as private sector-based fusion centers have emerged, usually organized along industry lines. Not surprisingly, there are levels of communication and cooperation among not only railroads, but also key shipper industries, such as the petrochemical industry.

The advent of the fusion center begins to connect the dots, but most importantly forges a level of cooperation that is far more robust than what existed previously. Mention was previously made of the railroad police, but their relationship with public sector law enforcement agencies was ad hoc and clearly inconsistent (Plant and Young 2007).

34.6 Future of Risk-Based Rail Security in the United States

Although change in approach is clearly in progress, both the railroads through their industry associations, the AAR and the American Short Line and Regional Railroad Association (ASLRRA), working in conjunction with the FRA and the Pipeline and Hazardous Material Safety Administration (PHMSA), have embarked on a risk assessment protocol that uses a 27-factor assessment tool known as the rail corridor risk management system (RCRMS) to evaluate risk and vulnerability. To ease explanation, the 27 variables can be subsumed into seven broad categories that consider (Gordon and Young 2015) the following:

- Rail shipments—the properties and quantities of the freight being transported
- Train operations—levels of maintenance, permitted speed, and overall transit time of a train from points A to B

- Route configuration—physical properties of the track structure including number of junctions, grade crossings, and various encumbrances
- Technology-based safety measures—the presence of trackside defect detection, wayside signaling, such as centralized train control, positive train control, and automatic equipment identification (AEI)
- Rail system network—includes traffic density, route length, whether freight shares the tracks with passenger trains, and relative congestion on the line that may pose potential delays
- Safety and incident response assets—availability and access time by first responders in proximity to not only the route, but also heavy recovery resources that are owned or contracted to others by the railroad
- Threats, risks, and vulnerabilities—the history of threats and exposure potential over a given route

A second initiative, closely related to RCRMS, is the requirement for railroads to disclose routes and volumes of hazardous materials that will be transported through both states and communities. Deemed sensitive security information, this has been a source of contention between the railroads and those several states that sought to widely publicize such information. The intent of the legislation was to improve response times, albeit that the initiative can be labeled *response*.

The third significant initiative, but one that falls under the heading of mitigation, is the improved standard for tank cars transporting flammable and explosive materials. With the events at Lac Megantic as its impetus, a new standard was promulgated requiring thicker steel for tank vessels, reinforced heads at either end of the car to prevent puncture from adjacent cars in the event of a derailment and protection of valves, fittings, and discharge connections to prevent shearing off should a rollover occur.

34.7 Infrastructure and Operations

When considering railroad security and from the perspective of commuters and intercity rail passengers, the infrastructure: track, culverts, bridges, tunnels, signals, and grade crossings (also known in Europe as level crossings) and their relationship to operations are relatively ignored. Why? Because US cities are adapted to accommodate automobiles (Rosenthal 2011) and are less costly to drive into and park, and people think more about roads, intersections, and traffic signals. The railroad physical environment is essentially "foreign" to them, and thus, anomalies, to include sabotage and tampering, are not readily recognized.

Trucking and the roadway infrastructure that they use are more familiar to the general public. The same is basically true for the other modes of bulk transportation that compete with rail, such as barge, pipeline, and, to some extent, intermodal.

Given the preceding statements, anomalies to the infrastructure, whether unintended, through a lack of or deferred maintenance, or intentionally introduced, will go unnoticed making exploitation easier. Taking this a step further, the maintenance of US railroads, unlike that observed in Europe and elsewhere in the world, is often performed to a lesser degree because of cost and impact on the profitability of a railroad. This impacts and reduces operating speeds making sabotage and compromise easier, as the faster a train goes the less the chance of compromise. Further, less than required track, bridge, and signal maintenance and often-resulting deteriorated conditions would make sabotage, exploitation, and compromise easier than if they were properly maintained and in a "good state of repair." The cost of maintenance, for all practical purposes, is the sole responsibility of the railroads in the United States as they are private companies, where in Europe and elsewhere in the world, maintenance is viewed to be subsidized. The reliance of rail travel over aviation and automobile in Europe exacerbates this phenomenon and can be applied to the transport of goods.

How can security be maximized given America's fixation on the automobile and aviation, as preferred modes of transportation, and lack of understanding of the rail environment? Infrastructure planning and design that emphasizes safety and security of the constructed facility and its impact on operations is the simplest and most effective approach. But the safety and security components of a project are often sacrificed because of budget constraints. As mentioned previously in this chapter, situational awareness among the railroad workers and passengers is important and supplements policing and technology. It should be noted that infrastructure protection and hardening should not preclude operational security, and operations should not compromise the safety and security measures built into the infrastructure.

How does a railroad ensure that the track, structure, and signals are of a quality to minimize the risk of compromise? The first is to ensure that the facilities are planned, designed, and constructed in accordance with sound engineering principles and construction practice, to include that of the FRA and the American Railway Engineering and Maintenance of Way Association. Next is to conduct periodic inspections to ensure that the facilities are in sound condition for the intended operations and that nothing has been placed or tampered with on the track, in a tunnel or on a bridge that could be used as a weapon, such as an IED, or tampered with, such as a signal approach circuit, to cause an accident or derailment. To ensure that the infrastructure is properly maintained, regular and periodic maintenance should be conducted. Why? Because poorly maintained track, structures, and signals could impact their structural and operational integrity making it easier to compromise. At stations, similar inspections and maintenance are recommended.

Recognizing that the planning, design, construction, and maintenance of infrastructure is the best way of minimizing risk, the DHS and other government agencies, as well as, stakeholders conduct formal or information security reviews of existing facilities and project designs. This is accomplished via security reviews, whereby the project sponsors, stakeholders, and involved government agencies, led by the DHS, identify design, to include redesign (of an existing facility), construction, and operational measures, such as standoff distances, to mitigate a terror attack. Assault planners are used to identify the likely threats and vulnerabilities of the transportation facility, which the project designers use to identify design solutions to mitigate the threat.

The exploitation of infrastructure could be made to look like an accident caused by or faulty maintenance, a deteriorated component of the infrastructure, or mechanical failure of rail equipment. An example of this premise at the time of this writing is the New Jersey Transit accident at Hoboken, New Jersey, station on September 29, 2016. A preliminary National Transportation Safety Board report provides a summary of the details of the investigation to date, but does not provide a preliminary analysis or state probable cause (National Transportation Safety Board 2016). With the reported acceleration of the train just prior to entering the station, one might conclude that human error could be the cause or a failure of one of the systems controlling the train, onboard or signal system. Until a cause is determined, one could speculate that the onboard train control systems had been tampered with, the signal system compromised, or the accident was intentional. The point here is that any accident could be created for radical purposes and appear to be "normal" to the "untrained eye."

Equipment safety inspections required by the AAR should also look for tampering with the rail cars and locomotives and the presence of IEDs or hidden compartments in rail cars. The latter could carry individuals or material that could pose a threat to the railroad or any location within the United States This is mainly focused on rail cars entering the United States. The inspections of the rolling stock by the train crews prior to departure and equipment maintainers can also provide additional layers of security.

The Customs and Border Protection (CBP) of DHS inspects rail cars, primarily at the border, looking for smuggling, drugs, and human trafficking. However, CBP personnel have been instructed to report observed anomalies that could compromise security even though it is outside the scope of their inspection. Further, the TSA works with CBP, as drug smuggling is suspected of funding terrorism.

34.8 Emergency Management Perspective

What constitutes an emergency or incident is one of the first things that must be considered when looking at the security of railroad operations. The FEMA *all-hazards* approach is the basis for the disaster life cycle; planning as part of and preparedness, response, recovery, and mitigation. *All-hazards* plans address the disaster life cycle from a common perspective and response activities, as well as addressing regionality and geographic-specific natural and human-made disasters, such as terror attacks.

One must first look at the types of disasters that are being considered and, in the case of terrorism, the intent to cause harm when maintaining and enhancing security as it applies to the disaster life cycle. Natural disasters are easily recognized as hurricanes, earthquake, floods, and tornados. Human-made disasters, such as an accidents, vandalism, or terrorism, are not as clearly recognized, as intent is the differentiator.

Security awareness is important when planning for the reconstitution of operations. Why? Because and in the case of a terrorist attack, individuals or organizations could take advantage of situations when preparing and "hunkering down" for a natural disaster or responding to a human-made disasters, especially one caused by terror intent. It is important to recognize not only the desire to exploit when hunkering down, but also the desire to inflict injury and damage in a second wave against the first responders and victims of the initial attack.

Let us look at this through two natural disasters: Hurricanes Rita and Ike on the Texas Gulf Coast. In 2005 and as Hurricane Rita approached and prior to landfall, the DHS advised that there was information regarding the desire of terrorists to compromise hazmat cars in rail yards in the forecasted path of the hurricane. Railroads in the impacted areas had to be notified and advised to take protective measures to secure their infrastructure and equipment even though it may put their employees in harm's way. This added dimension required railroads already impacted by Hurricane Katrina to manage and protect already crowded rail yards.

In 2008 and after Hurricane Ike made landfall, the Union Pacific Railroad (UP) had many of its nearly 1000 portable generators that powered grade crossing signals in the Houston metropolitan area stolen. It was determined by the railroad police that the generators were most likely stolen by local citizens for need, but there was the possibility that the theft could have been for sinister reasons to cause an accident between a train and a vehicle, as the UP in the immediate days after landfall was initially operating after the storm in a limited capacity with minimal signaling and "train order" operations. This stretched the resources of the railroad police and other departments to monitor other locations, such as bridges and tunnels, and replace the stolen generators. Could this situation provide an opportunity to compromise a bridge, tunnel, or grade crossing?

It is our opinion that emergency management and rail security is more focused on freight rail and especially hazmat and the damage an exploitation could have in conjunction with a natural disaster to the local environment. Freight rail in this situation could be used as a weapon or method of conducting an attack. Passenger rail, on the other hand, is more the target of an attack and not the "valued added" to a terrorist by compromising a hazmat rail car or train as a weapon or method of attack during or immediately after a natural disaster.

An attack on passenger rail will trigger a set and planned- for responses in an *all-hazards* approach to emergency management. FEMA, as well as DHS, TSA, and the FRA, has procedures and plans in place that address this. In addition to government plans, policies, and procedures, the industry itself has robust plans.

The industry plays a major role in security and emergency management indicative of the public/private partnership to security and the fact that the freight railroads are private companies, and Amtrak and commuter railroads are essentially and simply public/private partnerships in concept. Through the AAR and its member railroads, for example, training, planning, and emergency response are an integral part of an overall safety and security posture enjoyed by the industry. The ASLRRA and its member railroads have similar programs.

With regard to training, the AAR and ASLRRA offer the transportation rail incident preparedness and response (TRIPR) program to provide preincident planning and response training for incidents involving hazmat. TRIPR is intended to prepare first responders to handle incidents involving hazmat train incidents. Further and subsequent to the Lac Megantic disaster, the AAR has developed the RCRMS, which is an analytic method used to support railroads and determine safe and secure routing of trains with 20 or more cars of crude oil and other designated hazmat. It is a collaborative effort among DHS, and PHMSA, and FRA of the US Department of Transportation and employs 27 risk factors, to include local emergency response capabilities. An added value of the model is the ability of first responders along the route to be able to know what hazmat is involved and in what quantities and, thus, to plan and prepare accordingly should an accident occur. The hazmat transportation analytical risk model is a simplified version of RCRMS for the short line and regional railroads to analyze hazmat routing and is embraced by ASLRRA.

For passenger and commuter rail operations (and transit), American Public Transportation Association (APTA) has programs and initiatives to ensure that security measures are in place to protect resources, employees, and customers and are maintained and improved upon, and it recognizes that a disaster, whether natural and human-made, can occur anytime and anywhere. One such program is the emergency response preparedness program (ERPP), which is intended to support its constituents and stakeholders in emergency preparedness, response, and recovery from natural and human-made disasters, which can include evacuation and short-term transportation needs. The ERPP is just one of the APTA programs and initiatives that is focused on providing a safe and secure environment.

34.9 Conclusion

The bottom line in this discussion is that the US freight railroads are private companies and essentially rely on their own resources and, for the most part, funding for security and emergency response with overarching governmental requirements. Amtrak and the commuter railroads also rely on their own forces for the most part, but funding is through fare box revenues and subsidies. Although emergency preparedness and response are reactionary measures addressed by the railroads, planning and mitigation are proactive measures employed to minimize the impacts of a disaster. This is important in preventing an incident such as that experienced in Lac Megantic in 2013.

Passenger and commuter railroads are quasi-governmental, with many having private contract operators. Since they serve major metropolitan areas, including large terminals, such as New York City's Penn Station and Boston's South Station, they receive support from local and state governments and organizations operating the terminals. This is important in expediting emergency operations and response to a major disaster not unlike the train bombings at Atocha Station in Madrid in 2004. Since FEMA is the overarching agency for emergency management, there is a standard approach to emergency and disaster planning, preparedness, response, recovery, and mitigation that provides for relative uniformity in disaster management.

References

AAR (Association of American Railroads) (2016). *Class I Railroad Statistics*. https://www.aar.org/documents/railroad-statistics.pdf. Accessed October 15, 2016.

Canadian Transport Safety Board (2014). *Lac-Mégantic Runaway Train and Derailment Investigation Summary*. http://www.tsb.gc.ca/eng/rapports-reports/rail/2013/r13d0054/r13d0054-r-es.asp. Accessed November 7, 2015.

FEMA (Federal Emergency Management Agency) (2016). *FEMA Mission Areas*. https://www.fema.gov/mission-areas. Accessed October 15, 2016.

Gordon, G. (2011). Fixed facilities and infrastructure protection and passenger rail and rail transit security, *Proceedings of the American Railway Engineering and Maintenance of Way Association Annual Conference, Minneapolis, MN, September 18–21*.

Gordon, G., and Young, R. (2015). Railroad network redundancy: Operational flexibility, service continuity, and hazmat routing protection. *Proceedings of the Infrastructure Security Partnership, Society of American Military Engineers, Baltimore, MD*.

National Transportation Safety Board (2016). *Preliminary Report Railroad: New Jersey Transit (NJT) Train 1614*, DCA16MR011. October 13. http://www.ntsb.gov/investigations/AccidentReports/Pages/DCA16MR011_prelim.aspx. Accessed October 20, 2016.

Neffenger, P. (2016). *Surface Transportation Security: Protecting Passengers and Freight*. Written testimony of TSA Administrator Peter Neffenger for a Senate Committee on Commerce, Science, and Transportation hearing. https://www.dhs.gov/news/2016/04/06/written-testimony-tsa-administrator-senate-committee-commerce-science-and-transportation. Accessed September 30, 2016.

Plant, J., and Young, R. (2007). *Assessment to Terrorist Threats to the US Railway Network*, Boston, MA: Citizen for Rail Safety.

Rosenthal, E. (2011). Across Europe, irking drivers is urban policy, *New York Times*, June 26.

Staggers Rail Deregulation Act of 1980. 96th Congress, 49 USC 10101. http://www.gpo.gov/fdsys/pkg/statute-94/pdf/statute-94-Pg1895.pdf. Accessed November 2, 2015.

Wilson, J., Jackson, B., Eisman, M., Steinberg, P., and Riley, K. J. (2007). *Securing America's Passenger-Rail Systems*, Santa Monica, CA: RAND.

35

Introduction to IT Transformation of Safety and Risk Management Systems

Coen van Gulijk, Miguel Figueres-Esteban, Peter Hughes, and Andrei Loukianov

CONTENTS

35.1 Introduction: IT Transformation and BDRA

This chapter is an introduction to the information technology (IT) transformation of systems for safety management, risk analysis, and/or reliability, availability, maintainability, and safety (RAMS) analysis. The objective is to provide the reader with an overview of the formidable task of IT transformations in general and in particular for safety management on the Great Britain (GB) railways. The chapter can only scratch the surface of this transformation since it deals with selected elements of decades of research and development in the computer sciences. The authors believe that trying to describe the particulars of the transformation is out of scope for this book because it involves so many intricate peculiarities of computer science that it is infeasible to explain within the scope of this chapter.

This chapter contains a description of the domains that the IT transformation draws from. By doing so, this chapter conveys what kind of knowledge was assimilated by railway safety experts in GB to perform the IT transformation process for safety management systems on their railways. Despite the formidable effort this took, and will take in the future, we believe that the IT transformation of safety management systems is inevitable and will shape the future of safety and risk management.

This chapter does not contain treatises or new insights in RAMS applications. That means that RAMS standards such as EN 50126, 50128, 50129, or 61508 (CENELEC 2002; CENELEC 2010a; CENELEC 2010b; CENELEC 2014) are not treated in this chapter. It also does not treat computer-based safety case methods such as Hip-Hops (Papadopoulos et al. 2011) or dedicated software languages to deal with RAMS systems such as ScOLA or RailML (Issad et al. 2014; Nash et al. 2004). The chapter focuses solely on the development of enterprise systems that, at some point in the future, will provide the foundation for integrated safety systems.

The three most active protagonists of the IT transformation of UK railway safety systems are Network Rail with offering rail better information services (ORBIS), Railway Safety and Standards Board (RSSB) with safety management information system (SMIS+), and the University of Huddersfield (UoH) with big data risk analysis (BDRA). The efforts of RSSB and UoH are coordinated through a strategic partnership. RSSB is in the process of developing an industrywide software safety system called SMIS+, the university strengthens those efforts by exploring new technologies and applications in the BDRA research program. This work is written from the perspective of the BDRA research program. BDRA is short for big data risk analysis. Concisely, it is the application of big data techniques for safety analysis and safety management purposes. A BDRA management system is defined as an enterprise safety management system that performs the following:

- Extracts information from mixed data sources
- Processes it quickly to infer and present relevant safety management information
- Combines applications to collectively provide sensible interpretation
- Uses online interfaces to connect the right people at the right time

in order to

- Provide decision support for safety and risk management

This definition guides the development of BDRA systems that are of use to companies that work on the GB railways.

Section 35.2 provides safety delivery on the GB railways to introduce the complexity of the GB railway safety management system and its drive forward. Section 35.3 describes the digital railway enablers. Section 35.4 focuses on big data and the four building blocks for the BDRA research program. Section 35.5 describes two case studies related to BDRA: close call data and network text visualization. Section 35.6 discusses the IT transformation and the RAMS implications. Finally, Section 35.7 concludes the chapter and provides an outlook to the future.

35.2 Background and Driving Forces

Alongside the application of RAMS for the delivery of safe railways a number of mature, industrywide systems are present in the GB railways that deliver safety. These systems complement RAMS by focusing on operational aspects of safety delivery. The first is the taking safe decisions framework that provides a harmonized approach for dealing with normal operations and change management on the GB railways. The second is the safety risk model (SRM), a quantitative risk analysis model that calculates the overall level of risk on the GB railways.

35.2.1 Taking Safety Decisions

The UK rail industry has a common safety risk management framework that is used throughout the industry: the taking safe decisions framework (RSSB 2014a). It was developed by RSSB as a research project in the early 2000s to address issues with the harmonization of methods and clarity of obligations. The framework captures the legal and procedural obligations that all railway partners have to adhere to, ranging from the common safety method mandated by the European Union (EU) to national standards and best practices (Bearfield 2009; Commission Implementing Regulation (EU) 2013a, 2013b). Taking safe decisions is set up as a guidance document. It treats responsibilities and legal obligations for railway operators; the framework for risk management; contemporary GB safety analysis methods; and the SRM.

The legal obligations, like in any other country, are layered. On a European level, the common safety methods provide prescriptions. On a national level, the Health and Safety at Work Act (1974) demands that railway organizations safeguard their workers and members of the public. Other general safety legislation that applies to transport operators are the Occupiers' Liability Act 1984, which imposes a duty of care for trespassers; the Construction Regulations 2007 (design and management), which define legal duties for the safe operation of construction sites; and the Railways and Other Guided Transport Systems Regulations 2006, which set out requirements relating to safety management, including the need to maintain a safety management system; safety certification and authorization; risk assessment; transport operators' duty of cooperation; and safety-critical work. Taking safe decisions is not set up to describe the legal obligations in detail, but rather, it explains that the framework adheres to all prescribed legal obligations, so using it implies satisfying these legal obligations. However, it does not waive legal obligations, so the bottom line for the interpretation of the "so far as is reasonably practicable" principle in UK law is given as well.

The framework outline in Figure 35.1 shows the activities associated with *safety monitoring* (left), *analyzing and selecting options* (middle), and *making a change* (right). It also shows the associated activities for analyzing and selecting options: how to decide what to do when a problem or opportunity is identified. The reader can recognize activities that are required to meet general safety responsibilities under the Health and Safety at Work Act and the common safety methods. For analyzing the options and making a change, risk analysis principles based on the common safety methods are invoked by offering a choice between codes of practice, similar reference systems, and explicit risk estimations where the latter is the GB railway SRM.

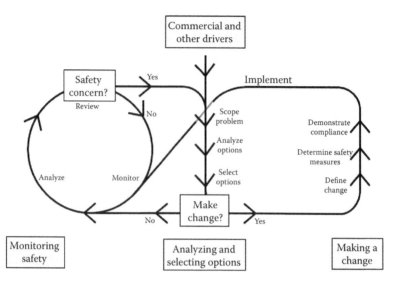

FIGURE 35.1
Taking Safety Decisions framework.

35.2.2 SRM

Preceding the taking safe decisions framework but of equal importance is the SRM for the GB railways. The SRM consists of a series of fault tree and event tree models representing 131 hazardous events that collectively capture the overall level of risk on the GB railways (Marsh and Bearfield 2008). Three major accident categories are used: train accidents, movement accidents, and nonmovement accidents. The models produce risk (Figure 35.1) in units of fatality and weighted injury per year (RSSB 2014b). Changes on the railways can be analyzed with the SRM to inform decision makers whether the fatality weighted injuries per year will go up or down depending on the changes they propose. In that sense, the SRM is a powerful tool to study the effect of changes on safety performance, but it is also used for continuous safety monitoring. A decisive strength of the SRM model is based on industry data. That is to say, the national incident reporting system for the GB railways, called SMIS, provides the data for base events in the fault trees.

The regulatory framework in GB, taking safe decisions, and the established good practice in the SRM support companies in the understanding of key hazards, accident causes, and control measures. They have ensured that the related risk is managed to an acceptable level.

Today, the relatively static SRM model is challenged. The SRM model is built from generic models that are based on network-wide average assumptions. It is difficult to extract risk profiles for locations and local circumstances. When such analyses are called for, a bespoke solution has to be designed every time. Also, the SRM does not explicitly include all control measures, making it difficult to use for the purposes of monitoring the effectiveness of individual controls. However, the technology exists to collect data of all types and align these with the risk control frameworks. This was tested in the GeoSRM model (Sadler et al. 2016). The GeoSRM model provides a step toward the application of big data techniques to deliver safety on the GB railways. It is on the back of this experience that a more ambitious IT transformation project was initiated: SMIS+.

35.2.3 Ready for the Next Step

The development of the taking safe decisions framework and GeoSRM are the last stages of a long process to gain trust in the industry to share information. Their development and cultivation by RSSB and their implementation in the industry greatly contributed to effective risk management on the UK railways. It provided a boost to safety reporting and sharing of information. A key milestone was reached in 2002 when the industry companies agreed to bring down barriers in their reporting systems and openly share all safety information. The principle that there is "no competition in safety" is one that is firmly understood and persists in the UK railway industry today. The availability of advanced risk analysis tools, a clear process description, and the level of collaboration in the industry enables the UK to progress into the next step: digital railway and digital safety system to match in the form of the SMIS+ project.

35.3 Digital Transformation on the Railway

Digital transformation is changing the world, and it has become a priority for the European Union. The Juncker Commission is paving the way for a "digital single market strategy" (European Commission 2015), which refers to a particular "digital single European railway area strategy." Although EU regulation already existed for the "digital railways" such as European Railway Traffic Management System (ERTMS) or the "telematics applications for passenger or freight," specific measures on data interoperability and automation can increase the safety, reliability, and performance of railways (Commission Regulation (EU) 2011, 2014). This is the aim of the "roadmap for digital railways" developed by the Community of European Railway and Infrastructure companies (2016) in collaboration with the International Rail Transport Committee (CIT), European Rail Infrastructure Managers (EIM) and the Union Internationale des Chemins de Fer (International Union of Railways) (UIC). The roadmap is based on four pillars, viz., the connectivity of railways, enhancing customer experience, increasing capacity, and open data platforms that should enable the competitiveness of digital railways. These pillars should allow the industry to take advantage of the digital transformation of railways and rapidly developing digital asset management strategies.

35.3.1 Digital Transformation in the United Kingdom

The digital transformation is represented in three strategic actions: ORBIS, SMIS+, and the BDRA.

35.3.1.1 ORBIS

ORBIS is one of the largest digital transformation programs created in Europe to achieve asset management excellence and support the infrastructure maintenance in the United Kingdom (Network Rail 2014). The program intends to improve the quality of asset information and to deliver efficient and safe decision-making tools. An example of these tools is the development of the rail infrastructure network model that provides (a) a visualization tool to see the locations of assets, the underlying asset information for each of them, and the environment surrounding the rail network and (b) an integrated network model in which a logical model provides asset attribute information and the relationships between the assets.

35.3.1.2 SMIS+

The GB railways have embarked on an ambitious project to deliver an IT transformation project for SMIS+, the accident database that feeds into the SRM model. It delivers a fundamentally new approach in the sense that it is based on an enterprise software platform. It incorporates a centralized safety reporting function to allow the reporting and analysis of the full range data in the database. SMIS+ is the program to deliver next-generation safety reporting technology for the GB railway industry (RSSB 2016), making it easier for people to collect information and extract intelligence. With SMIS+, RSSB and the UK railway industry are creating a completely new cloud-based online system exploiting commercial off-the-shelf, state-of-the-art, safety management software. This, in fact, is an IT transformation process where what has been used in the past, a static database with a rigid risk model, is replaced with a flexible risk-based data structure and new business intelligence software. Such a modern enterprise system connects users so it is easier to collaborate, and users can create their own reports more easily. The system also provides the opportunity for it to be fully developed to be a complete management system, customizable and evolvable to the management processes. It will be possible to be continually redesigned and configured by individual industry partners to meet their particular requirements while also providing centralized analysis capability for the railway.

Due to confidentiality considerations, the SMIS+ system cannot be described here. This chapter considers SMIS+ part of the larger agenda for the complete IT transformation of safety management systems that is aimed for in the BDRA research program. The main constituents of big data enterprise systems and analytical applications are described in Section 35.4.

35.4 BDRA

Big data is the label for methods and techniques that take advantage of data linkage for technological and nontechnological solutions to support business processes. The broad interpretation is that big data deals with huge volumes of a variety of data sources very quickly. In a narrower interpretation, it is the next step in the development of business process support and decision making, which is sometimes referred to as enterprise. In practice, big data deals with software-integration tools that combine structured and unstructured data sources to support control processes and decision-making processes. The analysis of big data requires a suite of methods that process the large amounts of data that are present in modern distributed information systems, such as the Internet.

BDRA systems are built upon four basic enablers, viz., data, ontology, visualization, and enterprise architecture. These enablers have to be integrated to produce the emergent behavior of BDRA, as illustrated in Figure 35.2.

35.4.1 Data

As suggested by its name, *data* is the basis of BDRA. Modern technological systems in the GB railways produce massive amounts of structured and unstructured data. For instance, a useful structured data source is the data produced by supervisory control and data acquisition systems (Daneels and Salter 1999), which consist of coded signals that use

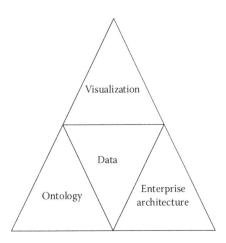

FIGURE 35.2
BDRA components.

Internet communication channels to remotely control equipment. These signals typically include error messages and information about the state of a piece of equipment. For safety and risk purposes, dedicated databases are maintained where error messages are stored for further analysis. In addition, RSSB maintains a large amount of unstructured data in the form of millions of reportable safety incidents from the railway. Although it is challenging to obtain safety knowledge from text such as close call reports (RSSB 2015), the combination of these data sources provides a powerful insight into how the operation and failure of equipment contribute to failures and accidents on the railway.

A huge technical concern is the way in which data are stored. Databases are commonly kept in large but slower "backing stores" or "disks." The backing store can be organized in different ways. For instance, it could be a file in the file system of an operating system. However, in more complex environments, it could be a database that provides more flexibility to store and access huge quantities of data. Traditional relational databases or Structured Query Language (SQL) databases depend on lookup tables for access. They have been very efficient when it comes to rapid and efficient access to data but remain efficient only for relatively "small" databases. When huge amounts of information have to be stored, individual disks or computers are too small or become unwieldy with relational databases (Hoffer, Venkataraman, and Topi 2016). A solution to bypass this problem is to omit the relational table by simply storing data in a system that, for lack of a better example, finds its analogy in an infinitely scalable library card catalog. In a library catalog, numerous pieces of information are stored on cards with label indexes in labeled boxes. Only a very basic index, usually alphabetic, states the approximate location of cards, but the system does not drill down to each exact card. Retrieving that information requires a query for that information, and a person and/or search engine finds the relevant card. A NoSQL database is similar in the sense that numerous pieces of digital information (often files of arbitrary type) are stored under an index label (identifying a unique information file), in an almost infinitely scalable database (Sadalage and Fowler 2013). Finding the right piece of information in the library catalog, as well as the NoSQL database, requires search efforts (a software query) to extract useful data. Such systems offer flexibility for the storage and use of data as well as speed for online applications because now, the application is no longer dependent on single disks and or computer processes anymore. For safety management systems, it means that different sources of information can be stored alongside each other and analyzed selectively and on demand. It provides flexibility for safety analysis in the sense that the analyst potentially has access to additional data for his/her risk problem, even if it is of a different type (numeric, visual, or text). It also allows for deeper scrutiny without changing or adding to data systems.

Software service solutions have to be built to exchange information with users and other databases. Such connections can be made in an enterprise service bus (ESB). An ESB routes, adopts, translates, and transfers information from the client or user to the appropriate answering service and/or database (Chappell 2004). Among others, an ESB monitors and controls routing of message exchange between services; it resolves conflicts in communicating service components; and it caters for services such as event handling, data transformation, mapping, message and event queuing, message sequencing, security, exception handling, and protocol conversion and enforces quality of the communication service.

The use of such technologies offers new opportunities for the development new risk analysis and risk management techniques, but a whole new suite of tools and skills has to be learned to deal with the technology.

35.4.2 Ontology

The railway system is a complex one due to its vast network and numerous interactions between subsystems. Each subsystem can be managed simultaneously by different organizations (e.g., infrastructure managers, operators, manufacturers, or maintainers). Each organization has its own structure (e.g., financial area, health and safety area, resources area, or selling area) formed by people of different expertise, skills, and competences that change over time and do not necessarily draw from the same jargon. So a passenger might be a ticket buyer for an economist, but equally, it might be an individual on a moving train for a safety expert (even if they did not pay their ticket). Thus, a railway system is a rich tapestry of different organizational knowledge bands created for different purposes and persons in a specific context. To date, no system exists that captures the knowledge associated with the railway system, but the technical systems to do so are available today.

Knowledge management (KM) is the discipline that deals with the interoperability of knowledge in different organizational contexts (Brewster, Ciravegna, and Wilks 2001). KM is specifically oriented toward capturing and storing organizational knowledge into a format that is useful for humans and computers alike. Capturing the interactions between parts of the railways is one of the greatest challenges in the BDRA program. The objective is to understand decision-making processes for risk analysis based on relevant safety knowledge from the railway system captured in databases. Another, more popular terminology for KM is ontology.

Ontology building is a common technique used by computer science to represent a common framework of understanding. Ontology is the systematic classification of domain knowledge that supports the use of different databases in a meaningful way: it can be compared to a search engine which holds the right search keys to produce results that are relevant to the human operator. The search keys are based on a repository of concepts and words that represent the knowledge structure of a specific domain. In this case, the domain is safety and risk for the GB railways; the concepts are the ways in which the components within the domain combine and interact to create the emergent behavior of the overall system.

Depending on the type of knowledge to represent, different levels of complexity of ontology can be required. Figure 35.3 shows a spectrum of ontologies depending on their complexity and descriptive power in computer science (Uschold and Gruninger 2004).

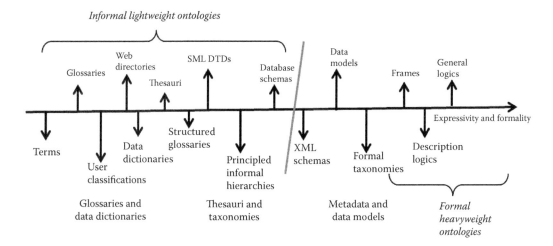

FIGURE 35.3
Different types of ontologies depending on their complexity. (After M. Uschold and M. Gruninger, *ACM SIGMOD Record*, 33, 58, 2004.)

To represent a knowledge domain, diverse lightweight ontologies such as vocabularies, on the left-hand side of Figure 35.3, can be created and linked to form heavyweight ontologies, on the right-hand side of Figure 35.3. Heavyweight ontologies tend to be very abstract which is desirable from a scientific point of view, but they tend to be unwieldy for practical computer solutions.

An ontology for BDRA is a systematic and explicit description of concepts and relationships for railway safety and risk (Van Gulijk and Figueres-Esteban 2016). It attempts to establish a common semantic framework among the current safety organizational knowledge to select data from different systems and enable data analysis. It means that it is not necessary to change the knowledge of each organization, just to describe and match it. That means that ontologies have to provide the semantics to obtain and use relevant data for railway safety.

Figure 35.4 illustrates how a semantic layer connects the data layer to the safety process. The semantic layer itself can be split into several layers as well. Safety lightweight

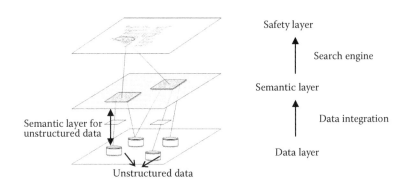

FIGURE 35.4
Semantic layers support analysis from unstructured data (e.g., text analysis), the data integration process from different sources (e.g., relational databases), and the search engine from safety operations (e.g., a bow tie).

ontologies are at the bottom of the semantic layer to access unstructured data. A second level of more complex semantic layer performs data integration from different sources and connects to the decision makers' interface. Since different ontologies are required for different purposes, the methodology to build ontologies has to be tailored to create ontologies of different degrees of expressiveness and specificity.

35.4.3 Visualization

Visualization has been developed from different disciplines (e.g. computer science, engineering, psychology or management sciences), its definition has been expressed in terms that might indicate different levels of abstraction and understanding depending on the discipline, generating conflicts and inconsistencies (Chen et al. 2009). Although we can find references related to the understanding and insight of data by means of visual perception to support the cognitive process in data analysis (Card, Mackinlay, and Shneiderman 1999), visualization has historically been divided into information visualization and scientific visualization. Scientific visualization focuses on visual techniques to depict scientific and spatial data, while information visualization focuses on abstract and nonspatial data. Visual analytics is a different proposition altogether: it focuses on interactive visual tools to analyze large datasets.

The term *visual analytics* arose around 2005, being defined as a combination of "… automated analysis techniques with interactive visualizations for an effective understanding, reasoning and decision making on the basis of very large and complex data sets" (Keim et al. 2008). VA is a multidisciplinary area that attempts to obtain insight from massive, incomplete, inconsistent, and conflicting data in order to support data analysis, but usually requires human judgment (Thomas and Cook 2005). Visual analytics explicitly combines four dimensions of big data (volume, variety, veracity, and value). Implicitly, the last attribute, velocity, may be considered a computational requirement to take into account in performing analysis and decision making. Therefore, it might be said that visual analytics is a variant of 'big data analytics' supported by interactive visualization techniques (Figueres-Esteban, Hughes, and Van Gulijk 2015). Visual Analytics has yielded five pillars where visualization can support data analysis tasks, viz., data management, data analysis/mining, risk communication, human–computer interaction, and information/scientific visualization (Figure 35.5). In the scope of this paper, data analysis is most relevant.

Data analysis (also called data mining or knowledge discovery) is the overall process of extracting valuable information from a collection of data (Grolemund and Wickham 2014). Since the beginning of data analysis, the collection of data and the methods for analyzing and interpreting data were central. Visual analytics, in support of data analysis, focused on large datasets and developing methods for semiautomated or fully automated analysis. Although these methods were initially developed for structured data, more recent efforts have been made to analyze semistructured and complex data.

In visual analytics, the automatic techniques have been addressed in two ways: supervised analysis and cluster analysis. The first technique designs analysis patterns based on methods and data that are known to solve the analysis (e.g., supervised by humans) while the second technique aims to extract structure from data without any human supervision or previous knowledge (Keim et al. 2008). Such techniques are successfully applied in bioinformatics, climate change, pattern identification, and spatial–temporal data mining. These tools support human reasoning from large datasets, and interactive visualization techniques are used for arriving at optimal results in analyses (Figure 35.6) (Endert et al. 2014; Keel 2006). That means that data mining in visual analytics is moving toward *visually controlled data mining* (Andrienko et al. 2010).

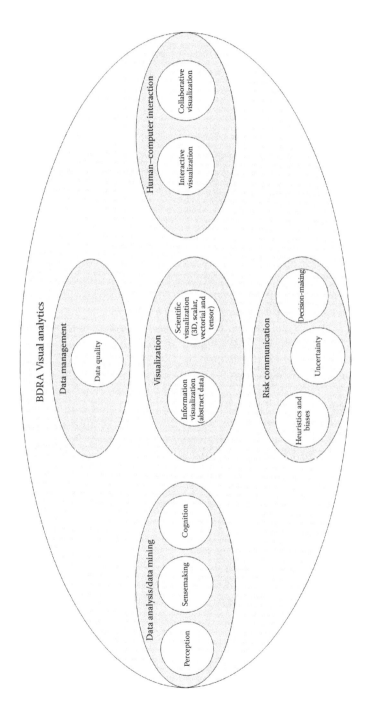

FIGURE 35.5
Main areas involved in BDRA visual analytics.

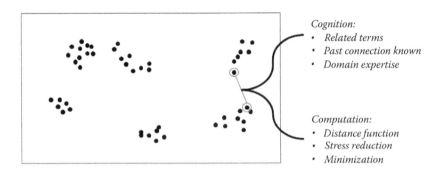

FIGURE 35.6
Computers can estimate a distance between two points based on features of data, but users with expertise domain can contradict those features and share this knowledge with the system by means of interactive visualization techniques. (After A. Endert et al., *Journal of Intelligent Information Systems*, 43, 411–35, 2014.)

35.4.4 Enterprise Architecture

Enterprise can be defined as any collection of organizations that have a common set of goals and/or aligned aims; architecture shows how components fit together, to build a service or to support a certain business capability (TOGAF 2006). Enterprise architecture was introduced to structure enterprise processes and to inform the design of software systems to support them. Enterprise architecture is the instrument to integrate the constituents of BDRA: data, ontology, and visualization. It is a holistic approach to system analysis that is specifically tailored to the needs of IT developers in the sense that it specifies the requirements of the software systems on a high level.

The Zachman enterprise architecture is one of the earliest and most enduring frameworks and was developed by John Zachman at IBM in 1980. The framework defines an enterprise as a two-dimensional classification matrix, based on the intersection of six communication questions (what, where, when, why, who, and how) with six rows according to data model transformation perspectives (transformation planner, business owner, system designer, system builder, and subcontractor) (see ZIFA Institute [1997]).

The enterprise architecture approach is very suitable to design and construct the backbone of a safety information system, in the sense that it provides a way to incorporate best practices of safety science in enterprise information systems management. The process stems from the same system analysis techniques that are used safety systems, and additional added value is found in understanding potential IT risks and better cost analysis. In this context, enterprise represents the interests of railway infrastructure owners or managers or the researchers in the BDRA research program (Sherwood, Clark, and Lynas 2009).

The first step of safety system design is the formulation of aims. On an industry level, it is any collection of organizations that have a common set of goals and/or the bottom line for the railway industry as a whole. The common safety methods are part of these goals, but they are supplemented with many others as discussed in the taking safety decisions framework in Section 35.2.1. In addition to that, there are goals about transport capability and finance. In IT, this forms the basis for an enterprise architecture that matches industry needs. On a BDRA research level, the requirements are less elaborate; it is usually the research objective with a specified analysis technique. The Zachman enterprise architecture method is a sufficiently comprehensive method for connecting high-level aims to business models, design, technical details, and IT systems. In that sense, Zachman is a top–down approach to the design of the system. The Zachman enterprise architecture method provides industry safety systems with clear business objectives, assigns ownership and responsibilities, the blueprint for organizational design and the framework for the IT systems. For BDRA, it provides the blueprint for the analysis and the technical requirements to achieve the research aims.

Metamodels are the bridge between the Zachman enterprise architecture and operational railway systems. Metamodel platforms provide an environment for the framework in which goals, models, rules, and constraints are formulated. In IT, they are used in a narrower sense: the metamodel entails a method to design ontologies and UML relationship charts, selection of programming languages, and development of semantic networks. In some respects, this is the bottom–up approach where designers try to capture all relevant knowledge to serve the enterprise architecture.

Some work on bridging big data and enterprise has been done by Loukianov (2012). The integrated view on technical, organizational and cultural factors in hazard operations was proposed by Ale et al. (2012). The high-level perspective on the alignment of information management and business strategy was suggested by Loukianov, Quirchmayr, and Stumptner (2012). Risk analysis on the basis of a provenance chain was developed (Viduto et al. 2014) and integrated into the Rolls-Royce equipment health management system, to improve the trust users place in system outputs and help understand associated risks of decisions recommended by the management system. The experience from such processes is now used to design BDRA systems.

35.5 BDRA Case Studies

35.5.1 Close Call Data

This case focuses on unstructured data from the close call database. A close call is a hazardous situation where the event sequence could lead to an accident if it had not been interrupted by a planned intervention or by random event (Figueres-Esteban, Hughes, and Van Gulijk 2016a). Network Rail workers and specific subcontractors within the GB railway industry are asked to report such events in the close call database. Close call reports are freeform text reports where users can describe a situation that, in their view, could have led to an accident. Providing a free text format for data entry allows the reporter to describe hazards in a rich way that would not be possible if data entry were constrained, for example, by selecting hazard types from a predefined list. The close call database receives about 150,000 entries each year which makes it impractical to manually review the records. Computer-based natural language processing (NLP) techniques have been developed to speed up the safety learning from this large body of data.

NLP is challenging because of the inherent ambiguity in written language, including the use of jargon, abbreviations, misspelling, and lack of punctuation. The processing of close call data by extracting information from free text involves the following processes (Hughes, Van Gulijk, and Figueres-Esteban 2015):

- Text cleansing, tokenizing, and tagging
- Ontology parsing and coding
- Clustering (creation of groups of records that are semantically similar)
- Information extraction

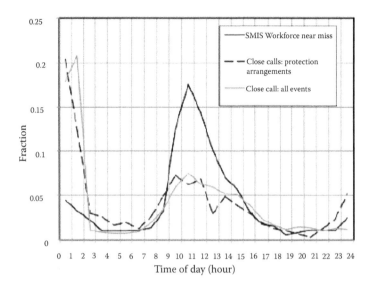

FIGURE 35.7
Frequencies of workforce incidents in SMIS+ and close call.

The method allowed for the automated identification of incidents with track workers. Reports from the SMIS+ system suggested that incidents with track workers took place more frequently between the hours of 11:00 and 15:00. The task for the close call analysis was to investigate whether the same pattern was present in the close call database. An automated search query was programmed to retrieve the protection/possession events in the close call database as function of time of day. The results were compared with track worker near miss events in the SMIS+ database.

The relative distributions of these events by time of day are shown in Figure 35.7, which shows that the incident database and close call reports follow similar trends during the day. However, further examination shows that the times at which reports are made for all close calls are similar to the times reports are made for protection arrangements, suggesting that there may be a reporting bias that interferes with the analysis. The high fraction of close call events between 00:00 and 01:00 is likely to be due to a default of the reporting system that sets the time stamp to 00:00 when the time of the incident is not entered by the person making the entry.

35.5.2 Network Text Visualization

Network text analysis is an alternative method to learn from close calls. It is a computer-based analysis that represents text as a graph. Words are represented as nodes, and their relationships, as edges. Interactive visualizations can support analysis to identify clusters related to risk areas. Figure 35.8 shows a cluster obtained from a sample of 500 close call records of three different types of risk scenarios. The cluster provides information about one of the risk types: level crossings. It shows railway workers reporting about level crossings. So without reading the close calls, risk types are easily identified. A more detailed description of the technique is described by Figueres-Esteban, Hughes, and Van Gulijk (2016a, 2016b).

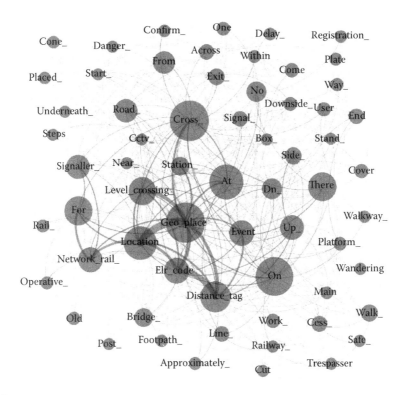

FIGURE 35.8
Subnetwork that represent a level crossing cluster. (From M. Figueres-Esteban et al., *Safety Science*, 89, 72–76, 2016.)

35.6 Discussion

The IT transformation is shaking the foundations of safety management systems. It is changing the way in which traditional safety management is delivered. IT systems are no longer reactive systems; they become part of flexible business systems that can respond to very quickly variations, which considerably increases the efficiency and the productivity of safety management. It also means a fundamental change in the conception of a safety management system.

An interesting point is how this digital transformation can impact the enterprise architectures involved in RAMS management. Business processes can be redefined to support and speed up the preliminary RAMS activities related to the life cycle of a system. Now the huge amounts of information produced by digital systems are not only just sources that might enhance maintenance by means of advanced monitoring systems, but they can be also used to define new systems that deal with challenging requirements, for instance, analyzing the behavior of similar systems in real environments for system identification (e.g., operational, environmental, or maintenance conditions) or improving the reliability and risk analysis (e.g., failures analysis).

IT transformation is a priority for the EU. New regulations have arisen to build an interoperable digital framework. Railways are building their own "digital market," and different actors are collaborating to make it possible. The UK railways is heavily involved in ambitious projects such as the ORBIS platform and the SMIS+/BDRA programs to deal with asset management and safety management, respectively.

The pillars of BDRA support in the core of future reliability, availability, and maintainability management. Large amounts of diverse data produced by the digital systems (structured and unstructured) have to be processed using big data techniques. Alternative types of data have to be stored alongside each other and crawled through by suitable NLP techniques to capture RAMS issues for the railways. KM systems based on railway RAMS ontologies have to be created in order to integrate information in database systems. Interactive visualizations have to allow data analysts to test their perceptual abilities to obtain insight from all types of available data and support the leaning process of KM systems. Finally, the elements have to be built under an enterprise architecture framework that defines the rules between the stakeholders.

35.7 Conclusion

This chapter introduces key elements of the IT transformation of safety and risk management systems. The elements are introduced from the hands-on experience in the UK railway perspective, but the IT transformation is taking place in many different countries; it is a global effort.

The chapter explained how the developments of railway safety management in the UK paved the way toward the digital transformation of safety management systems and how a select group of organizations approaches the problem. National harmonization of processes with the taking safe decisions framework, the national SMIS+ database, and a mature reporting culture provide the prerequisites for the transformation. ORBIS, SMIS+, and BDRA are on the way to exploit the potential and valuable lessons learned in these projects.

It is important to stress that the decision to make the IT transformation is not made by RAMS and safety experts. The IT transformation of railways and wider business systems is a global effort that is taking place today and leaves us almost with no choice but to follow. But this is not a tragedy. Two simple cases demonstrate that fundamentally, new ways of organizing data and analyzing it are emerging. They hold the promise of quick, efficient, and reliable RAMS decision making, even if it will take time to develop mature versions of those systems.

This chapter explains which are the most important elements of IT science that have to be considered. Experts in RAMS will eventually have to learn to work with such systems. Clearly, a formidable task is ahead of RAMS experts to enable the IT transformation of RAMS and safety systems. The experts do not need to keep up to date with the latest developments in computer science but need to understand some key components of it that will be important in the near future: data storage, KM, visual analytics, and enterprise architecture form the core. RAMS experts may feel that such topics are somewhat outside their comfort zone, but it is within their grasp to use them and to contribute to transformation projects. It also enables them to make a leap forward in safety and risk management for the railways, which after all, depends on informed decision making.

Acknowledgments

The work reported in this chapter was undertaken as part of the strategic partnership agreement between the University of Huddersfield and the Railway Safety and Standards Board.

References

Ale, B. J. M., D. M. Hanea, S. Sillem, P. H. Lin, C. Van Gulijk, and P. T. W. Hudson. 2012. Modelling risk in high hazard operations: Integrating technical, organisational and cultural factors. In *PSAM11 & ESREL 2012: 11th International Probabilistic Safety Assessment and Management Conference and the Annual European Safety and Reliability Conference*. Helsinki: Curran Associates.

Andrienko, G., N. Andrienko, H. Schumann, C. Tominski, U. Demsar, D. Dransch, J. Dykes, S. Fabrikant, M. Jern, and M. Kraak. 2010. Space and time. In *Mastering the Information Age Solving Problems with Visual Analytics*, edited by D. Keim, J. Kohlhammer, G. Ellis, and F. Mansmann, 57–86. Bad Langensalza, Germany: Eurographics Association.

Bearfield, G. J. 2009. Achieving clarity in the requirements and practice for taking safe decisions in the railway industry in Great Britain. *Journal of Risk Research* 12 (3–4): 443–53.

Brewster, C., F. Ciravegna, and Y. Wilks. 2001. Knowledge acquisition for knowledge management: Position paper. In *Proceeding of the IJCAI-2001 Workshop on Ontology Learning*, 121–34.

Card, S. K., J. D. Mackinlay, and B. Shneiderman. 1999. *Readings in Information Visualization: Using Vision to Think*. Vol. 1. San Francisco, CA: Morgan Kaufmann.

RSSB. 2015. *Close Call System*. http://www.closecallsystem.co.uk/.

CENELEC (European Committee for Electrotechnical Standardization). 2002. EN 61508-1:2002: *Functional Safety of Electrical/electronic/programmable Electronic Safety-Related Systems*. London: British Standards.

CENELEC. 2010a. EN 50126-1:1999: *The Specification and Demonstration of Reliability, Availability, Maintainability and Safety (RAMS). Part 1: Basic Requirements and Generic Process*. London: British Standards.

CENELEC. 2010b. EN 50129:2003: *Communication, Signalling and Processing Systems—Safety Related Electronic Systems for Signalling*. London: British Standards.

CENELEC. 2014. EN 50128:2011: *Communication, Signalling and Processing Systems. Software for Railway Control and Protection Systems*. London: British Standards.

Community of European Railway and Infrastructure. 2016. *A Roadmap for Digital Railways*. http://www.cer.be/publications/latest-publications/roadmap-digital-railways.

Chappell, D. 2004. *Enterprise Service Bus*. Sebastopol, CA: O'Reilly Media.

Chen, M., D. Ebert, H. Hagen, R. S. Laramee, R. Van Liere, K.-L. Ma, W. Ribarsky, G. Scheuermann, and D. Silver. 2009. Data, information, and knowledge in visualization. *Computer Graphics and Applications, IEEE* 29: 12–19.

Commission Implementing Regulation (EU). 2013a. *Regulation 402/2013 on the CSM for Risk Assessment and Repealing Regulation 352/2009*.

Commission Implementing Regulation (EU). 2013b. *Regulation No 1078/2012 on the CSM for Monitoring*.

Commission Regulation (EU). 2011. *Telematics Applications for Passenger Service—TAP*. TSI 454/2011.

Commission Regulation (EU). 2014. *Telematics Applications for Freight Service—TAF. TSI* 1305/2014.

Daneels, A., and W. Salter. 1999. What is SCADA? In *International Conference on Accelerator and Large Experimental Physics Control Systems*. Trieste.

Endert, A., M. S. Hossain, N. Ramakrishnan, C. North, P. Fiaux, and C. Andrews. 2014. The human is the loop: New directions for visual analytics. *Journal of Intelligent Information Systems* 43 (3): 411–35.

European Commission. 2015. *A Digital Single Market Strategy for Europe—COM(2015)*. Brussels: European Commission.

Figueres-Esteban, M., P. Hughes, and C. Van Gulijk. 2015. The role of data visualization in railway big data risk analysis. In *Safety and Reliability of Complex Engineered Systems*, edited by Luca Podofillini, Bruno Sudret, Bozidar Stojadinovic, Enrico Zio, and Wolfgang Kroger, 2877–82. London: Taylor & Francis Group.

Figueres-Esteban, M., P. Hughes, and C. Van Gulijk. 2016a. Visual analytics for text-based railway incident reports. *Safety Science* 89: 72–76.

Figueres-Esteban, M., P. Hughes, and C. Van Gulijk. 2016b. Ontology network analysis for safety learning in the railway domain. In *Proceedings of the 26th European Safety and Reliability Conference*, 6.

Grolemund, G., and H. Y. Wickham. 2014. A cognitive interpretation of data analysis. *International Statistical Review* 82 (2): 184–204.

Hoffer, J., R. Venkataraman, and H. Topi. 2016. *Modern Database Management*. Cranbury, NJ: Pearson Education.

Hughes, P., C. Van Gulijk, and M. Figueres-Esteban. 2015. Learning from text-based close call data. In *Safety and Reliability of Complex Engineered Systems*, edited by Luca Podofillini, Bruno Sudret, Bozidar Stojadinovic, Enrico Zio, and Wolfgang Kroger, 31–38. London: Taylor & Francis Group.

Issad, M., A. Rauzy, L. Kloul, and K. Berkani. 2014. SCOLA: A scenario oriented language for railway system specifications context formal model system specifications. In *Forum Académie Industrie de l'AFIS (Agence Française d'Ingénierie Système)*. Paris: HAL.

Keel, P. E. 2006. Collaborative visual analytics: Inferring from the spatial organization and collaborative use of information. In *IEEE Symposium on Visual Analytics Science and Technology 2006, VAST 2006*, 137–44. Cambridge, MA: Computer Science and Artificial Intelligence Laboratory, Massachusetts Institute of Technology.

Loukianov, A. 2012. Enterprise architecture and professional golf: Building a stochastic bridge. In *Proceedings to Pre-Olympic Congress on Sports Science and Computer Science in Sport*.

Loukianov, A., G. Quirchmayr, and M. Stumptner. 2012. Strategic alignment value-based analysis. In *Proceedings to IX International Symposium System Identification and Control Problems*.

Marsh, D. W. R., and G. Bearfield. 2008. Generalizing event trees using Bayesian networks. *Proceedings of the Institution of Mechanical Engineers, Part O: Journal of Risk and Reliability* 222 (2): 105–14.

Nash, A., D. Huerlimann, J. Schueette, and V. P. Krauss. 2004. RailML—A standard data interface for railroad applications. In *Computers in Railways IX*. Ashurst: Wessex Institute of Technology Press.

Network Rail. 2014. *Asset Management Strategy*. http://www.networkrail.co.uk/aspx/12210.aspx.

Papadopoulos, Y., M. Walker, D. Parker, E. Rüde, R. Hamann, A. Uhlig, U. Grätz, and R. Lien. 2011. Engineering failure analysis and design optimisation with HiP-HOPS. *Engineering Failure Analysis* 18 (2): 590–608.

RSSB (Railway Safety and Standards Board). 2014a. *Taking Safe Decisions—How Britain's Railways Take Decisions That Affect Safety*. http://www.rssb.co.uk/Library/risk-analysis-and-safety -reporting/2014-guidance-taking-safe-decisions.pdf.

RSSB. 2014b. *Safety Risk Model: Risk Profile Bulletin, Version 8.1*. http://www.rssb.co.uk/safety -risk-model/safety-risk-model/Documents/RPBv8-v1.1.pdf.

RSSB. 2016. *SMIS + Developing the New Safety Management Intelligence System*. http://www.rssb .co.uk/Library/risk-analysis-and-safety-reporting/2016-08-smis-plus-developing-the-new -safety-management-intelligence-system.pdf.

Sadalage, P. J., and M. Fowler. 2013. *NoSQL Distilled: A Brief Guide to the Emerging Word of Plyglot Persistence*. Cranbury, NJ: Pearson Education.

Sadler, J., D. Griffin, A. Gilchrist, J. Austin, O. Kit, and J. Heavisides. 2016. GeoSRM—Online geo-spatial safety risk model for the GB rail network. *IET Intelligent Transport Systems* 10(1): 17–24.

Sherwood, J., A. Clark, and D. Lynas. 2009. *Enterprise Security Architecture*. Hove, UK:SABSA Institute.

Thomas, J. J., and K. A. Cook. 2005. *Illuminating the Path: The Research and Development Agenda for Visual Analytics. IEEE Computer Society*. Vol. 54. IEEE Computer Society, Los Alamitos, CA.

TOGAF (The Open Group Architecture Framework). 2006. *The Open Group Architecture Framework*. San Francisco, CA.

Uschold, M., and M. Gruninger. 2004. Ontologies and semantics for seamless connectivity. *ACM SIGMOD Record* 33: 58.

Van Gulijk, C., and M. Figueres-Esteban. 2016. *Background of Ontology for BDRA*. Report 110-124.

Viduto, V., K. Djemame, P. Townend, J. Xu, S. Fores, L. Lau, M. Fletcher et al. 2014. Trust and risk relationship analysis on a workflow basis: A use case. In *Proceedings of the Ninth International Conference on Internet Monitoring and Protection. ICIMP 2014*. Paris.

ZIFA Institute. 1997. http://www.zifa.com/. Monument, USA: ZIFA.

36

Formal Reliability Analysis of Railway Systems Using Theorem Proving Technique

Waqar Ahmad, Osman Hasan, and Sofiène Tahar

CONTENTS

36.1 Introduction

In recent years, high-speed railway has been rapidly developed and deployed around the world including Germany, China, France, and Japan. The continuous endeavor to operate these trains at higher speeds has led to the development of high-speed railways into a new era. For instance, the high-speed railways in China had been operating at speeds of 300 km/h, but the introduction of the Beijing–Shanghai high-speed railway in June 2011 has further ushered China toward superhigh-speed trains that can operate at speeds of 380 km/h [1]. Due to the widespread coverage and continuous operation of the railway systems, the rigorous reliability analysis of these high-speed trains is a dire need. Moreover, a slight malfunctioning in the train components may cause undesirable delays at the arrival stations or even the loss of human lives in extreme cases.

Reliability block diagrams (RBDs) [2] are commonly used to develop reliability models for high-speed railway systems. Traditionally, these reliability models are analyzed by paper-and-pencil proof methods and simulation tools. However, the paper-and-pencil methods are prone to human errors for large systems, and it is often the case that many key assumptions that are essentially required for the analytical proofs are in the minds of the engineers and, hence, are not properly documented. These missing assumptions are thus not communicated to the design engineers and are ignored in system implementations, which may also lead to unreliable designs. On the other hand, there are numerous simulation tools available, such as ReliaSoft [3] and ASENT reliability analysis tool [4], that offer scalable reliability analysis compared to paper-and-pencil methods. However, these tools cannot ensure accurate analysis due to the involvement of pseudorandom numbers and numerical methods. Additionally, exhaustive verification of systems for all values of the variables is not possible.

To overcome the inaccuracy limitations of traditional techniques mentioned earlier, formal methods have also been proposed as an alternative for the RBD-based analysis using both state-based [5,6] and theorem-proving techniques [7]. The main idea behind the formal analysis of a system is to first construct a mathematical model of the given system using a state machine or an appropriate logic and then use logical reasoning and deduction methods to formally verify that this system model exhibits the desired characteristics, which are also mathematically specified using an appropriate logic. However, state-based approaches cannot be used for verifying generic mathematical expressions for reliability. On the other hand, theorem proving, which is based on the expressive higher-order logic (HOL) [8], allows working with a variety of datatypes, such as lists and real numbers and has been recently used to formalize commonly used RBDs [9] by leveraging upon the probability theory formalization in HOL [10]. This HOL-based RBD formalization provides the formally verified generic reliability expressions that can be used to carry out an accurate and rigorous reliability analysis of high-speed railway systems. In this chapter, we have utilized the recently proposed HOL formalization of RBDs [9] to conduct formal reliability analysis of a railway system designed for the Italian high-speed railways [11] consisting of several critical components, such as traction drive system, induction motors, converters, and transformers.

The rest of the chapter is organized as follows: Section 36.2 presents a review of the related work. Section 36.3 provides an overview of the proposed methodology that has been used to conduct formal reliability analysis of railway systems. To facilitate the understanding of the chapter for nonexperts in theorem proving, we present a brief introduction about theorem proving, the HOL theorem prover, and the formalization of probability and reliability theories in Section 36.4. This is followed by the description of our formalization of the RBD configurations in Section 36.5. The RBD-based formal reliability analysis of the Italian high-speed railway system is presented in Section 36.6, and finally Section 36.7 concludes the chapter.

36.2 Related Work

Many simulation tools, such as DNV-GL [12], ReliaSoft [3], and ASENT [4], support RBD-based reliability analysis and provide powerful graphical editors that can be used to construct the RBD models of the high-speed trains. These tools generate samples from the exponential or Weibull random variables to model the reliabilities of the individual system components. These samples are then processed by using computer arithmetic and numerical techniques in order to compute the reliability of the complete system. Although these software tools provide more scalable and quick analysis compared to paper-and-pencil based analytical methods, they cannot ascertain the absolute correctness of the system because of their inherent sampling based nature and the involvement of pseudorandom numbers and numerical methods.

Formal methods, such as Petri nets (PNs), have also been used to model RBDs [13] as well as dynamic RBDs [5] that are used to describe the reliability behavior of systems. PN verification tools, based on model checking principles, are then used to verify behavioral properties of the RBD models to identify design flaws [5,13]. Similarly, the probabilistic model checker *PRISM* [14] has been used for the quantitative verification of various safety and mission-critical systems, such as failure analysis for an industrial product development workflow [15], an airbag system [6], and the reliability analysis of a global navigation satellite system that enables an aircraft to determine its position (latitude, longitude, and altitude) [16]. However, due to the state-based models, only state-related property verification, such as deadlock checks, reachability, and safety properties, is supported by these approaches, i.e., we cannot verify generic reliability relationships for the given system using the approaches presented in the studies by Robidoux et al. [5], Norman and Parker [6], Signoret et al. [13], Herbert and Hansen [15], and Lu et al. [16].

A number of formalizations of probability theory are available in HOL (e.g., the studied by Mhamdi et al. [10], Hurd [17], and Hölzl and Heller [18]). Hurd's [17] formalization of probability theory has been utilized to verify sampling algorithms of a number of commonly used discrete [19] and continuous random variables [20,21] based on their probabilistic and statistical properties. Moreover, this formalization has been used to conduct the reliability analysis of a number of applications, such as memory arrays [22] and electronic components [23]. However, Hurd's formalization of probability theory only supports having the whole universe as the probability space. This feature limits its scope, and thus, this probability theory cannot be used to formalize more than a single continuous random variable, whereas in the case of reliability analysis of railways systems, multiple continuous random variables are required. The recent formalizations of probability theory by Mhamdi et al. [10] and Hölzl and Heller [18] are based on extended real numbers (including $\pm\infty$) and provide the formalization of Lebesgue integral to reason about advanced statistical properties. These theories also allow using any arbitrary probability space, a subset of the universe, and are thus more flexible than Hurd's formalization. Leveraging upon the high expressiveness of HOL and the inherent soundness of theorem proving, Mhamdi et al.'s [10] formalized probability theory has been recently used for the formalization of RBDs [9], including series [7], parallel [24], parallel–series [24], series–parallel [25], and k-out-*n* [26]. These formalizations have been used for the reliability analysis of many applications, including simple oil and gas pipelines with serial components [7], wireless sensor network protocols [24], logistic supply chains [25], and oil and gas pipelines [26]. Similarly, Mhamdi et al.'s probability theory has also been used for the formalization of commonly used fault tree (FT) gates, such as AND, OR, NAND, NOR, XOR, and NOT, and the probabilistic inclusion–exclusion principle [27]. In addition, the RBD and FT formalizations mentioned earlier have been recently utilized for availability analysis [28]. In this chapter, we utilize recently proposed HOL formalization of RBDs

[9] to carry out the formal reliability analysis of a railway system operated by the Italian high-speed railways.

36.3 Proposed Methodology

The proposed methodology for the formal reliability analysis of railway systems, depicted in Figure 36.1, allows us to formally verify the reliability expressions corresponding to the given *railway system description* and thus formally check that the given railway system satisfies its reliability requirements. The core component of this methodology is the HOL formalizations of the notions of probability, reliability, and RBDs.

The given railway system is first partitioned into segments, and the corresponding *RBD model* is constructed. This model can then be formalized in HOL using the core formalizations mentioned earlier, particularly the formalization of commonly used RBD configurations. The next step is to assign failure distributions, such as exponential and Weibull, to individual components of the given railway system. These distributions are also formalized by building upon the formalized probability theory and are used, along with the formal RBD model, to formalize the given reliability requirements as a proof goal in HOL. The user has to reason about the correctness of this *proof goal* using a theorem prover by building upon the core formalizations of probability and reliability theories. If all subgoals are discharged, then we obtain formally verified reliability expressions, which correspond to the given railway system and its reliability requirements of the given railway system. Otherwise, we can use the failing subgoals to debug the formal RBD model and proof goal, which represent the originally specified model and requirements, respectively, as depicted by the dotted line in Figure 36.1.

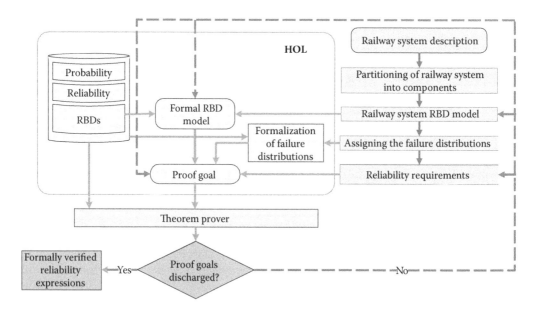

FIGURE 36.1
Methodology for formal railway system reliability analysis.

36.4 Preliminaries

In this section, we give a brief introduction to theorem proving and the HOL theorem prover to facilitate the understanding of the rest of the chapter.

36.4.1 Theorem Proving

Theorem proving [8] is a widely used formal verification technique. The system that needs to be analyzed is mathematically modeled in an appropriate logic, and the properties of interest are verified using computer-based formal tools. The use of formal logics as a modeling medium makes theorem proving a very flexible verification technique as it is possible to formally verify any system that can be mathematically described. The core of theorem provers usually consists of some well-known axioms and primitive inference rules. Soundness is assured as every new theorem must be created from these basic or already proven theorems and primitive inference rules. The verification effort of a theorem in a theorem prover varies from trivial to complex depending on the underlying logic [29].

36.4.2 HOL Theorem Prover

HOL [30] is an interactive theorem prover developed at the University of Cambridge, United Kingdom, for conducting proofs in HOL. It utilizes the simple type theory of Church [31] along with Hindley–Milner polymorphism [32] to implement HOL. HOL has been successfully used as a verification framework for both software and hardware as well as a platform for the formalization of pure mathematics.

The HOL core consists of only five basic axioms and eight primitive inference rules, which are implemented as meta language (ML) functions. The type system of the ML ensures that only valid theorems can be constructed. Soundness is assured as every new theorem must be verified by applying these basic axioms and primitive inference rules or any other previously verified theorems/inference rules.

In the work presented in this chapter, we utilize the HOL theories of Booleans, lists, sets, positive integers, *real* numbers, measure, and probability [10]. In fact, one of the primary motivations of selecting the HOL theorem prover for our work was to benefit from these built-in mathematical theories. Table 36.1 provides the mathematical interpretations of some frequently used HOL symbols and functions, which are inherited from existing HOL theories.

36.4.3 Probability and Reliability in HOL

Mathematically, a measure space is defined as a triple (Ω, Σ, μ), where Ω is a set, called the sample space; Σ represents a σ algebra of subsets of Ω, where the subsets are usually referred to as measurable sets; and μ is a measure with domain Σ. A probability space is a measure space (Ω, Σ, Pr), such that the measure, referred to as the probability and denoted by Pr, of the sample space is 1. In the HOL formalization of probability theory [10], given a probability space p, the functions `space`, `subsets`, and `prob` return the corresponding Ω, Σ, and Pr, respectively. This formalization also includes the formal verification of some of the most widely used probability axioms, which play a pivotal role in formal reasoning about reliability properties.

TABLE 36.1

HOL Symbols and Functions

HOL Symbol	Standard Symbol	Meaning
∧	*and*	Logical *and*
∨	*or*	Logical *or*
¬	*not*	Logical *negation*
::	*cons*	Adds a new element to a list
++	*append*	Joins two lists together
HD L	*head*	Head element of list *L*
TL L	*tail*	Tail of list *L*
EL n L	*element*	*n*th element of list *L*
MEM a L	*member*	True if *a* is a member of list *L*
λ x.t	λ *x.t*	Function that maps *x* to *t(x)*
SUC n	*n* + 1	Successor of a *num*
lim(λ n.f(n))	$\lim_{n \to \infty} f(n)$	Limit of a *real* sequence *f*

A random variable is a measurable function between a probability space and a measurable space. The measurable functions belong to a special class of functions, which preserves the property that the inverse image of each measurable set is also measurable. A measurable space refers to a pair (*S*, *A*), where *S* denotes a set and *A* represents a nonempty collection of subsets of *S*. Now, if *S* is a set with finite number of elements, then the corresponding random variable is termed as discrete; otherwise, it is known as a continuous random variable.

The probability that a random variable *X* is less than or equal to some value *t*, $Pr(X \le t)$ is called the cumulative distribution function (CDF), and it characterizes the distribution of both discrete and continuous random variables. The CDF has been formalized in HOL as follows [7]:

⊢ ∀ p X t. CDF p X t = distribution p X {y | y ≤ Normal t},

where the variables *p*: (α → *bool*)#((α → *bool*) → *bool*)#((α → *bool*) → *real*), *X*: (α → *extreal*), and *t*: *real* represent a probability space, a random variable, and a *real* number, respectively. The function Normal takes a *real* number as its input and converts it to its corresponding value in the *extended real* data type, i.e., it is the *real* data type with the inclusion of positive and negative infinity. The function distribution takes three parameters: a probability space *p*, a random variable *X*, and a set of *extended real* numbers and outputs the probability of a random variable *X* that acquires all values of the given set in probability space *p*.

Now, reliability *R(t)* is stated as the probability of a system or component performing its desired task over a certain interval of time *t*:

$$R(t) = Pr(X > t) = 1 - Pr(X \le t) = 1 - F_X(t), \tag{36.1}$$

where $F_X(t)$ is the CDF. The random variable *X*, in the preceding definition, models the time to failure of the system and is usually modeled by the exponential random variable with parameter λ, which corresponds to the failure rate of the system. Based on the HOL formalization of probability theory [10], Equation 36.1 has been formalized as follows [7]:

```
⊢ ∀ p X t. Reliability p X t = 1 - CDF p X t.
```

The series RBD, presented by Ahmad et al. [7], is based on the notion of mutual independence of random variables, which is one of the most essential prerequisites for reasoning about the mathematical expressions for all RBDs. If N reliability events are mutually independent, then

$$Pr\left(\bigcap_{i=1}^{N} A_i\right) = \prod_{i=1}^{N} Pr(A_i). \tag{36.2}$$

This concept has been formalized as follows [7]:

```
⊢ ∀ p L. mutual_indep p L = ∀ L1 n. PERM L L1 ∧
    1 ≤ n ∧ n ≤ LENGTH L ⇒
     prob p (inter_list p (TAKE n L1)) =
    list_prod (list_prob p (TAKE n L1))
```

The function `mutual _ indep` accepts a list of events L and probability space p and returns *True* if the events in the given list are mutually independent in the probability space p. The predicate PERM ensures that its two lists as its arguments form a permutation of one another. The function LENGTH returns the length of the given list. The function TAKE returns the first n elements of its argument list as a list. The function `inter _ list` performs the intersection of all sets in its argument list of sets and returns the probability space if the given list of sets is empty. The function `list _ prob` takes a list of events and returns a list of probabilities associated with the events in the given list of events in the given probability space. Finally, the function `list _ prod` recursively multiplies all elements in the given list of real numbers. Using these functions, the function `mutual _ indep` models the mutual independence condition such that for any 1 or more events n taken from any permutation of the given list L, the property $Pr\left(\bigcap_{i=1}^{N} A_i\right) = \prod_{i=1}^{N} Pr(A_i)$ holds.

36.5 Formalization of the Reliability Block Diagrams

Commonly used RBD configurations for the reliability analysis of the railway system include series, parallel, and a combination of both and are depicted in Figure 36.2. In this chapter, we present their formalization, which, in turn, can then be used to formally model the structures of a railway system in HOL and reason about their reliability, availability, and maintainability characteristics.

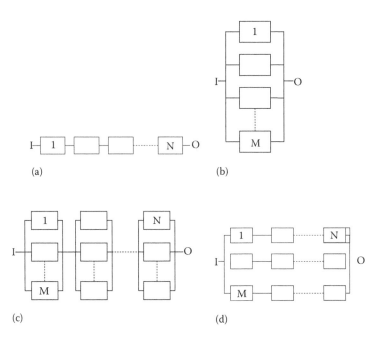

FIGURE 36.2
RBDs: (a) series; (b) parallel; (c) series–parallel; (d) parallel–series.

36.5.1 Formalization of Reliability Event

We describe the formally verified reliability expressions for the commonly used RBD configurations using reliability event lists, where a single event represents the scenario when the given system or component does not fail before a certain time. The HOL formalization of this concept is as follows:

Definition 36.1: ⊢ ∀ p X t.

```
rel_event p X t = PREIMAGE X {y | Normal t < y} ∩ p_space p
```

The function PREIMAGE takes two arguments, a function *f* and a set *s*, and returns a set, which is the domain of the function *f* operating on a given range set *s*. The function rel _ event accepts a probability space *p*; a random variable *X*, representing the failure time of a system or a component; and a real number *t*, which represents the time index at which the reliability is desired. It returns an event representing the reliable functioning of the system or component at time *t*.

Similarly, a list of reliability events can be derived by mapping the function rel _ event on each element of the given random variable list in HOL as follows:

Definition 36.2: ⊢ ∀ p L t.

```
rel_event_list p L t = MAP (λa. rel_event p a t) L,
```

where the HOL function MAP takes a function *f* and a list and returns a list by applying the function *f* on each element of the given list.

Now, we describe the formalization process by type abbreviating the notion of event, which is essentially a set of observations with type/$'a \rightarrow bool$ as follows:

```
type_abbrev ("event" , ``:'a ->bool'')
```

We then define a recursive datatype *rbd* in the HOL system as follows:

```
Hol_datatype 'rbd = series of rbd list |
                    parallel of rbd list |
                    atomic of 'a event'
```

An RBD can be either a series configuration, a parallel configuration, or an atomic event. The type constructors `series` and `parallel` recursively function on *rbd*-typed lists and thus enable us to deal with nested RBD configurations. The type constructor `atomic` is basically a typecasting operator between *event* and *rbd*-typed variables. Typically, a new datatype is defined in HOL as $(\alpha_1, \alpha_2, ..., \alpha_n)op$, where $(\alpha_1, \alpha_2, ..., \alpha_n)$ represent the arguments taken by the HOL datatype *op* [30]. For instance, the `atomic` type constructor is defined with the arbitrary type α, which is taken by the already defined type `events`. On the other hand, the type constructors `series` and `parallel` are defined without any arguments because the datatype `rbd` is not defined at this point.

We define a semantic function rbd _ struct (α *event* # α *event event* # (α *event* $\rightarrow$ *real*) $\rightarrow$ α *rbd* $\rightarrow$ α *event*) inductively over the *rbd* datatype. It extracts the corresponding event from the given RBD configuration as follows:

Definition 36.3: $\vdash (\forall p.$ rbd _ struct p (series []) = p _ space p) $\wedge$

```
(∀ xs x p.
  rbd_struct p (series (x::xs)) =
  rbd_struct p x ∩ rbd_struct p (series xs)) ∧
(∀ p. rbd_struct p (parallel []) = {}) ∧
(∀ xs x p.
  rbd_struct p (parallel (x::xs)) =
  rbd_struct p x ∪ rbd_struct p (parallel xs)) ∧
(∀ p a. rbd_struct p (atomic a) =  a)
```

The preceding function decodes the semantic embedding of an arbitrary RBD configuration by extracting the corresponding reliability event, which can then be used to determine the reliability of a given RBD configuration. The function rbd _ struct takes an *rbd*-typed list identified by a type constructor `series` and returns the whole probability space if the given list is empty and, otherwise, returns the intersection of the events that is obtained after applying the function rbd _ struct on each element of the given list in order to model the series RBD configuration behavior. Similarly, to model the behavior of a parallel RBD configuration, the function rbd _ struct operates on an *rbd*-typed list encoded by a type constructor `parallel`. It then returns the union of the events after applying the function rbd _ struct on each element of the given list or an empty set if the given list is empty. The function rbd _ struct returns the reliability event using the type constructor `atomic`.

In the subsequent sections, we present the HOL formalization of RBDs on any reliability event list of arbitrary length [24,25]. The notion of reliability event is then incorporated in the formalization while carrying out the reliability analysis of a real railway system, as it will be described in Section 36.6.

36.5.2 Formalization of Series Reliability Block Diagrams

The reliability of a system with components connected in series is considered to be reliable at time t only if all its components are functioning reliably at time t, as depicted in Figure 36.2a. If $A_i(t)$ is a mutually independent event that represents the reliable functioning of the ith component of a serially connected system with N components at time t, then the overall reliability of the complete system can be expressed as [33]

$$R_{\text{series}}(t) = Pr\left(\bigcap_{i=1}^{N} A_i(t)\right) = \prod_{i=1}^{N} R_i(t). \qquad (36.3)$$

Now using Definition 36.3, we can formally verify the reliability expression, given in Equation 3, for a series RBD configuration in HOL as follows:

Theorem 36.1: ⊢ ∀ p L. prob space p ∧

```
¬NULL L ∧ (∀x'. MEM x' L ⇒ x' ∈ events p) ∧
mutual_indep p L ⇒
(prob p (rbd_struct p (series (rbd_list L))) =
 list_prod (list_prob p L))
```

The first assumption, in Theorem 36.1, ensures that p is a valid probability space based on the probability theory in HOL [10]. The next two assumptions guarantee that the list of events L, representing the reliability of individual components, must have at least one event and the reliability events are mutually independent. The conclusion of the theorem represents Equation 36.3. The function rbd _ list generates a list of type rbd by mapping the function atomic to each element of the given event list L to make it consistent with the assumptions of Theorem 36.1. It can be formalized in HOL as

```
∀ L. rbd_list L = MAP (λa. atomic a) L.
```

The proof of Theorem 36.1 is primarily based on mutual independence properties and some fundamental axioms of probability theory.

36.5.3 Formalization of Parallel Reliability Block Diagrams

The reliability of a system with parallel connected submodules, depicted in Figure 36.2b, mainly depends on the component with the maximum reliability. In other words, the system will continue functioning as long as at least one of its components remains functional. If the event $A_i(t)$ represents the reliable functioning of the ith component of a system with N parallel components at time t, then the overall reliability of the system can be mathematically expressed as [33]

$$R_{\text{parallel}}(t) = Pr\left(\bigcup_{i=1}^{N} A_i(t)\right) = 1 - \prod_{i=1}^{N}(1 - R_i(t)). \qquad (36.4)$$

Similarly, by following the formalization approach of series RBD mentioned earlier, we can formally verify the reliability expression for the parallel RBD configuration, given in Equation 36.4, in HOL as follows:

Theorem 36.2: ⊢ ∀ p L.

```
prob_space p ∧ (∀x'. MEM x' L ⇒ x' ∈ events p) ∧
¬NULL L ∧ mutual_indep p L ⇒
 (prob p (rbd_struct p (parallel (rbd_list_L))) =
 1 - list_prod (one_minus_list (list_prob_p L)))
```

The preceding theorem is verified under the same assumptions as Theorem 36.1. The conclusion of the theorem represents Equation 36.4, where the function one _ minus _ list accepts a list of *real* numbers [$x1, x2, x3, ..., xn$] and returns the list of *real* numbers such that each element of this list is 1 minus the corresponding element of the given list, i.e., [$1 - x1, 1 - x2, 1 - x3, ..., 1 - xn$].

The preceding formalization described for series and parallel RBD configurations builds the foundation to formalize the combination of series and parallel RBD configurations. The type constructors `series` and `parallel` can take the argument list containing other *rbd* type constructors, such as `series`, `parallel`, or `atomic`, allowing the function rbd _ struct o yield the corresponding event for an RBD configuration that is composed of a combination of series and parallel RBD configurations.

36.5.4 Formalization of Series–Parallel Reliability Block Diagrams

If in each serial stage the components are connected in parallel, as shown in Figure 36.2c, then the configuration is termed as a *series–parallel structure*. If $A_{ij}(t)$ is the event corresponding to the proper functioning of the *j*th component connected in an *i*th subsystem at time index *t*, then the reliability of the complete system can be expressed mathematically as follows [33]:

$$R_{\text{series–parallel}}(t) = Pr\left(\bigcap_{i=1}^{N}\bigcup_{j=1}^{M}A_{ij}(t)\right) = \prod_{i=1}^{N}\left(1 - \prod_{j=1}^{M}\left(1 - R_{ij}(t)\right)\right). \tag{36.5}$$

By extending the RBD formalization approach presented in Theorems 36.1 and 36.2, we formally verify the generic reliability expression for series-parallel RBD configuration, given in Equation 36.6), in HOL as follows:

Theorem 36.3: ⊢ ∀ p L. prob _ space p ∧

```
(∀z. MEM z L ⇒ ¬NULL z) ∧
(∀x'. MEM x' (FLAT L) ⇒ x' ∈ events p) ∧
 mutual_indep p (FLAT L) ⇒
  (prob p
    (rbd_struct p ((series of (λa. parallel (rbd_list a))) L)) =
(list_prod of
  (λa. 1 - list_prod (one_minus_list (list_prob p a)))) L)
```

The first assumption in Theorem 36.3 is similar to the one used in Theorem 36.2. The next three assumptions ensure that the sublists corresponding to the serial substages are not empty, and the reliability events corresponding to the subcomponents of the parallel–series configuration are valid events of the given probability space p and are mutually independent. The HOL function FLAT is used to flatten the two-dimensional list, i.e., to transform a list of lists, into a single list. The conclusion models the right-hand side of Equation 36.5). The function of is defined as an infix operator [30] in order to connect the two *rbd*-typed constructors by using the HOL MAP function and thus facilitates the natural readability of complex RBD configurations. It is formalized in HOL as follows:

⊢ ∀ g f. f of g = (f o (λa. MAP g a))

36.5.5 Formalization of Parallel–Series Reliability Block Diagrams

If $A_{ij}(t)$ is the event corresponding to the reliability of the jth component connected in a ith subsystem at time t, then the reliability of the complete system can be expressed as follows:

$$R_{\text{parallel-series}}(t) = Pr\left(\bigcup_{i=1}^{M}\bigcap_{j=1}^{N}A_{ij}(t)\right) = 1 - \prod_{i=1}^{M}\left(1 - \prod_{j=1}^{N}\left(R_{ij}(t)\right)\right). \tag{36.6}$$

Similarly, the generic expression of the parallel–series RBD configuration, given in Equation 36.6, is formalized in HOL as follows:

Theorem 36.4: ⊢ ∀ p L. prob _ space p ∧

```
(∀z. MEM z L ⇒ ¬NULL z) ∧
(∀x'. MEM x' (FLAT L) ⇒ x' ∈ events p) ∧
 mutual_indep p (FLAT L) ⇒
(prob p
  (rbd_struct p ((parallel of (λa. series (rbd_list a))) L)) =
1 - (list_prod o (one_minus_list) of
  (λa. list_prod (list_prob p a))) L)
```

The assumptions of Theorem 36.4 are similar to those used in Theorem 36.3. The conclusion models the right-hand side of Equation 36.6.

To verify Theorems 36.3 and 36.4, it is required to formally verify various structural independence lemmas, for instance, given the list of mutually independent reliability events, an event corresponding to the series or parallel RBD structure is independent, in probability, with the corresponding event associated with the parallel–series or series–parallel RBD configurations.

36.6 Traction Drive System of High-Speed Trains

In order to illustrate the practical effectiveness of the RBD-based formal reliability analysis using theorem proving, we consider a multivoltage railway system, specifically designed for the Italian high-speed railways [11]. The overall railway system consists of three identical modules, i.e., A, B, and C, as depicted in Figure 36.3. Each module represents a traction drive system and two boogies that are composed of two bearings and one reduction gear. The most critical part in the railway system is the traction drive system because a slight malfunctioning in its key components may lead to train delay, affect the operation order, and endanger the safe operation of the train. A traction drive system in each module consists of a transformer, a filter, an inverter, two four-quardent converters and four induction motors that are connected with two boogies. The RBD diagram of the overall railway system is shown in Figure 36.3. The HOL formalization of the given train RBD is as follows:

Definition 36.4: ⊢ ∀ p T1 FQC1 FQC2 F1 I1 IM1
IM2 B1 IM3 IM4 B2 T2 FQC3 FQC4

F2 I2 IM5 IM6 B3 IM7 IM8 B4 T3 FQC5 FQC6 F3 I3 IM9 IM10 B5 IM11 IM12
B6.
railway_RBD p T1 FQC1 FQC2 F1 I1 IM1 IM2 B1 IM3 IM4 B2 T2 FQC3 FQC4
F2 I2 IM5 IM6 B3 IM7 IM8 B4 T3 FQC5 FQC6 F3 I3 IM9 IM10 B5 IM11 IM12
B6 =
rbd_struct p (parallel
  [MA_RBD T1 FQC1 FQC2 F1 I1 IM1 IM2 B1 IM3 IM4 B2;
  MB_RBD T2 FQC3 FQC4 F2 I2 IM5 IM6 B3 IM7 IM8 B4;
  MC_RBD T3 FQC5 FQC6 F3 I3 IM9 IM10 B5 IM11 IM12 B6])

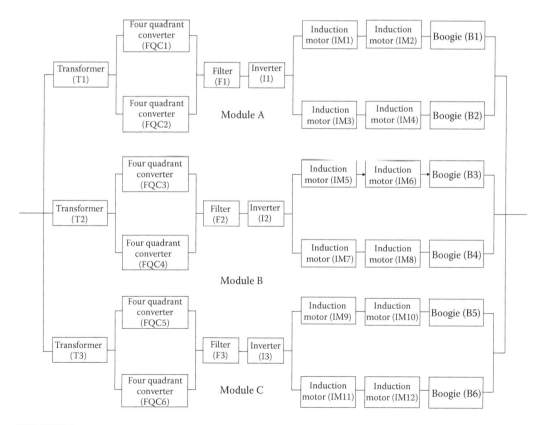

FIGURE 36.3
Railway system RBD.

where MA _ RBD, MB _ RBD, and MC _ RBD are RBDs corresponding to the train modules A, B, and C, as depicted in Figure 36.3. For instance, the HOL formalization of MA _ RBD is as follows:

Definition 36.5: ⊢ ∀ T1 FQC1 FQC2 F1 I1 IM1 IM2 B1 IM3 IM4 B2.

```
MA_RBD T1 FQC1 FQC2 F1 I1 IM1 IM2 B1 IM3 IM4 B2 =
series
 [atomic T1; parallel [atomic FQC1; atomic FQC2];
  series [atomic F1; atomic I1];
  parallel
   [series [atomic IM1; atomic IM2; atomic B1];
    series [atomic IM3; atomic IM4; atomic B2]]]
```

where T1, FQC1, FQC2, F1, I1, IM1, IM2, IM3, IM4, B1, and B2 are the events corresponding to transformer, four-quadrant converter, filter, inverter, induction motors and boogies, respectively.

In the same way, we can formalize the functions MB _ RBD and MC _ RBD in HOL, and the formal definitions can be found in the study by Ahmed [34]. It can be observed that the railway _ RBD does not seem to be directly mapping to any of the commonly used RBDs, which are described in Section 36.5. However, we can mathematically verify that this configuration is equivalent to one of those generic RBDs, i.e., the parallel–series RBD configuration in this case. The following lemma formally describes this relationship:

Lemma 36.1: ⊢ ∀ p T1 FQC1 FQC2 F1 I1 IM1 IM2 B1 IM3 IM4 B2 T2

```
FQC3 FQC4 F2 I2 IM5 IM6 B3 IM7 IM8 B4 T3 FQC5 FQC6 F3 I3 IM9 IM10 B5
IM11 IM12 B6.
 railway_RBD p T1 FQC1 FQC2 F1 I1 IM1 IM2 B1 IM3 IM4 B2 T2 FQC3
FQC4 F2 I2 IM5 IM6 B3 IM7 IM8 B4 T3 FQC5 FQC6 F3 I3 IM9 IM10 B5
IM11 IM12 B6 =
 rbd_struct p ((parallel of (λa. series (rbd_list a))))
[[IM1; IM2; B1; T1; F1; I1; FQC1];
 [IM1; IM2; B1; T1; F1; I1; FQC2];
 [IM3; IM4; B2; T1; F1; I1; FQC1];
 [IM3; IM4; B2; T1; F1; I1; FQC2];
 [IM5; IM6; B3; T2; F2; I2; FQC3];
 [IM5; IM6; B3; T2; F2; I2; FQC4];
 [IM7; IM8; B4; T2; F2; I2; FQC3];
 [IM7; IM8; B4; T2; F2; I2; FQC4];
 [IM9; IM10; B5; T3; F3; I3; FQC5];
 [IM9; IM10; B5; T3; F3; I3; FQC6];
 [IM11; IM12; B6; T3; F3; I3; FQC5];
 [IM11; IM12; B6; T3; F3; I3; FQC6]])
```

Each component of a railway system is exponentially distributed, as described by Dazi et al. [11], so we can express the reliability of the railway system, as shown in Figure 36.3, mathematically as follows:

$$R_{\text{railway\_system}} = 1 - \left(1 - e^{-(\lambda_{IM1}+\lambda_{IM2}+\lambda_{B1}+\lambda_{T1}+\lambda_{F1}+\lambda_{I1}+\lambda_{FQC1})t}\right)$$
$$\times \left(1 - e^{-(\lambda_{IM1}+\lambda_{IM2}+\lambda_{B1}+\lambda_{T1}+\lambda_{F1}+\lambda_{I1}+\lambda_{FQC1})t}\right)$$
$$\times \left(1 - e^{-(\lambda_{IM1}+\lambda_{IM2}+\lambda_{B1}+\lambda_{T1}+\lambda_{F1}+\lambda_{I1}+\lambda_{FQC2})t}\right)$$
$$\times \left(1 - e^{-(\lambda_{IM3}+\lambda_{IM4}+\lambda_{B2}+\lambda_{T1}+\lambda_{F1}+\lambda_{I1}+\lambda_{FQC1})t}\right)$$
$$\times \left(1 - e^{-(\lambda_{IM3}+\lambda_{IM4}+\lambda_{B2}+\lambda_{T1}+\lambda_{F1}+\lambda_{I1}+\lambda_{FQC2})t}\right)$$
$$\times \left(1 - e^{-(\lambda_{IM5}+\lambda_{IM6}+\lambda_{B3}+\lambda_{T2}+\lambda_{F2}+\lambda_{I2}+\lambda_{FQC3})t}\right)$$
$$\times \left(1 - e^{-(\lambda_{IM5}+\lambda_{IM6}+\lambda_{B3}+\lambda_{T2}+\lambda_{F2}+\lambda_{I2}+\lambda_{FQC4})t}\right) \quad (36.7)$$
$$\times \left(1 - e^{-(\lambda_{IM7}+\lambda_{IM8}+\lambda_{B4}+\lambda_{T2}+\lambda_{F2}+\lambda_{I2}+\lambda_{FQC3})t}\right)$$
$$\times \left(1 - e^{-(\lambda_{IM7}+\lambda_{IM8}+\lambda_{B4}+\lambda_{T2}+\lambda_{F2}+\lambda_{I2}+\lambda_{FQC4})t}\right)$$
$$\times \left(1 - e^{-(\lambda_{IM9}+\lambda_{IM10}+\lambda_{B5}+\lambda_{T3}+\lambda_{F3}+\lambda_{I3}+\lambda_{FQC5})t}\right)$$
$$\times \left(1 - e^{-(\lambda_{IM9}+\lambda_{IM10}+\lambda_{B5}+\lambda_{T3}+\lambda_{F3}+\lambda_{I3}+\lambda_{FQC6})t}\right)$$
$$\times \left(1 - e^{-(\lambda_{IM11}+\lambda_{IM12}+\lambda_{B6}+\lambda_{T3}+\lambda_{F3}+\lambda_{I3}+\lambda_{FQC5})t}\right)$$
$$\times \left(1 - e^{-(\lambda_{IM11}+\lambda_{IM12}+\lambda_{B6}+\lambda_{T3}+\lambda_{F3}+\lambda_{I3}+\lambda_{FQC6})t}\right).$$

In order to formally verify the preceding equation, we first formalize the notion of exponentially distributed random variable in HOL as follows:

Definition 36.6: ⊢ ∀ p X c. exp _ dist p X c =

∀ t. (CDF p X t = if 0 ≤ t then 1 - exp (-c * t) else 0)

The function exp _ dist guarantees that the CDF of the random variable X is that of an exponential random variable with a failure rate c in a probability space p. We classify a list of exponentially distributed random variables based on this definition as follows:

Definition 36.7: ⊢ (∀ p L. exp _ dist _ list p L [] = T) ∧

∀ p h t L. exp_dist_list p L (h::t) =
      exp_dist p (HD L) h ∧ exp_dist_list p (TL L) t

where the symbol T stands for logical *True*. The function exp _ dist _ list accepts a list of random variables L, a list of failure rates and a probability space p. It guarantees that all elements of the random variable list L are exponentially distributed with the corresponding failure rates, given in the other list, within the probability space p. For this purpose, it utilizes the list functions HD and TL, which return the *head* and *tail* of a list, respectively.

By using the definitions mentioned earlier, we can formally verify the reliability expression of the railway system, given in Equation 36.7, in HOL as follows:

Theorem 36.5: ⊢ ∀ p X _ T1 X _ FQC1 X _ FQC2 X _ F1 X _ I1
X _ IM1 X _ IM2 X _ B1 X _ IM3 X _ IM4

X_B2 X_T2 X_FQC3 X_FQC4 X_F2 X_I2 X_IM5 X_IM6 X_B3 X_IM7 X_IM8 X_B4 X_T3
X_FQC5 X_FQC6 X_F3 X_I3 X_IM9 X_IM10 X_B5 X_IM11 X_IM12 X_B6 C_FQC1
C_FQC2
C_F1 C_I1 C_IM1 C_IM2 C_B1 C_IM3 C_IM4 C_B2 C_T2 C_FQC3 C_FQC4 C_F2 C_I2
C_IM5 C_IM6 C_B3 C_IM7 C_IM8 C_B4 C_T3 C_FQC5 C_FQC6 C_F3 C_I3 C_IM9
C_IM10
C_B5 C_IM11 C_IM12 C_B6.
(A1): 0 ≤ t ∧ (A2): prob_space p ∧
(A3): in_events p [X_T1; X_FQC1; ···; X_B6] t ∧
(A4): mutual_indep p
      (rel_event_list p [X_T1; X_FQC1; ···; X_B6] t)) ∧
(A5): exp_dist_list p
      [X_T1; X_FQC1; ···; X_B6] [C_FQC1; C_FQC2; ···; C_B6] ⇒
   (prob p (railway_RBD p (rel_event p T1 t) (rel_event p FQC1 t)
      (rel_event p FQC2 t) ··· (rel_event p B6 t)) =
   1 - (list_prod o one_minus_list of
   (λa. list_prod (exp_func_list a t)))
    [[C_IM1;C_IM2;C_B1;C_T1;C_F1;C_I1;C_FQC1];
     [C_IM1;C_IM2;C_B1;C_T1;C_F1;C_I1;C_FQC2];
     [C_IM3;C_IM4;C_B2;C_T1;C_F1;C_I1;C_FQC1];
     [C_IM3;C_IM4;C_B2;C_T1;C_F1;C_I1;C_FQC2];
     [C_IM5;C_IM6;C_B3;C_T2;C_F2;C_I2;C_FQC3];
     [C_IM5;C_IM6;C_B3;C_T2;C_F2;C_I2;C_FQC4];
     [C_IM7;C_IM8;C_B4;C_T2;C_F2;C_I2;C_FQC3];
     [C_IM7;C_IM8;C_B4;C_T2;C_F2;C_I2;C_FQC4];
     [C_IM9;C_IM10;C_B5;C_T3;C_F3;C_I3;C_FQC5];
     [C_IM9;C_IM10;C_B5;C_T3;C_F3;C_I3;C_FQC6];
     [C_IM11;C_IM12;C_B6;C_T3;C_F3;C_I3;C_FQC5];
     [C_IM11;C_IM12;C_B6;C_T3;C_F3;C_I3;C_FQC6]])

In the preceding theorem, we have replaced the events that are associated with the railway components in function railway _ RBD with their corresponding random variable form by using Definition 36.1. It allows us to assign the exponential failure distribution with the random variables that correspond to the railway components. The assumptions A1 and A2 ensure that the time index must be positive, and p is a valid probability space. The assumptions ⊢ A3 and A4 guarantee that the events associated with the railway components are in events space p and mutually independent in the probability space p. The predicate in _ events takes a probability space p, a list of random variables, and a time index t and makes sure that each element in the given random variable list is in event space p. The last assumption (A5) ensures that the random variables that are exponentially distributed are assigned the corresponding failure rates. The conclusion of Theorem 36.5 models Equation 36.7. The proof of Theorem 36.5 utilizes Theorem 36.4, Lemma 36.1, and some fundamental axioms of probability theory.

The distinguishing features of the formally verified Theorem 36.5, compared to simulation-based reliability analysis of the railway system [11], include its generic nature and guaranteed correctness. All variables in Theorem 36.5 are universally quantified and can thus be specialized to obtain the reliability of any railway system for any given failure rates. The correctness of our results is guaranteed thanks to the involvement of a sound theorem prover in their verification, which ensures that all required assumptions for the validity of the results are accompanying the theorem. Unlike the work presented by Dazi et al. [11], the formally verified reliability result of Theorem 36.5 is sound and obtained through a rigorous reasoning process during the mechanization of their proofs. To the best of our knowledge, the benefits mentioned earlier are not shared by any other computer-based railway system reliability analysis approach.

36.7 Conclusion

The safe operation of high-speed trains has been the highest priority of railway companies around the world. However, their reliability analysis has been carried out using informal system analysis methods, such as simulation or paper-and-pencil, which do not ensure accurate results. The accuracy of the reliability results for railway systems is very critical since even minor flaws in the analysis could lead to the loss of many human lives or cause heavy financial setbacks. In order to achieve this goal and overcome the inaccuracy limitations of the traditional reliability analysis techniques, we propose to build upon the recent formalization of RBDs to formally reason about the reliability of high-speed railway systems using HOL theorem proving. As an application, we formally verified the reliability expressions of the a railway system designed for the Italian high-speed railways.

References

1. Liu, J., Li, S., Jiang, Y., and Krishnamurthy, M.: Reliability evaluating for traction drive system of high-speed electrical multiple units. In: *Transportation Electrification Conference and Expo*, Institute of Electrical and Electronics Engineers, Piscataway, NJ (2013) 1–6.
2. Trivedi, K. S.: *Probability and Statistics with Reliability, Queuing and Computer Science Applications*. John Wiley & Sons, Hoboken, NJ (2002).
3. ReliaSoft: http://www.reliasoft.com/ (2014).
4. ASENT: https://www.raytheoneagle.com/asent/rbd.htm (2016).
5. Robidoux, R., Xu, H., Xing, L., Zhou, M.: Automated modeling of dynamic reliability block diagrams using colored Petri nets. *IEEE Transactions on Systems, Man and Cybernetics, Part A: Systems and Humans* **40** (2) (2010) 337–351.
6. Norman, G., and Parker, D.: Quantitative verification: Formal guarantees for timeliness, reliability and performance. The London Mathematical Society and the Smith Institute (2014) http://www.prismmodelchecker.org/papers/lms-qv.pdf.
7. Ahmad, W., Hasan, O., Tahar, S., and Hamdi, M. S.: Towards the formal reliability analysis of oil and gas pipelines. *In: Intelligent Computer Mathematics*. Volume 8543 of LNCS. Springer, Berlin (2014) 30–44.
8. Gordon, M. J. C.: Mechanizing programming logics in higher-order logic. In: *Current Trends in Hardware Verification and Automated Theorem Proving*. Springer, Berlin (1989) 387–439.
9. Ahmed, W., Hasan, O., and Tahar, S.: Formalization of reliability block diagrams in higher order logic. *Journal of Applied Logic* **18** (2016) 19–41.
10. Mhamdi, T., Hasan, O., and Tahar, S.: On the formalization of the Lebesgue integration theory in HOL. In: *Interactive Theorem Proving*. Volume 6172 of LNCS. Springer, Berlin (2011) 387–402.

11. Dazi, G., Savio, S., and Firpo, P.: Estimate of components reliability and maintenance strategies impact on trains delay. In: *European Conference on Modelling and Simulation.* (2007) 447–452.
12. DNV-GL: http://www.dnvgl.com/oilgas/ (2015).
13. Signoret, J. P., Dutuit, Y., Cacheux, P. J., Folleau, C., Collas, S., and Thomas, P.: Make your Petri nets understandable: Reliability block diagrams driven Petri nets. *Reliability Engineering and System Safety* **113** (2013) 61–75.
14. PRISM: www.cs.bham.ac.uk/ dxp/prism (2015).
15. Herbert, L. T., and Hansen, Z. N. L.: Restructuring of workflows to minimise errors via stochastic model checking: An automated evolutionary approach. *Reliability Engineering and System Safety* **145** (2016) 351–365.
16. Lu, Y., Peng, Z., Miller, A. A., Zhao, T., and Johnson, C. W.: How reliable is satellite navigation for aviation? Checking availability properties with probabilistic verification. *Reliability Engineering and System Safety* **144** (2015) 95–116.
17. Hurd, J.: *Formal Verification of Probabilistic Algorithms.* PhD Thesis, University of Cambridge, Cambridge, UK (2002).
18. Hölzl, J., and Heller, A.: Three chapters of measure theory in Isabelle/HOL. In: *Interactive Theorem Proving.* Volume 6172 of LNCS. Springer, Berlin (2011) 135–151.
19. Hasan, O., and Tahar, S.: Formal verification of tail distribution bounds in the HOL theorem prover. Mathematical *Methods in the Applied Sciences* **32** (4) (2009) 480–504.
20. Hasan, O., and Tahar, S.: Formalization of the standard uniform random variable. *Theoretical Computer Science* **382** (1) (2007) 71–83.
21. Hasan, O., and Tahar, S.: Formalization of continuous probability distributions. In: *Automated Deduction.* Volume 4603 of LNCS. Springer, Berlin (2007) 2–18.
22. Hasan, O., Tahar, S., and Abbasi, N.: Formal reliability analysis using theorem proving. *IEEE Transactions on Computers* **59** (5) (2010) 579–592.
23. Abbasi, N., Hasan, O., and Tahar, S.: An approach for lifetime reliability analysis using theorem proving. *Journal of Computer and System Sciences* **80** (2) (2014) 323–345.
24. Ahmad, W., Hasan, O., and Tahar, S.: Formal reliability analysis of wireless sensor network data transport protocols using HOL. In: *Wireless and Mobile Computing, Networking and Communications,* Institute of Electrical and Electronics Engineers, Piscataway, NJ (2015) 217–224.
25. Ahmed, W., Hasan, O., and Tahar, S.: Towards formal reliability analysis of logistics service supply chains using theorem proving. In: *Implementation of Logics* Volume 40, EPiC Series in Computing. Suva, Fiji (2015) 111–121.
26. Ahmad, W., Hasan, O., Tahar, S., and Hamdi, M. S.: Formal reliability analysis of oil and gas pipelines. *Proceedings of the Institution of Mechanical Engineers, Part O: Journal of Risk and Reliability* (2017) 1–15.
27. Ahmed, W., and Hasan, O.: Towards the formal fault tree analysis using theorem proving. In: *Intelligent Computer Mathematics.* Volume 9150 of LNAI. Springer, Berlin (2015) 39–54.
28. Ahmad, W., and Hasan, O.: Formal availability analysis using theorem proving. In: *International Conference on Formal Engineering Methods.* Volume 10009 of LNCS. Springer, Berlin (2016) 1–16.
29. Harrison, J.: *Formalized Mathematics.* Technical Report 36, Turku Centre for Computer Science, Turku (1996).
30. Gordon, M. J., and Melham, T. F.: *Introduction to HOL: A Theorem Proving Environment for Higher Order Logic.* Cambridge University Press, Cambridge, UK (1993).
31. Church, A.: A formulation of the simple theory of types. *Journal of Symbolic Logic* **5** (1940) 56–68.
32. Milner, R.: A theory of type polymorphism in programming. *Journal of Computer and System Sciences* **17** (1977) 348–375.
33. Narasimhan, K.: Reliability engineering: Theory and practice. *The TQM Magazine* **17** (2) (2005) 209–210.
34. Ahmed, W.: *Formal Reliability Analysis of Railway Systems Using Theorem Proving Technique.* http://save.seecs.nust.edu.pk/train/ (2016).

37

Roles and Responsibilities for New Built, Extension, or Modernization of a Public Transport System: A Walk through the Life Cycle

Peter Wigger

CONTENTS

37.1 Introduction

For the example of a public transport system, the roles and responsibilities as well as the general activities throughout the life cycle of the system are described. Material from MODSafe research* [1] and the state-of-the-art life cycle concept as per EN 50126 [2] is therefore used, and for each life cycle phase, it will be described who has which

---

* MODSafe—Modular Urban Transport Safety and Security Analysis—is one of the latest projects in the European Transport sector under the Seventh Framework Programme (FP7) for Research and Technological Development of the European Union, performed from 2008 to 2012. The purpose of the MODSafe project was the development of a safety analysis and model reference for future urban guided transport projects. Even if the rail safety landscape in urban guided transport is highly diversified, the sector will benefit from some kind of harmonization. TÜV Rheinland was the project coordinator on behalf of the European Commission and furthermore responsible for the life cycle work package—in the person of Peter Wigger, the author of the chapter at hand. The raw material for the chapter at hand has been taken from the MODSafe deliverables respectively from TÜV Rheinland in-house sources. Refer to http://www.modsafe.eu for details [1].

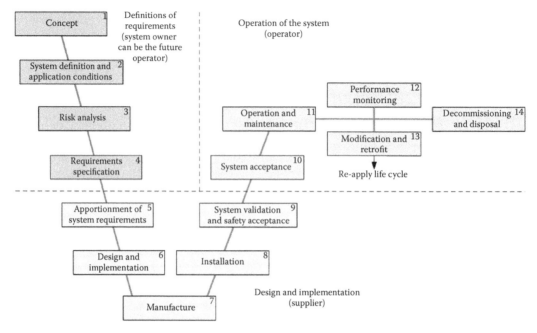

FIGURE 37.1
Key responsibilities.

responsibility for which activity, and how those interrelated to each other. Participants are, for example, the system owner, the future system supplier, the operator, the assessor, and the authority. Also, the interrelationship between system level aspects and the various subsystems as well as the principle structure of evidence documentation is discussed. Various diagrams will visualize the relationship. The general spilt of responsibilities is shown in Figure 37.1.

37.2 Life Cycle Phase 1: Concept

The concept phase (see Figure 37.2) includes successive studies such as the following:

- Prefeasibility studies including transport planning and traffic forecast setting up and justifying the proposed main characteristics of the project
- Feasibility studies and preliminary design studies
- Cost–benefit and financial analysis

Information is provided on the volumes of passengers to be used for relevant preliminary and, later, more detailed design studies as well as on the main characteristics of the interfaces with other transport modes.

The main responsibility for the concept phase is shared between the bodies in charge of the studies and the decision maker funding the project.

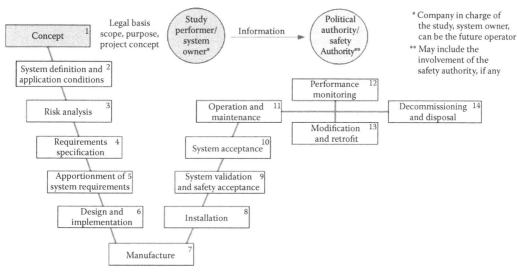

FIGURE 37.2
Life cycle phase 1—concept.

37.3 Life Cycle Phase 2: System Definition

In the early phase of system definition (see Figure 37.3), those comprehensive system safety concepts should be taken into account which might have an impact on the system decision and overall system design. These are the following:

- Basic operational concept (including grade of automation)
- Safety concept
- Safety concept in tunnels (if applicable)

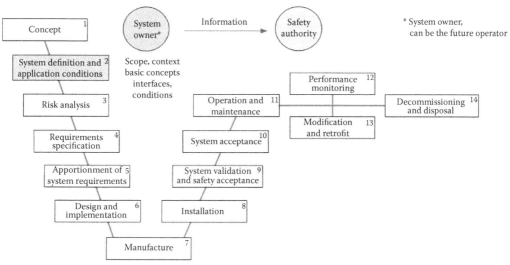

FIGURE 37.3
Life cycle phase 2—system definition.

- Evacuation and rescue concept
- Fire protection concept
- Environmental concept
- Concept for people with reduced mobility
- Integration of all different transport modes
- Interchanges with all transport modes including public transport internal interchanges

The main responsibility lies with the system owner in simple cases of project development, but may involve other actors depending on the project.

37.4 Life Cycle Phase 3: Risk Analysis

In particular, the risk analysis (see Figure 37.4) should include the following:

- Identification of the hazards and risks for human lives (users, staff, third party)
- Material and environmental injuries that can be caused by the system
- Consideration of events that cause hazards (trigger events)
- Consideration of accident scenarios (consequence analysis)
- Consideration of external events, which may cause internal failures of the system and consequently cause events as described earlier

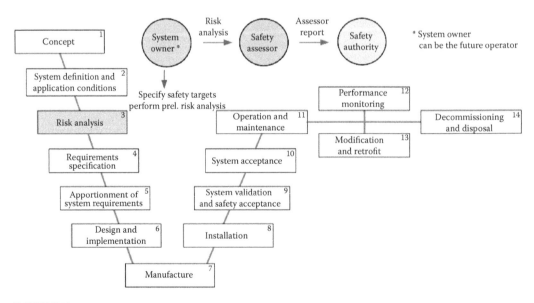

FIGURE 37.4
Life cycle phase 3—risk analysis.

- Consideration of external events, which directly influence the system and/or cause internal failures of the system and consequently cause preceding events
- Evaluation of risks caused by the system as a combination of severity and frequency of an accident (without putting certain measures in place to reduce the probability, but considering possible external events)
- Identification of mitigation measures (input safety requirements specification)

The main responsibility lies with the system owner, who might be checked and audited by the independent safety assessor.

37.5 Life Cycle Phase 4: Requirement Specification

In particular, the system requirement specification (see Figure 37.5) should describe the following:

- Type of system for which the application is planned
- Requirements with which the application should comply
- Standards, regulations, and operator's requirements which should be considered
- Safety requirements particularly derived from a risk analysis, with which the system should comply
- The interfaces by which the system is interacting with other systems: this should comprise technical components as well as user interfaces

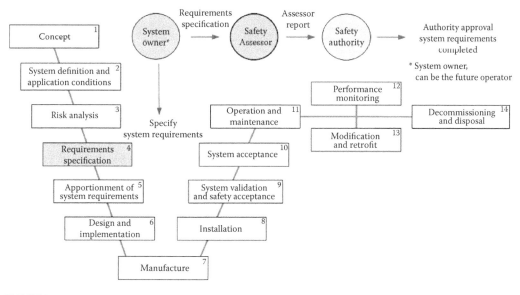

FIGURE 37.5
Life cycle phase 4—requirement specification.

- Performance and reaction times which the system should meet
- Methods of verification for the requirements after completion (requirements should have clear acceptance criteria)
- Separation of safety-related requirements from non-safety-related requirements with well-defined interfaces
- Specification of external measures necessary to achieve the requirements

The main responsibility lies with the system owner, who might be checked and audited by the independent safety assessor.

37.6 Life Cycle Phase 5: Apportionment of System Requirements

This phase (see Figure 37.6) describes the division of the overall system/system functions into subfunctions and the corresponding assignment of safety integrity requirements.

- Starting with the list of identified hazards from the initial risk analysis phase, and under consideration of the related hazard control measures and system requirements, tolerable hazard rates (THRs) can be assigned to individual safety related functions/subfunctions.
- Following the principle described by European Committee for Electrotechnical Standardization (CENELEC), THRs can be transferred to so-called safety integrity levels (SILs). Safety integrity correlates to the probability of failure to achieve required safety functionality.

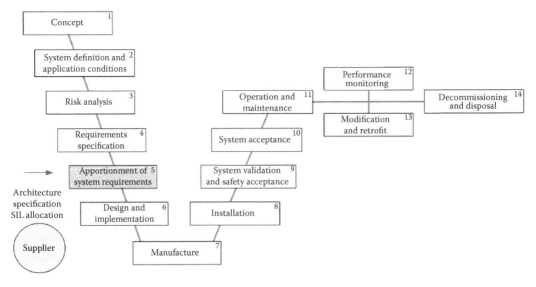

FIGURE 37.6
Life cycle phase 5—apportionment of system requirements.

- The concept of SILs is used as a means of creating balance between measures to prevent systematic faults and random failures.
- SIL is defined by a range of values for tolerable rates of hazardous events and defines measures to be implemented into the design and procedures to overcome/avoid systematic faults.

The main responsibility for the system architecture specification and safety requirement assignment process, and the apportionment of system requirements, lies with the supplier and is checked by the independent safety assessor.

A structured top–down approach should be followed, from risk analysis to system requirements on the overall transportation system level, to be refined to system functions, system architecture, and technical subsystems as shown in Figure 37.7. It is recommended to cluster the requirements, for example, into functional, safety, operational, environmental, and specific local application requirements in order to ease traceability into design and requirement traceability for compliance evidence purposes. Requirements should be refined to subsystems (e.g., signaling and rolling stock), using by the same clustering as mentioned earlier.

Further top–down refinement should be applied from subsystem level to subsystem components as shown in Figure 37.8 (e.g., signaling/communication-based train control [CBTC] onboard and trackside). Refer to the underlying life cycles in phase 6 as well.

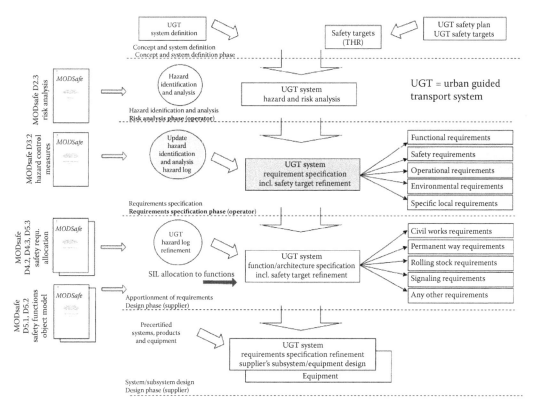

FIGURE 37.7
Structured top–down approach.

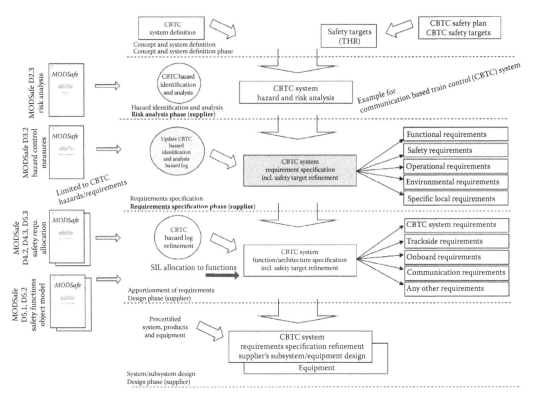

FIGURE 37.8
Structured top–down approach—example of CBTC system.

37.7 Life Cycle Phase 6: Design

Typical key processes, principles, and general requirements for design (see Figure 37.9) are as follows:

- Organization: The supplier should establish an organization as appropriate, ensuring compliance with the relevant requirements. This should, in particular, include a safety management and verification and validation (V&V) organization appropriate for the relevant safety integrity of the system to be designed and developed.

- Design and development process: The supplier should establish a design and development process as appropriate, ensuring compliance with the relevant system requirements. This should also include a process to resolve failures and incompatibilities.

- Quality: The supplier should establish a quality management system including staff qualification and training, in particular covering design and development.

- Environment: The supplier should implement processes and measures to cope with environmental requirements for design and development.

- Health and safety: The supplier should implement processes and measures to cope with health and safety requirements during design and development.

Figure 37.10 shows the typical subsystems and safety-related aspects of urban guided transport systems.

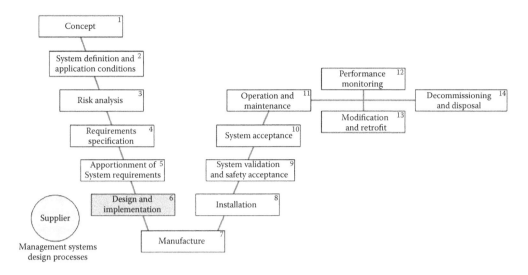

FIGURE 37.9
Life cycle phase 6—design.

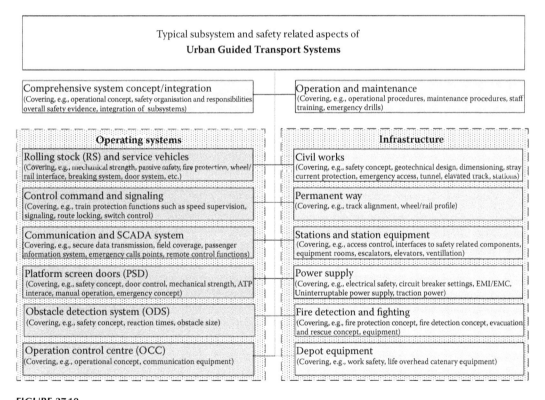

FIGURE 37.10
Typical subsystems.

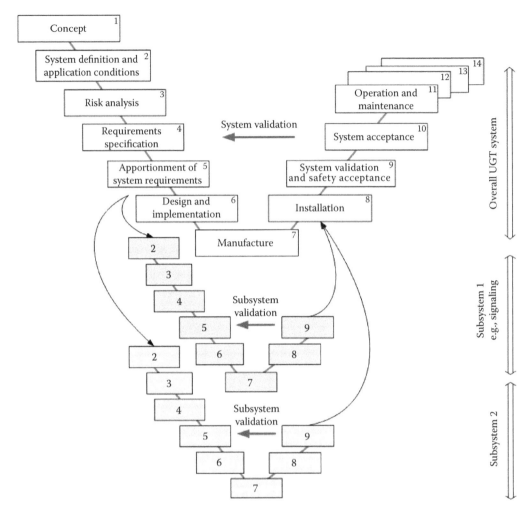

FIGURE 37.11
Life cycle phase 6—design—underlying subsystem life cycles.

In life cycle phase 6—design—the system design is usually split into multiple subsystems. Consequently, each subsystem is running its own design and development process, which can be seen as underlying life cycles as shown in Figure 37.11. Each subsystem—based on the system requirements assigned to the subsystem—may start with a subsystem internal risk analysis, followed by the subsystem requirement specification etc. Once the subsystem has been designed, manufactured and installed, and tested in the factory, it then contributes to the overall system life cycle in form of field installation and test.

37.8 Life Cycle Phase 7: Manufacture

Typical key processes, principles, and general requirements for manufacture from a life cycle approach perspective are as follows:

- Manufacturing process: The supplier establishes a manufacturing process as appropriate, ensuring compliance with the relevant design intent and requirements. It could also include a process to resolve failures and incompatibilities.
- Factory testing: The supplier establishes a factory test process as appropriate, ensuring compliance with the relevant test requirements. This also includes a process to resolve failures and incompatibilities.
- Quality: The supplier establishes a quality management system including staff qualification and training, in particular covering manufacture and factory processes.
- Environment: The supplier implements processes and measures to cope with environmental requirements for manufacture.
- Health and safety: The supplier implements processes and measures to cope with health and safety requirements during manufacture.

37.9 Life Cycle Phase 8: Installation

Typical key processes, principles, and general requirements for manufacture are as follows:

- Installation process: The supplier establishes a field installation process as appropriate, ensuring compliance with the relevant system integration intent and requirements. It could also include a process to resolve failures and incompatibilities.
- Field testing: The supplier establishes a field test process as appropriate, ensuring compliance with the relevant test requirements. This also includes a process to resolve failures and incompatibilities.
- Quality: The supplier establishes a quality management system including staff qualification and training, in particular covering field installation.
- Environment: The supplier implements processes and measures to cope with environmental requirements for installation.
- Health and safety: The supplier implements processes and measures to cope with health and safety requirements during installation.

37.10 Life Cycle Phase 9: System Validation and Safety Acceptance

Verification and validation applies on both UGT system level (ref. EN 50126) and subsystems, at least for the signaling and train control subsystem (ref. EN 50128 and EN 50129).

- This means that validation and safety acceptance applies first for the signaling and train control subsystem for the sole subsystem functions and second for the whole UGT system with respect to functional integration and interfaces incl. operational aspects.

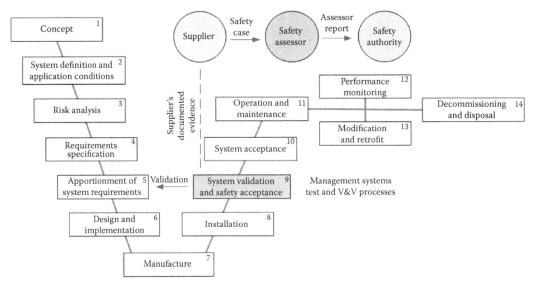

FIGURE 37.12
Life cycle phase 9—system validation and safety acceptance.

- Verification means the confirmation by examination and provision of objective evidence that the specified requirements have been fulfilled and is performed in each life cycle approach phase against the requirements of the previous life cycle approach phase.
- Validation means the confirmation by examination and provision of objective evidence that the particular requirements for a specific intended use have been fulfilled and are performed during life cycle approach phase 9 (system validation and safety acceptance) against the design intent as specified in life cycle approach phase 4 (requirements specification) as shown in Figure 37.12.

37.11 Life Cycle Phase 10: System Acceptance

EN 50126 "does not define an approval process by the safety regulatory authority" (ref. Scope section of the standard). The standard is applicable "for use by railway authorities and railway support industry" (ref. Scope)

- The approval process and system acceptance (see Figure 37.13) should be based on the life cycle approach.
- As an integral part, the CENELEC approach and process requires an appropriate safety management concept.
- Like the well-known quality management concept ISO 9001, the CENELEC safety management concept establishes the assumption that the level of safety of a

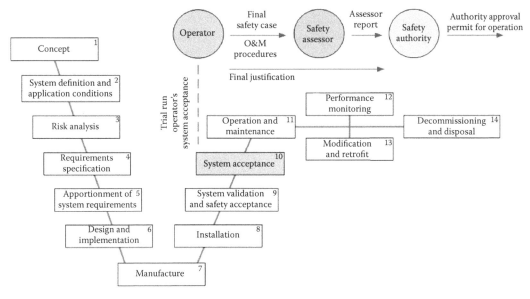

FIGURE 37.13
Life cycle phase 10—system acceptance.

complex product can seldom be proven simply by testing the finished product—especially with regard to electronic software and hardware components, this is impossible.

- Safety just like quality has to be an integral part of the design and production process in order to lead to products that possess a suitable level of safety.
- Furthermore, the required level of safety of a product is hugely dependent on the system it will be embedded in.
- The CENELEC safety management system therefore lends itself for combination with more prescriptive safety standard/regulation series, which may differ throughout the countries.

Following a clear top–down requirement refinement and traceability in life cycle phases 4–6, respective bottom–up requirement compliance evidence traceability should be applied in life cycle phases 7–10. This comes along with documented evidence in the form of a structured hierarchy of evidence documentation. Figure 37.14 shows a typical hierarchy of documentation for the signaling subsystem.

Following the bottom–up requirement compliance evidence approach, all subsystem evidence is feeding the overall transportation system level evidence documentation as underlying documentation. Figure 37.15 shows a typical hierarchy of documentation for the overall transportation system level. Also subsystems which have not consequently used and followed the railway application standards life cycle approach such as civil works can be linked as underlying evidence documentation in the form of civil works as build documentation. It is important that all subsystems clearly transfer to safety-related application conditions from subsystem to system.

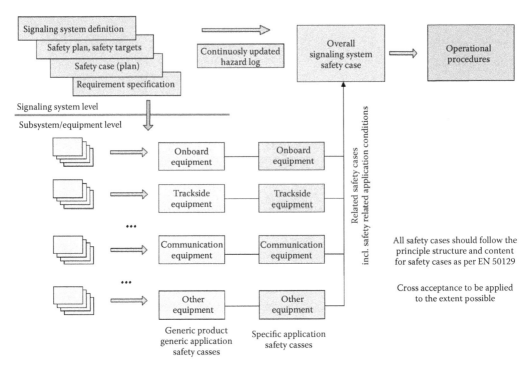

FIGURE 37.14
Hierarchy of documentation—example signaling system.

37.12 Life Cycle Phase 11: Operation and Maintenance

Typical key processes, principles, and general requirements for operation and maintenance (see Figure 37.16) from a life cycle approach perspective are as follows:

- Operations license: Commercial operation requires licensing by a safety authority. The license shall be available prior to start of operation.
- Operation plan: The operator develops a comprehensive operation plan for the complete system covering at least performance targets, operating strategies, handling of all safety relevant functions and systems, as well as key processes and staffing, etc.
- Maintenance plan: The operator/maintainer system develops a comprehensive maintenance plan for the complete system covering at least key reliability, maintainability, and availability (RAM) targets, maintenance conditions, maintenance strategies (preventive/corrective maintenance), facilities, logistics, costs, as well as key processes and staffing, etc.
- Operation procedures: The manual of procedures provide detailed instructions for the operation of the system such as, e.g., startup, shutdown, modes of operation and changing, vehicle dispatching, vehicle operation and function, controlling the system consoles, failure management, emergency response, passenger communications, service, and power distribution system management.

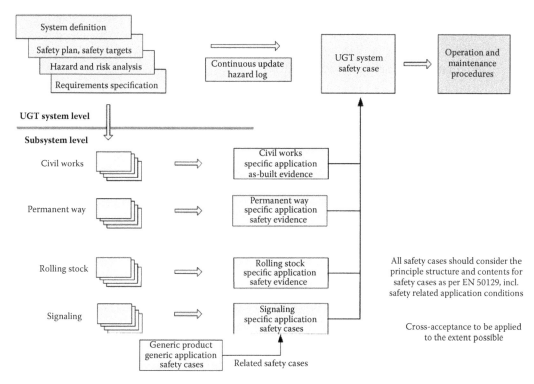

FIGURE 37.15
Hierarchy of documentation—overall transportation system level.

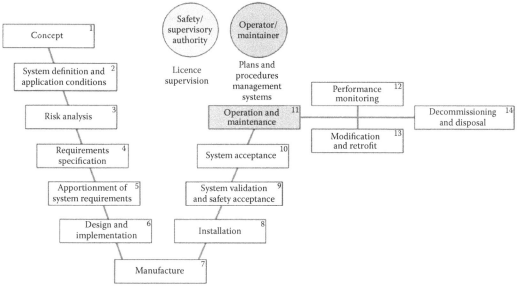

FIGURE 37.16
Life cycle phase 11—operation and maintenance.

- Maintenance procedures: The manual of maintenance procedures provide detailed maintenance instructions for the system such as, e.g., main steps, frequency and duration, necessary equipment and components, execution of work, results, documentation, and effort estimates.

- Driverless/unattended operation: In case of driverless/unattended operation, the operational concept and procedures reflect the system features (e.g., evacuation and rescue concept and safety concept in tunnels) and various functions (e.g., obstacle detection) that are to be performed by technical means (instead of by a driver). The initial version of the operational concept and procedures or at least the initial concepts with respect to driverless/unattended operation is available during the early life cycle phases 1–4.

- Quality and safety: The operator/maintainer establishes a performance and quality policy, a quality management system including staff qualification, and training. The quality system is maintained throughout the entire operation lifetime of the system/the entire life cycle phase 11 (which may last decades).

- Environment: The operator/maintainer establishes an environmental policy, an environmental management system to fulfill all legal and operational provisions, as well as emergency preparation and response procedure for the event of an environmental threat. The environmental management system is maintained throughout the entire operation lifetime of the UGT system/the entire life cycle approach phase 11 (which may last decades).

- Contingency: The operator creates and introduces a contingency plan before commencing the operation. A contingency plan has the objective to document and to instruct how to operate in an emergency case. The contingency plan is available during life cycle approach phase 10 at the latest since it is subject to approval and a prerequisite for licensing.

- Authority approval: The initial versions of the operation plan and maintenance plan or at least the initial concepts are available during the early life cycle phases 1–4. The operator's/maintainer's management systems, operation, and maintenance procedures are available during life cycle approach phase 10 at the latest since they are subject to safety authority approval and a prerequisite for licensing.

- Supervision: The supervisory authority supervises the operation and maintenance of the UGT system. This is an ongoing task and applies throughout the entire operation lifetime of the UGT system/the entire life cycle approach phase 11.

- Outsourcing: Operators can outsource (parts of) the operation and maintenance and functions. However, the operators' obligations to the supervisory authorities remain unchanged.

37.13 Life Cycle Phase 12: Performance Monitoring

Typical key processes, principles, and general requirements for performance monitoring (see Figure 37.17) from a life cycle approach perspective are as follows:

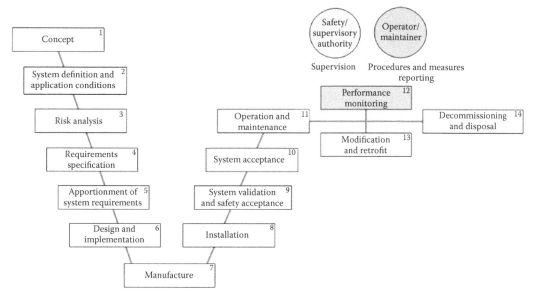

FIGURE 37.17
Life cycle phase 12—performance monitoring.

- Data establishment: The operator of the UGT system implements processes and measures for the regular collection of relevant performance/reliability, availability, maintainability, and safety (RAMS) data.

- Analysis: The operator of the UGT system implements a process for the regular analysis and statistical evaluation of the performance/RAMS data. A review of results in improvement/corrective actions for the further operation and maintenance of the system as needed.

- Record: The operator of the UGT system regularly records and reports on the results of the performance monitoring, including corrective actions, if any. The report is generally submitted to the supervisory authority.

37.14 Life Cycle Phase 13: Modification and Retrofit

Basically, the following types of modification and retrofit (see Figure 37.18) can be assumed:

- Complete shutdown of an existing UGT system to be substituted by a new one
- Extension of an existing UGT system, e.g., extension of a line or building of a new line
- Modernization/upgrade of a subsystem, e.g., renewal of the signaling and control system or upgrade to a CBTC system.
- Renewal of rolling stock, modernization, or new vehicles
- Upgrade of an existing UGT system operational mode, e.g., upgrade to driverless operation

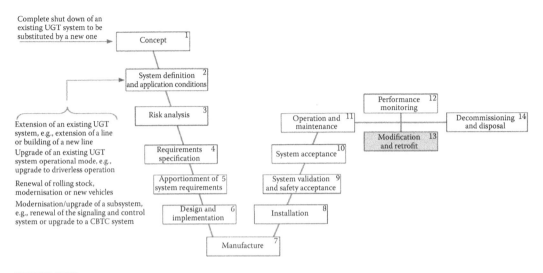

FIGURE 37.18
Life cycle phase 13—modification and retrofit.

37.15 Life Cycle Phase 14: Decommissioning and Disposal

Typical key processes, principles, and general requirements for decommissioning and disposal are as follows:

- Planning of the decommissioning and disposal and performance thereof
- Preparation of a safety plan and hazard and risk analysis and risk evaluation
- No RAM activities
- Waste management
- Often performed in combination with/as a prestep for modification and retrofit.

37.16 Summary and Outlook

The chapter at hand describes the life cycle roles and responsibilities and key activities for urban guided transport systems, following a structured approach ensuring completeness and traceability of safety and RAM aspects. One can expect this approach as reasonable suggestions for the future, aiming to contribute to the European drive for harmonization and to simplify the upgrade/modernization or new construction of urban guided transport systems. Cross acceptance of technical systems, processes, and approvals is one of the key attempts to benefit all parties involved, be it the manufacturers and suppliers, the operators, or the safety authorities.

References

1. TÜV Rheinland. http://www.modsafe.eu.
2. CENELEC (Comité Européen de Normalisation Électrotechnique). EN 50126: *Railway Applications—Specification and Demonstration of Reliability, Availability, Maintainability and Safety (RAMS)*. CENELEC, Brussels.

38

Holistic View on the Charm and Challenge of CENELEC Standards for Railway Signaling

Attilio Ciancabilla and Stephan Griebel

CONTENTS

38.1 Introduction

CENELEC is the European Committee for Electrotechnical Standardization and is responsible for standardization in the field of electrotechnical engineering. According to its website, "CENELEC prepares voluntary standards, which help facilitate trade between countries, create new markets, cut compliance costs and support the development of a Single European Market."

Figure 38.1 shows the CENELEC members (grey) and CENELEC affiliates (dark grey).

As CENELEC not only aims at creating market access at a European level, but also facilitates an international exchange of goods, it is vital that it positions itself with regard to the worldwide organization for standardization in the electrotechnical engineering field, i.e., the International Electrotechnical Commission (IEC). This was done in 1996 with the "Dresden agreement" with the objective of avoiding the duplication of effort and reducing time when preparing standards. A new version of this agreement is now available under the name of "Frankfurt agreement," issued in 2016.

Standards developed by CENELEC are officially recognized as "European Standards" by the European Union. The same applies to standards developed by the following:

- CEN, the regional mirror body to the international counterpart International Organization for Standardization (ISO)
- European Telecommunications Standards Institute, the regional mirror body to the international counterpart International Telecommunication Union, telecommunication standardization sector

FIGURE 38.1
CENELEC members and CENELEC affiliates.

Within CENELEC, the technical committee TC9X and its subcommittees have the following tasks:

- To prepare a single set of electrotechnical standards in the field of railway applications for Europe, in order to achieve a free market for goods and services inside Europe
- To mirror Technical Committee 9 of the IEC following the Dresden agreement

This is detailed in their business plan (CLC/TC9X/Sec/567C).

A challenge of developing standards for the railway field is that—although the topics are often similar to those covered in other industrial fields—the existing constraints of the coherence of the railway system have to be taken into account.

The Technical Committee for Railways TC9X is organized into three subcommittees SC9XA, SC9XB, and SC9XC, dealing with the railway system is a fully integrated system and its interfaces of signaling, rolling stock (RST), and fixed installations for energy supply.

The following sections will first highlight the basic principles for developing European standards for signaling in railways as well as its working environment.

Secondly, they will probe more profoundly into current developments in the following areas:

- Urban rail-related to signaling (nonmetro)
- Functional safety
- Compatibility between RST and train detection systems
- Information technology (IT) security

And, last but not least, a brief conclusion and outlook are presented.

38.2 Working Environment for and Basics of Development of CENELEC Standards

The relevant basics for railway signaling standardization are detailed in EN 45020 [1].

It holds that a "normative document on a technical subject, if prepared with the cooperation of concerned interests by consultation and consensus procedures, is presumed to constitute an acknowledged rule of technology at the time of its approval."

An acknowledged rule of technology is defined as a "technical provision acknowledged by a majority of representative experts as reflecting the state of the art," with *state of the art* being defined as the "developed stage of technical capability at a given time as regards products, processes and services, based on the relevant consolidated findings of science, technology and experience."

38.3 Areas for Standardization

38.3.1 Urban Rail Related to Signaling (Nonmetro)

Based on Mandate M/486 (for programming and standardization bodies in the field of urban rail), the so-called urban rail survey group with participants from various stakeholders in the field of urban rail identified the need for an EN standard to cover "signaling and other safety systems for trams, light rail and other nonmetro urban rail systems."

After an official enquiry among the national standardization committees for signaling, CENELEC SC9XA set up a survey group SGA14 to refine the scope of the future standard.

It was found that existing railway and highway standards may not appropriately address the need for standardization for some urban rail systems, particularly tram and light rail.

The urban rail survey group identified the need for a new standard to cover signaling on tramways and other urban rail systems which do not fall directly within either existing railway or highway standards. This would typically be the case for parts of systems which are along off-street alignment and which operate to "line of sight" or automatic block signaling with intermittent train control (i.e., as defined in IEC 62290 [2]: GOA0 and GOA1 with intermittent control systems).

The proposal of the survey group then states that such systems are in use in numerous member states within the European Union operating urban rail systems. They control points manage train movements along individual lines and prevent conflicts at junctions as well as at level crossings with road and pedestrian traffic. While adopting many of the functional requirements and safeguards used in standard traffic signal controllers, additional functionality is required and currently realized to fulfill the needs of urban rail.

The given justification of the need and the derived benefits of this future standard to be developed are directly taken from the "fiche" worked out by the urban rail survey group for the present standardization project as follows:

> A functional requirements specification for signaling and other safety systems for tram, light rail and cat 4-system is currently not in existence. The proposed new standard will not overlap with the scope and requirements of EN 62290 This standard might support system upgradeability, and support informed decisions when and how to make appropriate selection of safety systems depending on the characteristics of any particular network.

The standard should help avoid the repetition of inappropriate and sometimes unsuitable choices made in some places in the past and prevent over-engineered and expensive solutions being imposed. The standard will provide a common understanding and a common framework of expected functionality for both transport companies and the supply industry.

As of November 2015, the official working group for this new standard has been launched in the CENELEC Committee for Railway Signaling SC9XA.

38.3.2 Compatibility between Rolling Stock and Train Detection Systems: Track Circuits

The technological diversity of train detection equipment in Europe makes the definition of interference limits a complex task.

The basic conflict between RST and detection systems is due to the following:

- RST uses electrical switching power for traction and auxiliary systems, which results in high return currents at the fundamental frequency and its harmonics.
- Track circuits (TCs) use the rails to communicate between transmitter and receiver. TC signals are low power ones, compared to the return currents of the RST.

Combining RST and detection systems requires robust electromagnetic compatibility management in order to achieve a safe and reliable interface between the individual subsystems (RST, command and control and signaling [CCS] and Energy supply).

To meet the requirements for compatibility between train detection systems and RST in the future and to achieve interoperability and free movement within the European Union, it is necessary to define a "frequency management" including the complete set of interface requirements.

The train detection systems, track circuits, and axle counters are an integral part of the CCS trackside subsystem in the context of the rail interoperability directive. The relevant technical parameters are enumerated in the CCS and locomotives and passenger rolling stock Technical Specification for Interoperability and will be specified in the mandatory specification (index 77 of CCS TSI). In this specification, however, some parameters are still declared as "open points" and are being closed by the European Railway Agency (ERA). Meanwhile, a technical specification (CLC/TS 50238-2) has been published by CENELEC. It can be used for the interim period to ascertain conformity of individual train detection systems to the requirements of the TSIs. Another standard published by CENELEC is EN 50617-1:2015 that specifies the technical parameters of track circuits associated with the disturbing current emissions limits for RST in the context of interoperability defined in the form of a frequency management. The limits for compatibility between RST and track circuits currently proposed in this standard allow provision for known interference phenomena linked to traction power supply and associated protection (overvoltage, short-circuit current, and basic transient effects such as in-rush current and power cutoff). These effects are assessed using modeling tools that have been verified by the past European research project RAILCOM (electromagnetic compatibility between rolling stock and rail-infrastructure encouraging European interoperability).

The mask obtained is the result of a numerical simulation that, for each power system, has taken into account the characteristics of power supply system, the emissions of trains associated to the various types of electric drives employed any electrical resonances, the electrical characteristics of the track of electrical transients having a duration exceeding 3 s, and, in general, all the parameters which influence the emission of harmonic current by the RST.

The status quo of available compatibility limits across member states was established. In addition to that, a so-called frequency management, the emission limits for RST supplied under 50 Hz power systems, is visualized in the corresponding CENELEC standard.

38.3.3 Functional Safety

According to IEC 61508-1 [3], functional safety is that "part of the overall safety that depends on a system or equipment operating correctly in response to its inputs. Functional safety is achieved when every specified safety function is carried out and the level of performance required of each safety function is met."

The ability of a safety function to satisfactorily perform (under specific conditions) is known as "safety integrity" and is measured in safety integrity levels (SILs); see EN 50126:1999. The concept of SIL was primarily derived in the late 1980s: four levels, from 1 to 4, would define four sets of progressively more stringent requirements for software design. A natural extension was the equivalent definition of SIL for electronics hardware design and development. At the same time, the usage of microprocessors forced the move from a deterministic concept of risk management (inherent fail safety) to a probabilistic one, for which a *risk* was defined as the combination of "frequency" and "consequences" of a certain hazard.

In electronics, the frequency of a hazard was associated to the dangerous failure rate of the equipment performing the safety function. It was natural, at that point, to associate SIL, considered as an indicator of the integrity in respect to systematic faults (since more stringent requirements would bring to less residual possible bugs in the development of an electronic system), with failure rates that are an indicator of the integrity in respect to random faults.

As said earlier, the probabilistic approach considers risk as a combination of frequency and severity. The SIL is linked to frequency, but the correlation between frequency and severity that defines the maximum acceptable risk is not mandatorily defined in the CENELEC standards, since it is managed by the publications of the ERA.

The central standard for railway signaling, published by SC9XA, to cover these fundamental concepts of safety, evidence thereof, SILs, and the necessary measures of how to achieve it, is the CENELEC standard EN 50129:2003, *Railway Applications— Communication, Signalling and Processing Systems—Safety-Related Electronic Systems for Signalling* [4].

An official working group for a revision of EN 50129 was set up by the CENELEC Railway Signaling Committee SC9XA, the WGA15.

Together with a stringent and cooperative working procedure and a comprehensive list of precise tasks to tackle, this working group is now concentrating on various challenges such as the following:

- Integration of aspects of IT security in the framework of functional safety
- Treatment of programmable components
- Management of safety-related application conditions
- Revision of annex E containing the necessary measures for achieving a particular SIL for a function
- Cooperation with the new version of EN 50126, which is also in the process of revision in the overall CENELEC Railway Committee TC9X

38.3.4 IT Security

IT security requirements are rapidly growing in various branches of industry and are thus questioning current concepts and demanding answers for yet unsolved challenges.

On an international level, a variety of different standards and regulations on IT security already exist, such as, for example, the so-called common criteria or the international standard IEC 62443/ISA99. These documents emphasize that not only technical aspects are of importance to IT security, but ever more so additional factors such as operational circumstances, human aspects, management systems, and cultural influences.

CENELEC is looking at how the international standards and terminology about IT security could be adopted for the existing system of railway application standards in CENELEC for communication, signaling, and processing systems (EN 50129 [4], EN 50159 [5]).

Although both EN 50129 and EN 50159 already cover general issues such as access protection, answers are still missing when it comes to newly developing questions such as the relationship between safety and IT security. These questions have been particularly discussed in recent scientific publications.

This calls for a necessary framework that is a concept of how the international world of standards and terminology of IT security could be inserted into the existing railway application standards in CENELEC for communication, signaling, and processing systems:

- The framework shall predominantly deal with IT security aspects that are relevant to safety issues.
- The certificates for IT security should be part of the safety assessment and approval.
- The principles of risk analysis for safety issues could be adopted, such as the following:
  - Code of practice
  - Similar systems
  - Risk analysis
- The system could be compartmentalized according to IEC 62443 [6] into zones and conduits.
- IT security threats shall be treated in analogy with systematic faults, e.g., software faults, and are thus to be assessed purely qualitatively and not quantitatively.
- The IT security analysis shall take account of various parameters such as exposure of the assets, the capability of the attacker, or the possible safety impact.
- Various different IT security profiles could be defined.

38.4 Conclusion

Thus, CENELEC and its railway standardization body for railway signaling SC9XA is decisively contributing to achieve a free market for goods and services inside Europe by harmonizing requirements and thus fostering a framework in which innovation can thrive.

By dealing with cutting-edge issues, it provides the stakeholders, represented by the national committees for standardization, with the platform for translating existing approaches from standards into tailored standards for railway signaling.

In particular, in the field of IT security for functional safety, it will be necessary to align ever more closely with the IEC so as to facilitate cross-pollination and coherence with other fields of industry.

References

1. EN 45020:2006: *Standardization and Related Activities, General Vocabulary (ISO/IEC Guide 2:2004)*, Trilingual version EN 45020:2006, December 2006.
2. IEC 62290:2014: *Railway Applications—Urban Guided Transport Management and Command/Control Systems—Part 1: System Principles and Fundamental Concepts*, July 2014.
3. IEC 61508-1:2010: *Functional Safety of Electrical/Electronic/Programmable Electronic Safety-related Systems—Part 1: General Requirements*, April 2010.
4. EN 50129:2003: *Railway Applications—Communication, Signalling and Processing Systems—Safety-Related Electronic Systems for Signalling*, February 2003.
5. EN 50159:2010: *Railway Applications—Communication, Signalling and Processing Systems—Safety-Related Communication in Transmission Systems*, September 2010.
6. IEC 62443-2-1:2010: *Industrial Communication Networks—Network and System Security—Part 2-1: Establishing an Industrial Automation and Control System Security Program*, November 2010.

Appendix

A.1 Glossary

acceptance: The status achieved by a product, system or process once it has been agreed that it is suitable for its intended purpose.

accident: An unintended event or series of events resulting in loss of human health or life, damage to property or environmental damage.

application conditions: Those conditions which need to be met in order for a system to be safely integrated and safely operated.

approval: The legal act, often focused on safety, to allow a product, system or process to be placed into service.

assessment: Process to form a judgement on whether a product, system or process meets the specified requirements, based on evidence.

assessor: An entity that carries out an assessment.

assurance: Confidence in achieving a goal being pursued. Declaration intended to give confidence.

audit: A documented, systematic and independent examination to determine whether the procedures specific to the requirements

- comply with the planned arrangements,
- are implemented effectively and
- are suitable to achieve the specified objectives.

availability: The ability of a product to be in a state to perform a required function under given conditions at a given instant of time or over a given time interval assuming that the required external resources are provided.

collective risk: A risk, resulting from e.g. a product, process or system, to which a population or group of people is exposed.

commercial off-the-shelf software: Software defined by market-driven need, commercially available and whose fitness for purpose has been deemed acceptable by a broad spectrum of commercial users.

common cause failure: Failures of different items resulting from the same cause and where these failures are not consequences of each other.

compliance: A state where a characteristic or property of a product, system or process satisfies the specified requirements.

configuration management: A discipline applying technical and administrative direction and surveillance to identify and document the functional and physical characteristics of a configuration item, to control changes to those characteristics, to record and report change processing and implementation status and to verify compliance with specified requirements.

consequence analysis: To analyze the consequences of each hazard up to accidents and losses.

corrective maintenance: The maintenance carried out after fault recognition and intended to put a product into a state in which it can perform a required function.

designer: An entity that analyses and transforms specified requirements into acceptable design solutions that have the required safety integrity.

deterministic: Expresses that a behaviour can be predicted with certainty.

diversity: A means of achieving all or part of the specified requirements in more than one independent and dissimilar manner.

entity: A person, group or organisation that fulfils a role as defined in this standard.

equivalent fatality: An expression of fatalities and weighted injuries and a convention for combining injuries and fatalities into one figure for ease of evaluation and comparison of risks.

error: A discrepancy between a computed, observed or measured value or condition and the true, specified or theoretically correct value or condition.

fail-safe: A concept which is incorporated into the design of a product such that, in the event of a failure, it enters or remains in a safe state.

failure: The termination of the ability of an item to perform a required function.

failure mode: A predicted or observed manner in which the product, system or process under consideration can fail.

failure rate: The limit, if this exists, of the ratio of the conditional probability that the instant of time, T, of a failure of a product falls within a given time interval $(t, t + \Delta t)$ and the length of this interval, Δt, when Δt tends towards zero, given that the item is in an up state at the start of the time interval.

fault: The state of an item characterized by inability to perform a required function, excluding the inability during preventive maintenance or other planned actions.

function: A specified action or activity which may be performed by technical means and/or human beings and has a defined output in response to a defined input.

functional safety: The perspective of safety focused on the functions of a system.

generic product: Product (hardware and/or software) that can be used for a variety of installations, either without making any changes or purely trough the configuration of the hardware or the software (for example by the provision of application-specific data and/or algorithms).

hazard: A condition that could lead to an accident.

hazard analysis: An analysis comprising hazard identification, causal analysis and common cause analysis.

hazard log: The document in which hazards identified, decisions made, solutions adopted and their implementation status are recorded or referenced.

hazard rate: The rate of occurrence of a hazard.

implementation: The activity applied in order to transform the specified designs into their realisation.

independence (functional): Freedom from any mechanism which can affect the correct operation of more than one function as a result of either systematic or random failure.

independence (physical): Freedom from any mechanism which can affect the correct operation of more than one system/subsystem/ equipment as a result of random failures.

individual risk: A risk, resulting from e.g. a product, process or system, to which an individual person is exposed.

infrastructure manager: Any body or undertaking that is responsible in particular for establishing and maintaining railway infrastructure, or a part thereof, which may also include the management of infrastructure control and safety systems. The functions of the infrastructure manager on a network or part of a network may be allocated to different bodies or undertakings.

integration: The process of assembling the elements of a system according to the architectural and design specification, and the testing of the integrated unit.

life-cycle: Those activities occurring during a period of time that starts when the product, system or process is conceived and ends when the product, system or process is no longer available for use, is decommissioned and is disposed (if applicable).

logistic support: The overall resources which are arranged and organised in order to operate and maintain the system at the specified availability level at the required life-cycle cost.

maintainability: The ability of an item under given conditions of use, to be retained in, or restored to, a state in which it can perform a required function, when maintenance is performed under given conditions and using stated procedures and resources.

maintenance: The combination of all technical and administrative actions, including supervision actions, intended to retain a product in, or restore it to, a state in which it can perform a required function.

mission: An objective description of the fundamental task performed by a system.

mission profile: Outline of the expected range and variation in the mission with respect to parameters such as time, loading, speed, distance, stops, tunnels, etc., in the operational phases of the life-cycle.

negation: Enforcement of a safe state following detection of a hazardous fault.

negation time: Time span which begins when the existence of a fault is detected and ends when a safe state is enforced.

pre-existing software: All software developed prior to the application currently in question is classed as pre-existing software including commercial off-the-shelf software, open-source software and software previously developed but not in accordance with this European Standard.

preventive maintenance: The maintenance carried out at pre-determined intervals or according to prescribed criteria and intended to reduce the probability of failure or the degradation of the functioning of an item.

procedural safety: Aspects of the safety life-cycle which are governed by procedures and instructions (e.g. operational and maintenance procedures).

project management: The administrative and/or technical conduct of a project, including RAMS aspects.

project manager: An entity that carries out project management.

railway duty holder: The body with the overall accountability for operating a railway system within the legal framework.

railway undertaking: Means any public or private undertaking, whose activity is to provide transport of goods and/or passengers by rail on the basis that the undertaking must ensure traction; this also includes undertakings which provide traction only.

RAM plan: A documented set of time scheduled activities, resources and events serving to implement the organisational structure, responsibilities, procedures, activities, capabilities and resources that together ensure that an item will satisfy given RAM requirements relevant to a given contract or project.

RAMS management process: The activities and procedures that are followed to enable the RAMS requirements of a product or an operation to be identified and met. It provides a systematic and systemic approach to continually manage RAMS through the whole life-cycle.

reliability: The ability of an item to perform a required function under given conditions for a given time interval.

reliability growth: A condition characterised by a progressive improvement of a reliability performance measure of an item with time.

repair: Measures for re-establishing the required state of a system/sub-system/equipment after a fault/failure.

residual risk: Risk remaining once risk control measures have been taken.

restoration: Bring an item into a state where it regains the ability to perform its required function after a fault.

risk: The combination of expected frequency of loss and the expected degree of severity of that loss.

risk analysis: Systematic use of all available information to identify hazards and to estimate the risk.

risk assessment: The overall process comprising system definition, risk analysis and a risk evaluation.

risk based approach: In relation to safety, the risk based approach is a process for ensuring the safety of products, processes and systems through consideration of the hazards and their consequent risks.

risk evaluation: Procedure based on the Risk Analysis results to determine whether the tolerable risk has been achieved.

risk management: Systematic application of management policies, procedures and practices to the tasks of analysing, evaluating and controlling risk.

safe state: States in which the system is not allowed to fall back into a dangerous state if an additional failure occurs.

safety: Freedom from unacceptable risk to human health or to the environment.

safety authority: The body entrusted with the regulatory tasks regarding railway safety in accordance with the respective legal framework.

safety barrier: Any physical or non-physical means, which reduces the frequency of a hazard and/or a likely accident arising from the hazard and/or mitigates the severity of likely accidents arising from the hazard.

safety case: The documented demonstration that the product, system or process complies with the appropriate safety requirements.

safety function: A function whose sole purpose is to ensure safety.

safety integrity: Ability of a safety-related function to satisfactorily perform under all the stated conditions within a stated operational environment and a stated period of time.

safety integrity level: One of a number of defined discrete levels for specifying the safety integrity requirements of safety-related functions to be allocated to the safety-related systems.

safety management: The management structure which ensures that the safety process is properly implemented.

safety management process: That part of the RAMS management process which deals specifically with safety aspects.

safety plan: A documented set of time scheduled activities, resources and events serving to implement the organisational structure, responsibilities, procedures, activities, capabilities and resources that together ensure that an item will satisfy given safety requirements relevant to a given contract or project.

safety-related: Carries responsibility for safety.

software: An intellectual creation comprising the programs, procedures, rules, data and any associated documentation pertaining to the operation of a system.

software deployment: Transferring, installing and activating a deliverable software.

software maintenance: An action, or set of actions, carried out on software after deployment with the aim of enhancing or correcting its functionality.

system: Set of elements which interact according to a design, where an element of a system can be another system, called a subsystem and may include hardware, software and human interaction.

systematic failure: A failure due to errors, which cause the product, system or process to fail deterministically under a particular combination of inputs or under particular environmental or application conditions.

technical safety: That part of safety that is dependent upon the characteristics of a product, which derive from the system functional requirements and/or of the system design.

testing: The process of operating a product, system or process under specified conditions as to ascertain its behaviour and performance compared to the corresponding requirements specification.

traceability: Relationship established between two or more entities in a development process, especially those having a predecessor/successor or master/subordinate relationship to one another.

validation: Confirmation by examination and provision of objective evidence that the product, system or process is suitable for a specific intended use.

validator: An entity that is responsible for the validation.

verification: Confirmation by examination and provision of objective evidence that the specified requirements have been fulfilled.

verifier: An entity that is responsible for one or more verification activities.

A.2 Abbreviations

ALARP	as low as reasonably practicable
CBA	cost benefit analysis
CoP	code of practice
DRA	differential risk aversion
EFAT	equivalent fatality
EMC	electromagnetic compatibility
EMI	electromagnetic interference
FC	fault coverage
FMEA	failure mode and effects analysis
FMECA	failure mode, effects and criticality analysis
FRACAS	failure reporting analysis and corrective action system
FTA	fault tree analysis
GAME	globalement au moins equivalent
LAD	logistic and administrative delay

LCC	life-cycle cost
MACMT	mean active corrective maintenance time
MDT	mean down time
MEM	minimum endogenous mortality
MPMT	mean preventive maintenance time
MRT	mean repair time
MTBF/MOTBF	mean time between failures / mean operating time between failures
MTBM	mean time between maintenances
MTTR	mean time to restoration
MUT	mean up time
RAC	risk acceptance criteria
RAM	reliability, availability and maintainability
RAMS	reliability, availability, maintainability and safety
RC	repair coverage
SRAC	safety-related application conditions
THR	tolerable hazard rate
VPF	value of preventing a fatality

A.3 Typical Parameters and Symbols

In general any time-based parameter like MTBF can be converted/derived from the respective operated distance or operation cycles as well.

Definition and detailed guidance on mathematical treatment of RAM terms is given in EN 61703.

A.3.1 Reliability Parameters

TABLE A.1

Examples of Reliability Parameters

Parameter	Symbol	Dimension
Failure Rate	$\lambda(t)$	1/time, 1/distance, 1/cycle
Mean Up Time	MUT	time (distance, cycle)
Mean operating[a] Time To Failure (for non-repairable items)	MTTF	time (distance, cycle)
Mean operating[a] Time Between Failure (for repairable items)	MTBF	time (distance, cycle)
Mean Repair Time	MRT	time
Failure Probability	F(t)	dimensionless
Reliability (Success Probability)	R(t)	dimensionless

[a] According to EN 61703 and IEC 60050-191-2.

A.3.2 Maintainability Parameters

TABLE A.2

Examples of Maintainability Parameters

Parameter	Symbol	Dimension
Mean Down Time	MDT	time (distance, cycle)
Mean operating[a] Time Between Maintenance	MTBM	time (distance, cycles)
MTBM (corrective or preventive)	MTBM(c), MTBM(p)	time (distance, cycles)
Mean Time To Maintain	MTTM	time
MTTM (corrective or preventive)	MTTM(c), MTTM(p)	time
Mean Time To Restore	MTTR	time
Fault Coverage	FC	dimensionless
Repair Coverage	RC	dimensionless

[a] According to EN 61703 and IEC 60050-191-2.

"Operating" signifies the time, distance or cycles where the item under consideration is effectively in use.

A.3.3 Availability Parameters

TABLE A.3

Examples of Availability Parameters

Parameter	Symbol	Dimension
Availability	A	dimensionless
inherent	A i	
operational	A o	
Fleet Availability	FA	dimensionless
Schedule Adherence	SA	dimensionless or time

Under certain conditions, for instance constant failure rate and constant repair rate, the steady-state availability may be expressed by

$$A = \frac{MUT}{MUT + MDT} \leq 1.$$

$$A = \frac{MUT}{MUT + MDT} \leq 1 \ A = \frac{MUT}{MUT + MDT} \leq 1$$

with $0 \leq A \leq 1$ and generally has a value close to 1. Its complement is called *unavailability U*.

$$U = 1 - A = \frac{MDT}{MTBF + MDT} \geq 0 \ U = 1 - A = \frac{MDT}{MTBF + MDT} \geq 0.$$

$$U = 1 - A = \frac{MDT}{MTBF + MDT} \geq 0 \; U = 1 - A = \frac{MDT}{MTBF + MDT} \geq 0$$

Depending from the type of availability A to be considered, it has to be decided which fractions of MDT are relevant and therefore, taken into consideration for calculation. These fractions are to be defined.

Detailed guidance on calculations for systems with different properties and different repair characteristics is given in EN 61703.

The availability concept is illustrated by Figure A.1

For an item/system which is permanently in operation mode and no planned preventive maintenance is applied, MUT=MTBF holds. In this case MUT and MTBF can be used interchangeably for calculating the (steady state) operational availability. It may then be expressed by

$$A = \frac{MUT}{MUT + MDT} = \frac{MTBF}{MTBF + MDT} \leq 1$$

$$A = \frac{MUT}{MUT + MDT} = \frac{MTBF}{MTBF + MDT} \leq 1 \; A = \frac{MUT}{MUT + MDT} = \frac{MTBF}{MTBF + MDT} \leq 1$$

$A = \frac{MUT}{MUT + MDT} = \frac{MTBF}{MTBF + MDT} \leq 1$ MTBF is typically based on the time the system is in use (operated).

The parties involved should agree on the understanding of all the terms used (e.g. the MTBF time basis suitable for the specific application under consideration or which type of delay is taken into account). In case of contractual obligations it is highly recommended to stipulate the agreements.

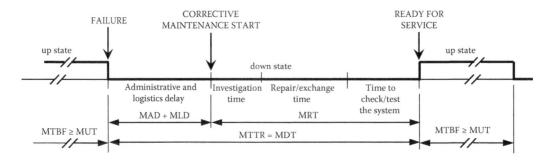

MTBF	mean (operating) time between failures	MAD	mean administrative delay
MUT	mean up time	MLD	mean logistics delay
MDT	mean down time	MTTR	mean time to restore (for corrective maintenance)
MRT	mean repair time		

Definitions can be found in IEC 60050-192 and EN61703.

FIGURE A.1
Availability concept and related terms.

The definitions of fleet availability (FA) as well as of schedule adherence (SA) are normally the subject of contractual negotiations. Therefore the elaboration of both parameters is not provided in this standard.

A.3.4 Logistic Support Parameters

TABLE A.4

Examples of Logistic Support Parameters

Parameter	Symbol	Dimension
Operation and Maintenance Cost	O&MC	money
Maintenance Cost	MC	money
Maintenance Man Hours	MMH	time (hours)
Logistic and Administrative Delay	LAD	time
Fault correction time		time
Repair time		time
Turn Around Time	TAT	time
Maintenance support performance		dimensionless
Employees for Replacement	EFR	number
Probability that Spare Parts are available (in Stock) when needed	SPS	dimensionless

A.3.5 Safety Parameters

TABLE A.5

Examples of Safety Performance Parameters

Parameter	Symbol	Dimension
Hazard rate	h(t)	1/time, 1/distance, 1/cycle
Probability of wrong-side failure	p_{WSF}	dimensionless
Active time to return to safe state	–	time

A.4 Risk Matrix Calibration and Risk Acceptance Categories

This section provides examples of application of a risk matrix and its parameters.

For the purpose of classifying any events Table A.6 and Table A.7 provide, in qualitative terms, typical categories of probability or frequency of occurrence of events and a typical description of each category for railways. Based on these typical categories, their numbers, and their numerical scaling (provided that numerical estimates are feasible) to be applied shall be defined by the railway duty holder, appropriate to the application under consideration.

Table A.8 and Table A.9 describe typical severity levels and the associated consequences. The number of severity levels and the consequences for each severity level to be applied shall be defined by the railway duty holder, appropriate for the application under consideration.

With regard to safety it is recommended that the relationship between injuries and fatalities ("equivalent fatalities" [efat] as defined in the Glossary), for the purpose of predicting and comparison only, be agreed with the railway duty holder based on the relevant legal framework.

Examples of such relationship may be:

- one equivalent fatality ≈ 1 fatality ≈ 10 major injuries ≈ 100 minor injuries or
- severe injuries and fatalities are fully equivalent or
- any other weighted equivalence.

By nature, the use of equivalent fatalities (efat) for prediction can not provide any certainty.

A.4.1 Frequency of Occurrence Levels

The following Table A.6 and Table A.7 give examples of frequency levels, adapted to a particular use.

The use of these particular examples is not mandatory, other classification can be used. More than one matrix can be used when this method is applied.

TABLE A.6

Frequency of Occurrence of Events with Examples for Quantification (Time Based)

Frequency Level	Description	Example of a Range of Frequency Based on Operation of 24h/day	Example of Equivalent Occurrence in a 30 Year Lifetime of 5000 h Operation/year
		Expected to happen	
Frequent	Likely to occur frequently. The event will be frequently experienced.	more than once within a period of approximately 6 weeks	more than about 150 times
Probable	Will occur several times. The event can be expected to occur often.	approximately once per 6 weeks to once per year	about 15 to 150 times
Occasional	Likely to occur several times. The event can be expected to occur several times.	approximately once per 1 year to once per 10 years	about 2 to 15 times
Rare	Likely to occur sometime in the system life-cycle. The event can reasonably be expected to occur.	approximately once per 10 years to once per 1000 years	perhaps once at most
Improbable	Unlikely to occur but possible. It can be assumed that the event may exceptionally occur.	approximately once per 1000 years to once per 100 000 years	not expected to happen within the lifetime
Highly improbable	Extremely unlikely to occur. It can be assumed that the event may not occur.	once in a period of approximately 100 000 years or more	extremely unlikely to happen within the lifetime

Note: This table is related to a single item (system/function/component). The examples given depend from the number of systems and/or the number of e.g. operational hours considered.

The expected mean time between two events is determined by the given reciprocal value of the frequency. For a frequency level bandwidth this formula has to be applied for its upper as well as its lower limit.

The expected occurrence or number of events in a time period is determined by the given time period multiplied by the given rate or frequency of occurrence. The time period has to be reduced if calendar time is not appropriate.

The considered time is an average and not necessarily a continuous time.

A distance based approach is given in Table A.7.

TABLE A.7

Frequency of Occurrence of Events with Examples for Quantification (Distance Based)

Frequency level	Description	Example of a Range of Frequency
F1	Likely to occur often	more than once every 5 000 km per train
F2	Will occur several times	once every 25 000 km per train
F3	Might occur sometimes	Once every 100 000 km per train

A.4.2 Severity Levels

The following tables give examples of severity levels, related to a particular use.

TABLE A.8

Severity Categories (Example Related to RAM)

RAM Severity Level	Description
Significant (immobilising failure)	A failure that: • prevents train movement or • causes a delay to service greater than a specified time • and/or generates a cost greater than a specified level
Major (service failure)	A failure that: • must be rectified for the system to achieve its specified performance and • does not cause a delay or cost greater than the minimum threshold specified for a significant failure
Minor	A failure that: • does not prevent a system achieving its specified performance and • does not meet criteria for Significant or Major failures

TABLE A.9

Severity Categories (Example 1 Related to RAMS)

Severity Level	Consequences to Persons or Environment	Consequences on Service/Property
Catastrophic	Many fatalities and/or extreme damage to the environment.	
Critical	Severe injuries and/or few fatalities and/or large damage to the environment.	Loss of a major system
Marginal	Minor injuries and/or minor damage to the environment	Severe system(s) damage
Insignificant	Possible minor injury	Minor system damage

TABLE A.10

Categories (Example 2 Related to Safety)

Severity Level	Consequences to Persons or Environment
S1	Many equivalent fatalities (likely more than about 10) or extreme damage to the environment.
S2	Multiple equivalent fatalities (likely less than about 10) or large damage to the environment.
S3	Single fatality or severe injury or significant damage to the environment.
S4	Minor injuries or minor damage to the environment
S5	Possible minor injury

TABLE A.11

Financial Severity Levels (Example)

Severity Level	Financial Consequences
SF1	The incident will incur people suing the company, severe impact to the public image of the company, and/or incur costs higher than 1 000 000 €.
SF2	The incident may have an impact on the public image of the company and/or incur costs higher than 100 000 €
SF3	The incident will not incur costs higher than 100 000 €

A.4.3 Risk Acceptance Categories

Risk acceptance categories allocated to identified risks allow classification for the purpose of decision making. Examples of risk acceptance categories are shown in Table A.12 and Table A.13.

TABLE A.12

Risk Acceptance Categories (Example 1 for Binary Decisions)

Risk Acceptance Category	Actions to be Applied
Unacceptable	The risk needs further reduction to be accepted
Acceptable	The risk is accepted provided adequate control is maintained

TABLE A.13

Risk Acceptance Categories (Example 2)

Risk Acceptance Category	Actions to be Applied
Intolerable	The risk shall be eliminated
Undesirable	The risk shall only be accepted if its reduction is impracticable and with the agreement of the railway duty holders or the responsible Safety Regulatory Authority.
Tolerable	The risk can be tolerated and accepted with adequate control (e.g. maintenance procedures or rules) and with the agreement of the responsible railway duty holders.
Negligible	The risk is acceptable without the agreement of the railway duty holders.

TABLE A.14

Risk Acceptance Categories (Example Related to Safety)

Frequency of Occurrence of an Accident (Caused by a Hazard)	Risk Acceptance Levels			
Frequent	Undesirable	Intolerable	Intolerable	Intolerable
Probable	Tolerable	Undesirable	Intolerable	Intolerable
Occasional	Tolerable	Undesirable	Undesirable	IIntolerableIIntolerable
Rare	Negligible	Tolerable	Undesirable	Undesirable
Improbable	Negligible	Negligible	Tolerable	Undesirable
Highly improbable	Negligible	Negligible	Negligible	Tolerable
	I	II	III	IV
	Severity of an accident (caused by a hazard)			

A.5 Applicable Standards

IEC 60050-192	International Electrotechnical Vocabulary – Dependability (February, 2015)
EN 614	Safety of machinery – Ergonomic design principles
IEC 60300-3-1	Dependability management – Part 3-1: Application guide – Analysis techniques for dependability – Guide on methodology
IEC 60300-3-2	Dependability management – Part 3-2: Application guide – Collection of dependability data from the field
IEC 60319	Presentation and specification of reliability data for electronic components
IEC 60605-2	Equipment reliability testing – Part 2: Design of test cycles
IEC 60605-3-1	Equipment reliability testing – Part 3: Preferred test conditions. Indoor portable equipment – Low degree of simulation
IEC 60605-4	Equipment reliability testing – Part 4: Statistical procedures for exponential distribution – Point estimates, confidence intervals, prediction intervals and tolerance intervals
IEC 60605-6	Equipment reliability testing – Part 6: Tests for the validity and estimation of the constant failure rate and constant failure intensity
IEC 60706-1	Guide on maintainability of equipment – Part 1: Sections 1, 2 and 3: Introduction, requirements and maintainability programme
IEC 60706-2	Maintainability of equipment – Part 2: Maintainability requirements and studies during the design and development phase
IEC 60706-3	Maintainability of equipment – Part 3: Verification and collection, analysis and presentation of data

IEC 60706-5	Maintainability of equipment – Part 5: Testability and diagnostic testing
IEC 60706-6	Guide on maintainability of equipment – Part 6: Section 9: Statistical methods in maintainability evaluation
IEC 60812	Analysis techniques for system reliability – Procedures for failure mode and effects analysis (FMEA)
IEC 61014	Programmes for reliability growth
IEC 61025	Fault tree analysis (FTA)
IEC 61070	Compliance test procedures for steady-state availability
IEC 61078	Analysis techniques for dependability – Reliability block diagram and boolean methods
IEC 61123	Reliability testing – Compliance test plans for success ratio
IEC 61160	Design review
IEC 61165	Application of Markov techniques
IEC 61508	Functional safety of electrical/electronic/programmable electronic safety-related systems
IEC 61703	Mathematical expressions for reliability, availability, maintainability and maintenance support terms
IEC 61709	Electric components – Reliability – Reference conditions for failure rates and stress models for conversion
IEC/TR 62380	Reliability data handbook – Universal model for reliability prediction of electronics components, PCBs and equipment

A.6 List of Methods

Some appropriate methods and tools for conducting and managing a RAMS programme are listed below. The choice of the relevant tool will depend on the system under consideration and the criticality, complexity, novelty etc., of the system.

1. An outline form of RAMS specification in order to assure assessment of all relevant RAMS requirements.

2. Procedures for formal design reviews with emphasis on RAMS, using some general and application specific check lists as appropriate, e.g.

   IEC 61160:2006   Design review

3. Procedures for performing "top down" (deductive methods) and "bottom up" (inductive methods) preliminary, worst case and in-depth RAM analysis for simple and complex functional system structures.

An overview of commonly used RAM analysis procedures, methods, advantages and disadvantages, data input and other requirements for the various techniques is given in:

IEC 60300-3-1   Dependability management – Part 3: Application guide – Section 1: Analysis techniques for dependability: Guide on methodology

The various RAM analysis techniques are described in separate standards, some of these are as follows:

IEC 60706   Guide on maintainability of equipment

IEC 60706-1   Part 1 – Sections 1, 2 and 3: Introduction, requirements and maintainability programme

IEC 60706-2   Part 2 – Maintainability requirements and studies during the design and development phase

IEC 60706-3   Part 3 – Verification and collection, analysis and presentation of data

IEC 60706-5   Part 5 – Testability and diagnostic testing

IEC 60706-6   Part 6 – Section 9: Statistical methods in maintainability evaluation

IEC 60812   Analysis techniques for system reliability – Procedures for failure mode and effects analysis (FMEA)

IEC 61025   Fault tree analysis (FTA)

IEC 61078   Analysis techniques for dependability – Reliability block diagram and boolean methods

IEC 61165   Application of Markov techniques

Availability of supportable statistical "RAM" data, for the components used in a design, (typically: failure rates, repair rates, maintenance data, failure modes, event rates, distribution of data and random events etc.) is fundamental to RAM analysis, e.g.

IEC 61709   Electronic components – Reliability – Reference conditions for failure rate and stress models for conversion

MIL-HDBK-217F Notice 2   Reliability Prediction for Electronic Systems

A number of computer programmes for system RAM analysis and statistical data analysis are also available.

IEEE 1633   IEEE recommended practice on Software Reliability

4. Procedures for performing hazard & safety/risk analysis

   Some of these are described in:

MIL-STD-882D	Standard Practise for System Safety
MIL-HDBK-764 (MI)	System Safety Engineering Design Guide For Army Materiel

   The same basic techniques and analysis methods listed for RAM (item 3), are also applicable for safety/risk analysis.

   Also see IEC 61508 Parts 1-7 under the general title "Functional safety of electrical/electronic/programmable electronic safety-related systems", consisting of the following parts:

IEC 61508-1:2010	Part 1: General requirements
IEC 61508-2:2010	Part 2: Requirements for electrical/electronic/ programmable electronic systems
IEC 61508-3:2010	Part 3: Software requirements
IEC 61508-4:2010	Part 4: Definitions and abbreviations
6.1.1 IEC 61508-5:2010	Part 5: Examples of methods for the determination of safety integrity levels
IEC 61508-6:2010	Part 6: Guidelines on the application of Parts 2 and 3
IEC 61508-7:2010	Part 7: Overview of techniques and measures

5. RAMS testing plans and procedures

   This step is in order to test the long-term operating behaviour of components, equipment or systems and to demonstrate compliance with the requirements. Furthermore RAMS analysis and test results are used to devise RAMS improvement programmes, e.g.

IEC 60605	Equipment reliability testing
IEC 60605-2	Part 2: Design of test cycles
IEC 60605-3-1	Part 3: Preferred test conditions. Indoor portable equipment – Low degree of simulation
IEC 60605-4	Part 4: Statistical procedures for exponential distribution – Point estimates, confidence intervals, prediction intervals and tolerance intervals
IEC 60605-6	Part 6: Tests for the validity of the constant failure rate or constant failure intensity assumptions
IEC 61014	Programmes for reliability growth
IEC 61070	Compliance test procedure for steady-state availability
IEC 61123	Reliability testing – Compliance test plan for success ratio IEC 61124 Reliability testing – Compliance tetst forn constant failure rate and constant failure intensity

Of greater importance is the assessment of RAMS data from the field (RAMS testing during operation), e.g.:

IEC 60300-3-2	Dependability management – Part 3: Application guide – Section 2: Collection of dependability data from the field
IEC 60319	Presentation of reliability data on electronic components (or parts)

6. Procedures/tools to perform LCC analysis (Life-cycle Cost)

Various computer programmes are available for LCC analysis

UNIFE Guideline for Lif-Cycle Cost for Railways (2001)

7. PHM

ISO 13372 Condition monitoring and diagnostics of machines '(2012)

IEEE P1656 Draft standard framework for prognosis and health management of electronic systems (to be submitted to ballot in June 2016)

Index

Page numbers followed f, t, and n indicate figures, tables, and notes, respectively.